Development, Function and Evolution of Teeth

Over the past 20 years there has been an explosion of information generated by scientific research. One of the beneficiaries of this has been the study of morphology, where new techniques and analyses have led to insights into a wide range of topics. Advances in genetics, histology, microstructure, biomechanics and morphometrics have allowed researchers to view teeth from new perspectives. However, up to now there has been little communication between researchers in the different fields of dental research. This book brings together for the first time overviews of a wide range of dental topics, linking genes, molecules and developmental mechanisms within an evolutionary framework. Written by leading experts in the field, this book will stimulate co-operative research in fields as diverse as palaeontology, molecular biology, developmental biology and functional morphology.

MARK F. TEAFORD is Professor of Anatomy at the Johns Hopkins University School of Medicine.
MOYA MEREDITH SMITH is Professor of Evolutionary Dentoskeletal Biology at the University of London, and is based at the Dental Institute of King's College London, Guy's Campus.
MARK W. J. FERGUSON is Professor of Basic Dental Sciences in the School of Biological Sciences at the University of Manchester.

Development, Function and Evolution of Teeth

Edited by Mark F. Teaford, Moya Meredith Smith and Mark W. J. Ferguson

 CAMBRIDGE
UNIVERSITY PRESS

CAMBRIDGE UNIVERSITY PRESS
Cambridge, New York, Melbourne, Madrid, Cape Town, Singapore, São Paulo

Cambridge University Press
The Edinburgh Building, Cambridge CB2 2RU, UK

Published in the United States of America by Cambridge University Press, New York

www.cambridge.org
Information on this title: www.cambridge.org/9780521570114

© Cambridge University Press 2000

First published 2000
This digitally printed first paperback version 2006

A catalogue record for this publication is available from the British Library

Library of Congress Cataloguing in Publication data
Development, function, and evolution of teeth / edited by Mark Teaford, Moya Meredith Smith & Mark W. J. Ferguson.
   p. cm.
1. Teeth – Physiology. 2. Teeth – Evolution. 3. Teeth – Molecular aspects.
I. Teaford, Mark, 1951– . II. Smith, Moya Meredith, 1955– . III. Ferguson, Mark W. J.
[DNLM: 1. Tooth – growth & development. 2. Tooth – anatomy & histology. WU 101 D489 2000]
QP88.6.D48 2000
573.3′56–dc21 99–39565 CIP
DNLM/DLC
for Library of Congress

ISBN-13 978-0-521-57011-4 hardback
ISBN-10 0-521-57011-5 hardback

ISBN-13 978-0-521-03372-5 paperback
ISBN-10 0-521-03372-1 paperback

Contents

Contributors

Barry K. Berkowitz GKT School of Biomedical Science Henriette Raphael House, Guy's Campus, London Bridge, London SE1 1UL, UK

P. M. Butler 23 Mandeville Court, Strode Street, Egham, Surrey TW20 9BU, UK

Mike I. Coates Biological Sciences, University College London, Gower Street, London WC1E 6BT, UK

M. C. Dean Evolutionary Anatomy Unit, Department of Anatomy and Developmental Biology, University College London, Gower Street, London WC1E 6BT, UK

George H. Dibdin MRC Dental Group, Dental School, University of Bristol, Lower Maudlin Street, Bristol BS1 2LY, UK

Alan G. Fincham Center for Cranofacial Molecular Biology, The University of Southern California, School of Dentistry, 2250 Alcazar Street, Los Angeles, CA, USA

M. Fortelius Division of Geology and Palaeontology, Department of Geology, University of Helsinki, PO Box 11, FIN-00014 University of Helsinki, Finland.

Peter Gaengler School of Dental Medicine, Conservative Dentistry and Periodontology, University of Witten-Herdecke, D-58448 Witten, Germany

J. P. Hunter Department of Anatomical Sciences, State University of New York at Stony Brook, Stony Brook, NY 11794–8081, USA

Ann Huysseune Instituut voor Dierkunde, University of Ghent, Ledeganckstraat 35, B-9000 Gent, Belgium

Jukka Jernvall Institute of Biotechnology, PO Box 56, 00014 University of Helsinki, Finland and Department of Anthropology, State University of New York at Stony Brook, Stony Brook, NY 11794, USA

Wighart von Koenigswald Institut für Paläontologie, Universität Bonn, Nussallee 8, Bonn D-53115, Germany

H. Lesot Institute de Biologie Medical Université Louis Pasteur, Strasbourg Faculté de Medicine, 11 Rue Humann, 67085 Strasbourg, France

Peter W. Lucas Department of Anatomy, University of Hong Kong, Li Shu Fan Building, 5 Sassoon Road, Hong Kong

Wen Luo Center for Craniofacial Molecular Biology, The University of Southern California, School of Dentistry, 2250 Alcazar Street, Los Angeles, CA, USA

Janet Moradian-Oldak Center for Craniofacial Molecular Biology, The University of Southern California, School of Dentistry, 2250 Alcazar Street, Los Angeles, CA, USA

Michael L. Paine Center for Craniofacial Molecular Biology, The University of Southern California, School of Dentistry, 2250 Alcazar Street, Los Angeles, CA, USA

Charles R. Peters Department of Anthropology and Institute of Ecology, University of Georgia, Athens, Georgia 30602, USA

John M. Rensberger Burke Museum of Natural and

Cultural History and Department of Geological Sciences, University of Washington, Seattle, WA 98195, USA

J. V. Ruch Institute de Biologie Medical Université Louis Pasteur, Strasbourg Faculté de Medicine, 11 Rue Humann, 67085 Strasbourg, France

P. Martin Sander Institut für Paläontologie, Universität Bonn, Nussallee 8, Bonn D-53115, Germany

Ivan J. Sansom School of Earth Sciences, University of Birmingham, Edgbaston, Birmingham B15 2TT, UK

Paul T. Sharpe Department of Craniofacial Development, UMDS, Guy's Hospital, London Bridge, London SE1 9RT, UK

R. Peter Shellis Division of Restorative Dentistry, Dental School, University of Bristol, Lower Maudlin Street, Bristol BS1 2LY, UK

B. Holly Smith Museum of Anthropology, 1109 Geddes, University of Michigan, Ann Arbor, MI 48109, USA

Moya Meredith Smith Department of Craniofacial Development, Dental Institute, Kings College London, Guy's Campus, London Bridge, London SE1 9RT, UK

A. J. (Tony) Smith School of Dentistry, Oral Biology, University of Birmingham, St Chad's Queensway, Birmingham B4 6NN, UK

Malcolm L. Snead Center for Craniofacial Molecular Biology, The University of Southern California, School of Dentistry, 2250 Alcazar Street, Los Angeles, CA, USA

David W. Stock Department of Anthropology, Pennsylvania State University, University Park, PA 16802, USA. *Present Address*: Department of Environmental, Population and Organismal Biology, University of Colorado, Boulder, CO80309-0334, USA

Mark F. Teaford Department of Cell Biology and Anatomy, Johns Hopkins University School of Medicine, Baltimore, MD 21205, USA

Irma Thesleff Institute of Biotechnology, PO Box 56, 00014 University of Helsinki, Finland

Kenneth M. Weiss Department of Anthropology, Pennsylvania State University, University Park, PA 16802, USA.

Margarita Zeichner-David Center for Craniofacial Molecular Biology, The University of Southern California, School of Dentistry, 2250 Alcazar Street, Los Angeles, CA, USA

Zhiyong Zhao Department of Anthropology, Pennsylvania State University, University Park, PA 16802, USA *Present address*: Department of Pediatrics, Yale University, New Haven, CT 06520, USA

Acknowledgements

This volume is a product of a symposium entitled 'Teeth: homeoboxes to function' held at the 4th International Congress of Vertebrate Morphology, at the University of Chicago in 1994. We wish to offer our sincere thanks to Sue Herring and Jim Hanken for inviting us to organize a session for the Congress. Robin Smith, from Cambridge University Press, provided enthusiastic guidance for this volume from the outset. Upon his leaving Cambridge Press, Tracey Sanderson skilfully (and patiently!) guided it through to completion. As is so often the case with a volume like this, many other people helped at various points along the way, so many that it is impossible to thank them all here. However, special thanks must go to Howard Farrell and Jayne Aldhouse for their expert editorial help; to innumerable reviewers for their expert insights and suggestions; and most of all, to each of the contributors, whose efforts and patience helped bring this volume successfully through its long gestation!

The original symposium would not have taken place without support from the Johns Hopkins University Department of Cell Biology & Anatomy and NIH support for the ICVM. Mark Teaford would also like to acknowledge the continued support of the National Science Foundation for his work on teeth. Moya Meredith Smith would also like to acknowledge numerous grants from the Royal Society for travel to Europe, USA and Australia and to exchange information with her colleagues in palaeontology, in particular for exchange visits to Lithuania and collaboration with Valia Karatjute-Talimaa. She would also like to thank the Dental School at Guy's Hospital and NERC for grants to pursue this work (GR/38543; GR3/10272), and to the Dean, Professor Ashley for his support of her research.

Finally, we could not have finished this project without continued encouragement, understanding, and insight from our families.

Genes, molecules and tooth initiation

Part one

1   Homeobox genes in initiation and shape of teeth during development in mammalian embryos

P. T. Sharpe

1.1.   Introduction

The past decade has seen remarkable advances in our understanding of the genetic control of embryonic development. We now know that developmental processes are initiated and controlled by interacting pathways of extracellular signalling molecules, receptors, intracellular signalling proteins and nuclear (transcription) factors. The different types of proteins (genes) that carry out these functions most often occur as members of families of related proteins characterised by possessing conserved amino acid motifs but which do not necessarily have similar functions in development. Thus for example, the transforming growth factor-beta (TGFβ) superfamily of secreted signalling proteins consists of a large family of proteins that share some homology with TGFβ, and in many cases share cell surface receptors (Kingsley, 1994). However, within this family different members have very different and specific functions in development. The bone morphogenetic protein Bmp-4, for example, probably has multiple functions as a signalling molecule in embryogenesis, including a role in lung morphogenesis, but targeted mutation of Bmp-4 (gene knock-out) shows a requirement for this protein for early mesoderm formation (Winnier et al., 1995; Bellusci et al.,1996). Bmp-7 on the other hand appears to have no direct role in mesoderm formation but is required for skeletal development (Luo et al., 1995). This illustrates a recurring theme in development where similar molecules have multiple functions, some of which overlap with other family members whereas others are unique. This almost certainly reflects the evolution of these gene families by gene duplication resulting in some shared and some unique functions.

The first of these families of developmental genes to be discovered and the one which has produced the greatest interest is that of the homeobox genes. The discovery of the homeobox as a small (180 bp) conserved region of DNA found in homeotic genes of flies (Drosophila) provided the springboard for all subsequent advances in this field (reviewed in Duboule, 1994). The discovery of the homeobox was an important milestone because it demonstrated that genes controlling fly development can also be present in vertebrates. This allowed progress in the understanding of vertebrate development to proceed at a far greater pace than was ever thought possible by cloning genes through nucleotide homology with fly genes.

Homeotic mutations in flies had long been suspected of holding important clues to understanding morphogenesis. Cells in an embryo differentiate into a limited number of specialised types, around 200 different cell types in mammals, and it is the arrangements of these cell types into defined structures with characteristic forms that is the main achievement of embryogenesis. The genetic control of morphogenesis of different structures was a mystery until molecular cloning became possible, followed by analysis of the expression, function and control of homeotic genes.

Homeotic mutations involve mutations in single genes that produce a phenotype where one body part of the fly is replaced by another. Thus the Antennapedia mutation is characterised by the development of legs in place of antennae on the fly's head. Eight of the homeotic genes identified in flies were cloned and found to be different except for one small 180 bp sequence towards the C terminus that was highly conserved. This sequence, or 'box', was named the homeobox and was also found to be present in other developmental genes in flies, most notably the segmentation genes fushi tarazu and engrailed (reviewed in Duboule, 1994).

The clues to the function of the homeobox came from structural analysis that showed that the homeodomain (60 amino acids) forms a helix-turn-helix structure which has DNA binding characteristics (Laughon and Scott, 1984, Sherperd et al., 1984). Homeoproteins are thus DNA-binding proteins that regulate gene transcription and as such may exert a hierarchical control function over the expression of genes required for morphogenesis of a particular structure.

The conservation of the homeobox sequence is not limited to flies, and since 1984 many vertebrate homeo-box genes have been cloned by nucleotide homology. However, there is a clear and important distinction between homeobox genes that are most closely related to homeotic genes and others that are not. Homeotic genes have unique organisational and functional characteris-tics in flies that distinguish them from other homeobox genes. Most significantly eight of the homeotic genes are clustered forming a complex (HOMC) and their linear order in the clusters is reflected in their anterior–poster-ior expression domains in fly embryos, a feature known as colinearity. Similarly the mammalian homeobox genes that most closely resemble homeotic genes, called Hox genes, also show the same feature of colinearity. In mammals there are 39 Hox genes arranged as four clus-ters on different chromosomes. Hox genes and HOMC genes probably had a single common evolutionary ancestor in a segmented worm and the Hox genes were duplicated during evolution of vertebrates (reviewed in Manak and Scott, 1994).

Hox gene expression is first detected in ectoderm and mesoderm cells in mammalian embryos during gastrula-tion (Gaunt et al., 1986). During organogenesis expression is found in the developing central nervous system and hindbrain, in the developing prevertebrae (somites) and more 5′ genes are also expressed in limb buds. Targeted mutation analysis has confirmed that Hox genes play a role in development of the axial skeleton, mutation of individual genes results in abnormal development of the axial skeleton at particular anterior–posterior levels. Thus for example a null mutation of Hoxb-4 results in partial homeotic transformation of the second cervical vertebra into a first cervical vertebra, whereas mutation of Hoxc-8 produces a change in a more posterior vertebra (LeMouellic et al., 1992; Ramirez-Solis et al., 1993).

A potentially important feature of Hox gene expres-sion for tooth development is the expression of the most 3′ Hox genes in cranial neural crest cells emanating from the developing hindbrain. Neural crest cells that form the branchial arches of the embryo migrate from distinct anterior–posterior positions in the hindbrain and caudal midbrain. Thus the cells of the first branchial arch which form the ectomesenchymal component of developing teeth are formed from neural crest cells that migrate from the rostral hindbrain (rhombomeres 1 and 2) and caudal midbrain. The anterior boundaries of expression of the most 3′ Hox genes have been shown to correspond to rhombomeric boundaries in the hindbrain and more significantly to be expressed in migrating neural crest cells at the appropriate axial level. Hoxb-2 expression, for example, has an anterior boundary at rhombomere 3/4 and is also expressed in neural crest cells that migrate from rhombomere 4 to populate the second branchial arch (Prince and Lumsden, 1994). This feature has been suggested as a mechanism of patterning the mesench-ymal cells of the branchial arches by positional specifica-tion of neural crest cells prior to their migration through the combination of Hox genes they express, referred to as the 'branchial Hox code'.

Since teeth in the first branchial arch were first thought to develop from neural crest cells contributed from rhombomeres 1 and 2 that populate this arch, and the neuroepithelium of the second arch expresses Hoxa-2, it was considered possible that tooth morphogenesis is patterned by the same branchial Hox code. However, it appears that this is not so, since rhombomere 1 does not express a Hox gene and although Hoxa-2 is expressed in rhombomere 2 it is unique in that this expression is not transferred to neural crest cells that migrate from this rhombomere. For both these reasons, expression of Hox genes has not been observed in the first branchial arch mesenchyme and so patterning of the first arch struc-tures cannot directly involve Hox genes (see also Chapter 11). Also, we now know that, in the mouse it is the non-segmental posterior midbrain crest which forms the ectomesenchyme of the mandibular molars (Imai et al., 1996). In the chick, and also the equivalent bones to those bearing teeth in the mouse, are formed from mid brain crest (Koentges and Lumsden, 1996). There are however many non-Hox homeobox genes that are expressed in the first branchial arch and also during tooth development. This chapter describes these expression patterns and proposes how some of these genes might function to control tooth morphogenesis via an 'odontogenic homeobox code' (Sharpe, 1995).

1.2.   Homeobox genes and tooth bud initiation

The first morphological sign of tooth development is a narrow band of thickened epithelium (primary epithelial band) on the developing mandible and maxilla that forms four zones, one in each quadrant. These bands specify the area of epithelium from which teeth are

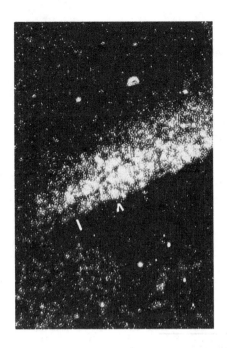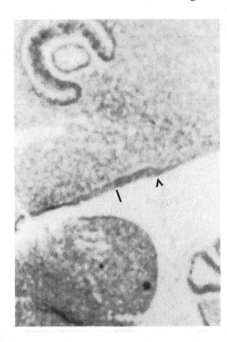

Figure 1.1. *In situ* hybridisation of a section in a dorsoventral plane with anterior at the top, of a day 10.5 gestation mouse embryo head showing *Msx-2* gene expression in the ectomesenchyme and oral epithelium in relation to the epithelial thickening. Left, dark field, right, light field. Section is sagittal (parallel to A/P axis, anterior to the top). Line, middle extent of epithelial band; arrowhead, lingual extent of epithelial band. Magnification × 200. (Courtesy of Bethan Thomas.)

capable of forming. The position of the bands thus determines and restricts the location within the mandible and maxilla of tooth development. Tooth buds form at discreet locations in these bands by secondary thickening of the epithelium and invagination into the underlying ectomesenchyme. The first important question in tooth development, therefore, is which cells provide the information that specifies the position of the oral epithelial thickenings? Although results from recombination experiments have largely supported the epithelium as the source of the initiation signals there is still some doubt as to whether it is the underlying neural crest-derived ectomesenchymal cells that are patterning the epithelial cells (Mina and Kollar, 1987; Lumsden, 1988; and reviewed by Ruch, 1995).

One of the interesting observations from the localisation of homeobox genes expressed in tooth development is that for many genes, expression is not restricted to either the dental epithelium or mesenchymal cells. This is particularly evident for the *Msx-2* homeobox gene, which shows spatially restricted expression in both epithelial and mesenchymal cells during tooth development (Figure 1.1; MacKenzie *et al.*,1992). More significantly, the early expression of *Msx-2* prior to tooth bud formation is also found in epithelium and mesenchyme. *Msx-2* is expressed in the distal tips of the mandibular and maxillary arch mesenchyme. *Msx-1*, a close relative of *Msx-2*, is expressed in a similar domain of ectomesenchyme as *Msx-2* but extends slightly more proximal then *Msx-2* (MacKenzie *et al.*, 1992). *Msx-1* is not expressed in the oral epithelium (MacKenzie *et al.*, 1991). These distinct early expression domains of *Msx-1* and *-2* in epithelium and mesenchyme suggest a possible role for these genes in initiation of the primary epithelial band (Sharpe, 1995).

Experimental evidence for a possible role of these genes in initiation of tooth development has come from *in vivo* experiments using targeted mutagenesis and also *in vitro* experiments involving explant cultures of early tooth germs. Targeted mutation of the *Msx-1* gene results in development of all teeth being arrested at the early bud stage (Satokata and Maas, 1994). Similarly mutations in MSX1 have been shown to be associated with tooth agenesis in humans (Vastardis *et al.*, 1996). Since *Msx-1* is expressed at high levels in the condensing mesenchyme at the bud stage, this suggests that *Msx-1* is required for a signalling pathway from bud mesenchyme to dental epithelium in tooth histogenesis but not initiation of the tooth bud. The recombination experiments have shown that following an initial signal (possibly Bmp-4 or Shh), from the thickened epithelium to the underlying mesenchyme, the direction of communication is then reversed and signals pass from the condensing mesenchyme to the epithelial bud (Figure 1.2A). The toothless phenotype of the *Msx-1* mutants implies that *Msx-1* regulates the expression of these signalling molecules. A possible candidate signalling molecule is Bmp-4, and in fact, the levels of *Bmp-4* expression in tooth bud mesenchyme of *Msx-1* mutants is reduced. Moreover, tooth development in *Msx-1* mutants can be rescued by the addition of beads coated in Bmp-4; thus there is a strong case for Msx-1 regulating *Bmp-4* in condensing

Epithelial thickening

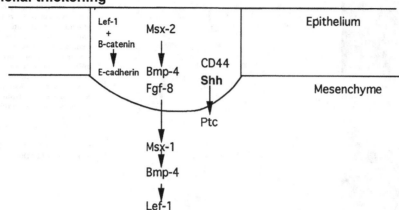

Figure 1.2. **A. Proposed signalling pathways involved in interactions between primary odontogenic epithelium thickening and ectomesenchyme at the epithelial band stage. B. Gene expression in the enamel knot precursor cells induced at the early bud stage by signals from the mesenchyme which may include Bmp-4.**

Tooth bud

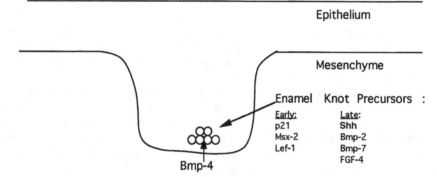

mesenchyme (Chen *et al.*, 1996). Although the *Msx-1* mutants give a clear tooth phenotype, the arrest does not occur at the initiation stage, i.e. tooth buds are produced. The earliest requirement for *Msx-1* would thus appear to be for mesenchymal to epithelial signalling at the bud stage. While it is possible that the targeted mutation is not a complete null (only the third helix is deleted in the mutant allele) it is more likely that *Msx-2* and/or *Dlx-2* may compensate for loss of *Msx-1* in tooth initiation. *Msx-2* targeted mutants have been generated and appear to have normal early tooth development. Significantly, however, tooth development in *Msx-1/Msx-2* double mutants is reported as being arrested earlier than the tooth bud stage and initiation may not occur at all in the absence of both genes (Maas *et al.*, 1996). Thus, there is strong *in vivo* data that supports the role of these genes in initiation. The possible role of *Dlx-2* in tooth development is discussed below.

Lef-1 is a transcription factor which is a member of

the HMG-box family of DNA-binding proteins, of which the best known is the mammalian sex-determining gene SRY. Lef-1 is very closely related to another protein, Tcf-1, and these genes probably appeared by duplication of a single ancestor gene (Gastrop, *et al.*, 1992). *Lef-1* and *Tcf-1* show a very similar pattern of expression during tooth development, and both are expressed in T-lymphocytes. The function of both genes has been studied by targeted gene disruption where, surprisingly, the phenotypes produced are quite different. *Lef-1* mutant mice have tooth development arrested at the bud stage (similar to the *Msx-1 -/-* mice) and show no major defects in T-lymphocytes (van Genderen *et al.*, 1994). *Tcf-1* mutants have normal tooth development but severely impaired T-cell function (Verbeek *et al.*, 1995). One possible hypothesis is that the archetypal gene is expressed and required for tooth development and that *Tcf-1*, duplicated from *Lef-1* and acquired a novel function in T-cells but which is not required for tooth development. *Lef-1* is expressed in secondary epithelial thickenings and in condensing mesenchyme of the tooth bud in a similar domain to *Msx-1*. Because tooth development in *Lef-1* mutants is arrested at the bud stage, it was originally believed that *Lef-1* would be required for initiating the signalling pathway of molecules from condensing mesenchyme to the epithelium of the tooth bud similar to *Msx-1*. However, detailed recombination experiments using *Lef-1* mutant and wild-type epithelium and mesenchyme have shown that the *Lef-1* is required in the early thickened epithelium for tooth development (Kratochwil *et al.*, 1996). Thus a recombination of *Lef-1* mutant epithelium with wild-type mesenchyme does not give tooth development, whereas the reciprocal recombination of mutant mesenchyme and wild-type oral epithelium allows normal tooth development *in vitro*. Moreover, the requirement for *Lef-1* in the thickened epithelium is transient, since mutant mesenchyme, cultured with wild-type epithelium and then dissociated and recultured with mutant epithelium, gives rise to normal tooth development, indicating that the signals regulated by *Lef-1* have been initiated in the mutant mesenchyme by the wild-type epithelium. Interestingly *Tcf-1* expression is barely detectable in epithelium but mesenchymal expression overlaps that of *Lef-1* indicating why *Tcf-1* may not be able to compensate for loss of *Lef-1* function in tooth bud development. The possibility that expression of Lef-1 in epithelium may be sufficient to induce tooth development has come from ectopic expression experiments in transgenic mice using a keratin 14 promoter which produced ectopic teeth (Zhou *et al.*, 1995).

The genetic pathway in which *Lef-1* belongs is starting to be elucidated, providing important insights into the control of bud stage tooth development. Lef-1 interacts with β-catenin in the cell cytoplasm, resulting in transport of the Lef-1 protein into the nucleus. β-catenin is a component of the cell adhesion molecule E-cadherin, the expression of which is localised to tooth epithelial thickenings (Kemler, 1993, Behrens *et al.*, 1996, Huber *et al.*, 1996). β-catenin has been shown to be downstream of Wnt signalling in *Xenopus* and since some Wnt genes colocalise with Lef-1 and E-cadherin in tooth epithelial thickenings (A. McMahon, personal communication) a pathway involving Wnt regulation of *Lef-1* expression resulting in activation of E-cadherin with β-catenin as an intermediate may be important in early tooth development (Yost *et al.*, 1996; Molenaar *et al.*, 1996; Luning *et al.*, 1994).

The localisation of other signalling molecules and transcription factors in tooth bud epithelial thickenings demonstrates that the Lef-1 pathway is not the only one of importance. The high levels of Shh gene expression in secondary tooth bud epithelial thickenings suggest an

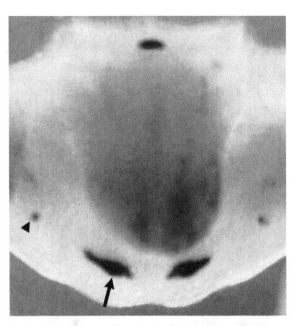

Figure 1.3. **Whole mount** *in situ* **hybridisation of a day-13.5 mandible showing expression of** *Shh* **in the epithelium of the developing incisor (arrow) and molar (arrowhead) tooth germs viewed from the oral aspect.**

important yet undetermined role for this pathway in odontogenesis (Figure 1.3) (Bitgood and McMahon, 1995; Koyama *et al.*, 1996; Sharpe, unpublished).

1.3. Patterning of tooth position and shape

Mammalian teeth have characteristic shapes for each position in the jaws. The shape and position are important for dietary requirements and have evolved and diversified for particular specialised feeding functions. Incisors are conical, or chisel-shaped and located at the front of the jaws, where they are used not only for obtaining and cutting food, but also for grooming or defence functions. Molars are triangular, rectangular and multicuspid in shape and are located towards the back of the jaws aand are for processing food, either by cutting, grinding or crushing. Variations on these basic shapes have evolved for specialised diets such as carnassial teeth in carnivores which are a sectorial adaptation of tritubercular molars. The correct pattern of the dentition is thus essential for animal survival. The dentition of any animal species is as unique as its DNA and since patterns are inherited the developmental mechanisms that direct pattern formation must be genetically controlled.

The importance of Hox genes in development of the

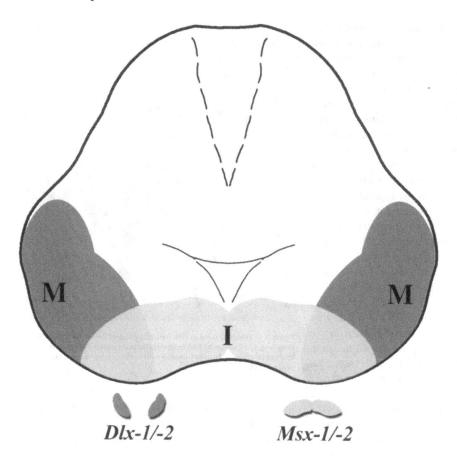

Figure 1.4. **Simplified model of the 'odontogenic homeobox gene code' of dental patterning based on the expression of *Msx-1, Msx-2, Dlx-1* and *Dlx-2* in the developing mandible. Overlapping domains of homeobox gene expression in the ectomesenchyme of the mandibular process can specify the shape of the tooth that develops at a particular position. Oral aspect, anterior to the top; I, incisor field; M, molar field.**

axial skeleton has been discussed. But it is possible that homeobox gene expression in the first branchial arch may be involved in patterning the development of first branchial arch structures such as teeth. A molecular model of patterning the dentition, based on the expression of several homeobox genes in neural crest-derived ectomesenchyme has been proposed and termed the 'odontogenic homeobox gene code' (Sharpe, 1995). A simplified form of this model is illustrated in Figure 1.4 where the expression of homeobox genes, such as *Dlx-1, Dlx-2, Msx-1* and *Msx-2* are proposed to specify the development of tooth germs into either molars or incisors. *Dlx-1* and *Dlx-2* are homeobox genes belonging to a family of seven or eight genes related to the *Drosophila Distalless* gene which is involved in appendage development. *Dlx-1* and *Dlx-2* are located within 10 kb of each other on mouse chromosome 2 and their expression in mandibular and maxillary mesenchyme is almost indistinguishable (Bulfone *et al.*, 1993). Both genes are predominately expressed in the proximal ectomesenchyme of the mandible and maxilla in the area where molars will develop prior to the start of tooth development (E10). Expression is absent in more distal mesenchyme where incisors will develop. The expression of *Dlx-1* and *Dlx-2* is complemented by that of *Msx-1*

and *Msx-2*, which are predominantly expressed in distal mesenchyme and not in proximal areas. In order to test this model, mice embryos with targeted disruption of the *Dlx-1* and *Dlx-2* genes using homologous recombination in ES cells have been analysed. Mice with null mutations in either gene have normal tooth development. It is only when double mutations are created, i.e. mice with null mutations in both genes, that tooth development is affected. *Dlx-1/-2 -/-* embryos have normal upper and lower incisors and lower molars but do not develop any upper molars (Qiu *et al.*, 1997; Thomas *et al.*, 1997).

Since the odontogenic mesenchyme expressing these homeobox genes is neural crest derived, this model predicts that populations of cranial neural crest cells are specified as odontogenic and further regionally specified as maxilla/mandible/molar/incisor. The failure of maxillary molar teeth to develop in the *Dlx-1/-2* double mutant embryos thus supports the odontogenic homeobox code model for patterning of maxillary molar tooth development and suggests that *Dlx-1* and *Dlx-2* are required for the specification of a subpopulation (maxillary molar) of odontogenic neural crest cells. The normal development of mandibular molars implies that tooth patterning in the lower and upper jaws is controlled independently, a feature that was not originally predicted by the model but one that has interesting implications for mechanisms of evolution of dental patterns.

Goosecoid (Gsc) was originally envisaged as part of the odontogenic homeobox code but has since been shown to be important for mandibular skeletal, but not tooth, development. Targeted null mutations in the Gsc gene produce craniofacial defects that resemble first arch syndrome in humans, but the development of teeth in these mice is normal (Rivera-Perez *et al.*, 1995; Yamada *et al.*, 1995).

Gsc has been shown to be upregulated in *Xenopus* mesoderm formation by the secreted signalling protein activin. Activin is a member of the TGFβ superfamily of growth factor-like signalling molecules that have wide-ranging functions in embryogenesis. Targeted null mutation of the activin-βA gene in transgenic mice was predicted to result in defects in mesoderm formation. However, mesoderm formation in activin-βA mutants was found to be normal but significantly, the major phenotype was abnormalities in craniofacial development, particular tooth development. Activin-βA -/- mice develop no incisor teeth and no mandibular molars but development of maxillary molars is always normal. Thus null mutations in a molecule (activin-βA) produce a tooth patterning phenotype. In common with the Dlx-1/2 mutations, activin-βA is not required for development of all teeth but only for initiation of incisors and mandibular molars. Unlike the Dlx-1/2 mutations, activin-βA affects incisor tooth development to the same extent in the mandible and maxilla and this may either indicate that incisor development is controlled differently to molar development or that activin is involved in regulating different pathways in the mandible and maxilla (Matzuk *et al.*, 1995).

Other homeobox genes are expressed during tooth development and some of these may also contribute to the odontogenic homeobox code. Other members of the Dlx family such as *Dlx-5* and *Dlx-6* are expressed in proximal regions of the developing mandible but not the maxilla and may form part of the code that is required for mandibular molar development. *Barx-1* is a homeobox gene that is expressed in the mesenchyme in areas where molars form but not incisor mesenchyme (Tissier-Seta *et al.*, 1995).

1.4.   Regulation of tooth shape

Once the spatial information is provided to specify development of a tooth germ into an incisor or molar, genes must be activated that control the shaping process to produce cusps (morphogenesis). The physical processes that direct cusp development have classically been suggested to involve the differentiation of the stellate reticulum and differential cessation of mitosis in the dental epithelium. The importance of differentiated transient epithelial structures, enamel knots, has recently gained significance. The enamel knot was originally identified in the 1920s in molar cap stage tooth germs as a transient, small group of epithelial cells immediately above the condensing mesenchyme forming the dental papilla. A role for the enamel knot in cusp formation was proposed by Orban (1928) and Butler (1956) who suggested that enamel knot cells act as a local restraint causing post-mitotic internal enamel epithelium to inflect at the site of the future first cusp and the external enamel epithelium to dimple as the swelling pressure of the developing stellate reticulum separated the external and internal enamel epithelia everywhere else in the tooth germ.

Msx-2 was the first gene whose expression was localised to enamel knot cells and it was proposed that *Msx-2* expression provided a molecular link between tooth initiation and shape (MacKenzie *et al.*, 1992). Subsequently the expression of several more genes has been shown to be restricted to enamel knot epithelial cells in the tooth bud and the origin of enamel knot cells traced back to a few epithelial cells at the tip of molar tooth buds. Expression of the genes for the secreted factors Shh, Bmp-2, Bmp-4, Bmp-7 and Fgf-4 are all localised in enamel knot cells (Chapter 2, Figure 1.2B) suggesting that this structure acts as a signalling centre similar to both the AER and the zone of polarising activity (ZPA) in limb development (Vaahtokari *et al.*, 1996). Expression of these genes has also identified the formation of secondary enamel knots in regions corresponding to the future cusp tips and it is proposed that the first (primary) enamel knot acts as a signalling centre to direct secondary knot formation which functions to control local epithelial cell proliferation rates. Enamel knot cells have a transient existence and rapidly undergo apoptosis which is probably controlled in part by early expression of p21 in the enamel knot cells (I. Thesleff, personal communication).

The specification of tooth patterning by the odontogenic homeobox code suggests that one way the code acts is by the odontogenic mesenchyme cells that condense at the base of the epithelial tooth bud communicating to the epithelial tooth bud cells to initiate enamel knot cell differentiation at specific positions (see Chapter 2).

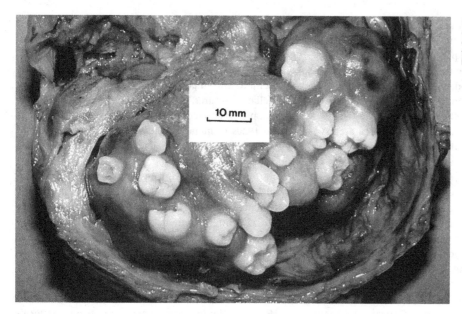

Figure 1.5. **Development of ectopic teeth in a human ovarian teratoma. (Courtesy of the Anatomy Museum, Guy's Hospital.)**

1.5.   Ectopic tooth development

The correct development of teeth in the right place at the right time clearly involves many different interacting cellular and molecular processes. It comes as something of a surprise, therefore, to see perfectly formed teeth develop at ectopic sites. One of the most common sites of ectopic teeth formation is in ovarian teratomas (Figure 1.5). Teratomas are defined as germ cell tumours and since the normal development of ovaries and testes does not involve any contribution from cells of neural crest origin it is difficult to reconcile the development of teeth in these structures. Since other neural crest-derived organs such as pigmented hair also form in these teratomas it seemed likely that cells of neural crest origin are present. Alternatively since germ cells are pluripotent it is possible that in certain circumstances these may contribute to tooth mesenchyme, and that the neural crest cell phenotype, although only present normally in developmental stages, is strongly expressed in this abnormal development of these germ cells, neural crest being a vertebrate synapomorphy. Whatever the cellular origin of these teeth the fact remains that their shape, mineralisation, etc. is perfectly normal and clearly the mechanisms operating in normal tooth development in the oral cavity are working in this ectopic site. If it were only possible to study the formation of these ectopic teeth during the early developmental stages they could teach us a great deal about the molecular control of normal tooth development.

Summary

Homeobox genes are involved in the genetic control of many different developmental processes, including organogenesis, in invertebrate and vertebrate embryogenesis. Many different homeobox genes are expressed in the developing orofacial region and during tooth development in mammalian embryos and the roles of several of these genes is beginning to be elucidated. Homeobox genes were discovered in *Drosophila* where mutations in certain homeobox genes, namely homeotic genes, produced changes in embryonic patterning resulting in transformations of body parts. The potential role of homeobox genes in positional information of cells in organogenesis has led to investigation of the possible role of homeobox genes expressed in the developing orofacial region of mammalian embryos in patterning for tooth position and shape.

Acknowledgements

Work in the author's laboratory is supported by the Medical Research Council and the Human Frontier Science Programme.

References

Behrens, J., von Kries, J. P., Kühl, M., Bruhn, L., Wedlich, D., Grosschedl, R. and Birchmeier, W. (1996). Functional interaction of β-catenin with the transcription factor LEF-1. *Nature*, **382**, 638–642.

Bellusci, S., Henderson, R., Winnier, G., Oikawa, T. and Hogan, B. L. M. (1996). Evidence form normal expression and targeted misexpression that bone morphogenetic protein-4 (Bmp-4) plays a role in mouse embryonic lung morphogenesis. *Development*, **122**, 1693–1702.

Bitgood, M. J. and McMahon, A. P. (1995). *Hedgehog* and *Bmp* genes are coexpressed at many diverse sites of cell–cell interaction in the mouse embryo. *Developmental Biology*, **172**, 126–138.

Bulfone, A., Kim, H. J., Puelles, L., Porteus, M. H., Grippo, J. F. and Rubenstein, J. L. R. (1993). The mouse *Dlx-2 (Tes-1)* gene is expressed in spatially restricted domains of the forebrain, face and limbs in midgestation mouse embryos. *Mechanisms of Development*, **40**, 129–140.

Butler, P. M. (1956). The ontogeny of molar pattern. *Biological Reviews*, **31**, 30–70.

Chen, Y., Bei, M., Woo, I., Satokata, I. and Maas, R. (1996). *Msx1* controls inductive signalling in mammalian tooth morphogenesis. *Development*, **122**, 3035–3044.

Duboule, D. (1994). *Guidebook to the Homeobox Genes*. Oxford: Oxford University Press.

Gastrop, J., Hoevenagel, R., Young, J. R. and Clevers, H. C. (1992). A common ancestor of the mammalian transcription factors RCF-1 and TCF-1 alpha/LEF-1 expressed in chicken T cells. *European Journal of Immunology*, **22**, 1327–1330.

Gaunt, S. J., Miller, J. R., Powell, D. J. and Duboule, D. (1986). Homeobox gene expression in mouse embryos varies with position by the primitive streak stage. *Nature*, **324**, 662–664.

Gehring, W. J. (1987). Homeo boxes and the study of development. *Science*, **236**, 1245–1252.

van Genderen, C., Okamura, R., Fariñas, I., Quo, R. G., Parslow, R. G., Bruhn, L. and Grosschedl, R. (1994). Development of several organs that require inductive epithelial–mesenchymal interactions is impaired in LEF-1 deficient mice. *Genes and Development*, **8**, 2691–2703.

Huber, O., Korn, R., McLaughlin, J., Ohsugi, M., Herrmann, B. G. and Kemler, R. (1996). Nuclear localization of β-catenin by interaction with transcription factor LEF-1. *Mechanisms of Development*, **59**, 3–10.

Imai, H., Osumi-Yamashita, N.,Ninomiya, N. and Eto, K. (1996). Contribution of early-emigrating midbrain crest cells to the dental mesenchyme of mandibular molar teeth in rat embryos. *Developmental Biology*, **176**, 151–165.

Kemler, R. (1993). From cadherins to catenins: cytoplasmic protein interactions and regulation of cell adhesion. *Trends in Genetics*, **9**, 317–321.

Kingsley, D. M. (1994). The TGF-B superfamily: new members, new receptors, and new genetic tests of function in different organisms. *Genes and Development*, **8**, 133–146.

Koentges, G. and Lumsden, A. (1996). Rhombencephalic neural crest segmentation is preserved throughout craniofacial ontogeny. *Development*, **122**, 3229–3242.

Koyama, E., Yamaai, T., Iseki, S., Ohuchi, H., Nohno, T., Yoshioka, H., Hayashi, Y., Leatherman, J. L., Golden, E. B., Noji, S. and Pacifici, M. (1996). Polarizing activity, *Sonic Hedgehog*, and

tooth development in embryonic and postnatal mouse. *Developmental Dynamics*, **206**, 59–72.

Kratochwil, K., Dull, M., Fariñas, I., Galceran, J. and Grosschedl, R. (1996). *Lef-1* expression is activated by Bmp-4 and regulates inductive tissue interactions in tooth and hair development. *Genes and Development*, **10**, 1382–1394.

Laughon, A. and Scott, M. P. (1984). Sequence of a *Drosophila* segmentation gene: protein structure homology with DNA-binding proteins. *Nature*, **310**, 25–31.

LeMouellic, H., Lallemand, Y., and Brûlet, P. (1992). Homeosis in the mouse induced by a null mutation in the *Hox-3.1* gene. *Cell*, **69**, 251–264.

Lumsden, A. G. S. (1988). Spatial organisation of the epithelium and the role of neural crest cells in the initiation of the mammalian tooth germ. *Development* (Supp.), **103**, 155–169.

Lüning, C., Rass, A., Rozell, B., Wroblewski, J. and Öbrink, B. (1994). Expression of E-cadherin during craniofacial development. *Journal of Craniofacial Genetics and Developmental Biology*, **14**, 207–216.

Luo, G., Hofmann, C., Bronckers, A. L. J. J., Sohocki, M., Bradley, A. and Karsenty, G. (1995). Bmp-7 is an inducer of nephrogenesis, and is also required for eye development and skeletal patterning. *Genes and Development*, **9**, 2808–2820

Maas, R. L., Chen, Y., Bei, M., Woo, I. and Satokata, I. (1996). The role of Msx genes in mammalian development. *Annals of the New York Academy of Sciences*, **785**, 171–181.

MacKenzie, A., Leeming, G., Jowett, A., Ferguson, M. W. J. and Sharpe, P. T. (1991). The homeobox gene *Hox-7.1* has specific regional and temporal expression patterns during early murine craniofacial embryogenesis, especially tooth development *in vivo* and *in vitro*. *Development*, **111**, 269–285.

MacKenzie, A. L., Ferguson M. W. J. and Sharpe P. T. (1992). Expression patterns of the homeobox gene *Hox-8* in the mouse embryo suggest a role in specifying tooth initiation and shape. *Development*, **115**, 403–420.

Manak, J. R. and Scott, M. P. (1994). A class act: conservation of homeodomain protein functions. In *The Evolution of Developmental Mechanisms*, eds. M. Akam, P. Holland, P. Ingham and G. Wray pp. 61–77. Cambridge: The Company of Biologists.

Matzuk, M. M., Kumar, T. R., Vassaill, A., Bickenbach, J. R., Roop, D. R., Jaenisch, R. and Bradley, A. (1995). Functional analysis of activins during mammalian development. *Nature*, **374**, 354–356.

Mina M. and Kollar E. J. (1987). The induction of odontogenesis in non-dental mesenchyme combined with early murine mandibular arch epithelium. *Archives of Oral Biology*, **32**, 123–127.

Molenaar, M., van de Wetering, M., Oosterwegel, M., Peterson-Maduro, J., Godsave, S., Korinek, V., Roose, J., Destree, O. and Clevers, H. (1996). XTcf-3 transcription factor mediates beta-catenin-induced axis formation in *Xenopus* embryos. *Cell*, **86**, 391–399.

Orban, B. (1928). *Dental Histology and Embryology*. Chicago: Rogers Printing.

Prince, V. and Lumsden, A. (1994). *Hoxa-2* expression in normal and transposed rhombomeres: independent regulation in the neural tube and neural crest. *Development*, **120**, 911–923.

Qiu, M., Bulfone, A., Ghattas, I., Menses, J. J., Sharpe, P. T., Presley, R., Pedersen, R. A. and Rubenstein, J. L. R. (1997). Role of the *Dlx* homeobox genes in proximodistal patterning of the branchial arches: mutations of *Dlx-1*, *Dlx-2* and *Dlx-1* and *-2* alter morphogenesis of proximal skeltal and soft tissue structures derived from the first and second arches. *Developmental Biology*, **185**, 165–184.

Ramirez-Solis, R., Zheng, H., Whiting, J., Krumlauf, R. and Bradley, A. (1993). *Hoxb-4 (Hox 2.6)* mutant mice show homeotic transformation of a cervical vertebra and defects in the closure of the sternal rudiments. *Cell*, **73**, 279–294.

Rivera-Perez, J. A., Mallo, M., Gendron-Maguire, M., Grindley, T. and Behringer, R. R. (1995). goosecoid is not an essential component of the mouse gastrula organiser but is required for craniofacial and rib development. *Development*, **121**, 3005–3012.

Ruch, J. V. (1995). Tooth crown morphogenesis and cytodifferentiations: candid questions and critical comments. *Connective Tissue Research*, **32**, 1–8.

Satokata, I. and Maas, R. (1994). Msx1-deficient mice exhibit cleft palate and abnormalities of craniofacial and tooth development. *Nature Genetics*, **6**, 348–356.

Sharpe, P. T. (1995). Homeobox genes and orofacial development. *Connective Tissue Research*, **32**, 17–25.

Sherperd, J. C. W., McGinnis, E., Carrasco, A. E., DeRobertis, E. M. and Gehring, W. J. (1984). Fly and frog homeo domains show homology with yeast mating type regulatory proteins. *Nature*, **310**, 70–71.

Thomas, B. L., Tucker, A. S., Qiu, M., Ferguson, C. A., Hardcastle, Z., Rubenstein, J. L. R. and Sharpe, P. T. (1997). Role of *Dlx-1* and *Dlx-2* genes in patterning of the murine dentition. *Development*, **124**, 4811–4818.

Tissier-Seta, J.-P., Mucchielli, M.-R., Mark, M., Mattei, M.-G. and Brunet, J.-F. (1995). *Barx1*, a new mouse homeodomain transcription factor expressed in cranio-facial ectomesenchyme and the stomach. *Mechanisms of Development*, **51**, 3–15.

Vaahtokari, A., Åberg, T., Jernvall, J., Keränen, S. and Thesleff, I. (1996). The enamel knot as signalling centre in the developing mouse tooth. *Mechanisms of Development*, **54**, 39–43.

Vastardis, H., Karimbux, N., Guthua, S. W., Seidman, J. G. and Seidman, C. E. (1996). A human MSX1 homeodomain missense mutation causes selective tooth agenesis. *Nature Genetics*, **13**, 417–421.

Verbeek, S., Izon, D., Hofhuis, F., Robanus-Maandag, E., te Riele, H., van de Wetering, M., Oosterwegel, M., Wilson, A., MacDonald, H. R. and Clevers, H. (1995). An HMG-box-containing T-cell factor required for thymocyte differentiation. *Nature*, **374**, 70–74.

Winnier, G., Blessing, M., Labosky, P. A. and Hogan, B. L. M. (1995). Bone morphogenetic protein-4 is required for mesoderm formation and patterning in the mouse. *Genes and Development*, **9**, 2105–2116.

Yamada, G., Mansouri, A., Torres, M., Stuart, E. T., Blum, M., Schultz, M., De Robertis, E. M. and Gruss, P. (1995). Targeted mutation of the murine *goosecoid* gene results in craniofacial defects and neonatal death. *Development*, **121**, 2917–2922.

Yost, C., Torres, M., Miller, J. R., Huang, E., Kimelman, D., and Moon, R. T. (1996). The axis-inducing activity, stability, and subcellular distribution of β-catenin is regulated in *Xenopus* embryos by glycogen synthase kinase 3. *Genes and Development*, **10**, 1443–1454.

Zhou, P., Byrne, C., Jacobs, J. and Fuchs, E. (1995). Lymphoid enhance factor 1 directs hair follicle patterning and epithelial cell fate. *Genes and Development*, **9**, 570–583.

2 Return of lost structure in the developmental control of tooth shape

J. Jernvall and I. Thesleff

2.1. Introduction

Perhaps one of the main attractions, both scientifically and aesthetically, of mammalian teeth is their diversity. Among other things, differently shaped teeth are a good reminder of both evolutionary flexibility and precision of developmental control mechanisms. In this respect, mammalian teeth offer a good opportunity to use the fossil record to test our models of development. However, although the diversity of tooth shapes has been described in great detail and their evolutionary history has been relatively well reconstructed, our current knowledge of general mechanisms controlling tooth shape development is limited. As a first approximation, a great deal of mammalian molar tooth diversity derives from different combinations of cusps (Chapter 19). Cusp number, size and shape can vary from one tooth to the next and often cusps are connected by crests (lophs). Even this 'gross tooth' diversity constitutes a considerable challenge to a developmental biologist wishing to study the development of tooth shapes. Indeed, the list of students of comparative tooth development is extensive and extends back in time to Richard Owen (e.g. Owen, 1845; Leche, 1915; Bolk, 1920–22; Butler, 1956; Gaunt, 1961).

While the enthusiasm to study comparative ontogeny of different tooth shapes appears to have slowly declined, molecular studies of mouse tooth development have flourished (see Weiss, 1993; Thesleff et al., 1995; Thesleff and Sahlberg, 1996 for reviews). As a subject of study, the diversity of biological shapes has been transposed by the diversity of genes and their products. In practical terms this implies that, of total mammalian tooth diversity, we are left with the dentition of a mouse. This does not imply that mouse teeth are somehow uninteresting.

Rather, mouse, the standard laboratory model of mammalian development, has to represent all the rest of mammalian evolutionary diversity. This has of course prevented studies on, for example, the developmental control of premolars, which mice lack.

Recently, however, the problem of biological diversity has resurfaced among developmental biologists. This is, rather paradoxically, because new molecular discoveries have demonstrated an increasing number of developmental regulatory processes that are shared among different organs and organisms (which has also revived interest in homology, see Gilbert et al., 1996). Indeed, now the question is how to define a tooth using developmental parameters of regulatory processes. And this is where the development of cusps becomes important. If we simplify matters and define teeth as a set of cusps that have sufficient developmental plasticity to facilitate morphological evolution, we have restated the research programme to specify the developmental basis of morphological diversity (or evolutionary developmental biology, see Gilbert et al., 1996). As a great deal of mammalian molar tooth diversity comprises different arrangement, size and shape of cusps, a great many specific developmental mechanisms are presumably facilitating this diversity. So what is, developmentally speaking, a tooth cusp anyway? As a partial answer, we shall discuss in this chapter the role of the 'epithelial enamel knot' in tooth shape development. The enamel knot is an embryological structure whose origins go back to classical descriptive studies, but which has lain dormant in the literature for several decades. Recent discoveries concerning gene expression patterns, however, have suggested that, in a developing tooth, the enamel knot is very far from being dormant.

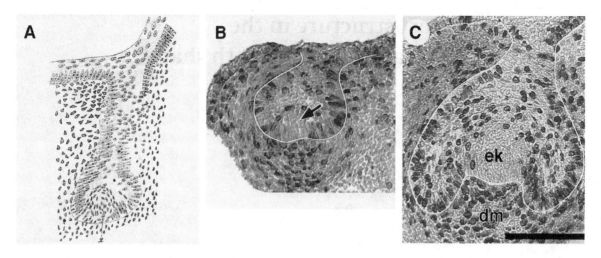

Figure 2.1A–C. **Transverse sections of tooth germs. A drawing (A) of late bud-stage bottlenose whale tooth germ (drawing from Ohlin, 1896) showing the characteristic appearance of the enamel knot cell cluster (x). The bud- (B) and cap-staged (C) first mouse lower molar histological sections showing the formation of the primary enamel knot. The first sign of the enamel knot (arrow in B) is marked by the cessation of the cell proliferation (unlabelled cells, proliferating cells have dark nuclei). The basement membrane separating epithelium and mesenchyme is marked with a white line. ek, enamel knot; dm, dental mesenchyme. Scale bar represents 100 μm in B and C.**

2.2.   Found and lost: the enamel knot

Enamel knots can be seen in several old descriptions of tooth development, starting more than a century ago (e.g. Ohlin, 1896). In these drawings of histological sections, the enamel knot is a clump of cells, usually in the centre of the tooth germ (Figure 2.1A). The packing and flattening of the cells in a histological section gives the enamel knot its characteristic look, hence names like 'spherical bodies' (Tims, 1901), 'Kugelkörper' (Heinick, 1908) or 'cellanhopningar' (Ohlin, 1896) have been used to describe the appearance of the tooth germs with the enamel knot ('a halved cabbage head' being quite accurate; Jernvall, 1995). Although usually recognized, their possible significance for tooth development was not considered. The enamel knot enjoyed a more prominent place in the literature during the first three decades of the twentieth century. Functional interpretations for the role of the enamel knot in tooth development include functioning as a reservoir of cells for the rapidly growing enamel organ, in particular for the stellate reticulum above the inner enamel epithelium (Butler, 1956). It has also been suggested to cause the epithelial-mesenchymal interface to fold by pushing or acting as a restraint (Ahrens, 1913). However, although the problem of its significance or even its existence never seems to quite subside, the enamel knot was gradually reduced to a developmental curiosity, possibly originating as a histological artifact (reviewed in Butler, 1956; also Mackenzie et al., 1992). The notion that the enamel knot is actually a widespread structure in developing teeth was proposed by Westergaard and Ferguson (1987) where they reported

enamel knots in crocodilians. The authors of this chapter also admit to ignoring the enamel knot for a long time. For us, the rediscovery started with the cells of the enamel knot itself.

2.3.   Cells different from all others: the enamel knot found again

The enamel knot is formed during early tooth morphogenesis in the centre of the tooth germ epithelium (Figure 2.1). It forms at the late bud stage and this also marks the beginning of tooth shape development. Moreover, the cells of the enamel knot cease to proliferate (Jernvall et al., 1994). This was our initial discovery (or rediscovery, there was disagreement in the literature on this matter, see Butler, 1956 for review), which drove our attention to this area of the tooth germ.

Before the actual cell division into two daughter cells, the genome has to be duplicated (S-phase of a cell cycle). The proliferating cells in the tooth germ can be localized by mapping cells that are replicating their DNA and incorporating a labelled DNA-base analogue. However, the standard histological sections used have the problem of leaving the three-dimensional associations of cell proliferation and tooth shape to an investigator's imagination. Therefore, because one mouse tooth germ can consist of up to a hundred sections, we decided early on to use three-dimensional reconstructions of serial sections to relate the cellular level processes to morphology. After the reconstructions, the

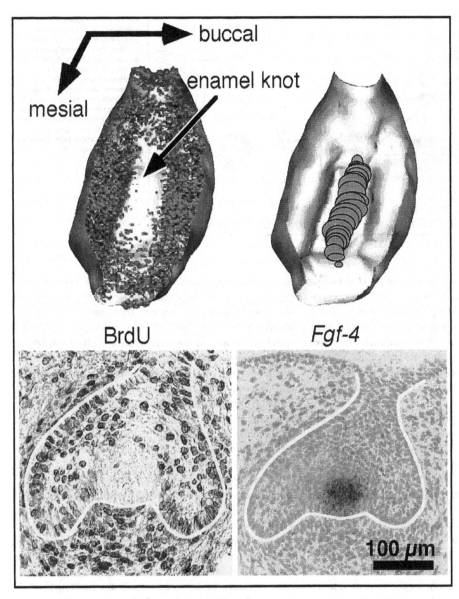

Figure 2.2. **A three-dimensional reconstruction of a mouse cap-stage first lower molar. The basement membrane side of the inner enamel epithelium is viewed from above at an oblique angle. Note how cell proliferation patterns (as incorporated BrdU) and expression pattern (detected messenger RNA) of fibroblast growth factor 4 gene (*Fgf-4*) are inverted: only the non-proliferating enamel knot cells express *Fgf-4*. Below corresponding transverse histological sections. The basement membrane is marked with a white line. (Adapted from Jernvall, 1995.)**

enamel knot was impossible to miss (Figure 2.2). Morphogenesis of the tooth crown begins as lateral growth of the tooth bud epithelium forms a characteristic 'cap'. All teeth, incisors as well as molars, pass through this stage. The enamel knot sits along the midline of this cap stage growth and one easily gets the impression that the tip of the tooth bud is fixed and grows laterally instead. Moreover, the enamel knot is present only during the formation of the tooth germ cap stage. The knot disappears and the cells of the inner enamel epithelium resume proliferation, and the whole life span of the enamel knot is less that 2 days in a mouse molar (Jernvall, 1995). The transition of a tooth bud to a cap offers us a very clear

example of allometric growth which happens around the enamel knot.

The return of the enamel knot from oblivion was finally clinched by the discovery that their cells have a distinct gene expression pattern that differs from the rest of the tooth germ. First, Mackenzie *et al.*, (1992) found that the transcription factor Msx-2 had an expression pattern largely restricted to the cells of the enamel knot in a cap staged tooth. Second, fibroblast growth factor 4 (*Fgf-4*) was reported to be expressed in a specific manner in a tooth germ (Niswander and Martin, 1992). Particularly the expression of *Fgf-4* turned out to be remarkably restricted to the cells of the enamel knot

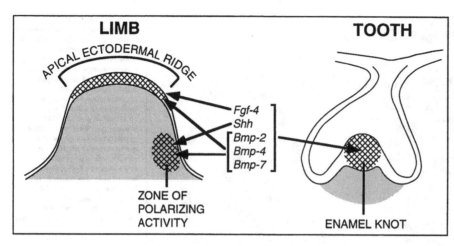

Figure 2.3 **Schematic comparison of expression patterns of some signalling molecules in a limb bud and in a cap-staged tooth. Limb and tooth mesenchyme are marked in grey. While different sets of signalling molecules are expressed in the limb bud in one mesenchymal and one epithelial domain, in the teeth these signalling molecules are expressed in the epithelial enamel knot.**

(Jernvall *et al.*, 1994). So much so, that now we had three markers for the enamel knot cells; the histology, cessation of cell proliferation and *Fgf-4* expression (Figure 2.2). While this should have been enough to convince most embryologists of the existence of the enamel knot, it also resurrected the question about its functions during tooth morphogenesis. However, because of the new tools in molecular biology, we now had a little bit more to work with.

Fgf-4 expression by the non-proliferating cells of the enamel knot is intriguing because secreted FGF-4 protein is a known mitogen in vertebrate limbs stimulating cells of the limb mesenchyme to divide (Niswander and Martin, 1993; Vogel and Tickle, 1993). Our experiments confirmed this to be the case for both the dental epithelium and mesenchyme *in vitro* (Jernvall *et al.*, 1994). However, these results implied a slightly paradoxical situation for the cells of the enamel knot itself; the same cells that do not divide, are manufacturing a protein that stimulates cells to divide. On the other hand, this could fulfill the minimum requirement for the control of tooth bud folding and formation of the cap stage: the enamel knot may cause the unequal growth of the epithelium by concurrently remaining non-proliferative and by stimulating growth around it (Jernvall *et al.*, 1994; Jernvall, 1995; Vaahtokari *et al.*, 1996a). The combination of the non-dividing enamel knot cells and surrounding dividing cells could create packing of cells around the knot resulting in folding and downgrowth of the epithelium around it. While this kind of 'physical pushing' has been argued to play an important role in the initiation of cusp development (e.g., Butler, 1956; Osborn, 1993), molecular evidence supporting it had been lacking.

But genes never function alone. Vertebrate limb development is an excellent example of the multigene cascades involved in shape development. In the limb bud, many signalling molecules are present in two separate signalling tissues: the apical ectodermal ridge (AER) and the zone of polarizing activity (ZPA), which have been shown to control proximodistal growth and anteroposterior patterning, respectively (Tabin, 1991). *Fgf-4*, which is expressed in the tooth germ in the enamel knot, is restricted to the posterior AER in the limbs (Niswander and Martin, 1992; Suzuki *et al.*, 1992). Other known signalling molecules include sonic hedgehog (*Shh*) and bone morphogenetic proteins (*Bmp-2, -4, -7*). In the limb buds, *Shh* is expressed only in the ZPA (Echelard *et al.*, 1993) while the *Bmps* are expressed both in the AER and in the ZPA (Francis *et al.*, 1994; Lyons *et al.*, 1995). Positive feedback loops operate between the AER and the ZPA, and for example, once induced, *Shh* expression in the ZPA is maintained by FGF-4 signal from AER (Laufer *et al.*, 1994; Niswander *et al.*, 1994).

Contrasting with limbs, there seems to be only one signalling centre in the cap stage tooth germ, the enamel knot. And, at least on the molecular level, the enamel knot performs the duties of both limb signalling centres (Figure 2.3). *Fgf-4*, *Shh* and *Bmp-2, -4*, and *-7* are all expressed in the enamel knot (Vaahtokari *et al.*, 1996b). This very centralized expression pattern of several signalling molecules also explains why the role of the enamel knot has been difficult to demonstrate experimentally. There has been no way to 'dissect' the roles of different centres because there is only one present. Hence, one gets an individual tooth or no tooth.

The similarities between different signalling centres are not only limited to signalling molecules. Also the same transcription factors are expressed in the enamel knot and limb signalling centres. The already mentioned homeobox-containing *Msx-2* gene (Mackenzie *et al.*, 1992) is expressed both in the limb AER and in the enamel knot. Transcription factors function like cell's internal switches that can cause the cell to progress to the next developmental phase (differentiation, mitosis) and to manufacture new sets of signalling molecules. These signalling molecules, on the other hand, may trigger the neighbouring cells to proceed to the next developmental phase, which may or may not be the same as that of the signalling cells. Also the signalling cells themselves may respond to a signal called autocrine induction (in contrast to paracrine – between populations of cells). The exact biochemical role of *Msx-2* is not known, but it appears to be associated with programmed cell death (apoptosis) (Graham *et al.*, 1994). Recent findings indicate that the signalling of apoptosis is another universal molecular cascade in the embryo. Both BMP-4 and *Msx-2* have been shown to be involved in apoptosis of interdigital tissue in chick limbs and rhombomeres (Graham *et al.*, 1994; Zou and Niswander, 1996). While *Msx-2* transcripts are present in the cells that are going to go into apoptosis, the actual (autocrine) signalling molecule that mediates apoptosis is BMP-4 (Graham *et al.*, 1994; Zou and Niswander, 1996). And, perhaps no surprise, the disappearance of the enamel knot happens via apoptosis (Vaahtokari *et al.*, 1996a) and *Bmp-4* expression marks the death of its cells as well (Jernvall *et al.*, 1998).

While the enamel knot cells appear distinctly different from all the other cells of the tooth germ, no 'enamel knot genes' have been discovered that would put the enamel knot apart from other embryonic signalling centres and their molecular cascades. Indeed, the enamel knot nicely demonstrates how different organs can be made using quite similar developmental processes or modules (Wray, 1994; Gilbert *et al.*, 1996, Raff, 1996). These developmental modules are homologous as processes even though the structures themselves (such as a limb and a tooth) are not homologous (Bolker and Raff, 1996; Gilbert *et al.*, 1996). Roth (1988) proposed that new morphological structures can result from 'genetic piratism'. The signalling functions and the apoptotic pathways may have risen only once (or a few times, convergence and parallelism are rampant in the fossil record, e.g., Hunter and Jernvall, 1995), but these molecular cascades have been co-opted (pirated) repeatedly to generate new structures. Whereas the enamel knot may not have any specific genes of its own, the *combination* of active signalling molecules and transcription factors may well be unique (Figure 2.3). Overtly simplifying, we can formulate that limb AER + ZPA = EK (the enamel knot). That is, the molecular identity of the enamel knot is the sum of molecular activities of two limb signalling centres. Evolutionarily speaking though, the enamel knot may represent a signalling centre for general epithelial-mesenchymal organs (e.g. similar cascades are likely to be found in feathers, Nohno *et al.*, 1995), and limb signalling centres may have evolved with the endoskeleton resulting from a partly pirated developmental pathway from an epithelial–mesenchymal centre.

2.4.   Back to cusps: how many enamel knots are there?

A reasonable hypothesis is that the enamel knot may be important for tooth shape development. Indirect evidence of this is the discovery that multicusped teeth have more than one enamel knot (Jernvall *et al.*, 1994). The classical enamel knot that forms at the end of the bud stage correlates with the beginning of tooth crown shape development. The tooth germ grows around it and the knot sits at the tip of the folding. At least for rodent incisors and crocodilian teeth, this is the one and only enamel knot present. But in teeth with several cusps (e.g. mouse and opossum molars; Figure 2.4), secondary enamel knots form later at the tips of each cusp (Jernvall *et al.*, 1994; Jernvall, 1995). The secondary knots can be detected at the onset of, or even slightly prior to, the visible development of the cusps (Figure 2.4). Therefore, a secondary enamel knot 'marks the cusp' and once cusp development is initiated, the knot disappears (Jernvall *et al.*, 1994; Jernvall, 1995). This, again, happens via apoptosis (Vaahtokari *et al.*, 1996a).

The existence of secondary enamel knots has been even more elusive than the first (primary) enamel knot. Actually, early workers sometimes recognized them, but no distinction was made between the primary and secondary knots. And the secondary knots can look quite similar to the primary knot: the cells are packed, postmitotic, and they express *Fgf-4* gene (Jernvall *et al.*, 1994). The secondary enamel knots are not, however, as different from the rest of the tooth germ as the primary enamel knot. While the expression of many signalling molecules is tightly restricted to the primary enamel knot, during the formation of secondary enamel knots, the expression domains of these signals are broader (Vainio *et al.*, 1993; Bitgood and McMahon, 1995). Indeed, only *Fgf-4* remains restricted to enamel knots throughout the odontogenesis. This may reflect a more restricted molecular function for secondary enamel knots as compared with the primary enamel knot.

The short lifespan of individual enamel knots may mean that the control for cusp development is needed only intermittently. In order for a cusp to develop, it has to be initiated, and the initiation of a cusp appears to be the most important event in its development. The sequence of cusp formation closely follows the evolutionary appearance of the cusps, and usually the first cusp

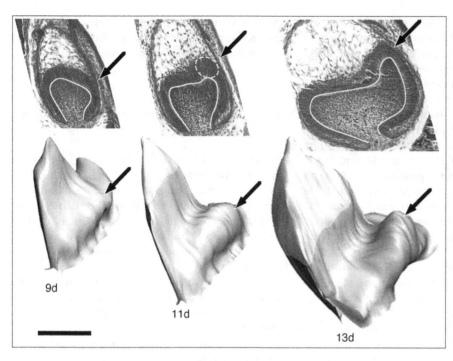

Figure 2.4. **Three-dimensional reconstructions of the opossum (*Monodelphis domestica*) first lower molar development (obliquely lingual view, mesial is toward the viewer) showing the formation of metaconid (arrows). In the histological section of the 11-day-old tooth (11d), the secondary enamel knot is visible as a maximum of 90 µm thickening (average 68 µm) of the inner enamel epithelium (white dashed circle). No secondary enamel knot is detected in the 9- (9d) or 13-day-old (13d) tooth germs (maximum thickness of the metaconid area inner enamel epithelium = 53 µm and average thickness = 40 µm). Note the mineralization of protoconid tip (marked as white) by the time of the initiation of metaconid. The basement membrane, from which the 3-D reconstructions were made, is marked with a white line in the histological sections. Scale bar represents 200 µm.**

(the primary cusp) to develop is the protoconid in the lower, and the paracone in the upper teeth (e.g. Butler, 1956; Luckett, 1993). However, because cusps grow from the tip down, the order of initiation of cusp development is the main determinant of relative cusp size and phylogenetically 'recent' but large cusps can start to form prior to phylogenetically older cusps (Berkovitz, 1967; Butler, 1956, 1967). Therefore, for a cusp to become the right size, its development has to be initiated at the right moment. The control of cusp initiation by the enamel knots is perhaps all that is required for the cusps to form. So, a very simplistic view of tooth shape development has basically two variables; where and when. The placement of secondary enamel knots determines the cusp pattern, and their timing determines the relative height of the cusps.

One complication is the function of the primary enamel knot in the initiation of cusp development. Does it initiate the paracone on the upper molars and protoconid on the lower molars, the phylogenetically oldest cusps? Currently we do not really know this. The primary enamel knot establishes the whole crown area (including talonid, Figure 2.4), and the temporal initiation of the first cusp is initiated from the 'remnants' of the primary enamel knot. Thus, it is possible that the enamel knot of the primary cusp is actually a portion of the primary knot. This is an issue that can be helped with careful mapping of different molecular markers.

2.5.   Developmental control of tooth shape?

We certainly do not yet know what controls tooth shape development. Even in itself, tooth shape is a rather elusive concept. Tooth shape is not the same as tooth family (incisor, canine, premolar or molar). Tooth family is defined by relative position, and individual teeth can evolve shapes that are found in adjacent tooth families (e.g. molarization of premolars in many ungulates, 'premolarization of molars' in seals). The existence of different tooth families manifests developmental decoupling of these sets of teeth, and may be as old as heterodonty itself. The determination of tooth families may be under homeobox-containing genes which may create a patchwork of overlapping gene expression domains (for Dlx genes see Chapters 1, 11). The actual determination of tooth shape during development is probably a downstream process from the specification of tooth family. The primary enamel knot can be considered as one candidate for the control of actual tooth shape. Its timely appearance, and cellular functions suggest that it is needed for tooth morphogenesis. Indeed, mutant mice with experimentally inactivated transcription factors *Msx-1* and *Lef-1* (Satokata and Maas, 1994; van Genderen *et al.*, 1994) have their tooth development arrested to the bud stage and no enamel knot forms. Both these genes are normally expressed in the tooth bud (*Msx-1* is needed in the mesenchyme, *Lef-1* in the epithelium; Satokata and

Maas, 1994; Kratochwil *et al.*, 1996). Evidently, the primary knot is needed for the tooth morphogenesis to take place. On the other hand, the appearance of the secondary enamel knots would mark the actual process of crown shape development. How are the secondary enamel knots created? The signalling molecules of the primary enamel knot make it tempting to hypothesize that the molecular signals diffusing from the primary knot determine the locations and activation times for each secondary knot. It is also possible that the secondary enamel knots form at boundaries of different gene expression domains and the elucidation of this will require careful analysis of gene expression patterns. However, there are at least two problems.

First, this theory appears to contradict the prevailing concept that the mesenchymal dental papilla determines the shape of teeth. The tissue recombination studies have shown that the dental papilla mesenchyme has an instructive role in regulating the shape of teeth. During cap stage, the molar dental papilla directs molar tooth development when associated with incisor enamel organ epithelium and vice versa (Kollar and Baird, 1969, 1970). However, because these experiments were done using tissues from the same species (mouse), it is difficult to determine whether mesenchyme is instructing the tooth type only (incisor/molar) or also the individual species-specific tooth cusp patterns. Therefore, in these recombinants the dental papilla may induce in the epithelium the formation of a primary enamel knot that is tooth-type specific and which subsequently directs the formation of secondary enamel knots and cuspal pattern inherent to the dental epithelium. Alternatively, it is possible that the mesenchymal dental papilla alone, without epithelial influence is capable of determining the placement of secondary enamel knots. Hence, it still remains to be demonstrated how the mesenchymal dental papilla and the primary enamel knot interact in the regulation of the secondary enamel knots and subsequently the tooth shape.

The second problem is that in most molar teeth, the primary knot is long gone by the time the last secondary knots appear and cusps form (Figure 2.4). A primary knot-centric model on tooth morphogenesis would thus predict that the cusp patterns are determined early during morphogenesis. This could also mean an increasing relaxation in the control of later developing cusps. This, of course, usually means the smallest cusps, and while it is generally known that small cusps tend to be variable in number, it is not really known if this is a result of increasing variability in the later developing cusps. A second possibility is that the secondary enamel knots affect the formation of each other. This could be via signalling molecules (e.g. *Fgf-4*), and each secondary

knot could create an inhibition field around it preventing a new secondary knot from forming too close (Jernvall, 1995). Indirect evidence for 'enamel knot inhibition' comes from a mouse molar. The first lower molar of a mouse is, by any standards, a small tooth (1.5 mm long). It, however, has six (some fused) cusps that are equally high. This creates a developmental challenge. To be able to grow several equally high cusps on a very small tooth, there are essentially two options. A tooth can initiate cusp development at a very small size and then gradually add more cusps as is seen in larger molars (Jernvall, 1995). The other option is to initiate the cusps at a larger size, but almost simultaneously. It is the latter option we observe in a mouse molar. This mouse molar has a large primary enamel knot that disappears quickly (24 h), and within 2 days all the main cusps are initiated (Jernvall, 1995). The gradual development of cusps may not be an option for a mouse tooth because the tooth germ would have to be extremely small (40 cells in length) when the cusp development would start and small disturbances in growth would have proportionally larger effects on the final form.

Summary

One can be fairly certain that, like other aspects of tooth development including initiation and cell differentiation, cusp development is regulated by epithelial–mesenchymal interactions. Conceivably, the formation of secondary enamel knots is affected by the underlying mesenchyme, and the mesenchyme is reciprocally affected by the epithelium and the enamel knots. It is this interplay of different tissues and signalling centres that controls shape development. In order for the correct tooth shape to form, cusp development has to be initiated in the right place at the right time. The enamel knots are probably mediators of this process. But the exact mechanism as to how the formation of the enamel knots is controlled, as well as the regulation of other aspects of tooth shape, such as tooth crown height, are still waiting to be discovered. And some of the evidence is perhaps still dormant in the literature.

Acknowledgements

The *Monodelphis* sections were kindly made available by Dr Kathleen Smith, Durham, NC, USA.

References

Ahrens, K. (1913). Die Entwicklung der menschlichen Zahne. *Arbeiten der anatomisches Institut, Wiesbaden*, **48**, 169–266.

Berkovitz, B. K. B. (1967). The dentition of a 25–day pouch-young specimen of *Didelphis virginiana* (Didelphidae: Marsupialia). *Archives of Oral Biology*, **12**, 1211–1212.

Bitgood, M. J. and McMahon, A. P. (1995). *Hedgehog* and *Bmp* genes are coexpressed at many diverse sites of cell–cell interaction in the mouse embryo. *Developmental Biology*, **172**, 126– 158.

Bolk, L. (1920–1922). Odontological essays. *Journal of Anatomy*, **55**, 138–186, 219–234; **56**, 107– 136; **57**, 55–75.

Bolker, J. A. and Raff, R. A. (1996). Developmental genetics and traditional homology. *Bioessays*, **18**, 489–494.

Butler, P. M. (1956). The ontogeny of molar pattern. *Biological Reviews*, **31**, 30–70.

Butler, P. M. (1967). Dental merism and tooth development. *Journal of Dental Research* (Suppl. 5), **46**, 845–850.

Echelard, Y., Epstein, D. J., St-Jacques, B., Shen, L., Mohler, J., McMahon, J. A. and McMahon, A. (1993). *Sonic hedgehog*, a member of a family of putative signalling molecules, is implicated in the regulation of CNS and limb polarity. *Cell*, **75**, 1417–1430.

Francis, P. H., Richardson, M. K., Brickell, P. M. and Tickle, C. (1994). Bone morphogenetic proteins and a signalling pathway that controls patterning in the developing chick limb. *Development*, **120**, 209–218.

Gaunt, W. A. (1961). The development of the molar pattern of the golden hamster (*Mesocricetus auratus* W.), together with a reassessment of the molar pattern of the mouse (*Mus musculus*). *Acta Anatomica*, **45**, 219–251.

van Genderen, C., Okamura, R. M., Farinas, I., Quo, R. G., Parslow, T. G. and Bruhn, L. (1994). Development of several organs that require inductive epithelial-mesenchymal interactions is impaired in LEF-1–deficient mice. *Genes and Development*, **8**, 2691–2703.

Gilbert, S. F., Opitz, J. M. and Raff, R. A. (1996). Resynthesizing evolutionary and developmental biology. *Developmental Biology*, **173**, 357–372.

Graham, A., Francis-West, P., Brickell, P. and Lumsden, A. (1994). The signalling molecule BMP4 mediates apoptosis in the rhombencephalic neural crest. *Nature*, **372**, 684–686.

Heinick, P. (1908). Über die Entwicklung des Zahnsystems von Castor fibre L. *Zoolische Jahrbüchern, (Anat.)*, **XXVI**, 355–402.

Hunter, J. P. and Jernvall, J. (1995). The hypocone as a key innovation in mammalian evolution. *Proceedings of the National Academy of Sciences, USA*, **92**, 10718–10722.

Jernvall, J. (1995). Mammalian molar cusp patterns: developmental mechanisms of diversity. *Acta Zoologica Fennica*, **198**, 1–61.

Jernvall, J., Kettunen, P., Karavanova, I., Martin, L. B. and Thesleff, I. (1994). Evidence for the role of the enamel knot as a control center in mammalian tooth cusp formation: non-dividing cells express growth stimulating *Fgf-4* gene. *International Journal of Developmental Biology*, **38**, 463–469.

Jernvall, J., Åberg, T., Kettunen, P., Keränen, S. and Thesleff, I. (1998). The life history of an embryonic signaling center: BMP-4 induces p21 and is associated with apoptosis in the mouse tooth enamel knot. *Development*, **125**, 161–169.

Kollar, E. J. and Baird, G. R. (1969). The influence of the dental papilla on the development of tooth shape in embryonic mouse tooth germs. *Journal of Embryology and Experimental Morphology*, **21**, 131–148.

Kollar, E. J. and Baird, G. R. (1970). Tissue interactions in embryonic mouse tooth germs. II. The inductive role of the dental papilla. *Journal of Embryology and Experimental Morphology*, **24**, 173–186.

Kratochwil, K., Dull, M., Farinas, I., Galceran, J. and Grosschedl, R. (1996). *Lef1* expression is activated by BMP-4 and regulates inductive tissue interactions in tooth and hair development. *Genes and Development*, **10**, 1382–1394.

Laufer, E., Nelson, C. E., Johnson, R. L., Morgan, B. A. and Tabin, C. (1994). *Sonic hedgehog* and *Fgf-4* act through a signalling cascade and feedback loop to integrate growth and patterning of the developing limb bud. *Cell*, **79**, 993–1003.

Leche, W. (1915). Zur Frage nach der stammesgeschichtlichen Bedeutung des Milchgebisses bei den Säugetieren. *Zoologischen Jahrbüchern* (syst.) **XXXVIII**,, 275–370.

Luckett, W. P. (1993). Ontogenetic staging of the mammalian dentition, and its value for assessment of homology and heterochrony. *Journal of Mammalian Evolution*, **1**, 269–282.

Lyons, K. M., Hogan, B. L. and Robertson, E. J. (1995). Colocalization of BMP-7 and BMP-2 RNAs suggests that these factors cooperatively mediate tissue interactions during murine development. *Mechanisms of Development* 50:71–83.

Mackenzie, A., Ferguson, M. W. J. and Sharpe, P. T. (1992). Expression patterns of the homeobox gene, *Hox-8*, in the mouse embryo suggest a role in specifying tooth initiation and shape. *Development*, **115**, 403–420.

Niswander, L. and Martin, G. R. (1992). *Fgf-4* expression during gastrulation, myogenesis, limb and tooth development in the mouse. *Development*, **114**, 755–768.

Niswander, L. and Martin, G. R. (1993). FGF-4 and BMP-2 have opposite effects on limb growth. *Nature*, **361**, 68–71.

Niswander, L., Jeffrey, S., Martin, G. R. and Tickle, C. (1994). A positive feedback loop coordinates growth and patterning in the vertebrate limb. *Nature*, **371**, 609–612.

Nohno, T., Kawakami, Y., Ohuchi, H., Fujiwara, A., Yoshika, H. and Noji, S. (1995). Involvement of the *Sonic hedgehog* gene in chick feather formation. *Biochemical and Biophysical Research Communications*, **206**, 33–39.

Ohlin, A. (1896). Om tandutvecklingen hos hyperoodon. *Bihang till Kongliga Svenska Vetenskaps-Akademiens Handlingar*, **Band 22** (Afd IV, No 4), 1–33.

Osborn, J. W. (1993). A model simulating morphogenesis without a morphogen. *Journal of Theoretical Biology*, **165**, 429–445.

Owen, R. (1845). *Odontography*. London: Hippolyte Baillière.

Raff, R. A. (1996). *The Shape of Life*. chicago: University of Chicago Press.

Roth, V. L. (1988). The biological basis of homology. In *Ontogeny and Systematics*, ed. C. J. Humphries, pp. 1–26. New York: Columbia University Press.

Satokata, I. and Maas, R. (1994). *Msx1*-deficient mice exhibit cleft palate and abnormalities of craniofacial and tooth development. *Nature Genetics*, **6**, 348–356.

Suzuki, H. R., Sakamoto, H., Yoshida, T., Sagimura, T., Terada, M. and Solursh, M. (1992). Localization of *Hst1* transcripts to the apical ectodermal ridge in the mouse embryo. *Developmental Biology*, **150**, 219–222.

Tabin, C. J. (1991). Retinoids, homeoboxes, and growth factors: toward molecular models for limb development. *Cell*, **66**, 199–217.

Thesleff, I. and Sahlberg, C. (1996). Growth factors as inductive signals regulating tooth morphogenesis. *Seminars in Cell and Developmental Biology*, **7**, 185–193.

Thesleff, I., Vaahtokari, A. and Partanen, A.-M. (1995). Regulation of organogenesis. Common molecular mechanisms regulating the development of teeth and other organs. *International Journal of Developmental Biology*, **39**, 35–50.

Tims, H. W. M. (1901). Tooth genesis in the Caviidae. *Journal of the Linnean Society (Zoology)*, **28**, 261–290.

Vaahtokari, A., Åberg, T. and Thesleff, I. (1996a). Apoptosis in the developing tooth: association with an embryonic signalling center and suppression by EGF and FGF-4. *Development*, **122**, 121–126.

Vaahtokari, A., Åberg, T., Jernvall, J., Keränen, S. and Thesleff, I. (1996b). The enamel knot as a signalling center in the developing mouse tooth. *Mechanisms of Development*, **54**, 39–43.

Vainio, S., Karavanova, I., Jowett, A. and Thesleff, I. (1993). Identification of BMP-4 as a signal mediating secondary induction between epithelial and mesenchymal tissues during early tooth development. *Cell*, **75**, 45–58.

Vogel, A. and Tickle, C. (1993). FGF-4 maintains polarizing activity of posterior limb bud cells *in vivo* and *in vitro*. *Development*, **119**, 199–206.

Weiss, K. M. (1993). A tooth, a toe, and a vertebra: the genetic dimensions of complex morphological traits. *Evolutionary Anthropology*, **2**, 121–134.

Westergaard, B. and Ferguson, M. W. J. (1987). Development of dentition in *Alligator mississippiensis*. Later development in the lower jaws of embryos, hatchlings and young juveniles. *Journal of Zoology*, **212**, 191–222.

Wray, G. A. (1994). Developmental evolution: new paradigms and paradoxes. *Developmental Genetics.*, **15**, 1–6.

Zou, H. and Niswander, L. (1996). Requirement for BMP signalling in interdigital apoptosis and scale formation. *Science*, **272**, 738–741.

3 Molecules implicated in odontoblast terminal differentiation and dentinogenesis

J. V. Ruch and H. Lesot

J. V. Ruch and H. Lesot

3.1. Introduction

Histological organization, cytological features as well as functional aspects confer specificity to the odontoblasts. These post-mitotic, neural crest-derived cells are most often aligned in a single layer at the periphery of the dental pulp and secrete the organic components of pre-dentin-dentin, including collagens and non-collagenous proteins. The functional odontoblasts, with one or more cytoplasmic processes extending into the predentin-dentin are connected by junctional complexes which probably do not completely seal the inter-odontoblastic spaces. The dentin extracellular matrix shows great similarities to bone matrix. However Smith and Hall (1990) strongly supported the phylogenetic emergence of distinct osteogenic and dentinogenic cell lineages.

Odontoblast terminal differentiation, allowing for dentinogenesis, represents the end point of a developmental sequence initiated during specification of neural crest-derived connective cell phenotypes and including patterning of the dental cells in the developing arches. Tooth class-specific histomorphogenesis leads to tooth-specific spatial distribution of functional odontoblasts (see Jernvall, 1995; Ruch, 1985, 1987, 1995a; Ruch et al., 1995 for reviews). Interestingly the analysis of the fate of uniparental mouse cells in chimeras indicated that the odontoblasts have a maternal origin (Fundele and Surani, 1994); thus genetic imprinting might affect odontogenesis.

Chapter 5 of this book examines the origin and early evolution of dentine, and Chapter 6 examines pulpo-dentinal interactions in the development and repair of dentine. Both include discussions of the initial developmental aspects of the odontoblast cell lineage. Here we concentrate exclusively on the physiological terminal differentiation of these cells. As far as reactionary and reparative dentins are concerned see Lesot et al., 1994; Smith et al., 1995; Tziafas, 1995.

3.2. Cytological aspects of odontoblast terminal differentiation

The terminal differentiation of odontoblast is initiated at the tip of the cusps, progressing in an apical direction and occurs in each cusp according to a specific temporo-spatial pattern (Figure 3.1). The preodontoblasts located adjacent to the basement membrane, beneath the inner dental epithelium, show no special orientation and contain few cisternae of rough endoplasmic reticulum, only small Golgi apparatus and few mitochondriae (Garant and Cho, 1985). These preodontoblasts probably withdraw sequentially from the cell cycle (Ruch, 1990). During the last division of the mouse preodontoblast, the mitotic spindle lies perpendicular to the epithelial–mesenchymal junction (Ruch, 1987) and the daughter cell in contact with the basement membrane will overtly differentiate. These cells demonstrate rapid hypertrophy and polarization: the nuclei take up an eccentric, basal, position; the supra-nuclear Golgi complex develops and cisternae of granular endoplasmic reticulum flatten and become parallel to the long axis of the cell (Figures 3.2, 3.3). Polarization of these cells is accompanied by an apical accumulation of actin, actinin and vinculin (Kubler et al., 1988) as well as a redistribution of vimentin (Lesot et al., 1982). Cytokeratin 19 has also been found in differentiating odontoblasts (Webb et al., 1995) and functional odontoblasts express the intermediate filament nestin (Terling et al., 1995). The important role of the

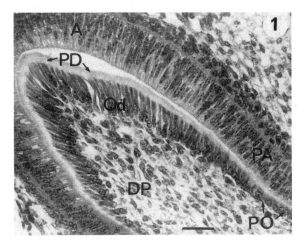

Figure 3.1. **Frontal section of a day-18.5 (p.c.) first lower mouse molar (main cusp) illustrating the gradient of odontoblast terminal differentiation. Od, functional odontoblasts; PD, predentin; PO, preodontoblasts; PA, preameloblasts; DP, dental papilla. (From Ruch** *et al.,* **1995 with permission.)**

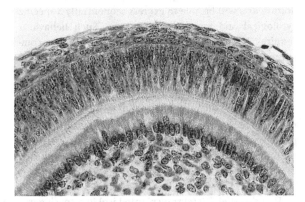

Figure 3.2. **Histological aspect of functional odontoblasts.**

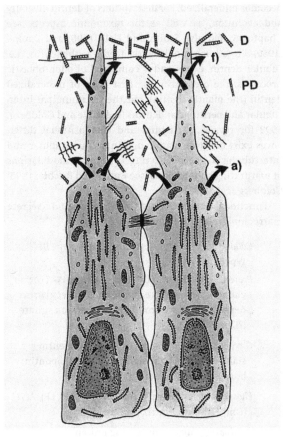

Figure 3.3. **Schematic representation of functional odontoblasts. PD, predentin; D, dentin. The arrows indicate the levels of secretion of matrix components: a) base of odontoblast process, f) mineralization front (From Ruch** *et al.,* **1995.)**

cytoskeleton in odontoblast differentiation was recently stressed in transgenic mice which express desmin under the direction of the vimentin promoter (Berteretche *et al.,* 1993). Young odontoblasts are characterized by numerous processes oriented toward the basement membrane. Fully differentiated odontoblasts are 50–60 μm long columnar cells which are formed by the cell body demonstrating polarized distribution of the organelles and one odontoblastic process containing a well-developed cytoskeleton and exocytotic and endocytotic vesicles (Garant and Cho, 1985; Goldberg and Takagi, 1993; and references therein). During the terminal, functional differentiation, the frequency of interodontoblastic junctions, comprising tight and gap junctions as well as zonulae adherens, increases (Linde and Goldberg, 1993 and references therein). The existence of a permeability barrier in the odontoblastic layer has recently been discussed (Bishop and Yoshida, 1992). The old question of von Korff

fibres, i.e. the presence of collagen fibres in between functional odontoblasts, is not yet definitively resolved (Ten Cate, 1989; Bishop *et al.,* 1991; Ohsaki and Nagata, 1994; Sawada and Nanci, 1995; Yoshiba *et al.,* 1994, 1995). The extent of the odontoblast processes within the dentinal tubule is also the subject of some controversy (Holland, 1976; Sigal *et al.,* 1984; Meyer *et al.,* 1988; Byers and Sugaya, 1995).

Age-related changes exist such that human odontoblasts, after long dentinogenic activity, constitue a nonrenewing cell population. Changes in the shape of the nucleolus and degenerative changes have been reported (Couve, 1986).

3.3.   **Functional aspects**

Odontoblasts are involved in production of predentindentin matrix in an extracellular framework that becomes mineralized. For descriptions of dentin diversity and evolution, as well as for taxonomic aspects, see Chapter 5, and also Smith and Hall (1990) and Baume (1980). In all mammals with teeth, orthodentin, i.e. tubular dentin, exists and is composed of odontoblastic processes, the predentin, the first layer of mineralized dentin (the mantle dentin) and the circumpulpal inter-tubular and peritubular dentins (see Linde and Goldberg, 1993 for review). Structural and compositional differences exist between mantle dentin and peritubular and intertubular dentins reflecting functional modulations of maturating odontoblasts (see Butler and Ritchie, 1995; Goldberg *et al.*, 1995).

Functional odontoblasts synthesize and secrete matrix molecules including:

Collagens: Type I, type I trimer, type V, type III (?), type VI

Proteoglycans: decorin (PG II), biglycans (PG 1) other chondroitin-4 sulfate containing proteoglycans, dermatan sulfate proteoglycan, keratan sulfate (?), perlecan (?)...

Glycoproteins/sialoproteins: osteonectin, dentin sialoprotein(s), bone sialoprotein, osteopontin, bone acidic glycoprotein 75

Phosphoproteins: phosphophorins (HP, MP, LP), AG1 (Dmp1, dentin matrix protein 1)

γ-Carboxyglutamate-containing proteins: osteocalcin, matrix GLA protein

Lipids, a minor part of dentin organic components, comprise phospholipids (phosphatidyl-choline, phosphatidylethanolamine), cholesterol, cholesterol esters and triacylglycerols...

Serum derived proteins including α_2HS-glycoprotein, albumin, immunoglobulins are also present in predentin/dentin.

For recent reviews see Ruch (1995a), and Wöltgens *et al.* (1995).

Some of these dentin components may or may not be dentin specific and expressed exclusively by odontoblasts (e.g. dentin sialoprotein(s), dentin phosphoprotein(s) and Dmp1 (Butler, 1995; George *et al.*, 1995).

Transcripts for type IV collagenase MMP-2 have also been detected in functional odontoblasts. This enzyme might be involved in basement membrane degradation (Heikinheimo and Salo, 1995). Porcine predentin-dentin contains metalloproteinases: gelatinases and proteogly-canases (Fukae *et al.*, 1994).

Finally, functional odontoblasts play an important role in calcium transport. Higher calcium concentration in predentin compared with the dental pulp (almost three times higher in the former) implies the existence of an ion-concentrating mechanism across the odontoblast layer (Linde and Lundgren, 1995 and references therein). Odontoblast calcium transport mechanisms probably imply Ca^{2+}-activated ATPase, Na^+/Ca^{2+} exchanger, calcium channels of L^-, N^- and T types (see Linde and Goldberg, 1993). Ca^{2+}-binding proteins like calmodulin, the 28 kDa calbindin and parvalbumin (Hubbard *et al.*, 1981; Taylor, 1984; Magloire *et al.*, 1988; Berdal *et al.*, 1989) and annexins III to VI (Goldberg *et al.*, 1991) have been shown to be present in functional odontoblasts.

3.4.   Control of odontoblast terminal differentiation

Odontoblasts overtly differentiate according to a cusp-specific temporospatial pattern. That means that preo-dontoblasts will be able to express sequentially a specific cytological and metabolic phenotype. Such behaviour requires both epigenetic signalling and specific com-petence.

3.4.1.   Epigenetic signalling

To the best of our knowledge, preodontoblasts never give rise to a functional odontoblast layer when dental pa-pillae, isolated by treatment with proteolytic enzymes, are either cultured *in vitro* in conventional media or grafted *in vivo* (Kollar and Baird, 1970; Ruch and Karcher-Djuricic, 1971; Kollar, 1972; Ruch *et al.*, 1973).

Huggins *et al.* (1934) transplanted isolated tooth germ tissues in heterotopic positions and suggested that the inner dental epithelium plays a role in odontoblast dif-ferentiation. Koch (1967), analysing cultures of transfilter associations of enamel organs and dental papillae, con-cluded that epithelial-mesenchymal interactions were necessary for the functional differentiation of odonto-blasts. Both experimental investigations performed during the past 25 years and descriptive data demon-strate that stage- and space-specified preameloblasts control odontoblast terminal differentiation. This inter-action is mediated by the basement membrane, a dynamic asymmetric interface, acting both as a specific substrate and as a reservoir for paracrine factors (for reviews see Ruch, 1995a; Ruch *et al.*, 1995 and references therein).

Putative role of matrix molecules and substrate adhesion molecules

Fibronectin. This glycoprotein(s), of mesenchymal origin is associated with the basement membrane, surrounds preodontoblasts and accumulates at the apical pole of polarizing odontoblasts.

When investigating the ability of fibronectin to interact with plasma membrane proteins transferred to polyvinylidenedifluoride membrane, we observed an interaction with three high molecular weight proteins present in membranes prepared from dental mesenchyme but not from the enamel organ (Lesot et al., 1985). An immunological approach allowed study of one of these proteins with an apparent molecular weight of 165 kDa. Two monoclonal antibodies directed against this protein were used to investigate its localization and function. The 165 kDa protein and fibronectin transitorily accumulated at the apical pole of polarizing odontoblasts (Lesot et al., 1990, 1992).

A monoclonal antibody that recognized an extracellular epitope of the 165 kDa proteins, was found to interfere specifically with the organization of microfilaments, to have no effect on microtubules, and to block odontoblast elongation and polarization. At later stages, the maintenance of odontoblast polarization no longer required fibronectin–165 kDa protein interaction (Lesot et al., 1988). At this stage, both fibronectin and the 165 kDa protein tended to disappear from the apical pole of odontoblasts while the formation of junctional complexes, including tight junctions, zonulae adherens and gap junctions, increased (Iguchi et al., 1984; Bishop, 1985; Callé, 1985). The molecular mechanisms supporting the role of the 165 kDa protein in the reorganization of microfilaments during odontoblast polarization still remain unknown. No direct interactions could be detected between the membrane protein and either α-actinin, vinculin or talin (Fausser et al., 1993a, b). Either direct interactions are mediated by a ligand different from those we tested, or the strength of the interaction was too weak to be maintained in the experimental conditions used. Another possibility could be that there is no direct interaction, but instead a transduction pathway (Lesot et al., 1994).

Several complementary approaches indicate that the 165 kDa protein is not a member of the integrin family. The 165 kDa protein is a monomer and its interaction with fibronectin is calcium independent, thus differing from integrin β3IIb. A large 175 kDa C-terminal fragment of fibronectin, including the RGD sequence, did not interact with the membrane protein. This observation was confirmed by the fact that GRGDS peptides did not interfer with odontoblast differentiation, although the 165 kDa protein plays a major role in this process. The protein would thus differ from β1α5, β1αv, β3αIIb, β3αv and β6αv integrins which are all RGD dependent (Hynes, 1992). The 165 kDa protein was found to interact with a 62 kDa proteolytic fragment of fibronectin comprising the collagen-binding domain and the first type III repeat of fibronectin (Figure 3.4). No integrin interacts with this region of fibronectin (Hynes, 1992; Lesot et al., 1992). The 165 kDa protein is a minor constituent among membrane proteins and attempts are now being made to get more information by cloning the corresponding gene. Most interestingly it was demonstrated that cell attachment to superfibronectin (multimers) is mediated partially by receptors with properties distinct from those of integrins (Moria et al., 1994). However, preodontoblasts of isolated dental papillae cultured on fibronectin-coated filters never became functional: fibronectin apparently is not involved in triggering the chain reaction involved in odontoblast terminal differentiation.

Enamel proteins. Sawada and Nanci (1995) documented a general co-distribution of enamel proteins with fibronectin and suggested that proteins of the enamel family might act as mediators of events leading to the terminal differentiation of odontoblasts. Ruch et al. (1989) also demonstrated that a monoclonal antibody produced against mouse dental papilla cells (MC22–45D) strongly stained the enamel and that this antibody was capable of inhibiting odontoblast terminal differentiation in vitro, suggesting a functional role of enamel proteins in odontoblast differentiation.

Integrins. As far as integrins are concerned, immunofluorescent staining of frozen sections of mouse tooth germ with specific antibodies revealed positive staining of the inner dental epithelium corresponding to β4 and α6 chains of integrins while no reaction was observed when using anti-β1 antibodies (Lesot et al., 1993, Meyer et al., 1995, Ruch et al., 1995). An oscillation of β5 integrin mRNA expression during initial tooth morphogenesis, and the presence of β5 transcripts in the inner dental epithelium during odontoblast differentiation were observed (Yamada et al., 1994). Salmivirta et al. (1996) documented the presence of integrin 1 subunit in basement membrane areas and in mesenchymal cells throughout mouse molar development.

Syndecans. Syndecan-1, a cell surface heparan-sulfate proteoglycan, which acts as a receptor for several matrix molecules including tenascin (Bernfield et al., 1992), also binds growth factors, particularly FGFs. Dental papilla cells express syndecan-1 (Thesleff et al., 1987). Syndecan-1, interacting with tenascin, may play a role in the condensation of the mesenchymal cell in the tooth germ rather than during odontoblast terminal differentiation (Salmivirta et al., 1991). Fibroglycan (syndecan-2), however, is expressed by differentiating odontoblasts (Bai et al., 1994).

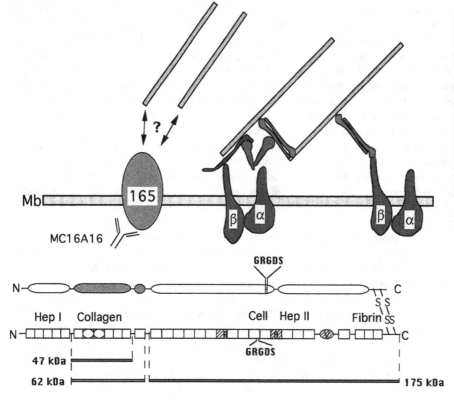

Figure 3.4. **Schematic representation of microfilaments-membrane (Mb) – fibronectin interaction. Fibronectin interacts with integrin by means of the GRGDS sequence and with the 165 kDa membrane protein by means of a N-terminal 62 kDa peptide including the collagen-binding domain and the first type III repeated unit. Neither the collagen-binding domain alone (47 kDa) nor the large 175 kDa protein interacted with the 165 kDa protein. Although fibronectin – 165 kDa interaction interferes with microfilament organization neither, actin, nor α-actinin nor vinculin directly interacted with the 165 kDa protein.**

Implications of growth factors

Immunolocalization and/or *in situ* hybridization of growth factors and/or specific transcripts (although in some cases providing somewhat conflicting data) suggest intervention of the following growth factors in odontoblast terminal differentiation:

> Insulin-like growth factor (IGF) (Joseph *et al.*, 1993); αFGF, βFGF (Gonzales *et al.*, 1990; Cam *et al.*, 1992; Niswander and Martin, 1992; Martin *et al.*, 1998); int.2 (Wilkinson *et al.*, 1989).
>
> Nerve growth factor (NGF) (Byers *et al.*, 1990; Mitsiadis and Luukko, 1995 and references therein).
>
> TGFβ superfamily: TGFβ1, 2 and 3 (Lehnert and Akhurst, 1988; Cam *et al.*, 1990; D'Souza *et al.*, 1990; Pelton *et al.*, 1990, 1991; Millan *et al.*, 1991; Wise and Fan, 1991; Thesleff and Vaahtokari, 1992; Heikinheimo *et al.*, 1993a); and BMP$_{2,4,6}$ (Heikinheimo *et al.*, 1993b; Vainio *et al.*, 1993; Heikinheimo, 1994). Transcripts for activin βA subunit are abundant in day-17 mouse molar preodontoblasts, and transcripts for follistatin are present in the corresponding preameloblasts (Heikinheimo *et al.*, 1997).

Members of the epidermal growth-factor (EGF) family may be involved during tooth initiation and morphogenesis rather than during odontoblast differentiation (Partanen and Thesleff, 1987; Cam *et al.*, 1990; Kronmiller *et al.*, 1991; Hu *et al.*, 1992; Heikinheimo *et al.*, 1993b *but see also* Topham *et al.*, 1987).

Both growth hormone and IGF-I have effects on developing tooth germs *in vitro* and on the cells of odontogenesis therein. Growth hormone appears to affect odontogenic cell proliferation and subsequent differentiation, equivalent to the influence of foetal calf serum. IGF-I strongly promotes the differentiation and development of odontoblasts and of their differentiated cell functions in the form of dentinal matrix formation as well as promoting significant volumetric growth (Young *et al.*, 1995). Exogenous bFGF also stimulated mouse molar and incisor development *in vitro* (Marrtin *et al.*, 1998).

For further functional investigations, isolated dental papillae were grown *in vitro*. When isolated dental papillae were embedded in agar-solidified culture medium, the cusp pattern could be maintained *in vitro* (Bègue-Kirn *et al.*, 1992). Trypsin-isolated day-17 mouse molar dental papillae (containing only preodontoblasts) were cultured for 6 days in semi-solid agar medium containing EDTA-

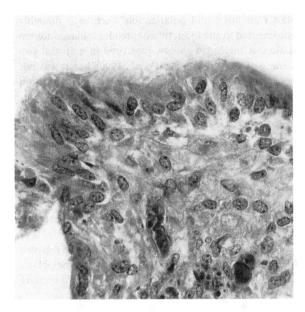

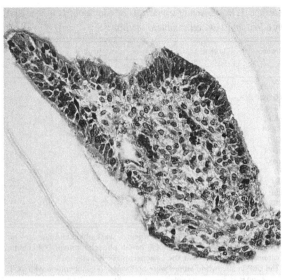

Figure 3.6. **Section of day-17 first lower mouse dental papilla (main cusp) cultured in semi-solid medium containing a mixture of BMP2, 3 and 4. Most of the preodontoblasts differentiated into functional odontoblast-like cells**

Figure 3.5. **Section of day-17 first lower mouse dental papilla (main cusp) cultured in semi-solid medium containing TGFβ3 combined with heparin. Extended differentiation of functional odontoblast-like cells is obvious.**

soluble dentin matrix fractions separated by ion exchange chromatography (Smith and Leaver, 1979). The total unpurified EDTA-soluble fraction neither promoted nor maintained odontoblast differentiation. However, dentin matrix fractions retained on DEAE (diethylaminoethyl)-cellulose (Bègue-Kirn et al., 1992) were observed to initiate the differentiation of odontoblast-like cells at the periphery of the explants. Furthermore, in these conditions the normal pattern of a gradient of odontoblast differentiation could be seen in the dental papillae which was initiated at the tips of the main cusps and progressed in an apical direction. Collagen type I, decorin and biglycan (but not fibronectin) were found to be present in the extracellular matrix which accumulated at the secretory pole of the elongated cells (Lesot et al., 1994). The active components present in the EDTA-soluble fraction of dentin could be retained on heparin-agarose columns and might include growth factors such as TGFβs, BMPs and IGFs (Finkelman et al., 1990) and indeed the addition of a blocking anti-TGFβ antibody (Dasch et al., 1989) to the culture medium abolished the biological effects of the active dentin matrix fraction in odontoblast differentiation (Bègue-Kirn et al., 1992). These data prompted investigations on the effects of growth factors on odontoblast differentiation in vitro.

Effects of members of the TGFβ family on differentiation of odontoblasts in vitro.   In the presence of recombinant

TGFβ1, BMP2 or BMP4 (added alone to the agar-solidified culture medium), the preodontoblasts of isolated dental papillae never overtly differentiated. When combined with heparin (or fibronectin) these growth factors had positive, albeit differential effects (Bègue-Kirn et al., 1992, 1994; Ruch et al., 1995): TGFβ1–heparin (like EDTA-soluble fractions of dentin) induced gradients of cytological and functional differentiation; whereas BMP2 (or 4)-heparin allowed funtional differentiation in restricted areas, i.e. at the tips of the cusps. Heparin alone had no effect. Recombinant TGFβ3 combined with heparin induced extended functional differentiation of odontoblast-like cells (Figure 3.5). Similar results were obtained in the presence of a mixture of BMP2, 3 and 4 prepared by Murata et al. (1997) (Figure 3.6).

In collaboration with K. Heikinheimo (Heikinheimo et al., 1998), we analysed the effects of recombinant human activin A and its binding protein follistatin. Follistatin combined with heparin triggered the differentiation of odontoblast-like cells in the dental papilla culture model. In this context, it is also interesting to note that recent results with activin-βA subunit and follistatin-deficient mice demonstrated major defects in the dentition (Matzuk et al., 1995a, b).

IGF1.   Transcripts for IGF1 have been detected neither in preameloblasts-ameloblasts nor in preodontoblasts-odontoblasts (Bègue-Kirn et al., 1994). However, in our in vitro dental papilla cultures, exogenous IGF1, combined with

Table 3.1. *Expression of transcripts by odontoblasts in vivo and by odontoblast-like cells induced in vitro*

Probes	In vivo	In vitro		
		TGF 1–HN	BMP2–HN	IGF1–HN
TGFβ1	++	+	+	–
TGFβ2	~	–	–	–
TGFβ3	++	+	~	~
BMP2	++	+	+	+
BMP4	+	++	++	++
IGF1	–	–	–	–
msx1	++	~	–	~
msx2	~	+	+	+
FN	–	–	–	–
ON	++	++	+	+
BSP	++	++	+	+
DSP	++	+	ND	ND
DPP	++	+		

++ strong positive signal; + positive signal; ~ weak ubiquitous signal; – no signal; BSP, bone sialoprotein; DPP, dentin phosphoprotein; DSP, dentin sialoprotein; FN, fibronectin; ON, osteonectin; HN, heparin
*The DSP in situ hybridizations were performed in collaboration with W. T. Butler and H. H. Ritchie. For details see Bègue-Kirn *et al.*, 1994.

heparin, induced extended cytological differentiation of odontoblast-like cells without apical matrix deposition (Bègue-Kirn *et al.*, 1994).

By means of *in situ* hybridization, we compared the expression of several transcripts of odontoblast-like cells induced *in vitro* and by *in vivo* differentiating odontoblast (Bègue-Kirn *et al.*, 1994). The transcriptional phenotype of odontoblasts induced by TGFβ1 or BMP2 (combined with heparin) is very similar to that of *in vivo* differentiating odontoblasts (Table 3.1.). On the other hand, IGF1 did not affect TGFβs' transcriptions, which might be correlated with the absence of matrix deposition by IGF1–induced odontoblast-like cells.

Low affinity receptors. Several growth factors, including bFGF and members of the TGFβ family, bind with rather low affinity to various cell surface proteoglycans, non-signalling receptors that present the ligand to high-affinity, signalling receptors. Proteoglycans may also play an important role in activation of the signalling receptors (see Schlessinger *et al.*, 1995 for review).

Low-affinity receptors for TGFβs include decorin, β-glycan and biglycan. Decorin, has been localized at the epithelio-mesenchymal junction, where it was found to change its localization during odontoblast differentiation: it surrounded polarizing odontoblasts and later accumulated at the apical pole of functional odontoblasts (Ruch *et al.*, 1995). This change in the distribution of decorin could be compared to that of fibronectin. Since the two molecules have been reported to interact (Schmidt *et al.*, 1991), their redistribution might be correlated and it would be of importance to investigate whether decorin indeed plays an active role in odonto-

blast elongation and polarization. Decorin is probably not involved in the induction of preodontoblast differentiation: it might, in theory, intervene in a spatial patterning of TGFβ (Yamaguchi *et al.*, 1990), but it was not detected at the dental epithelio-mesenchymal junction in front of preodontoblasts.

Biglycan is also present at the epithelio-mesenchymal junction (Ruch *et al.*, 1996). Latent TGFβ-binding protein (LTBP), another TGFβ-binding protein, potentially involved in the activation of TGFβ (Flaumenhaft *et al.*, 1993), as well as the latency associated peptide (LAP), have been localized in the dental basement membrane (Ruch *et al.*, 1995; Cam *et al.*, 1997) and might play a more important role than decorin in early steps of odontoblast differentiation.

Most interestingly, the transcripts for the β-glycan (receptor TGFβ type III) were localized in preameloblasts facing preodontoblasts that were able to differentiate (Ruch *et al.*, 1995).

Signalling receptors. TGFβ family members signal through complexes of type I and type II serine-threonine kinase receptors (Massagué *et al.*, 1994). Hardly anything is known about the expression of these receptors in differentiating odontoblasts. Recently, however, transcripts for TGFβ RII were reported to be expressed in young preodontoblasts (Wang *et al.*, 1995). Heikinheimo *et al.* (unpublished data) demonstrated the expression of TGFβ RII, and Act.-RIIA in differentiating odontoblasts.

Polarizing odontoblasts express neurotrophin receptors: gp75 and members of the Trk protein tyrosine kinase receptor family (Mitsiadis and Luukko, 1995). Growth hormone receptor (Zhang *et al.*, 1992), FGF receptor 1 (Peters *et al.*, 1992) and PDGF receptor (Hu *et al.*, 1995) are also expressed.

The accumulated data from *Drosophila*, *Caenorhabditis elegans* and vertebrates suggest that Notch signalling plays a functional role in the differentiation of uncommitted cells (Artavanis-Tsakonas *et al.*, 1995). Expression of Notch 1, 2 and 3 during mouse odontogenesis was correlated with determination of ameloblast cell fate (Mitsiadis *et al.*, 1995a).

Vitamins A and D

The developing teeth are targets for retinoid induced teratogenesis *in vivo* (Knudsen, 1967) and *in vitro* (Hurmerinta *et al.*, 1980; Mark *et al.*, 1992; Bloch-Zupan *et al.*, 1994b). However, despite hints of possibilities (Berdal *et al.*, 1993; Bloch-Zupan *et al.*, 1994a), the role of retinoic acid, if any, during odontoblast differentiation remains to be elucidated, as does the role of functional heterodimers formed by vitamin D and retinoic acid receptors.

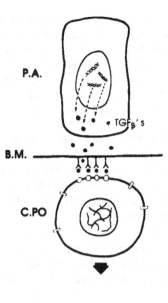

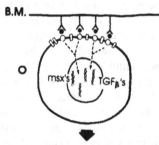

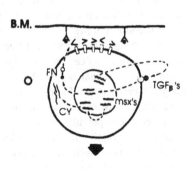

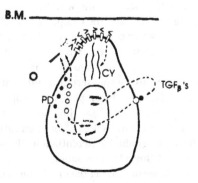

3.4.2.   Current hypotheses and questions

At the cellular level, as a schematic working hypothesis (Figure 3.7) we might suggest the following: one or several (partially redundant) epithelially derived member(s) of the TGFβ family are eventually activated and trapped by basement membrane-associated low-affinity receptors. These non-signalling receptors will present the ligand(s) to high-affinity signalling cell membrane receptors expressed by competent preodontoblasts. Diffusion of the growth factors will be limited and the interaction with signalling receptors occurs according to a definite geometry.

BMPs are secreted as mature, active, protein dimers and might interact with follistatin and/or with matrix molecules like chordin and noggin (Aubin, 1996). On the other hand, TGFβs are secreted as inactive molecules and interact with LAP and LTBP. These complexes further interact with low-affinity matrix receptors. The proteolytic activation probably implies plasmin, cathepsin D and/or thrombospondin (Feige et al., 1996). Cathepsin D has been localized both in odontoblasts and ameloblasts (Andujar et al., 1989).

The transduced downstream signals regulate the transcription of genes encoding for proteins belonging to the TGFβ family and homeoproteins such as msx2. The msx2 transcription factor might, in turn, modulate the expression of genes involved in cytoskeleton assembly. Calpains might play a role (Hanson et al., 1995). Endogenous members of the TGFβ family could regulate matrix production including transitory synthesis of fibronectin which interacts with the 165 kDa, non-integrin, membrane receptor and thus co-regulates processes involved in cell elongation and polarization. It has been shown also that TGFβ can regulate microtubule stabilization (Gundersen et al., 1994). Members of the TGFβ superfamily also upregulate the transcription of the nuclear proto-oncogenes c-jun and c-fos. Interestingly, differentiating and functional odontoblasts express these genes (Byers et al., 1995; Kitamura and Terashita, 1997). c-Fos protein forms the AP-1 transcription factor with Jun family, and genes of proteins expressed by odontoblasts including alkaline phosphatase, osteocalcin and collagens types I and III have AP-1 binding sites (Katai et al., 1992). The raf-1 proto-oncogene is also strongly expressed by odontoblastic and subodontoblastic cells (Sunohara et al., 1996). Clearly, the mechanisms involved in odontoblast differentiation are very complex and synergistic

Figure 3.7. **Members of the TGF family (•) secreted by preameloblasts (PA) are eventually activated and trapped by basement membrane (BM)-associated low-affinity receptors (LAR) having an epithelial or mesenchymal origin. TGFβs interact according to a specific geometry with signalling receptors (SR) expressed at the surface of competent preodontoblasts (C.PO).**

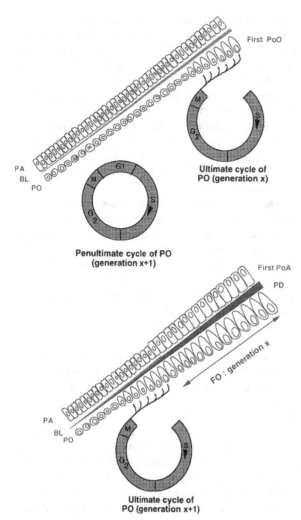

Figure 3.8. **Stage-space specific preameloblasts (PA) instruct the neighbouring preodontoblasts (PO) to differentiate along a particular pathway and in a coordinated sequential order giving rise to the gradient of terminal differentiation. The known facts suggest both the existence of an intrinsic timing mechanism (leading to the expression of specific competence by preodontoblasts) and that division counting might be involved. PO of successive, juxtaposed generations (x, x+1, etc.) might sequentially withdraw from the cycle and overtly differentiate. The counting of mitoses could be initiated early during odontogenesis in such a way that all PO might be able to express specific competence after the same number of cell cycles (Ruch, 1990). PoO, post-mitotic, polarizing odontoblasts; BL, basement membrane; PD, predentin; PoA, post-mitotic, polarizing ameloblasts; FO, functional odontoblasts. (From Ruch et al., 1955.)**

interactions with other endogenous or circulating growth factors (IGFs, FGFs) probably exist.

This model does not explain the gradients of odontoblast differentiation. Once more, the terminal differentiation of odontoblasts controlled by preameloblasts occurs in each cusp according to a specific temporospatial pattern: the first overtly differentiating cells are located at the tip of the cusps and progressively further

functional cells emerge towards the basal (apical) parts. Preameloblasts facing differentiating odontoblasts express several of the growth factors putatively involved in the control of odontoblast differentiation: TGFβ1, 2, 3, BMP2, 4, 6, follistatin (Bègue-Kirn et al., 1994, Heikinheimo et al., 1997). Schematically, at least two, not mutually exclusive possibilites can be evoked:

1.   The epithelially derived signal (simple or combinatorial) progressively reaches an operational level from the top to the base of the cusp, i.e. there exists a spatial gradation of inducibility and an even distribution of competent preodontoblasts.

2.   During initiation of odontoblast differentiation, the operational signal has an even distribution and the responsive competent preodontoblasts emerge sequentially in an apical direction.

Most of the in vivo, in situ hybridization data concerning the expression of different transcripts by the preameloblasts involved in odontoblast terminal differentiation support the second possibility. The results of our dental papilla culture experiments demonstrate that a uniform distribution of active growth factors combined with heparin inside the solidified medium allow, as in vivo, the gradual differentiation of odontoblasts.

Histological investigations, combined with ^{3}H-thymidine radioautography of in vitro cultured heterochronal enamel organ-dental papilla recombinations (Ruch and Karcher-Djuricic, 1971; Ruch et al., 1976), suggest that the competence of preodontoblasts to respond to specific epigenetic signals (triggering terminal differentiation) requires a minimal number of cell cycles. Terminal differentiation cannot be anticipated. On the other hand, supplementary cell cycles do not hamper terminal differentiation and may allow regulative phenomena. The gradual emergence of competent preodontoblasts might also be related to cell kinetics according to Ruch (1990, 1995b): an intrinsic timing mechanism could imply division counting, and preodontoblasts of juxtaposed cell generations might sequentially withdraw from the cell cycle (Figure 3.8). The specific competence expressed by post-mitotic daughter cells might result either from the expression of a particular combination of signalling receptors and/or of specific transductional and/or post-transductional steps. During terminal differentiation of odontoblasts, junctional complexes develop and particularly the frequency of gap junctions increases allowing for direct intercellular communication that might affect the gradual differentiation. However, the implantation of thin plexiglas slides in cultured mouse molars does not seem to affect the formation of gradients

(Bradamante and Ruch, unpublished data). Nevertheless functional investigations concerning cell adhesion molecules (CAMs and cadherins) and proteins related to junctional complexes are required.

Another important question has to be stressed: we do not know if withdrawal from the cell cycle and initiation of polarization and funtional differentiation are co-regulated. We suggest that the expression of a preodontoblast-specific competence implies intrinsic control mechanisms. However, withdrawal from the cell cycle is controlled by a stage-space-specific inner dental epithelium (Ruch and Karcher-Djuricic, 1971) and members of the TGFβ family might be involved. The control of the expression of G1 phase regulators – cyclins, cyclin dependent kinases and inhibitors – has to be investigated. The cyclin-dependent kinase-inhibitor p21 has been shown to be strongly expressed in post-mitotic differentiating and functional odontoblasts (and ameloblasts) (Bloch-Zupan et al., 1998). Further studies have to focus on the interplay of differentiation signals and cell cycle controlling factors.

Recently, several culture systems have been described for the in vitro maintenance of odontogenic cells (Nakashima, 1991; MacDougall et al., 1995 and references therein). MacDougall et al. (1995) realized temperature sensitive Simian virus 40 large T antigen immortalization of mouse dental papilla cell cultures and established a clonal 'odontoblast' cell line. These cells showed high constitutive expression of dentin phosphoprotein, type I collagen and alkalin phosphatase and will allow isolation of regulatory factors involved in dentin protein gene expression. It will be of great interest to investigate if these cells in appropriate culture conditions are able to express the integrated cyto-functional odontoblast phenotype. If this were the case, such cell lines would be of more general interest.

3.5.   Conclusion

As stated by Noden (1983): 'cells of the neural crest must resolve two problems: phenotypic commitment, i.e. what to become, and spatial programming, i.e. where to do it'. When and how the odontogenic cells are committed is not known but odontoblasts will do it at locations defined by tooth-specific morphogenesis and due both to delivery of epithelially derived signals and time-space-dependent expression of particular competence. Implying the expression of signalling receptors, odontoblasts share many metabolic aspects with osteoblasts and cementoblasts, but probably express some specific dentin extracellular matrix components. However, their most striking specificity resides in their unique cyto-functional behaviour, i.e. odontoblasts are post-mitotic cells that assume some features of epithelial cells and secrete extracellular predentin-dentin components at their apical pole.

Thanks to the application of adequate, current molecular technologies, the respective role and nature of endogenous, cell kinetic related maturation (competence) and epigenetic signalling will hopefully be unravelled.

Summary

Odontoblasts are post-mitotic, neural crest-derived cells that overtly differentiate according to tooth-specific temporospatial patterns and secrete predentin-dentin collagenous and non-collagenous components. Neither the timing nor the molecular mechanisms of odontoblast specification are known and the problem of their patterning in the developing jaws is far from being solved. On the other hand, some significant strides have been made concerning the control of their terminal differentiation. Fibronectin interacting with a 165 kDa, non-integrin, membrane protein intervenes in the cytoskeletal reorganization involved in odontoblast polarization and then terminal differentiation can be triggered in vitro by immobilized members of the TGFβ family. Histological aspects and the transcriptional phenotypes (transcripts of TGFβs, BMPs, msxs, IGF1, fibronectin, osteonectin, bone sialoprotein genes) are very similar in vivo and in vitro. In vivo members of the TGFβ superfamily secreted by preameloblasts, trapped and activated by basement membrane-associated components, might initiate odontoblast terminal differentiation. The time-space-regulated expression of low-affinity and signalling receptors has to be investigated to determine both the specific geometry of epigenetic signalling and the particularity of preodontoblast competence.

Acknowledgements

The original work reported in this chapter was mainly supported by INSERM, CNRS, MRT, Fondation pour la Recherche Médicale and EEC. We are grateful to all our present and former co-workers, and to Dr A. J. Smith for critical reading of the manuscript.

For additional information see Ruch (1998).

References

Andujar, M., Hartmann, D. J., Caillot, G., Ville, G. and Magloire, H. (1989). Immunolocalization of cathepsin D in dental tissues. *Matrix*, 9, 397–404.

Artavanis-Tsakonas, S., Matsuno, K. and Fortini, M. E. (1995). Notch signalling. *Science*, 268, 225–232.

Aubin, B. M. L. (1996). Bone morphogenetic proteins: multifunctional regulators of vertebrate development. *Genes and Development*, 10, 1580–1594.

Bai, X. M., von der Schuren, B., Cassiman, J. J., van den Berghe, H. and David, G. (1994). Differential expression of multiple cell surface heparan sulfate proteoglycans during embryonic tooth development. *Journal of Histochemistry and Cytochemistry*, 42, 1043–1053.

Baume, L. J. (1980).The biology of pulp and dentine. *Monographs in Oral Science*, 8, 1–220.

Bègue-Kirn, C., Smith, A. J., Ruch, J. V., Wozney, J. M., Purchio, A., Hartmann, D. and Lesot, H. (1992). Effects of dentin proteins, transforming growth factor 1 (TGF1) and bone morphogenetic protein 2 (BMP2) on the differentiation of odontoblast in vitro. *International Journal of Developmental Biology*, 36, 491–503.

Bègue-Kirn, C., Smith, A. J., Loriot, M., Kupferle, C., Ruch, J.V. and Lesot H. (1994). Comparative analysis of TGFs, BMPs, IGF, msxs, fibronectin, osteonectin and bone sialoprotein gene expression during normal and in vitro-induced odontoblast differentiation. *International Journal of Developmental Biology*, 38, 405–420.

Berdal, A., Nanci, A., Balmain, N., Thomasset, M., Breaier, A., Cuisinier-Gleizes, P. and Mathieu, M. (1989). Immunolocalization and dosing of calbindins during amelogenesis in normal and vitamin D deficient rat. In *Tooth Enamel V*, ed. R.W. Fearnhead, pp.154–162. Yokohama: Florence Publishers.

Berdal, A., Hotton, D., Pike, J. W., Mathieu, H. and Dupret, J. M. (1993). Cell and stage specific expression of vitamin D receptor and calbinding genes in rat incisor: regulation by 1,25-dihydroxyvitamin D3. *Developmental Biology*, 155, 172–179.

Bernfield, M., Kokenyesi, R., Kato, M., Hinkes, M., Spring, J., Gallo, R. and Lose, E. (1992). Biology of the syndecans. *Annual Review of Cell Biology*, 8, 333–364.

Berteretche, M. V., Dunia, I., Devilliers, G, van der Kemp, A., Pieper, F., Bloemendal, H., Benedetti, E.L. and Forest, N. (1993). Abnormal incisor-tooth differentiation in transgenic mice expressing the muscle-specific desmin gene. *European Journal of Cell Biology*, 62, 183–193.

Bishop, M. A. (1985). Evidence for tight junctions between odontoblasts in the rat incisor. *Cell Tissue Research*, 239, 137–140.

Bishop, M. A. and Yoshida, S. (1992). A permeability barrier to lanthanum and the presence of collagen between odontoblasts in pig molars. *Journal of Anatomy*, 181, 29–39.

Bishop, M. A., Machotra, M. and Yoshida, S. (1991). Interodontoblastic collagen (von Korff fibres) and circumpulpul dentin formation: an ultra thin serial section study in the rat. *American Journal of Anatomy*, 191, 67–73.

Bloch-Zupan, A., Decimo, D., Loriot, M., Mark, M. P. and Ruch, J.V. (1994a). Expression of nuclear retinoic acid receptors during mouse odontogenesis. *Differentiation*, 57, 195–203.

Bloch-Zupan, A., Mark, M. P., Weber, B. and Ruch, J. V. (1994b). In vitro effects of retinoic acid on mouse incisor development. *Archives of Oral Biology*, 39, 891–900.

Bloch-Zupan, A., Leveillard, T., Gorry, P. and Ruch, J.V. (1998). Expression of p21 WAF1/CIP1 during mouse odontogenesis. *European Journal of Oral Sciences*, 106, 104–111.

Butler, W. T. (1995). Dentin matrix proteins and dentinogenesis. *Connective Tissue Research*, 33, 381–387.

Butler, W. T. and Ritchie, H. (1995). The nature and functional significance of dentin extracellular matrix proteins. *International Journal of Developmental Biology*, 39, 169–180.

Byers, M. R. and Sugaya, A. (1995). Odontoblast processes in dentin revealed by fluorescent Di-I. *Journal of Histochemistry and Cytochemistry*, 43, 159–168.

Byers, M. R., Schattman, G. C. and Bothwell, M. (1990). Multiple functions for NGF receptor in developing, aging and injured teeth are suggested by epithelial, mesenchymal and neural immunoreactivity. *Development*, 109, 464–471.

Byers, M. R., Mecifi, K. B. and Iadarola M. J. (1995). Focal c-fos expression in developing rat molars, correlations with subsequent intradental and epithelial sensory innervation. *International Journal of Developmental Biology*, 39, 181–189.

Callé, A. (1985). Intercellular junctions between human odontoblasts. A freeze-fracture study after demineralization. *Acta Anatomica*, 122, 138–144.

Cam, Y., Neumann, M. R. and Ruch, J. V. (1990). Immunolocalization of TGFβ1 and EGF receptor epitopes in mouse incisors and molars. Demonstration of in vitro production of transforming activity. *Archives of Oral Biology*, 35, 813–822.

Cam, Y., Neumann, M. R., Oliver, L., Roulais, D., Janet, T. and Ruch, J. V. (1992). Immunolocalization of acidic and basic fibroblast growth factors during mouse odontogenesis. *International Journal of Developmental Biology*, 36, 381–389.

Cam, Y., Lesot, H., Colosetti, P. and Ruch, J. V. (1997). Distribution of transforming growth factor beta 1-binding proteins and low-affinity receptors during odontoblast differentiation in mouse. *Archives of Oral Biology*, 42, 385–391.

Couve, E. (1986). Ultrastructural changes during the life cycle of human odontoblasts. *Archives of Oral Biology*, 31, 643–651.

Dasch, J. R., Pace, D. R., Weagell, W., Inenaga, D. and Ellingworth, L. (1989). Monoclonal antibodies recognizing transforming growth factor β. *Journal of Immunology*, 142, 1536–1541.

D'Souza, R. N., Happonen, R. P., Ritter, N. M. and Butler, W. T. (1990). Temporal and spatial patterns of transforming growth factor β1 expression in developing rat molars. *Archives of Oral Biology*, 35, 957–965.

Fausser, J. L., Staub, A., Ungewickell, E., Ruch, J. V. and Lesot H. (1993a). Absence of interactions between the 165 kDa fibronectin-binding protein involved in mouse odontoblast differentiation and vinculin. *Archives of Oral Biology*, 38, 537–540.

Fausser, J. L., Ungewickell, E., Ruch, J. V. and Lesot H. (1993b). Interaction of vinculin with clathrin heavy chain. *Journal of Biochemistry, Tokyo*, 114, 498–503.

Feige, J. J., Quirin, N. and Souchelnitsky, S. (1996). TGF, un peptide biologique sous contrôle: formes latentes et mécanismes d'activation. *Médecine/Sciences*, 12, 929–939.

Finkelman, R. D., Mohan, S., Jennings, J. C., Taylor, A. K., Jepsen, S. and Baylink, D. J. (1990). Quantitation of growth factors IGF-I, SGF/IGF-II and TGF-β in human dentin. *Journal of Bone and Mineral Research*, 5, 717–723.

Flaumenhaft, R., Abe, M., Sato, Y., Miyazono, K., Harpel, J.,

Heldin, C. and Rifkin, D. B. (1993). Role of latent TGFβ in cocultures of endothelial and smooth muscle cells. *Journal of Cell Biology*, **120**, 995–1002.

Fukae, M., Tanabe, T. and Yamada, M. (1994). Action of metallo-proteinases on porcine dentin mineralization. *Calcified Tissue International*, **55**, 426–435.

Fundele, R. H. and Surani, M. A. (1994). Experimental embryological analysis of genetic imprinting in mouse development. *Developmental Genetics*, **15**, 515–522.

Garant, R. H. and Cho, M. I. (1985). Ultrastructure of odontoblast. In *The Chemistry and Biology of Mineralized Tissues*, ed. W. T. Butler, pp. 22–33. Birmingham, AL: EBSCO Media.

George, A., Silberstein, R. and Veis, A. (1995). In situ hybridization shows Dmp1 (AG1) to be developmentally regulated dentin-specific protein produced by mature odontoblasts. *Connective Tissue Research*, **33**, 389–394.

Goldberg, M. and Takagi, M. (1993). Dentine proteoglycans: composition, ultrastructure and functions. *Histochemical Journal*, **25**, 781–806.

Goldberg, M., Feinberg, J., Rainteau, D., Lécolle, S. Kaetzel, M. A., Dedman, J. R. and Weinman, S. (1991). Annexins I-VI in secretory ameloblasts and odontoblasts of rat incisor. *Journal de Biologie Buccale*, **18**, 289–298.

Goldberg, M., Septier, D., Lécolle, S., Chardin, H., Quintana, M. A., Acevedo, A. C., Gafni, G., Dillouya, D., Vermelin, L., Thonemann, B., Schmalz, G., Bissal-Mapahou, P. and Carreau, J. P. (1995). Dental mineralization. *International Journal of Developmental Biology*, **39**, 93–110.

Gonzales, A. M., Buscaglia, M., Ong, M. and Baird, D. (1990). Distribution of basic fibroblast growth factor in the 18-day rat fetus: localization in the basement membranes of diverse tissues. *Journal of Cell Biology*, **110**, 753–765.

Gundersen, G. G., Kim, I. and Chapin, C. J. (1994). Induction of stable microtubules in 3T3 fibroblasts by TGFβ and serum. *Journal of Cell Science*, **107**, 645–659.

Hanson, B., Gritli-Linde, A., Ruch, J. V. and Linde, A. (1995). Calpains in developing murine dental tissue. *Connective Tissue Research*, **22**, 92.

Heikinheimo, K. (1994). Stage-specific expression of decapenta-plegic-Vg-related genes 2, 4 and 6 (bone morphogenetic proteins 2, 4 and 6) during human tooth morphogenesis. *Journal of Dental Research*, **73**, 530–597.

Heikinheimo, K. and Salo, T. (1995). Expression of basement membrane type IV collagen and type IV collagenase (MMP-2 and MMP-9) in human fetal teeth. *Journal of Dental Research*, **74**, 1226–1234.

Heikinheimo, K., Happonen, R. P., Miettinen, P. J. and Ritvos, O. (1993a). Transforming growth factor β2 in epithelial differentiation of developing teeth and odontogenic tumors. *Journal of Clinical Investigation*, **91**, 1019–1027.

Heikinheimo, K.; Voutilainen, R., Happonen, R. P. and Miettinen, P. J. (1993b). EGF receptor and its ligands, EGF and TGFβ in developing and neoplastic odontogenic tissues. *International Journal of Developmental Biology*, **37**, 387–396.

Heikinheimo, K., Bègue-Kirn, C., Ritvos, O., Tuuri, T. and Ruch, J. V. (1997). The actin binding protein follistatin is expressed in developing murine molar and induces odontoblast-like cell differentiation *in vitro*. *Journal of Dental Research*, **76**, 1625–1636.

Heikinheimo, K., Bègue-Kirn, C., Ritvos, O., Tuuri, T. and Ruch, J. V. (1998). Activin and bone morphogenetic protein (BMP) signalling during tooth development. *European Journal of Oral Sciences*, **106**, 167–173.

Holland, G. T. (1976). The extent of the odontoblast process in the rat. *Journal of Anatomy*, **121**, 133–149.

Hu, J. C. C., Sakakura, Y., Sasano, Y., Shum, L., Bringas, P., Werb, Z. and Slavkin, H. C. (1992). Endogenous epidermal growth factor regulates the timing and pattern of embryonic mouse molar tooth morphogenesis. *International Journal of Developmental Biology*, **36**, 505–516.

Hu, J. C. C., Zhang, C. and Slavkin, H. C. (1995). The role of platelet-derived growth factor in the development of mouse molars. *International Journal of Developmental Biology*, **39**, 393–945.

Hubbard, M. J., Bradley, M. P., Kardos, T. B. and Forrester, L. T. (1981). Calmodulin-like activity in a mineralizing tissue: the rat molar tooth germ. *Calcified Tissue International*, **33**, 545–548.

Huggins, C. R., McCarrol, M. D. and Dahlberg, A. A. (1934). Transplantation of tooth germ elements and the experimental heterotopic formation of dentin and enamel. *Journal of Experimental Medicine*, **60**, 199–210.

Hurmerinta, K., Thesleff, I. and Saxen, L. (1990). *In vitro* inhibition of mouse odontoblast differentiation by vitamin A. *Archives of Oral Biology*, **25**, 385–393.

Hynes, R. O. (1992). Integrins: versatility, modulation, and signalling in cell adhesion. *Cell*, **69**, 11–25.

Iguchi, Y., Yamamura, T., Ichikawa, T., Hachimoto, S., Horiuchi, T. and Shimono, M. (1984). Intercellular junctions in odontoblasts of the rat incisor studied with freeze-fracture. *Archives of Oral Biology*, **29**, 487–497.

Jernvall, J. (1995). Mammalian molar cusp pattern: developmental mechanisms of diversity. *Acta Zoologica Fennica*, **198**, 1–61.

Joseph, B. K., Savage, N. W., Young, W. G., Gupta, G. S., Breier, B. H., and Waters, M. J. (1993). Expression and regulation of insulin-like growth factor-1 in the rat incisor. *Growth Factors*, **8**, 267–275.

Katai, H., Stephenson, J. D., Simkevich, C. P., Thompson, J. P. and Raghow, R. (1992). An AP-1 like motif in the first intron of pro1CI) collagen is a critical determinant of its transcriptional activity. *Molecular and Cellular Biochemistry*, **118**, 119–129.

Kitamura, C. and Terashita M. (1997). Expression of c-jun and jun-B proto-oncogenes in odontoblast during development of bovine tooth germs. *Journal of Dental Research*, **76**, 822–830.

Knudsen, P. A. (1967). Tooth germs in non-exencephalic sibling of mouse embryos with exencephaly induced by hypervitaminosis A. *Acta Odontologica Scandinavica*, **25**, 669–676.

Koch, W. E. (1967). *In vitro* differentiation of tooth rudiments of embryonic mice. I. Transfilter interaction of embryonic incisor tissues. *Journal of Experimental Zoology*, **165**, 155–170.

Kollar, E. J. (1972). The development of the integument: spatial, temporal and phylogenetic events. *American Zoologist*, **12**, 125–135.

Kóllar, E. J. and Baird, G. (1970). Tissue interactions in embryonic mouse tooth germs. II. The induction role of dental papilla. *Journal of Embryology and Experimental Morphology*, **24**, 173–184.

Kronmiller, J. E., Upholt, W. B. and Kollar, E. J. (1991). EGF antisens oligodeoxynucleotides block murine odontogenesis *in vitro*. *Developmental Biology*, **147**, 485–488.

Kubler, M. D., Lesot, H. and Ruch, J.V . (1988). Temporo-spatial distribution of matrix and microfilament components during odontoblast and ameloblast differentiation. *Roux's Archives of Developmental Biology*, **197**, 212–220.

Lehnert, S. A. and Akhurst, R. J. (1988). Embryonic expression patterns of TGFβ type 1 RNA suggests both paracrine and autrocrine mechanisms of action. *Development*, **104**, 263–273.

Lesot, H., Meyer, J. M., Ruch, J. V., Weber, K. and Osborn, M. (1982). Immunofluorescent localization of vimentin, prekeratin and actin during odontoblast and ameloblast differentiation. *Differentiation*, **21**, 133–137.

Lesot, H., Karcher-Djuricic, V., Mark, M., Meyer, J. M. and Ruch, J. V. (1985). Dental cell interaction with extracellular matrix constituents: Type I collagen and fibronectin. *Differentiation*, **29**, 176–181.

Lesot, H., Karcher-Djuricic, V., Kubler, M. D. and Ruch, J. V. (1988). Membrane–cytoskeleton interactions: inhibition of odontoblast differentiation by a monoclonal antibody directed against a membrane protein. *Differentiation*, **37**, 62–72.

Lesot, H., Kubler, M. D., Fausser, J. L. and Ruch, J. V. (1990). A 165 kDa membrane antigen mediating fibronectin–vinculin interaction is involved in murine odontoblast differentiation. *Differentiation*, **44**, 25–35.

Lesot, H., Fausser, J. L., Akiyama, S. K., Staub, A., Black, D., Kubler, M. D. and Ruch, J.V. (1992). The carboxy-terminal extension of the collagen-binding domain of fibronectin mediates interaction with a 165 Kda membrane protein involved in odontoblast differentiation. *Differentiation*, **49**, 109–118.

Lesot, H., Bègue-Kirn, C., Kubler, M. D., Meyer, J. M., Smith, A. J., Cassidy, N. and Ruch, J. V. (1993). Experimental induction of odontoblast differentiation and stimulation during reparative processes. *Cells and Materials*, **3**, 201–217.

Lesot, H., Smith, A. J., Tziafas, D., Bègue-Kirn, C., Cassidy, N. and Ruch, J. V. (1994). Effects of biologically active molecules on dental tissue repair: comparison of reactionary and reparative dentinogenesis with the induction of odontoblast differentiation in vitro. *Cells and Materials*, **4**, 199–218.

Linde, A. and Goldberg, M. (1993). Dentinogenesis. *Critical Reviews in Oral Biology and Medicine*, **4**, 679–728.

Linde, A. and Lundgren, T. (1995). From serum to the mineral phase. The role of odontoblast in calcium transport and mineral formation. *International Journal of Developmental Biology*, **39**, 213–222.

MacDougall, M., Thiemann, F., Ta, H., Hsu, P., Song Chen, L. and Snead, M. L. (1995). Temperature sensitive simian virus 40 Large T antigen immortalization of murine odontoblast cell cultures: establishment of clonal odontoblast cell line. *Connective Tissue Research*, **33**, 419–425.

Magloire, H., Joffre, A., Azerad, J. and Lawson, D. E. M. (1988). Localization of 28 kDa calbindin in human odontoblasts. *Cell Tissue Research*, **254**, 341–346.

Mark, M. P., Bloch-Zupan, A. and Ruch, J. V. (1992). Effects of retinoids on tooth morphogenesis and cytodifferentiations in vitro. *International Journal of Developmental Biology*, **36**, 517–526.

Marrtin, A., Unda, F. J., Bègue-Kirn, C., Ruch, J. V. and Arechaga, J. (1998). Effects of αFGF, βFGF, TGF1, and IGF1 on odontoblast differentiation. *European Journal of Oral Sciences*, **106**, 117–121.

Massagué, J., Attisano, L. and Wrana, J. L. (1994). The TGF-β family and its composite receptors. *Trends in Cell Biology*, **4**, 172–178.

Matzuk, M. M., Kumar, T. R., Vassalli, A., Bickenbach, J. R., Roop, D. R., Jaenisch, R. and Bradley, A. (1995a). Functional analysis of activins during mammalian development. *Nature*, **374**, 354–356.

Matzuk, M. M., Lu, N., Vogel, H., Selheyer, K., Roop, D. R. and Bradley, A. (1995b). Multiple defects and perinatal death in mice deficient in follistatin. *Nature*, **374**, 360–363.

Meyer, J. M., Staubli, A. and Ruch, J.V. (1988). Ultrastructural duality of extracellular fibrillar components of the odontoblast layer in the mouse molar. *Archives of Oral Biology*, **33**, 25–31.

Meyer, J. M., Ruch, J. V., Kubler, M. D., Kupferle, C. and Lesot, H. (1995). Cultured incisors display major modifications in basal lamina deposition without further effect on odontoblast differentiation. *Cell Tissue Research*, **279**, 135–147.

Millan, F. A., Denhez, F., Kondaian, P. and Akhurst, R.J. (1991). Embryonic gene expression patterns of TGFβ1, 2 and 3 suggest different developmental functions in vivo. *Development*, **111**, 131–144.

Mitsiadis, T. and Luukko, K. (1995). Neutrotophins in odontogenesis. *International Journal of Developmental Biology*, **39**, 195–202.

Mitsiadis, T. A., Lardelli, M., Lendahl, U., and Thesleff, I. (1995a). Expression of Notch 1, 2 and 3 is regulated by epithelial-mesenchymal interactions and retinoic acid in the developing mouse tooth and associated with determination of ameloblast cell fate. *Journal of Cell Biology*, **130**, 407–418.

Mitsiadis, T., Muramatsu, T., Muramatsu, H. and Thesleff, I. (1995b). Midkine, a heparin-binding growth/differentiation factor is regulated by retinoic acid and epithelial–mesenchymal interactions in the developing mouse tooth and affects cell proliferation and morphogenesis. *Journal of Cell Biology*, **129**, 267–281.

Moria, A., Zhang, Z. and Ruoslahti, E. (1994). Superfibronectin is a functionally distinct form of fibronectin. *Nature*, **367**, 193–196.

Murata, M., Tsujigiwa, H., Kinuta, Y., Liu, G. R., Lesot, H., Haikel, Y., Ruch, J. V. and Nagai, N. (1997). Bone morphogenetic protein (BMP) stimulates cytological and functional differentiation of preodontoblasts in vitro. *Journal of Hard Tissue Biology*, **6**, 105–113.

Nakashima, M. (1991). Establishment of primary cultures of pulp cells from bovine permanent incisors. *Archives of Oral Biology*, **36**, 655–663.

Niswander, L. and Martin, G. R. (1992). FGF-4 expression during gastrulation, myogenesis, limb and tooth development in the mouse. *Development*, **114**, 755–768.

Noden, D. M. (1983). The role of the neural crest patterning of avian cranial skeletal, connective and muscle tissues. *Developmental Biology*, **96**, 144–165.

Ohsaki, Y. and Nagata, K. (1994). Type III collagen is a major component of mouse molar root. *Anatomical Record*, **240**, 308–313.

Partanen, A. M. and Thesleff, I. (1987). Localization and quantitation of ^{125}I-epidermal growth factor binding in mouse embryonic tooth and other embryonic tissues at different developmental stages. *Developmental Biology*, **120**, 186–197.

Pelton, R. W., Dickinson, M. E., Moses, H. L. and Hogan, B. L. M. (1990). In situ hybridization analysis of TGFβ3 RNA expression during mouse development: comparative studies with TGFβ1 and 2. *Development*, **110**, 609–620.

Pelton, R. W., Saxen, B., Jones, M., Moses, H. L. and Gold, J. I. (1991). Immunohistochemical localization of TGFβ1, TGFβ2 and TGF3 in the mouse embryo: expression patterns suggest multiple roles during embryonic development. *Journal of Cell Biology*, **115**, 1091–1105.

Peters, K. G., Werner, S., Chen, G. and Williams, L. T. (1992). Two FGF receptor genes are differentially expressed in epithelial and mesenchymal tissues during limb formation and organogenesis in the mouse. *Development*, **114**, 233–243.

Ruch, J. V. (1985). Odontoblast differentiation and the formation of the odontoblast layer. *Journal of Dental Research*, **64**, 489–498.

Ruch, J. V. (1987). Determinisms of odontogenesis. *Cell Biology Reviews*, **14**, 1–112.

Ruch, J. V. (1990). Patterned distribution of differentiating dental cells: facts and hypotheses. *Journal de Biologie Buccale*, **18**, 91–98.

Ruch, J. V (ed). (1995a). Odontogenesis. *International. Journal of Developmental Biology*, **39**, 298 pp.

Ruch, J. V. (1995b). Tooth crown morphogenesis and cytodifferentiations: candid questions and critical comments. *Connective Tissue Research*, **32**, 1–8.

Ruch, J. V. (1998). Odontoblast commitment and differentiation. *Biochemistry and Cell Biology*, **76**, 923–938.

Ruch, J. V. and Karcher-Djuricic, V. (1971). Mise en évidence d'un rôle spécifique de l'épithélium adamantin dans la différenciation et le maintien des odontoblastes. *Annales d'Embryologie et Morphogenise*, **4**, 359–366.

Ruch, J. V., Karcher-Djuricic, V., and Gerber, R. (1973). Les déterminismes de la morphogenèse et des cytodifférenciations des ébauches dentaires de souris. *Journal de Biologie Buccale*, **1**, 45–56.

Ruch, J. V., Karcher-Djuricic, V. and Thiebold, J. (1976). Cell division and cytodifferentiation of odontoblasts. *Differentiation*, **5**, 165–169.

Ruch, J. V., Zidan, G., Lesot, H., Kubler, M. D., Smith, A. J., Mathews, J., Andujar, M. and Magloire, H. (1989). A rat monoclonal antibody MC22–45D directed against a mouse enamel component inhibits terminal differentiation of odontoblasts in vitro. In *Tooth Enamel V*, ed. R. W. Fearnhead, pp. 242–246. Yokohama: Florence Publishers.

Ruch, J. V., Lesot, H. and Bègue-Kirn, C. (1995). Odontoblast differentiation. *International Journal of Developmental Biology*, **39**, 51–68.

Ruch, J. V., Lesot, H., Bègue-Kirn, C., Cam, Y., Meyer, J. M. and Bloch-Zupan, A. (1996). Control of odontoblast differentiation: current hypotheses. In *Proceedings of the International Conference on Dentin/Pulp Complex 1995*, ed. M. Shimono, T. Maeda, H. Suda and K. Takahashi, pp. 105–111. Tokyo: Quintessence Publishing.

Salmivirta, M., Elenius, K., Vainio, S., Hofer, V., Chiquet-Ehrismann, R., Thesleff, I. and Jalkanen, M. (1991). Syndecan from embryonic tooth mesenchyme binds tenascin. *Journal of Biological Chemistry*, **266**, 7733–7739.

Salmivirta, K., Gullberg, D., Hirsch, E., Altruda, F. and Ekblom, P. (1996). Integrin subunit expression associated with epithelial–mesenchymal interactions during murine tooth development. *Developmental Dynamics*, **205**, 104–113.

Sawada, T. and Nanci, A. (1995). Spatial distribution of enamel proteins and fibronectin at early stages of rat incisor tooth formation. *Archives of Oral Biology*, **40**, 1029–1038.

Schlessinger, J., Lax, I. and Lemmon, M. (1995). Regulation of growth factor activation by proteoglycans: what is the role of the low-affinity receptors. *Cell*, **83**, 357–360.

Schmidt, G., Haisser, H. and Kresse, H. (1991). Interaction of the small proteoglycan decorin with fibronectin. Involvement of the sequence NKISK of the core protein. *Biochemical Journal*, **280**, 411–414.

Sigal, M. J., Aubin, J. E., Ten Cate, A. R. and Pitaru, S. (1984). The odontoblast process extend to the dentino enamel junction: an immunocytochemical study of rat dentin. *Journal of Histochemistry and Cytochemistry*, **32**, 872–877.

Smith, A. J. and Leaver, A. G. (1979). Non-collagenous components of the organic matrix of rabbit incisor dentin. *Archives of Oral Biology*, **24**, 449–454.

Smith, A. J., Cassidy, N., Perry, H., Bègue-Kirn, C., Ruch, J. V. and Lesot, H. (1995). Reactionary dentinogenesis. *International Journal of Developmental Biology*, **39**, 273–280.

Smith, M. M., and Hall, B. K. (1990). Development and evolutionary origins of vertebrate skeletogenic and odontogenic tissues. *Biological Reviews*, **65**, 277–373.

Sunohara, M., Tanzawa, H., Kaneko, Y., Fuse, A. and Sato, K. (1996). Expression patterns of raf-1 suggest multiple roles in tooth development. *Calcified Tissue International*, **58**, 60–64.

Taylor, A. N. (1984). Tooth formation and the 28 000 Dalton vitamin D-dependent calcium-binding protein: an immunocytochemical study. *Journal of Hisotchemistry and Cytochemistry*, **32**, 159–164.

Ten Cate, A. R. (1989). Dentinogenesis. In *Oral Histology, Development Structure and Function*, third edition, ed. A. R. Ten Cate, pp. 139–156. St Louis: Mosby.

Terling, C., Rass, A., Mitsiadis, T. A., Fried, K., Lendahl, U. and Wroblewski, J. (1995). Expression of the intermediate filament nestin during rodent tooth development. *International Journal of Developmental Biology*, **39**, 947–956.

Thesleff, I. and Vaahtokari, A. (1992). The role of growth factors in determination and differentiation of the odontoblastic cell lineage. *Proceedings of the Finnish Dental Society*, **88**, 357–368.

Thesleff, I., Kokenyesi, R., Kato, S. and Chiquet-Ehrismann, R. (1987). Changes in the distribution of tenascin during tooth development. *Development*, **101**, 289–296.

Topham, R. T., Chiego, D. J., Gattone, V. H., Hinton, D. A. and Klein, R. M. (1987). The effect of epidermal growth factor on neonatal incisor differentiation in mouse. *Developmental Biology*, **124**, 532–543.

Tziafas, D. (1995). Basic mechanisms of cytodifferentiation and dentinogenesis during dental pulp repair. *International Journal of Developmental Biology*, **39**, 281–290.

Vainio, S., Karavanova, I., Jowett, A. and Thesleff, I. (1993). Identification of BMP-4 as a signal mediating secondary induction between epithelial and mesenchymal tissues during early tooth development. *Cell*, **75**, 45–58.

Wang, Y. Q., Sizeland, A., Wang, X. F. and Sasson, D. (1995). Restricted expression of type II TGF receptor in murine embryonic development suggests a central role in tissue modeling and CNS patterning. *Mechanisms of Development*, **52**, 275–289.

Webb, P. P., Mokham, B. J., Ralphs, J. R. and Benjamin, M. (1995). Cytoskeleton of the mesenchymal cells of the rat dental papilla and dental pulp. *Connective Tissue Research*, **32**, 71–76.

Wilkinson, D. G., Bhatt, S. and McMahon, A. P. (1989). Expression pattern of the FGF-related proto-oncogene *int*-2 suggests multiple role in fetal development. *Development*, **105**, 131–136.

Wise, G. E. and Fan, W. (1991). Immunolocalization of transforming growth factor beta in rat molars. *Journal of Oral Pathology and Medicine*, **20**, 74–80.

Wöltgens, J. H. M., Bronckers, A. L. J. J., and Lyaruu, D. M. (eds.) (1995). Fifth international conference on tooth morphogenesis and differentiation. *Connective Tissue Research*, **32, 33**, 554 pp.

Yamada, S., Yamada, K. M. and Brown, K. E. (1994). Integrin regulatory switching in development: oscillation of 5 integrin mRNA expression during epithelial–mesenchymal interactions in tooth development. *International Journal of Developmental Biology*, **38**, 553–556.

Yamaguchi, Y., Mann, D. M. and Ruoslahti, E. (1990). Negative regulation of transforming growth factor by the proteoglycan decorin. *Nature*, **346**, 281–284.

Young, W. G., Ruch, J. V., Stevens, M. R., Bègue-Kirn, C., Zhang, C. Z., Lesot, H. and Waters, M. J. (1995). Comparison of the effects of growth hormone, insulin-like growth factor-1 and fetal calf serum on mouse odontogenesis in vitro. *Archives of Oral Biology*, **40**, 789–799.

Yoshiba, N., Yoshiba, K., Iwaku, M., Nakamura, H. and Ozawa, H. (1994). A confocal laser scanning microscopic study of the immunofluorescent localization of fibronectin in the odontoblast layer of human teeth. *Archives of Oral Biology*, **39**, 395–400.

Yoshiba, N., Yoshiba, K., Nakamura, H., Iwaku, M. and Ozawa, H. (1995). Immunoelectron microscopic study of the localization of fibronectin in the odontoblast layer of human teeth. *Archives of Oral Biology*, **40**, 83–89.

Zhang, C. Z., Young, W. G. and Waters, M. J. (1992). Immunocytochemical localization of growth hormone receptor in maxillary teeth. *Archives of Oral Biology*, **37**, 77–84.

4 Enamel biomineralization: the assembly and disassembly of the protein extracellular organic matrix

A. G. Fincham, W. Luo, J. Moradian-Oldak, M. L. Paine, M. L. Snead and M. Zeichner-David[1]

4.1. Introduction

The component tissues of teeth include an outer enamel cover over underlying dentin. Enamel is almost completely inorganic mineral, so much so that it has been said that teeth emerge into the mouth as preformed fossils (Boyde, 1996). While this notion certainly reflects the biological function of teeth as elements of the masticatory apparatus, it regrettably and erroneously implies that the formation of teeth is a simple or passive affair. This is not the case and the dentition represents a remarkable and elegant biological adaptation of dermal appendages whose emergence coincides with vertebrate radiation. This is not surprising since teeth provide for increased nutrition through enhanced food processing, as well as providing formidable weapons of offense and defense, and hence, animals bearing teeth might well be predicted to compete more efficiently in a specific niche.

Tooth development has been studied in detail for several decades because it provides a useful model for instructive interactions occurring between dissimilar germ layers. These epithelial–mesenchymal interactions are responsible for the majority of vertebrate organogenesis, including hair, pancreas, mammary gland, salivary gland, thymus, vibrissae, and others. We will not include a detailed analysis of such interactions in this review except where they relate to control of amelogenin gene expression. The reader is directed to several excellent reviews that have recently appeared (Weiss et al., 1994; Thesleff et al., 1995; Zeichner-David et al., 1995; Maas and Bei, 1997). We are cognizant of several exceptional treatments of the field of biomimetics, biomineralization and enamel formation over the past several years and readers are directed to these scholarly works for alternative views of enamel formation (Fearnhead, 1989; Lowenstam

and Weiner, 1989; Archibald and Mann, 1993; Mann, 1993; Slavkin and Price, 1992; Deutsch et al., 1995a,b; Robinson et al., 1995; Simmer and Fincham, 1995; Simmer and Snead, 1995). We have omitted entirely any discussion of dentin formation and touch only briefly upon a newly identified gene associated with root formation and then only because it is also expressed by ameloblasts. Rather than attempt to reinterpret the existing data yet again, we have purposely attempted to confine this review to a narrow perspective of enamel gene expression as it relates to assembly and disassembly of the organic enamel extracellular matrix. It is this assembly of proteins that directs the formation of the mineral phase, and for that reason, we believe that an enhanced understanding of the supramolecular assembly of enamel proteins holds answers to the physiologic function of proteins within this unique vertebrate tissue.

The scaffolding within a living cell is composed of a myriad of proteins, including the microtubules, microtubule-associated proteins, cytoskeletal elements, actin, and others which are bridged to integral membrane components by various linkage proteins. The intracellular assembly of these components occurs through highly regulated pathways which are predominately dependent upon high-energy sources (ATP, GTP). A series of kinases and phosphatases utilize the trinucleotides to catalyze the transfer or removal of phosphate groups to acceptor proteins of the cytoskeleton. These post-translational events often result in striking changes in conformation of the protein resulting in negative or positive modulation in their capacity to participate in

---

[1] Authors are listed alphabetically and each contributed equally to the manuscript.

assembly and disassembly of the higher-ordered structure. The assembly of the enamel matrix occurs extracellularly, without the benefit of energy sources or the exquisite regulation gleaned through a cascade of regulatory phosphorylation–dephosphorylation events (Mosher et al., 1992). Rather, the enamel organic matrix must assemble without direct contiguous cellular intervention. Enamel matrix assembly more closely matches the paradigm of assembly of the extracellular basement membrane, another multiple protein member structure, which assembles solely by virtue of information contained within the constituent proteins themselves. Unlike the basement membrane, enamel does not remodel, nor does it even remain in contact with the cells that synthesized the enamel proteins: the ameloblast retracts from the forming matrix with concomitant mineral deposition. The proteins within the basement membrane are identified as containing multiple domains each with a unique function required for interaction among constituent proteins. For example, the protein laminin contains multiple domains each with a unique contribution towards recognizing and interacting with various collagens through self-assembly to form highly ordered structures with distinctive biological activities.

Dental enamel serves as a model to study protein interactions leading to matrix assembly and biomineralization. Enamel is the hardest, most mineralized tissue in the vertebrate body where the mineral phase is dominated by calcium hydroxyapatite (HAP), which is organized into the largest known biologically formed HAP crystallites. During enamel formation, ameloblasts synthesize and secrete enamel proteins into the extracellular space. Crystallite geometry in biomineralized tissues is guided by protein assemblies occurring within an extracellular matrix, yet the molecular basis for the assembly of the pre-enamel organic matrix (being referred to in this chapter as the enamel organic extracellular matrix) remains vague. Information relating to nucleators of enamel biomineralization and the sites of initial nucleation is non-existent. The subsequent events that result in an almost total replacement of the organic phase by the inorganic hydroxyapatite crystallites are better described, but knowledge of the underlying mechanisms is still in its infancy!

Enamel is first formed in organic matrix devoid of mineral. The enamel matrix is composed principally of proteins, with the organization of the matrix components occurring outside the cell, implying that all the biologically relevant information required for supramolecular assembly of constituent parts resides in the primary structure of the protein itself. While a number of proteins have been reported as being involved with the enamel organic matrix, only three enamel-specific genes have been cloned and partly characterized. Presently, the known organic constituents of the enamel extracellular matrix (ECM) are the amelogenins (Termine et al., 1980; Snead et al., 1983) and the enamelins including tuftelin (Deutsch et al., 1991). The recent cloning of ameloblastin (Krebsbach et al., 1996), also known as amelins by Cerny and co-workers (Cerny et al., 1996) or sheathlin by Hu and co-workers (J. P. Simmer and C.-C. Hu, personal communication) has introduced a third class of ameloblast-specific proteins. With the identification of several 'low abundance' non-amelogenin proteins (Fukae and Tanabe, 1987a,b), as well as other proteins that are potentially involved in post-translational modification of enamel matrix proteins, such as sulfated enamel matrix proteins (Chen and Smith, 1994; Smith et al., 1995a,b), there may even yet be additional gene(s) and the protein(s) they encode potentially involved in enamel organic extracellular matrix biogenesis. Enamel proteins are expressed exclusively in developing enamel and are subsequently removed during mineral phase maturation, with an amelogenin-specific protease being recently described (Moradian-Oldak et al., 1994b). Clearly the amelogenins are the best characterized of the 'enamel' proteins.

The essential role for amelogenin during enamel organic extracellular matrix assembly has been shown by several independent lines of investigation, thus implying that all the biologically relevant information required for supramolecular assembly of the enamel organic extracellular matrix resides primarily in protein structure itself. Early speculations (Snead et al., 1989) that alterations in amelogenin expression would be identified as the pathogenic event in several types of human *amelogenesis imperfecta* have now been documented (Lagerström et al., 1990, 1991; Aldred et al., 1992; Diekwisch et al.,1993; Lagerström-Fermer et al., 1995; Lyngstadaas et al., 1995). Aldred and his team of detectives (Aldred et al., 1992) undertook a molecular dissection of the gene(s) responsible for the *amelogenesis imperfecta* phenotype in a kindred. Rather than identifying a unique mutation conferring messenger instability, inappropriate splice junctions selection, or other underlying alterations in the requisite protein domains for enamel biogenesis as suggested by Snead et al. (1989) they instead discovered a previously unknown locus mapping to the long arm of the X-chromosome. A direct role for amelogenin-directed control of crystallite growth comes from the experiments of Doi and colleagues (Doi et al., 1984) who used purified amelogenin protein to direct mineral deposition confined to the ends of hydroxyapatite seed crystals. Recently, additional experimental support for self-assembly of enamel proteins comes from *in vitro* experiments in which highly purified bacterial-expressed M180 amelo-genin was shown to assemble into 'nanospheres', which can also be observed during *in vivo* enamel formation (Moradian-Oldak *et al.*, 1994a; Fincham *et al.*, 1995a,b). The principles that regulate enamel formation appear to be congruent for all mammalian species examined to date, including mouse and man. For all of these reasons, we have chosen to focus our attention on the assembly of the enamel organic extracellular matrix that results in a unique extracellular micro-milieu, which is generally believed to regulate the initiation, propagation, termina-tion and maturation of enamel hydroxyapatite crystal-lites during mineral phase maturation and the regulated disassembly of the matrix through selective degradation of protein components (Moradian-Oldak *et al.*, 1994b).

4.2.   The enamel proteins

4.2.1.   Overview
Most papers written in the field of tooth development described the proteins present in enamel as belonging to one of two classes of proteins: amelogenins or enamelins. Amelogenins comprise 90% of the enamel proteins and are the better characterized of the two enamel protein classes. The other remaining proteins present in the enamel extracellular matrix are grouped into the enam-elin class of enamel proteins. In contrast to amelogenins, detailed knowledge for these 'other' enamel proteins has remained rather sparse. The major factor contributing to the lack of information concerning these proteins has been their under-representation within the forming enamel extracellular matrix.

The term 'enamelin' was originally introduced by Mechanic (1971) to describe the neutral soluble proteins present in mature enamel. However, Eastoe (1965) had been referring to these same neutral soluble proteins as 'amelogenins' since 1965 and suggested that the term enamelin be reserved for the proteins remaining in the mature enamel. Termine and co-workers (1980) used the term enamelin operationally to describe a class of proteins that are associated with the mineral phase of enamel. Proteins were extracted using a sequential dis-sociative method with 4 M guanidine hydrochloride, followed by removal of the mineral with 4 M guanidine hydrochloride containing 0.5 M EDTA to solublilize the enamelins. The EDTA solubilized extract contained proteins of relatively higher molecular weight than amelogenins, which were phosphorylated and had an acidic nature. The enamelin proteins had an amino acid composition different than that for the amelogenins, being rich in glycine, serine, glutamic acid, and aspartic

acid residues. The enamelin proteins more closely re-sembled the low-cystine keratin proteins present in the oral epithelium than the amelogenins. No physiologic function has yet been demonstrated for the enamelin proteins, however, because of their acidic characteristics, and, by analogy to other mineralizing systems, it was suggested that their function is to serve as nucleators for hydroxyapatite crystallite formation (Termine *et al.*, 1980).

Originally, it was believed that enamelins represented as much as 10% of the total enamel proteins (Termine *et al.*, 1980, Fincham *et al.*, 1982a); however, several studies demonstrated the presence of many other proteins present in the enamel extracellular matrix such as albumin, other serum proteins (Okamura, 1983; Zeichner-David *et al.*, 1987; Menanteau *et al.*, 1987; Lime-back *et al.*, 1989; Strawich and Glimcher 1990), serine proteases (Suga, 1970; Carter *et al.*, 1984, 1989; Crenshaw and Bawden, 1984; Menanteau *et al.*, 1986; Moradian-Oldak *et al.*, 1994b), Ca$^+$-dependent proteases (Moradian-Oldak *et al.*, 1994b), glycoconjugates (Menanteau *et al.*, 1988) and sulfated glycoconjugates (Goldberg and Septier, 1986; Kogaya and Furuhashi, 1988, Chen *et al.*, 1995) (see Figure 4.1). Although some of these proteins might fit in one of the criteria previously referred to as enamelin (e. g. based either upon solubility, or presence in mature enamel, or high molecular weight, or anionic nature), their individual physiologic function(s) remain unknown and they should not be grouped under the term enamelin. It has become more usual to refer to these diverse proteins as the 'non-amelogenin enamel proteins', retaining the name enamelin for proteins closely associated the hydroxyapatite crystallite formation.

4.2.2.   Anionic enamel proteins
The difficulty in obtaining pure anionic enamel proteins in sufficient amounts has been a major drawback in obtaining their complete characterization. Initially, several investigators attempted to isolate and to purify enamelins from bovine (Ogata *et al.*, 1988; Menanteau *et al.*, 1988), porcine (Fukae *et al.*, 1980; Limeback 1987), human (Deutsch *et al.*, 1984; Zeichner-David *et al.*, 1987) and rabbit (Zeichner-David *et al.*, 1988,1989) tissues. At-tempts to produce enamelin-specific polyclonal antibo-dies (Zeichner-David *et al.*, 1987; Deutsch, 1989) failed to distinguished one class of enamel proteins from the other; possibly due to contamination of enamelins with other proteins, such as amelogenin aggregates (Belcourt *et al.*, 1983; Limeback and Simic, 1990), serum-derived proteins (Limeback *et al.*, 1989; Strawich and Glimcher, 1990) and/or shared epitopes between the amelogenins and enamelins (Zeichner-David *et al.*, 1989). Preparations

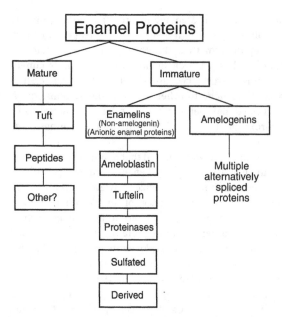

Figure 4.1. 'Organizational chart' for dental enamel proteins. Immature enamel contains both amelogenins and non-amelogenins. The enamelin fraction (e. g. non-amelogenin, anionic enamel protein) is some 10% of the secretory stage enamel matrix, and comprises several unrelated proteins; ameloblastin, tuftelin(s) and proteinases. Tuftelin may be the precursor of the insoluble 'tuft protein' found in the mature enamel (Robinson *et al.*, 1975). Recent advances in molecular biology and the cloning the genes that encode these enamel proteins makes any nomenclature rapidly archaic. This organization is subject to revision as new genes are discovered and their role in enamel biomineralization uncovered.

of monoclonal antibodies against bovine enamelin were able to recognize only high molecular weight enamel proteins (Shimokawa *et al.*, 1984). However, these antibodies failed to cross-react with secretory ameloblast cells (Rosenbloom *et al.*, 1986; Deutsch *et al.*, 1987), and these biochemical studies have so far yielded only partial amino acid sequences. Zeichner-David *et al.* (1988) published the N-terminal amino acid sequence for a rabbit enamel extracellular matrix derived protein as LPLPPHPGHPG. The rabbit sequence was identical to the N-terminal conserved sequence of amelogenins from all other mammalian species examined to date, except for having an N-terminal leucine residue instead of methionine. An anti-peptide antibody directed against the rabbit sequence identified the oligopeptide epitope in high molecular weight anionic enamel proteins obtained from several other species, suggesting a conserved function for this domain in both proteins (Zeichner-David *et al.*, 1989). Alternatively, the epitope may not have been identified in a high molecular weight anionic enamel protein, but rather in an amelogenin aggregate as suggested by several other investigators (see discussion of Zeichner-David *et al.*, 1989), or the

epitope defined by the anti-peptide antibody used in these experiments may be shared between enamelin and amelogenin.

Tanabe *et al.* (1990) reported a partial N-terminal sequence for a 32 kDa phosphorylated bovine non-amelogenin as LXQVPGRIPPGYGRPPTP. The protein that contained this amino acid sequence was eventually found to be the processed product of a larger protein, referred to as the 89 kDa enamelin (Fukae *et al.*, 1996) with the N-terminal sequence, MPMQMPRMPGFSPKR-EPM, that is rich in methionine residues. Strawich and Glimcher (1990) reported four N-terminal amino acid sequences, presumably for bovine enamelins: (a) MPLPPXXG; (b) FPYDGFAGY; (c) LKQPSSXXAQ and (d) YQQPPWHIIHKKIPGFF, one of which, sequence d, also contained a portion of the amelogenin amino acid sequence. Due to the difficulty of extracting pure enamelins from the enamel extracellular matrix, some investigators have suggested that enamelins do not exist and what some investigators have been describing as enamelins are actually albumin and/or amelogenin aggregates (Limeback and Simic, 1989; Strawich and Glimcher, 1990). Couwenhoven *et al.*, (1992) have shown that messenger RNA encoding albumin can not be identified in preparations of developing mouse enamel organs, suggesting that any albumin recovered from developing enamel is not due to its expression by ameloblasts. Chen and colleagues (1995) have shown that if enamel organs are freeze-dried prior to their removal from the jaw, so as to prevent disruption of the vasculature, albumin is not recovered in protein preparations of the resulting enamel. Collectively, this data suggests that if albumin is postulated to play a functional role during enamel organic matrix assembly it must be biosynthesized at a site distant from the enamel organ and then transported back into the enamel organ; such a pathway has not yet been demonstrated.

Tuftelins

Another group of non-amelogenin proteins are the 'tuft' proteins (Robinson *et al.*, 1975, 1989a). These are insoluble proteins present in the enamel tufts radiating from the dentin—enamel junction of mature teeth. The tuft proteins have an amino acid composition rich in glutamic acid, aspartic acid, leucine and serine (Robinson *et al.*, 1975) resembling slightly the composition of the keratins. Antibodies directed against freeze-dried tuft protein indicated that tuft proteins share epitopes with keratins, recognizing proteins in the 55–57 kDa and 25 kDa range. The anti-tuft protein antibody also cross-reacted with a protein present in secretory organelles of the ameloblast cells, as well as in the multivesicular structures of the stratum granulosum of the foot pad

(Robinson et al., 1989a). These data attest to the secretory nature of tuft proteins and their synthesis by ameloblast cells. Because of the location, amino acid composition similarity and antibody cross-reactivity of tuft protein with keratins, it was suggested that the tuft proteins represent ancestral (e. g. primitive) ameloblast secretory products (Robinson et al., 1975). It should be noted that cross-reactivity with keratins has also been reported for antibodies raised against proteins extracted from dentin extracellular matrix and bone matrix (Lesot et al., 1988). The function of the tuft proteins remains unknown.

Advancement in the field came through the application of a molecular biology approach when Deutsch and colleagues (1991) reported the isolation and characterization of a cDNA clone for a novel acidic enamel protein that they termed 'tuftelin'. Although the clone, one of nine identified, was isolated by screening a bovine cDNA tooth library with a monoclonal enamelin antibody (Deutsch et al., 1987), the deduced amino acid sequence resembled the composition of the tuft proteins (Robinson et al., 1975, 1989a,b), thus the name tuftelin. There is, however, a difference in size between the conceptual translation of the cDNA versus the protein recovered from forming enamel. This could be accounted for by ameloblast-directed post-translational modifications that are predicted to retard the protein mobility on SDS gels. Thus proteins recovered from developing teeth may demonstrate an apparent larger molecular weight. Additional information on tuftelin genes from other species is now forthcoming (e. g., Zeichner-David et al., 1993; Dodds et al., 1996) and may aid in the identification of protein domains conserved among species and thus direct investigators towards protein domains of physiologic relevance.

Using reverse transcription coupled to polymerase chain amplification (RT-PCR), Zeichner-David et al. (1993) reported the initial expression of tuftelin mRNAs in the developing mouse molar at the bud stage of tooth development (embryonic day 13 (E13), vaginal plug is day 0), whereas amelogenins mRNAs are first detected at E15. The identification of the tuftelin protein in these developing teeth was performed by immunohistochemistry and in contrast, the tuftelin protein was not apparent until the cap stage (E17), while the amelogenin proteins are first immunodetected at the newborn stage of development (Snead et al., 1985; Zeichner-David et al., 1994). Mouse tuftelin is first expressed at a developmental stage in which the basal lamina is still present, the inner enamel epithelial cells are not yet polarized but rather are still dividing (Zeichner-David et al., 1994, 1995). The discrepancy between the detection of transcripts and the detection of translation product might represent a difference in the sensitivity of the two methods employed. PCR

is an extremely sensitive technique, whereas immunodetection requires many more molecules for positive detection. Another interpretation of these results is that the observed differences might represent a physiological stage of ameloblast proto-differentiation, a situation in which very low levels of transcripts are expressed, as suggested by studies reported by Couwenhoven and Snead (1994). The function of tuftelin remains unknown. However, because of its anionic character, its localization to the dentin–enamel junction (Deutsch et al., 1991; Zeichner-David et al., 1994,1995) and its expression prior to amelogenins, it has been suggested that tuftelin protein might serve as nucleator for hydroxyapatite crystal formation (Deutsch et al., 1991, 1994; Zeichner-David et al., 1995). Tuftelin has been localized to human chromosome 1q21–31 using the fluorescence in situ hybridization (FISH) method (Deutsch et al., 1994). The tuftelin gene remains a promising candidate for the inherited disturbances to enamel formation known as amelogenesis imperfecta, that have an autosomal mode of inheritance. To date, no linkage to chromosome 1 has been demonstrated for amelogenesis imperfecta.

4.2.3. The 'other' enamel proteins: non-amelogenin/non-enamelin

Ameloblastin

In 1996 a new member of the non-amelogenin class of enamel proteins was simultaneously characterized by three different groups of investigators, two groups using rat incisors and one group using porcine teeth. In the USA, Krebsbach et al. (1996) named it 'ameloblastin'; in Sweden, Cerny et al. (1996) named it 'amelin', and in a joint study between investigators in Japan and the USA, C.-C. Hu and colleagues (personal communication) named it 'sheathlin' (see also Snead 1996). For the purpose of this discussion, the term 'ameloblastin' is used. All groups reported the presence of two transcripts, resulting from alternative splicing, with deduced amino acid sequences indicating proteins of approximately 422–395 amino acids for one and 380–324 amino acids for the other, rich in proline, glycine and leucine with isoelectric points of ranging from 6.7–5. 2. The homology between the rat and porcine proteins is 77%. There is also significant homology to the N-terminal sequence of a 15 kDa porcine protein first identified as a 'sheath' protein that accumulates at the rod boundaries (Uchida et al., 1995). Using an antibody produced against a recombinant ameloblastin, Krebsbach and co-workers (1996) identified two bands; one of 70 kDa and the other of 20 kDa, detected by Western immunostaining using proteins extracted from the rat enamel extracellular matrix. This antibody also localizes ameloblastin to Tome's process, the highly specialized plasma membrane component of secretory ameloblast cells, suggesting that ameloblastin may be involved in linking the ameloblast to the enamel extracellular organic matrix. The ameloblastin molecule has a 'DGEA' domain (Cerny et al., 1996), which has been identified in collagen type I as a recognition site for alpha 2 beta 1 integrin (Staatz et al., 1991). A trombospondin-like cell adhesion domain, 'VTKG' (Yamada and Kleinman, 1992), is also present. These data have prompted the suggestion that ameloblastin, by analogy to the role of collagen in other mineralizing tissues, might serve to provide interactions between ameloblasts and the enamel extracellular matrix (Cerny et al., 1996).

Recent data indicated that Hertwig's epithelial root sheath (HERS) cells, also express ameloblastin mRNA during root development, although the protein has yet to be detected in similar sites during root formation (Fong et al., 1996). The function of this protein, if any, in root formation is also unknown at this time. The presence of enamel proteins in the murine acellular cementum has been previously postulated by Slavkin and colleagues (1988) using antibodies directed against mouse amelogenin oligopeptides (anti-peptide antibodies) or human enamelins. Since ameloblastin has a 25% similarity to amelogenin (particularly to the N-terminal region LPPHPGHPGYI of mouse amelogenin to which the anti-peptide antibody was produced), it is possible that the results described by Slavkin et al. (1988) of an 'amelogenin-like' protein in developing tooth roots were due to the presence of cross-reacting epitopes in ameloblastin. Using solution hybridization techniques, Cerny and colleagues (1996) determined that 5% of the non-amelogenin mRNAs is represented by ameloblastin. A similar frequency of ameloblastin clones was found by Krebsbach et al. (1996). Ameloblastin has been localized to mouse chromosome 5 (Krebsbach et al., 1996) which is syntenic to human chromosome 4. Recently, a family with autosomal dominant *amelogenesis imperfecta* has been linked to human chromosome 4q (Forsman et al., 1994) thus raising the possibility that ameloblastin might be the gene responsible for the enamel defect observed in this family. A physiologic role for the ameloblastin protein in tooth development remains unknown.

The expression of ameloblastin mRNA in the developing mouse molar is first detected at E14 by RT-PCR (Simmons et al., 1996). Couwenhoven and Snead (1994) showed that when the enamel organ epithelia is isolated from the dental ectomesenchyme and placed in culture, continued differentiation to the ameloblast phenotype requires the presence of a reconstituted (Matrigel) basement membrane gel. When these investigators placed isolated epithelia derived from dental lamina (E12), bud stage (E13) or early cap stage (E14) in culture, only the cap

stage-derived enamel organ epithelium expressed amelogenin in prolonged tissue culture within reconstituted basement membrane gels; the earlier stages of odontogenic epithelia did not express amelogenins. These experiments suggest that the instructive signal(s) that controls amelogenin transcription occurs prior to, or during, the early cap stage of development. These studies also suggest that the inducer(s) for tuftelin and ameloblastin transcription may be different than those for amelogenin.

Sulfated enamel proteins

Another class of proteins present in the enamel extracellular matrix are the 'sulfated enamel proteins' (Smith et al., 1995a,b). These proteins are present in such small amounts that their presence can only be detected by radioactive labeling of the sulfate groups. Within six minutes of an intravenous injection of [35]SO4, a 65 kDa protein can be identified in enamel homogenates that has less than a 30 minute half-life, after which, the 65 kDa protein disappears to be replaced by two proteins of ~49 and ~25 kDa. These proteins are destroyed within 1–2 hours (Smith et al., 1995a,b). The acidic nature of the sulfated enamel proteins (pI = 4.8–5.6) suggests that they belong to the family of anionic enamel proteins. The sulfated anionic proteins could be one or more of the previously described anionic enamel proteins recovered by EDTA with guanidine HCl. However, this appears unlikely, since the previously described proteins can be recovered from the mature enamel extracellular matrix, whereas the sulfated proteins disappear shortly after they are synthesized. The role of the sulfated enamel proteins during enamel formation is unknown at this time.

Serum- and saliva-derived proteins

Other proteins identified in the enamel may be derived from serum and these would include albumin, alpha-2 HS glycoprotein, gamma-globulin, and fetuin. Additional contaminants may be derived from saliva, and include the proline-rich salivary protein (Strawich and Glimcher, 1990). The presence of these proteins in the enamel and their physiologic role, if any, during enamel biomineralization has been controversial. Some investigators believed that their presence is an artifact of blood contamination when the teeth are dissected (Chen et al., 1995), while other investigators have suggested a physiologic role for albumin, including: (1) controlling the rate of hydroxyapatite crystal formation by inhibiting its growth (Robinson et al., 1995); and (2) acting as calcium binding or releasing proteins (Menanteau et al., 1987). To date, the serum and salivary derived proteins have not been shown to be synthesized by the ameloblast cells, but rather appear to be absorbed onto the mineralized matrix due to their high affinity for apatite (Menanteau et al., 1987; Chen et al., 1995).

Proteinases

Our knowledge of the proteinases present in the enamel extracellular matrix has been greatly amplified in the past few years. Several classes of proteinases, including serine proteases and metalloproteases have been identified and characterized in forming enamel (Crenshaw and Bawden, 1984; Carter et al., 1989; DenBesten and Heffernan, 1989; Moradian-Oldak et al. 1994b; 1996a,b). The role of these gene products is discussed in section 5, a section devoted to the disassembly of the organic enamel matrix.

The 1990s have started the era of the 'other enamel proteins' with the molecular cloning of tuftelin by Deutsch and colleagues (1991), as well as the cloning of ameloblastin by three independent groups (Kresbach et al., 1996; Cerny et al., 1996; C.-C. Hu and J. P. Simmer, personal communication). A significant shift in strategy so as to employ more molecular biology approaches, has resulted in the removal of a previous barrier to the study the non-amelogenin enamel proteins, that being their relatively low concentration in forming enamel. The resulting progress has been remarkable. The present state of our knowledge for the newly discovered enamel proteins of tuftelin and ameloblastin, has almost reached the same knowledge level as that obtained for amelogenins in over a decade. While there will probably be (many?) more enamel proteins discovered in the near future, one major question that is still awaiting an answer is: What is their function in the assembly and disassembly of the enamel organic matrix that directs biomineralization?

4.3. Control of amelogenin gene expression

An archetypal characteristic of the multicellular organism is the exquisite capacity to regulate the site, timing and amount of a particular protein biosynthesized by constituent cells. This property exclusively accounts for the diversity of tissues that comprise the vertebrate body plan: each tissue having a unique physiologic capacity that reflects the function of proteins selected for expression by their constitutive cells. Regulation of protein expression can occur at several levels of cellular complexity. At a coarse level, transcriptional control dominates the regulated expression of amelogenin. However, post-transcriptional controls should not be overlooked since they can play important roles, in-cluding: regulation of availability of messenger RNA for translation; regulation of alternative splicing of amelogenin isoforms that contribute to organic matrix assembly; and degradation of messenger RNA species to prevent translation. In addition, several levels of translation and post-translational control including phosphorylation, vesicular transport, and premature degradation of translated proteins prior to export into the enamel extracellular space should be considered. For the purpose of this review, we shall focus on regulation at the transcriptional level, that is, regulation that controls the amount of messenger RNA synthesized by the ameloblast.

A popular and powerful approach to understanding regulated gene expression during development is the analysis of cis and trans acting elements that lie upstream of the coding region and serve to regulate tissue specific patterns of gene expression. This experimental approach is commonly referred to as 'promoter bashing'. The fruits of such an experimental approach are the identification of the transcription factors (e. g., proteins) and the DNA sequence in the promoter that bind the protein in a mutually dependent, sequence-specific manner. The promoter segment of DNA contains DNA elements in the form a unique sequence of nucleotides, that provide a template upon which is built a stable assembly of transcription factors. Since only cells determined to a specific fate contain the required transcription factor proteins, expression of certain genes is restricted to such differentiated cells. It is this assembly of transcription factors that acts to modulate tissue-specific expression of the RNA transcript by a specific tissue

Amelogenin is normally expressed only by ameloblasts and then in a precise temporal and spatial pattern consistent with the embryonic construction of a tooth. Because of this exquisite level of regulation it has long been held that only by identifying the DNA elements within the promoter, and the transcription regulatory proteins that bind the DNA elements, can an understanding of amelogenin gene expression be achieved at a mechanistic level. The fruits of 'amelogenin promoter bashing' would include an understanding of the DNA elements and transcription binding proteins that confer temporal and spatial control over the larger issue of tooth development. These can be difficult experiments to carry out and to interpret, but they provide significant insight into the mechanisms that regulate an essential underlying cellular function of ameloblasts. Understanding the regulation of amelogenin expression may also provide insight for developmental biologists and evolutionary biologists by providing the molecular participants that operate during development. In as much as the fossil record reflects the structure produced by a developmental regulatory mechanism, scientists attempting to unravel the evolution of tooth structure may also gain insight into how changes in regulated gene expression can account for variation in tooth form. Alternatively, DNA elements in the promoter are also capable of binding protein factors that repress or silence the expression of a specific gene in a tissue that does not normally express it.

Silencing has been the outcome realized by Adeleke-Stainback and colleagues (1995), who have analyzed fragments of the 3500 nucleotide (nt) putative promoter for the bovine X-chromosomal amelogenin gene. These investigators fused either the 1500 nt or 1900 nt fragment of the promoter to a construct containing the SV40 viral promoter coupled to the reporter gene, chloramphenicol transferase (CAT), to give either a pCAT BAX 1.5 or a pCAT BAX 1.9 construct, respectively. Either construct was able to silence expression of the viral promoter when transfected into HeLA cells. Moreover, this silencing was observed when the bovine X-chromosomal amelogenin promoter fragment was cloned in reverse orientation. The investigators concluded that silencing is an important part of the mechanism(s) used to achieve cell-specific expression of amelogenin by ameloblasts.

In contrast to the silencing, Chen and co-workers (1994) demonstrated that the same 3500 nt putative promoter for the bovine X-chromosomal amelogenin gene directed expression of the beta-galactosidase gene exclusively to secretory ameloblasts as determined by transgenic mouse studies. Transgenic animal technology represents the culmination of several advances in molecular biology and animal husbandry, disparate techniques that have been linked to provide an extremely powerful approach to gene function analysis (see also Hogan *et al.*, 1986; Ignelzi *et al.*, 1995). In this approach, fertilized mouse eggs are injected with a 'foreign' gene, the transgene, in this case formed by joining the amelogenin promoter to a reporter gene, so that expression of the reported gene can be easily followed. In some cases the injected egg will develop into a mouse that can be identified by molecular hybridization methods to contain the transgene. These are called 'founder mice' and in some cases, these animals will have the transgene included in their germ line cells and hence future generations of mice bearing the transgene can be produced by simply breeding appropriate parental lines. In analyzing their transgenic lines, Chen and co-workers (1994) showed that the bovine amelogenin promoter responds to mouse-derived transcription factor proteins synthesized by differentiated mouse ameloblasts. This same 3500 nucleotides of X-chromosomal bovine amelogenin promoter can contain DNA elements capable of suppressing expression when transfected into non-ameloblast cells.

We focused upon the mouse, since there is only one amelogenin locus residing on the X-chromosome (Chapman *et al.*, 1991). Snead and coworkers (1996) cloned the putative mouse amelogenin promoter, an upstream region of regulatory DNA and produced a transgene by coupling it to the easily detected luciferase reporter protein. Based upon evidence gained by immunohistochemical detection using a monospecific anti-luciferase antibody in transgenic mice, ameloblasts along the tip of the major buccal cusp of mandibular first molars first express detectable levels of luciferase at the new born stage of development. These transgenic animals thus express the luciferase gene exclusively in ameloblasts, in a temporal and spatial pattern that matches the expression profile for endogenous amelogenin expression. This finding suggests that the 2263 nt piece of DNA from the mouse X-chromosomal amelogenin contains all of the requisite regulatory elements that result in a canonical pattern for the temporal and spatial pattern of amelogenin expression. The finding of appropriate temporal and spatial regulation for the transgene suggests that it may be possible to identify the specific DNA elements and members of the hierarchical cascade of signals, including growth factor pathways and transcription factors, which are involved with regulating amelogenin transcription Thus, the nucleotide sequence for this piece of DNA can be examined for the presence and function of important regulatory hierarchical elements that have been suggested to be essential to tooth formation and enamel biogenesis. DNA sequence comparison between the bovine X-chromosomal amelogenin promoter with the murine X-chromosomal amelogenin promoter suggest that several areas of DNA sequence similarity immediately adjacent to the start site of transcription, but that areas further upstream are less similar between species (Figure 4.2). Shared nucleotide sequences in this region of the promoter suggests that the minimal basal promoter for amelogenin, and hence the transcription factors that bind to them, may be similar between these two species. If this is correct, then a correlate is that the DNA upstream of this area of shared sequence may provide additional sites for transcription factor binding leading to the silencing observed by Adeleke-Stainback *et al.* (1995), or the facilitation observed by Snead and co-workers (1996).

The genetic engineering of mice bearing the murine X-chromosomal amelogenin promoter coupled to the sensitive reporter gene, luciferase, as produced by Snead and co-workers, will make possible the identification and testing of putative regulatory elements. Towards this end, several independent lines bearing the amelogenin promoter-luciferase transgene have been established (Snead *et al.*, 1996), as determined by Southern blot

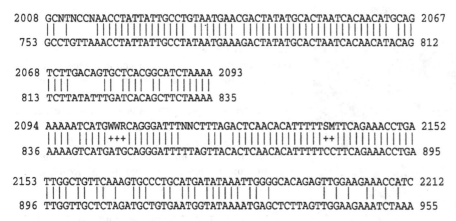

```
2008  GCNTNCCNAACCTATTATTGCCTGTAATGAACGACTATATGCACTAATCACAACATGCAG  2067
      || |       |||||||||||||||  ||||||| ||||||||||||||||||||||||| |||
 753  GCCTGTTAAACCTATTATTGCCTATAATGAAAGACTATATGCACTAATCACAACATACAG  812


2068  TCTTGACAGTGCTCACGGCATCTAAAA  2093
      ||||     || ||||| ||  |||||||
 813  TCTTATATTTGATCACAGCTTCTAAAA  835


2094  AAAAATCATGWWRCAGGGATTTNNCTTTAGACTCAACACATTTTTSMTTCAGAAACCTGA  2152
      ||||  |||||+++||||||||||     |||  ||||||||||||||||++|||||||||||
 836  AAAAGTCATGATGCAGGGATTTTTAGTTACACTCAACACATTTTTCCTTCAGAAACCTGA  895


2153  TTGGCTGTTCAAAGTGCCCTGCATGATATAAATTGGGGCACAGAGTTGGAAGAAACCATC  2212
      ||||  || || |   |||  || |||| |||||| || |          ||| ||  | | |
 896  TTGGTTGCTCTAGATGCTGTGAATGGTATAAAATGAGCTCTTAGTTGGAAGAAATCTAAA  955


2213  GGATCAAGCATSCCTGAGCTTCAGACAG  2240
      |  |||||||||+||||| ||||||||||
 956  GAATCAAGCATCCCTGAGTTTCAGACAG  983
```

Figure 4.2. **Comparison of the mouse and bovine X-chromosomal amelogenin promoters. The area shown is proximate to the transcription start site and suggests a highly conserved grouping of nucleotides. The sequences that lie further 5′-upstream are far less similar and may contain as yet unidentified regulatory domains. These upstream regulatory domains may account for the variation that investigators have encountered in analyzing the regulation of the amelogenin gene in transgenic animals or by transient transfection assays.**

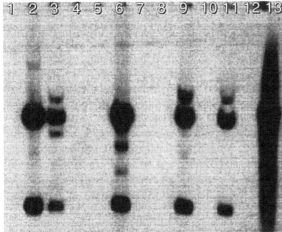

Figure 4.3. **Confirmation of transgene status by Southern blot analysis of the F⁰ mice. The genotype was analyzed for the presence of the amelogenin promoter-luciferase transgene by hybridization. Positive hybridization indicates integration of the transgene into the genome, with the intensity and number of bands suggestive of multiple integration sites and/or variable copy number for the transgene. Lanes 1 through 12 contain 10 µg of EcoR1 digested DNA from individual fetuses; lane 13 contains 100 ng of the luciferase template that was also used as a positive control.**

analysis of F⁰ pups (Figure 4.3). Expression of the reporter gene was shown to be under control of the putative amelogenin promoter with transgene expression limited to the ameloblasts lining the surface of the major cusp of the first mandibular molar at the newborn stage of development (Figure 4.4). This pattern of temporally and spatially restricted expression for the amelogenin gene is consistent with the pattern of expression for the endogenous amelogenin gene at the newborn stage of development (Snead et al., 1988). Moreover, maxillary and mandibular first molars show the pattern of expression

required for effective articulation of opposing cusp surfaces as is observed for the canonical amelogenin gene (Figure 4.4). There were no sites of transgene expression identified within non-dental craniofacial tissues: transgene expression was restricted exclusively to cells of the ameloblast lineage. Snead and co-workers (1996) concluded that the 2263 nt amelogenin promoter is capable of responding to inductive signals arising between dissimilar germ layers that result in the temporal and spatial regulation of the amelogenin gene (Snead et al., 1996). These transgenic animals will provide the reagents for analyzing the transcription factors required for amelogenin expression using a simple mating strategy. For example, if a specific transcription factor is a candidate protein believed to be involved in regulating amelogenin gene expression and if a gain of function transgenic mouse line is available for that transcription factor (e. g., mice overexpression of the transcription factor), then it would be a simple matter to breed such an animal to the mice bearing the amelogenin-luciferase transgene. Animals bearing both transgenes might be expected to show an increase in the luciferase-reported gene activity if the transcription factor manifests a facilitory role in regulation of amelogenin expression. Alternatively, the animals bearing both transgenes might show reduced luciferase reporter gene activity as the transcription factor inhibits amelogenin expression. Such whole animal experiments are an important adjunct to the typical promoter bashing strategies that use cell lines and selected portions of the promoter in an attempt to identify transcription binding sites and their role in gene regulation of amelogenin. Thus in vitro cell line analysis and whole animal experiments should complement each other in delineating critical pathways for amelogenin regulation.

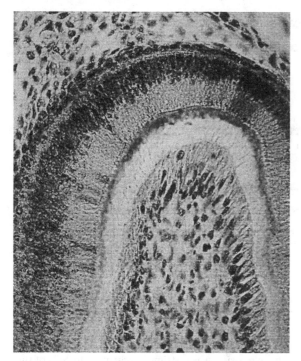

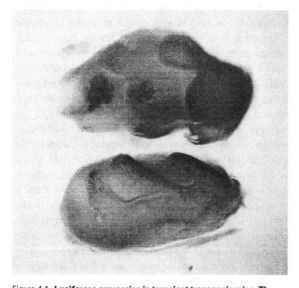

Figure 4.4. Luciferase expression in transient transgenic mice. The two panels show the expression of the luciferase reporter protein a consequence of appropriate temporal and spatial regulation exerted on the amelogenin promoter. The upper panel is a 5 μm tissue section that shows luciferase expression exclusively in ameloblasts within the first mandibular molar at the newborn stage of development. The lower panel is a whole mount that shows the complementary expression of luciferase between maxillary first and mandibular first molars. Note the expression of luciferase is under the control of the amelogenin promoter and results in expression only on the major cusp tips with a typical gradient of expression from the more mature cusp tip towards the less mature ameloblasts nearer the dentin–enamel junction (base of cusp). This pattern of expression is identical to the endogenous amelogenin gene and suggest that the promoter contains all of the necessary control elements for developmentally regulated amelogenin expression. Luciferase was detected using a monospecific anti-luciferase antibody.

4.4.   Self-assembly of enamel proteins

4.4.1.   Overview

The amelogenin proteins, first characterized by Eastoe (1965), are the principal tissue-specific proteins of the extracellular organic matrix of developing dental enamel (Termine *et al.*, 1980). Amelogenins are generally hydrophobic and enriched in glutamine, proline, histidine and leucine, with molecular weights (before processing), ranging from 5 to 22 kDa. For a detailed presentation of amelogenin compositions, hydrophilicity plots, etc. (see table 2 in Simmer and Snead, 1995).

Amelogenins are secretory products of genes generally located on the sex chromosomes (Lau *et al.*, 1989) and are specifically expressed in the secretory stage of enamel development (Skobe *et al.*,1995) by ameloblast cells of the dental inner enamel epithelium. In the developing mouse molar, expression of amelogenin protein is first detected at birth (Snead *et al.*, 1988), apparently following expression of the more hydrophilic enamelins (Slavkin *et al.*, 1988; Zeichner-David *et al.*, 1993). Further, it has now been established by the cloning of amelogenin mRNAs, that multiple amelogenins are expressed through alternative splicing from the amelogenin gene(s) (Gibson *et al.*, 1991, 1992; Lau *et al.*, 1992; also see Simmer and Snead 1995, and references therein). In the case of the mouse, nine such constructs have now been identified (J. P. Simmer, personal communication), although details of the timing and levels of expression of these amelogenins remain to be clarified.

The distinction among the proteins of developing enamel was initially based on procedural criteria, in which it was found that the extraction of scrapings of developing bovine enamel matrix with guanidine hydrochloride readily solubilized the amelogenin fraction while a 'mineral-bound' enamelin fraction was only solubilized by a subsequent extraction–demineralization of the residue with guanidine-EDTA (Termine *et al.*, 1980). This procedural definition of these two matrix protein components has been widely used but, due to the incomplete nature of the separation procedure, may have led to some ambiguities. More recently other components of the enamel protein matrix have been described which may reasonably be attributed to the enamelin fraction [e. g., 'ameloblastin' (Krebsbach *et al.*, 1996), 'tuftelin or tuft' protein (Deutsch *et al.*, 1991, 1995*a*,*b*) and 'ameloproteinase-I' (Moradian-Oldak *et al.*, 1996*b*). An tentative approach to the organization this protein data is presented in the schematic chart of Figure 4.1. It should be noted that nomenclature in this field is undergoing constant revision as new genes are cloned and the physiologic function of the protein they encode is identified.

During the secretory stage of amelogenesis, the amelogenin proteins comprise the major matrix component (Termine *et al.*, 1980) and may reasonably be assumed to occupy the bulk of the matrix volume. During this stage of enamel development, enamel mineral (carbonated hydroxyapatite or possibly octacalcium phosphate) occurs as extremely long mineral 'ribbons', some 2–4 nm by 5–10 nm in width and thickness. These structures can be seen to originate at, or close to, the dentin enamel junction (Diekwisch *et al.*, 1994) and appear to extend throughout the full thickness of the matrix with their ends impinging onto the cell membrane of the ameloblast Tome's processes (Daculsi *et al.*, 1984). Subsequent to the formation of these initial enamel ribbons within the secretory phase, a striking transformation of the developing enamel occurs known as the maturation phase. The maturation phase can be identified in a region of the developing tooth as a 'chalky' band (Robinson *et al.*, 1981*a*). In this region it has been shown that a massive loss of matrix protein and water is accompanied by rapid growth of the initial mineral crystallites in both thickness and width, leading eventually, in the mature enamel prism, to crystal–crystal fusion.

Following the initial description of the amelogenin protein in the early 1960s, several attempts were made to gain information on its secondary and tertiary structures, against a background of biophysical studies of the fibrous proteins (Astbury, 1961). Generally, data obtained from X-ray diffraction suggested the presence of components with a poorly defined beta or cross-beta conformation (Bonar *et al.*, 1965; Fincham *et al.*, 1965), although subsequent work (Traub *et al.*, 1985 suggested that this diffraction pattern might in fact originate from the enamelin component of the matrix. Other studies of developing enamel protein preparations (mostly bovine) revealed evidence for some form of reversible assembly of the protein into supramolecular structures (Katz *et al.*, 1965; Mechanic, 1971). These observations appeared to correlate with the report (Nikiforuk and Simmons, 1965) that enamel protein preparations exhibit a temperature-sensitive and a pH-sensitive reversible coacervation, presumably involving protein to protein hydrophobic interactions. The establishment of the primary amino acid sequence for amelogenins from a range of species (mouse, rat, human, pig, cow) has shown them to be highly conserved across evolution, with a moderately hydrophobic amino terminal motif and a hydrophilic carboxyl-terminal sequence (Snead *et al.*, 1985; Simmer and Snead, 1995). Such an organization of the primary structure has led to suggestions that orientation of amelogenin monomers by intermolecular associations of their hydrophilic and hydrophobic motifs may lead to supramolecular assemblies as were reported in the chro-

Figure 4.5. **Atomic force microscopy of recombinant M179 amelogenin showing characteristic spherical structures on a nanoscale, the 'nanospheres'. Each sphere is approximately 20 nm in diameter and would correspond to approximately 100 amelogenin molecules.**

matographic and sedimentation experiments noted above (Snead *et al.*, 1985; Fincham *et al.*, 1991*a*,*b*).

4.4.2.   Amelogenin assembly into nanospheres

Following the cloning of the mouse 180–residue amelogenin (Snead *et al.*, 1983) a recombinant mouse amelogenin expressed in *Escherichia coli* bacteria has become available (Simmer *et al.*, 1994). The expressed protein ('M179') contains 179 amino acid residues (the N-terminal methionine being cleaved), but lacks the single phosphoserine residue that has been shown to be present in other native amelogenins (Fincham *et al.*, 1994). Recent studies of the properties of this recombinant amelogenin (Fincham *et al.*, 1994, 1995*b*; Moradian-Oldak *et al.*, 1994*a*) have shown that, in solution under physiological conditions (e. g. pH 7.4, [I] = 0.02M) the protein appears to self-assemble to generate quasi-spherical structures ('nanospheres') of some 18–20 nm diameter (see Fincham *et al.*, 1995*a*,*b*; Figure 4.5). Further, examination of transmission electron microscope images of developing mouse enamel have shown that the initial ribbon crystallites are spaced some 20 nm apart (Diekwisch *et al.*, 1994) and that electron-lucent structures, comparable to those observed in the *in vitro* preparations of the recombinant protein, occur as beaded rows of nanospheres aligned with and between these earliest crystallites (Fincham *et al.*, 1995). These observations have led to the hypothesis

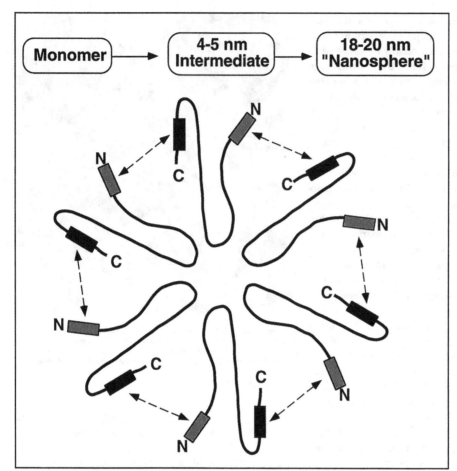

Figure 4.6. **Hypothetical assembly of amelogenin monomers to form 4–5 nm size Intermediate Structures. Six amelogenin monomers are folded to permit protein–protein interactions between carboxyl and amino-terminal domains, as described by Paine and Snead (1997). Arrows indicate intermolecular interactions. The folded carboxyl-terminal sequence is externalized, permitting proteolytic cleavage as described by Moradian-Oldak et al. (1994) causing disruptions of the assembly.**

that, following secretion by the ameloblast cells, the amelogenin protein of the developing enamel self-assemble to form structures that function to control both initial crystallite spacing and their c-axial extension to form the immensely long ribbons seen in the secretory stage enamel. Thus, while the amelogenin monomers may fail to retain stable secondary or tertiary structures (as evidenced by the nuclear magnetic resonance studies of Termine and Torchia, 1980), supramolecular assembly to generate the quaternary-level nanospheres may create the functional amelogenin unit that directs and controls initial enamel matrix ultrastructure. Subsequent to this nanosphere-mediated phase, which creates an assembly of long thin crystallite ribbons probably extending through the full thickness of the secretory matrix, it is envisaged that the activation of proteinase(s) within the matrix initially cleaves carboxyl-terminal regions of the exposed amelogenins of the nanospheres causing modification, or collapse of the structures, translocation of protein from the matrix, and permitting the maturation-stage thickening of the mineral crystallites.

In other studies employing a molecular biological approach utilizing the two-hybrid system (Fields and Song, 1989), Paine and Snead (1997) have demonstrated that the amelogenin monomer appears to have two distinct domains that may facilitate intermolecular interactions. Such a finding is consistent with the requirements of a self-assembly model in that it permits the generation of multimeric forms by a chain-wise linking of interactive sites (Figure 4.6). Evidence for some form of bead-like assemblages in the developing enamel matrix is not new, although the interpretation in terms of amelogenin function is novel. Smales (1975) reported helical structures in demineralized specimens of developing rat enamel, and inspection of his published electron micrographs reveals structures comparable to those recently described by Fincham and Moradian-Oldak (1996). Also, both Bai and Warshawsky (1985) and Robinson et al. (1981b) have presented freeze-fracture images of developing enamel showing aligned rows of beaded structures comparable in size to the recombinant amelogenin nanospheres.

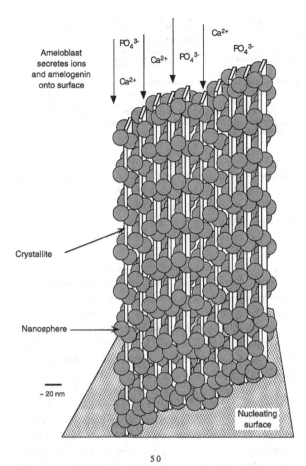

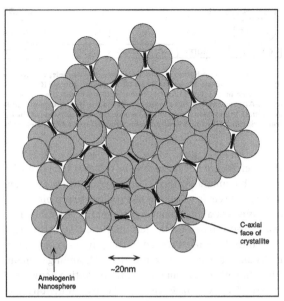

Figure 4.8. **'Vertical view' of structure shown in Figure 4.7. Ribbon-like initial crystallites are enclosed in amelogenin nanospheres leaving only their c-axial faces open to mineral accretion. As the crystallites extend in the c-axial direction, additional amelogenin is secreted by the ameloblast, self-assembles and packs around the extending crystallite.**

Figure 4.7. **Hypothetical assembly of amelogenin nanospheres and mineral crystallites during the secretory phase of amelogenesis. Initial, secretory stage, enamel crystallites extend from the dentin–enamel junction (nucleating surface) and are enclosed in amelogenin nanospheres which inhibit lateral crystal growth. Nanospheres (18–20 nm in diameter) may self-assemble from smaller 'Intermediate Structures' (see Figure 4.6). Ions transported by the ameloblast cells cause crystallite growth on the 'exposed' c-axial faces. Subsequent (maturation phase) proteolytic processing of the amelogenin disrupts the structure, now permitting lateral crystal growth.**

Most recently, further analyses of dynamic light scattering data for recombinant amelogenin (M179) have shown the presence of bimodal distributions of polymeric structures suggesting the presence of both 18–20 nm nanospheres, as previously described (Moradian Oldak et al., 1994a) together with a population of 4–5 nm structures which are computed to be comprised of some 4–6 amelogenin monomers. These observations suggest that the assembly of the nanospheres may occur through such smaller intermediate structures (Figure 4.6).

In summary, the amelogenin proteins have been shown to be highly conserved proteins exhibiting a bipolar hydrophobic–hydrophilic primary structure. Biophysical and structural studies have shown that amelo-genin does not exhibit a stable secondary or tertiary conformation, but provides evidence for a supramolecular assembly. Recent studies of recombinant amelogenins employing a range of imaging techniques and molecular biological approaches have established that the proteins interact to generate supramolecular assemblies (nanospheres). *In vivo* studies of developing enamel suggest that amelogenin nanospheres are the functional components of the matrix, serving both to space, orient and control initial (secretory phase) enamel biomineralization (Figures 4.7, 4.8).

4.4.3.   Identification of protein domains directing self-assembly

One of our current interests has been to define the molecular basis for the protein–protein interactions which are believed to be required for the assembly and disassembly of the enamel extracellular matrix. To accomplish this task we have chosen to use the sensitive yeast two-hybrid system (Fields and Song, 1989), which reflects the ability of two proteins to interact under physiologic conditions that approximate those observed in the enamel organic matrix milieu (Aoba and Moreno, 1988). Yeast cells will perform most post-translational modifications of transgene proteins, allowing for glycosylation and/or phosphorylation reactions to occur, thus closely mimicking the mammalian modifications.

Figure 4.9. **Yeast two-hybrid system for testing the capacity of two proteins to interact. See text for details of the assay and its application. A. Hybrid protein formed from the GAL4 DNA-binding domain (GAL4-BD) and amelogenin bind to the GAL1 upstream activating sequence (GAL1 UAS) but cannot activate reporter transcription without the activation domain. B. A hybrid protein formed from the GAL4 activation domain (GAL4-AD) and amelogenin cannot localize to the UAS and cannot independently activate transcription of the reporter without the binding domain. C. Interaction between the two amelogenin proteins of the hybrid proteins reconstitutes GAL4 functions and results in the transcription of the *lacZ* reporter, gene, proportional to the amelogein–amelogenin interaction.**

However, we are not suggesting by this that the yeast will form a biomineral, rather, that the yeast will provide an appropriate environment to test protein to protein interactions essential to enamel organic matrix assembly. The assay itself is based on the fact that the GAL4 transcription control elements consist of two physically separable molecular domains: one acting as a DNA-binding domain while the other functions as the transcription activation domain (Figure 4.9). Normally, the two functions are located on the same molecule, but it has been shown that a functional molecule can be assembled *in vivo* from separated domains. In this case, the DNA-binding domain and the activator domain for GAL4 transcription activator has been cloned separately into two different expression vectors, each of which can accept a target gene (gene A and gene B) to form fusion proteins. Both hybrid proteins are targeted to the nucleus by nuclear localization signals. If protein A interacts with protein B, the DNA-binding domain of GAL4 will be tethered to its transcription activation domain and the activity of the transcription complex reconstituted. Expression of a reported gene, such as *lacZ*, downstream from the GAL4-binding site is used as a colorimetric assay to indicate interaction between protein A with protein B.

Prior to the development of the yeast two-hybrid system, protein–protein interactions were generally studied using biochemical techniques such as cross-linking, co-immunoprecipitation and co-fractionation by chromatography. These alternative techniques are largely not applicable to enamel matrix proteins because of their aberrant physical-chemical properties which lead to their aggregation, and for this reason we choose the yeast two hybrid system.

The yeast hybrid system can further be used to define functional binding domains responsible for protein–protein interactions. This involves isolating protein regions of interest, cloning these smaller peptides in the hybrid vectors, and assaying for binding activity. Just as the transcriptional activating genes such as GAL4 can be separated into two domains with distinct functional activities (a DNA-binding and a transcriptional activating domain), the yeast two-hybrid assay allows the investigator to 'map' minimal domains that define the basis for protein–protein interactions. Critical amino acids within each domain can also be defined (Mosteller *et al.*, 1994), and library screening procedures (e.g., Chevray and Nathans, 1992) illustrate the range of protein interactions that are yet to be discovered using this system.

4.4.4.   Self-assembly properties of enamel proteins

We have used the yeast two-hybrid system to demonstrate that protein to protein interactions occur during enamel organic matrix assembly, by demonstrating and characterizing protein to protein interactions in the yeast two-hybrid assay. Full length cDNA for the two mouse amelogenin isoforms, M180 and the leucine-rich amelogenin protein (LRAP) (Lau *et al.*, 1992), for bovine tuftelin (Deutsch *et al.*, 1991) and for rat ameloblastin (Krebsbach *et al.*, 1996) have been cloned into the vector bearing the DNA-binding domain, and into the vector bearing the activation domain, producing enamel protein hybrids fused to a portion of the GAL4 complex. The amelogenin constructs have been generated without the pre-pro-leader sequence ensuring that any interactions observed are not due to hydrophobic interactions occurring in the amelogenin pre-pro-leader domain. We predicted that members of the three classes of enamel proteins would be capable of interacting with one another in a consistent and reproducible manner and would interpret such results to suggest that enamel protein interactions are specific in nature and not strictly due exclusively to mass aggregation. Interaction between amelogenin isoforms, tuftelin to tuftelin and amelogenin isoforms to tuftelin have been tested. We found tuftelin interacts with itself (Paine *et al.*, 1996), that amelogenin interacts with itself and at least one isoform (LRAP), but tuftelin and the amelogenins fail to interact with each other. Further, ameloblastin fails to interact with either itself, amelogenin or tuftelin. Results for interactions amongst the three cloned enamel matrix proteins have been tabled along with positive control combinations (Table 4.1).

An alternative approach to defining enamel matrix proteins is to catalogue the genes that participate in enamel matrix assembly. Perhaps all of the genes expressed exclusively by ameloblasts have already been

Table 4.1. *Interactions as detected by the yeast two-hybrid system*

GAL4 DNA-binding domain fusion	GAL4 activation domain fusion	Interaction
Amelogenin	Amelogenin	+
Amelogenin	LRAP	Weak positive
LRAP	LRAP	−
Tuftelin[a]	Tuftelin	++
Ameloblastin	Ameloblastin	−
Amelogenin	Ameloblastin	−
Amelogenin	Tuftelin	−
Tuftelin	Ameloblastin	−
p53[b]	SV40 large T antigen	+
H-ras[c]	CDC25	+++

Note: Relative strength of interaction indicated by a positive signal. Interactions [a](Paine *et al.*, 1996) [b](Li and Fields, 1993) and [c](Mosteller *et al.*, 1994) have been reported previously. Positive controls were the tumor suppressor protein p53 co-transformed with the SV40 large T-antigen (Lane and Crawford, 1979; Li and Fields, 1993), and H-ras (wild type) co-transformed with the CDC25 protein of *Saccharomyces cerevisiae* (Mosteller *et al.*, 1994; Mosteller *et al.*, 1995). Negative controls involve each of the enamel protein–hybrid constructs co-transformed with either pPC86 (for the pPC97 hybrids) or pPC97 (for the pPC86 hybrids). Blue color (indicating a positive interaction) developed within the first hour for H-ras co-transformed with the CDC25 protein, and over a 2-hour period for tuftelin co-transformed with itself. Blue color developed for all the other positive combinations over an approximate 8–10 hour period. A very pale blue color for amelogenin co-transformed with LRAP was present after 24 hours. No blue color developed over the 24 hours for reactions involving ameloblastin, or leucine-rich amelogenin protein (LRAP) co-transformed with itself, or amelogenin co-transformed with tuftelin, and these combinations have been recorded as a negative result. Negative control combinations showed no blue color development over the 24 hours.

cloned, but this appears unlikely. Recent attempts to broaden our understanding of enamel specificity have used a random or 'shotgun' approach to sequence genes from a tooth-derived cDNA library (Matsuki *et al.*, 1995). This investigation resulted in the identification of a number of novel genes including ameloblastin (Krebsbach *et al.*, 1996). An independent approach to ameloblast-derived cDNA library screening is based on the hypothesis that known enamel proteins may interact with other classes of proteins, such as chaperones, kinases and proteinases and that enamel matrix proteins may also interact with other enamel matrix (structural) proteins. In this scenario, amelogenin protein could be used as the probe (or 'bait') to screen a tooth-derived cDNA expression library in an attempt to find proteins essential to enamel matrix assembly and subsequent biomineralization (see Figure 4.9). For example, amelogenin could be cloned into one of the yeast two-hybrid vectors (typically the GAL4 DNA-binding domain vector) and then used to screen for unknown cDNAs that exhibit amelogenin-binding ability. In this case, the cDNA encoding an unknown protein capable of interacting with amelogenin can be selectively screened and isolated for further characterization.

Information gained from the above experimental ap-

proaches could be used establish a database to design enamel proteins with defects in domains responsible for the protein to protein interaction property. These defective proteins could be expressed in transgenic animals and are predicted to result in defects in enamel matrix assembly that will lead to enamel crystallite defects. For example, coupling the expression of the cDNA bearing such an assembly defect with the putative amelogenin promoter (Chen *et al.*, 1994; Park *et al.*, 1994; Snead *et al.*, 1996) would 'drive' the expression of the mutated cDNA (or any other gene such as growth factors) to express exclusively in ameloblasts. We predict that this defective protein might create a dominant-negative effect when synthesized and secreted, thereby altering the assembly of the remaining enamel proteins. Additional information may be derived from the yeast two-hybrid system regarding assembly domains for other enamel proteins that can be confirmed in a transgenic animal model. Based on the demonstrated ability to define protein-binding domains in amelogenin and tuftelin (Paine *et al.*, 1996) that are essential to enamel organic matrix assembly, it is expected that transgenic animal phenotypes will include traits common to the hereditary enamel defect *amelogenesis imperfecta*. Aldred demonstrated that some cases of *amelogenesis imperfecta* are due to mutation of the amelogenin gene structure (Aldred *et al.*, 1992) that is predicted to alter protein-protein interactions during enamel formation. Using the yeast two-hybrid system we can identify and design specific defects in enamel proteins that will phenocopy naturally occurring defects in man.

Enamel biogenesis requires that enamel proteins be synthesized and secreted into the extracellular space by ameloblasts and that these proteins self-assemble into a matrix that can support hydroxyapatite crystallite formation. Despite striking similarities in amelogenin amino acid sequences between mammalian species, alternative splicing of amelogenin yields protein products which can include or exclude amino acid subsets and hence structural domains of the mouse M180 amelogenin. These alternatively spliced amelogenins may alter the stoichiometry among matrix components to favor self-assembly. Proteins other than tuftelin or ameloblastin in the enamel extracellular organic matrix may participate in enamel organic matrix self-assembly by interacting with the amelogenin components to provide alternative or additional regulatory mechanisms for self-assembly. The previous absence of a laboratory-based assay for enamel protein interactions which may be of functional significance to the molecular mechanism for assembly of the enamel organic matrix has limited progress in this area. To help remedy this situation, we are extending the yeast two-hybrid experimental strategy to

Table 4.2. *A summary of previous reports on enamel proteinases studied during the last 10 years showing different proteinases found in: secretory-stage enamel (SE), late maturation-stage enamel (LME), developing enamel (DE) and maturation-stage enamel (ME)*

Source	Molecular weight range	Classification	Optimum pH	Effect on amelogenin	Reference
Bovine SE	ND	Serine-proteinase (Ca-dependent)	ND	ND	Crenshaw and Bawden 1984
Pig SE	62–130	Metallo-proteinase	–	Inactive (based on enzymograms)	Overall and Limeback 1988
Pig LME	31–36	Serine-proteinase	7.4 and 5.5	Active (based on enzymograms)	
Porcine DE	29	Serine-proteinase	6–8	ND	Carter *et al.*, 1989
Rat SE	38–40 50–80	–	–	–	Smith *et al.*, 1989*a*
Rat ME	29	Serine-proteinase	8	Active (based on enzymograms)	
Bovine SE	45–66	Ca-dependent	–	Not active	DenBesten and Hefferman, 1989
Rat SE	43–92	Ca-dependent	–	Not active	
Rat ME	28–33	No Calcium	ND	Active	
Bovine	30	No Calcium	5.5–6	Cleaves TRAP	Sasaki *et al.*, 1991*a*
Porcine outer layer	76	Metallo-proteinase	8	Cleaves C-terminus	Tanabe *et al.*, 1992
Porcine inner layer	30–34	ND	7.3	Cleaves 1/3 of the molecule at N-terminus	

analyses of other proteins of the enamel extracellular matrix which may be important for enamel self-assembly. Finally, we will use recombinant methodology to define the minimal protein domain(s) responsible for these interactions. The information gained from these studies will enhance our understanding of the mechanism for enamel biomineralization. The data will also provide the intellectual insight to design transgenic mice designed to perturb enamel biogenesis at the genetic level.

4.5.   **Matrix disassembly**

4.5.1.   Overview

One of the fundamental steps towards understanding enamel biomineralization is the investigation into the mechanisms by which enamel proteins are proteolytically cleaved and finally removed from enamel extracellular matrix. Removal of the extracellular matrix proteins, particularly amelogenins, is concomitant with the incremental growth of carbonate hydroxyapatite crystals resulting in maturation of a highly ordered mineralizing system, enamel.

It has been three decades since the presence of pro-teolytic activity in developing enamel was first demonstrated by digestion of gelatin in a photographic emulsion substrate (Suga, 1970). Subsequently, several investigators have studied enamel proteinases from bovine, pig, rat and mouse using more specific methods such as enzymography. A summary of reports on enamel proteinases during the past 10 years is provided in Table 4.2. Examination of that data indicates that while there is a general agreement that both serine proteinases and metalloproteinases are present in the extracellular matrix during enamel maturation, there has been less than uniform agreement concerning their specific activities against amelogenin, pH optima and calcium requirement of the proteinases. The main difficulty in isolation and purification of enamel proteinases is due to their extremely low abundance in the enamel extracellular matrix, while another restrictive limitation has been the close association of the proteinases to their substrate, amelogenin, and the added complexity that amelogenin tends to aggregate during purification steps.

Recently, Moradian-Oldak *et al.* (1994*a*, 1996*a,b*) used a recombinant murine amelogenin substrate (Simmer *et al.*, 1994) , M179, to examine the specificity of enamel proteinase fractions isolated from both developing bovine and pig enamel. This was justified based upon the high level of sequence conservation (more than 80%) for amelogenins between the species studied (pig, human, mouse, cow and rat) (Snead et al., 1985, Shimokawa et al., 1987, Yamakoshi et al., 1989, Salido et al., 1992, Bonass et al., 1994). The mouse amelogenin termed M179 contains 179 amino acid residues and is identical in amino acid sequence to the native mouse amelogenin M180 (Lau et al., 1992), except for the following. First, M180 is post-translationally modified by a single phosphorylation of serine16 (Fincham et al., 1995a) which is lacking in M179. Second, the bacteria-derived processing events remove the first methionine from the N-terminus, leaving M179. Further support of our approach is provided by the recent data on the self-assembly process by which M179 forms monodisperse macromolecular assembles (Moradian-Oldak et al., 1994b, Fincham et al., 1994, 1995a,b). Therefore, we believe that M179 is an appropriate model substrate for our proteolysis studies since it is essentially authentic mouse amelogenin.

The strategy for the preparation of proteinase fractions was based on their isolation by size-exclusion chromatography and their further fractionation by either reversed-phase HPLC techniques (Moradian-Oldak et al., 1994a) or ammonium sulfate precipitation (Moradian-Oldak et al, 1996a). Two major fractions were defined; one contained the 'the high molecular weight proteinase' or fraction 'A' and the other contained the 'low molecular weight proteinases (30 kDa)' or fraction 'B' (Moradian-Oldak et al., 1996a). The fraction A, enriched in the high molecular weight proteinases, could be further resolved by electrophoresis into two principal bands, one of 60 kDa and one of 68 kDa using a gelatin containing acrylamide gel.

4.5.2. Carboxyl-terminus processing of amelogenins in vitro and in vivo

Incubation of the substrate M179 with fraction A at 37 °C generated a major proteolytic product with a N-terminal amino acid sequence identical to that of the substrate M179. Mass spectroscopic analysis of the product confirmed that it resulted from the cleavage of the M179 recombinant protein at Pro^{168} indicating that cleavage occurred at the first proline encountered starting from the carboxyl-terminal region of M179. The activity of the fraction A proteinase was found to be optimal at pH 8. Systematic incubation experiments with different protease inhibitors and divalent cations indicated that the proteinase in fraction A is a calcium-dependent metalloproteinase that cleaved M179 within the carboxyl-terminal region (Moradian-Oldak et al., 1994a).

In a recent study of the bovine- and porcine-leucine-rich amelogenin polypeptides (LRAP), Fincham and Moradian-Oldak (1996) found the carboxyl-terminal sequence to be LPELPLEAWP in both species, contrary to earlier reports (Fincham et al., 1981). It is known that in the bovine case, the LRAP molecule arises by the post-secretory processing of a precursor of 59 amino acid residues which is generated by the expression of an alternatively-spliced mRNA (Gibson et al., 1991). This LRAP precursor has a carboxyl-terminal sequence identical to the recombinant M179 protein used in this study and we suggest that the in vivo proteolytic processing of the bovine LRAP-precursor occurs through the action of a proteinase identical to that described by Moradian-Oldak et al. (1994b).

Previous reports have suggested that the carboxyl-terminal processing of the amelogenin occurred by a specific scission at the Ala^{166}-Trp^{167} residues of the conserved carboxyl-terminal region (Fincham and Moradian-Oldak, 1996; Tagaki et al., 1984). In an attempt to examine potential initial pathway(s) of amelogenin proteolysis in enamel extracellular matrix in vivo Fincham and Moradian-Oldak (1996) made comparisons between five species: mouse, pig, bovine, rat and human, using mass spectrographic analysis of the lyophilized proteins in the acetic extract. Fragments larger than 12 kDa were analyzed and most of the sequences identified were found to be derived by carboxyl-terminal processing rather than processing from the amino-terminus. In the case of mouse and rat, strong evidence was obtained for sequential proteolytic processing from the carboxyl-terminus. We therefore speculate that the carboxyl-terminal processing may occur through exo-peptidase activity rather than a specific endo-peptidase recognizing the Ala-Trp or Ala-Pro motifs. A comparison between the five species showed a fragment uniquely identified as being derived by the processing of the parent amelogenin to the first proline residue from the carboxyl-terminal. This is a result of the cleavage of 11 residues of the hydrophilic carboxyl terminal sequence as demonstrated in the in vitro proteolytic experiments using M179 as substrate. Carboxyl terminal cleavage of an amelogenin isolated from porcine developing enamel by a 76 kDa proteinase has also been demonstrated by Tanabe et al. (1992). The proteinase in this study appeared to be completely inhibited by benzamidine and EDTA. Tanabe et al. (1992) suggested that the 76 kDa proteinase cleaves the secreted porcine amelogenin of 173 residues (e. g. the '25 kDa amelogenin') to form a 148 residue amelogenin (the '20 kDa amelogenin').

4.5.3. Complete degradation of amelogenin by a serine proteinase, ameloproteinase-I

In order to investigate the activity of the 30 kDa proteinase present in the low molecular weight fraction B against amelogenin, the fraction B was further purified using the ammonium sulfate precipitation technique and assayed in the *in vitro* proteolytic experiments using M179 as a substrate (Moradian-Oldak *et al.*, 1996*a*). As Sasaki *et al.* (1991*a*) had previously described a purified enzyme (30 kDa) with a specific activity against bovine 28 kDa amelogenin that generated a tyrosine rich amelogenin polypeptide (TRAP) molecule (see Table 4.2), we were encouraged to see that incubation of M179 with the proteinase fraction resulted in major products whose amino acid analysis showed no tyrosine content indicating cleavage of the TRAP from the amino-terminal region of the parent molecule. Difficulties in isolating a major amelogenin product after the initial incubation experiment suggested that, subsequent to the cleavage event liberating TRAP, further degradation of the remaining amelogenin substrate was occurring, an observation that further supports the amelogeninase activity of the 30 kDa proteinase (see also DenBesten and Heffernan, 1989). The action of the 30 kDa proteinase on amelogenin M179 resulted in the production of small polypeptides whose molecular masses varied between 300 and 2000 Da (unpublished data). Systematic incubation experiments in the presence and absence of different proteinase inhibitors indicated that this proteinase is a serine proteinase and this enzymatic activity was found to have an optimum at pH 6.

In order to obtain a partial amino acid sequence of the 30 kDa serine proteinase responsible for amelogenin degradation it was necessary for us to develop a two-dimensional gel electrophoresis system appropriate for zymogram analyses (Moradian-Oldak *et al.*, 1996*a*). Based on the two-dimensional zymograms, the area corresponding to the proteinases was identified and probed by immunoblot assay for contaminating amelogenin using an amelogenin antibody. It was demonstrated that the proteins focused in this area are not amelogenins. Hence, the amelogenin-specific proteinases were completely separated from co-purifying amelogenins by isoelectric focusing (IEF). Three closely resolving spots on the two-dimensional zymogram that appeared to correspond to the 30 kDa proteinase were recovered by transfer to a polyvinylidene difluoride (PVDF) membrane for amino acid sequence determination. Each of these spots yielded an identical amino-terminal amino acid sequence, with residue identity extending to at least the first 15 residues. This sequence identity indicates that the three areas seen on two dimensional zymograms are in fact the same protein, but derived from different levels of post-translational modification. A search of the protein database was conducted, using the identity of the 15 amino acid residues obtained for the proteinase as the query sequence. These 15 amino acid residues were found within the sequence of the '32 kDa enamelin protein' and the '89 kDa enamelin' protein previously reported by

Tanabe *et al.* (1990) and Fukae *et al.* (1996). Based upon the sequence identity for the first 15 amino acid residues obtained for the amelogenin-specific proteinase, the 32 kDa enamelin is in actuality an enamel proteinase, whose name is proposed as 'Ameloproteinase-I'. The 89 kDa enamelin (Uchida *et al.*, 1991) can be recognized as the precursor from which the 32 kDa enamelin (Uchida *et al.*, 1991) that harbors the amelogenin-degrading activity, as suggested by Moradian-Oldak and co-workers (1996*b*). The physiologic function of a protein derived from the enamel organic matrix is biochemically defined for the first time: ameloproteinase-I is responsible for the progressive degradation of amelogenin during enamel maturation.

4.5.4. Biological significance of the action of enamel proteinases

The unique characteristics of enamel proteinases studied (molecular weights, isoelectric points, calcium and pH dependence), their appearance at specific stages of amelogenesis (Smith *et al.*, 1989*a*,*b*, Robinson *et al.*, 1982, 1983), and their specific activity with respect to amelogenin are supportive evidence for the idea that proteinases may be directly involved in controlling some of the critical steps of enamel biomineralization. Additional evidence for this view may well be related to the cases of *amelogenesis imperfecta*, a diverse group of hereditary disorders characterized by a variety of developmental enamel defects including hypoplasia and hypomineralization. It has been suggested (Wright *et al.*, 1992) that enamel protein content may be increased in all the major *amelogenesis imperfecta* groups, with the hypocalcified and hypomaturation types, appearing to have greatest retained abundance of enamel proteins. It has been proposed that amelogenin may be retained in hypomaturation *amelogenesis imperfecta* resulting in deficient crystallite growth and thus hypomineralization of the enamel (Wright *et al.*, 1992). Retention of amelogenin could result for a variety of reasons, including defective proteinase activity.

The *in vitro* experiments described above have indicated that developing enamel contains at least two different classes of proteinases which appear to have selective action cleavage of the amelogenin molecule. Further, it has been suggested (Moradian-Oldak *et al.*, 1994*b*) that the specificity of cleavage and the susceptibility of amelogenin molecules to the proteinases may well be controlled *in vivo* by local changes in pH (Sasaki *et al.*, 1991*b*) and calcium concentration resulting from the formation of apatite crystals. A proteinase of about 60–68 kDa molecular weight that cleaves the hydrophilic C-terminal region of amelogenin appeared to be a calcium-dependent metalloproteinase with a pH optimum of 8. On the other hand, the lower molecular weight proteinase (ameloproteinase I, previously known as the 32 kDa enamelin), was characterized as a serine proteinase which is optimally active at pH 6. This proteinase further degraded the TRAP molecule and it appeared to cause further degradation of M179 amelogenin. In contrast, ameloprotease-I appears to be a unique enzyme in terms of substrate specificity. It is important to mention that the amino acid sequence of ameloproteinase-I (Fukae *et al.*, 1996) is not similar to other known serine proteinase sequences (i. e., trypsin, chymotrypsin, subtlisin).

4.6.   A model for enamel biomineralization

These recent experimental data together with the existing knowledge on the appearance and abundance of enamel proteinases during enamel maturation (Smith *et al.*, 1989a,b) support the following assumptions that the hydrophilic carboxyl-terminal segment is cleaved at an early stage of enamel formation when the extracellular matrix is enriched in higher molecular weight proteins. The cleavage event may interrupt the nanosphere self-assembly process of amelogenin molecules resulting in structural modification of the initially secreted amelogenin matrix (Fincham *et al.*, 1995a,b). Cleavage of the hydrophilic segment may also result in reduced affinity of amelogenin molecules to the hydroxyapatite crystals (Aoba and Moreno, 1989). At a later stage of enamel development, the TRAP segment is cleaved and may perform additional function such as preventing crystal–crystal fusion. At the maturation stage of amelogenesis, when the low molecular weight proteinases are abundant (Smith *et al.*, 1989b), massive degradation of amelogenin takes place to provide the appropriate space for the oriented growth and maturation of carbonated apatite crystals, the hallmark of enamel (Figures 4.7, 4.8).

Based upon our observations, a model for enamel matrix-mediated biomineralization can be formulated and predicts that the first step in amelogenin regulation of crystallite formation is the binding of the carboxyl-tail to the crystallite surface, perhaps poisoning a specific crystallite surface for further growth (Aoba *et al.*, 1989). Following proteolytic cleavage of the amelogenin 'tail', the amelogenin would leave the crystallite surface and through interactions regulated by two protein domains then interact with other amelogenins to provide appropriate spacing between growing crystallites through the formation of nanospheres. With progressive cleavage of the amelogenin proteins, or through the introduction of increasing stoichiometric amounts of LRAP, crystallites could expand in girth as the relative diameters of the

nanospheres decreased. Consistent with this model is the prediction that LRAP (and other alternately spliced amelogenins) should accumulate in the more mature regions of developing enamel in which crystallite growth is restricted to width expansion. This predicted outcome has been observed in developing bovine teeth (Fincham *et al.*, 1982b).

Summary

Biomineralization is the process by which a micro-environment is created by cells that serves to regulate the initiation, propagation, maturation and termination of inorganic crystallites of a precise chemical composition and shape. The effect of biomineralization shapes our world by altering the mineral distribution of the seas and on the land. Similarly, analysis of the skeletal remains of extinct and extant organisms has contributed significantly to our present understanding of evolution and evolutionary medicine (Williams and Nesse, 1991). Recent advances in the field of biomineralization may be seen to be centered around the discipline of biomimetics (Archibald and Mann, 1993; Mann, 1993), and the attendant creation of biopolymers with sufficient surface organization or three-dimensional constraints to mimic the process of naturally occurring interactions designed by Nature to achieve mineral deposition. To date, biomimetics has focused on creating novel assemblies of repeated elements that can direct crystallite formation. In biomineralization systems, cells biosynthesize macromolecules that direct and control crystal growth (Lowenstam and Weiner, 1989). Remarkable scientific insight and creativity has gone into the design of materials that can mimic the naturally occurring assembly of proteins responsible for directing biomineralization in teeth, shells and bones. Regrettably, while progress has been made through empirical approaches to the design of new polymer assemblies that can mimic biosynthesized protein assemblies, the biomimetic design has not been based on knowledge gained from an understanding of the naturally occurring substrates. An overarching goal of our research group is to identify and characterize the molecular events, contributing proteins and protein domains that lead to the assembly and disassembly of the enamel proteins. It is provocative to speculate that such knowledge gained from Nature may be used to produce artificial matrices *in vitro* that will direct mineral deposition. Such knowledge can then be coupled to protein engineering and recombinant DNA methods in order to create a matrix-mediated self-organizing assembly of biopolymers and bioceramics that can be exploited in the development of future dental restorative materials (Heuer *et al.*, 1992).

This review describes recent events in the field of enamel biomineralization from a genetic, molecular and cellular perspective. We focus on critical aspects of the proteins component and the cellular events that direct the self-assembly of the enamel organic matrix, a process that results in a protein structure that is competent to undergo replacement by the enamel mineral phase. We include the application of recombinant DNA approaches which are used to express and recover amelogenin proteins (i.e., recombinant amelogenin) from bacteria. We demonstrate the capacity of amelogenin to self-assembly into higher order structure called 'nanosphere', *in vivo* and *in vitro*. A novel genetic approach employing the yeast two-hybrid assay system is described and used to identify the enamel protein domains that appear to be required for protein self-assembly. The identification and characterization of a specific enamel proteinase that has the capacity to selectively and progressively degrade amelogenin proteins, resulting in the progressive breakdown of the enamel organic matrix is described. Using transgenic animal techniques, we describe a means to identify DNA elements responsible for regulating amelogenin gene expression during development. We present a rationale joining these disparate approaches of gene expression and self-assembly domains of enamel proteins in order to create a transgenic mouse model mimicking *amelogeniesis imperfecta*. A model of enamel formation is presented in an attempt to link these disparate ideas to the formation of the organic enamel extracellular matrix and its replacement by the enamel mineral phase.

Acknowledgements

The authors thank Harold C. Slavkin DDS, for his never ending enthusiasm and capacity for research into the mechanisms of biomineralization that has touched each of our professional lives. His vision for the future of dental research materialized in the formation of the Center for Craniofacial Molecular Biology at the University of Southern California School of Dentistry, a site where scientists from many different disciplines unite their expertise in order to advance knowledge of craniofacial development. Dr Slavkin is presently serving in his new capacity as the Director of the National Institute of Dental Research. The authors wish to thank the following individuals for their participation in the production of this work: Jim Simmer for the LRAP cDNA; Danny Deutsch for the tuftelin cDNA; Paul Krebsbach for the ameloblastin cDNA; Pierre M. Chevray and Daniel Nathans for the plasmids and yeast strains used in the yeast two-hybrid investigation; Daniel Broek and Raymond D. Mosteller provided hybrid plasmids H-ras and CDC25. Many colleagues contributed to critical discussions of the ideas contained in this review and we thank each of them. Deans Howard M. Landesman and William Crawford have provided a nurturing and stable environment for the development of these ideas. This work was supported the United States Public Health Service, National Institutes of Health, National Institute of Dental Research grants DE 06988, DE08678, DE02848, and HL60390.

References

Adeleke-Stainback, P., Chen, E., Collier, P., Yuan, Z.-A., Piddington, R., Decker, S., Rosenbloom, J. and Gibson, C. W. (1995). Analysis of the regulatory region of the bovine X-chromosomal amelogenin gene. *Connective Tissue Research*, **32**, 115–118.

Aldred, M. J., Crawford, P. J. M., Roberts, E. and Thomas N. S. T. (1992). Identification of a nonsense mutation in the amelogenin gene (AMELX) in a family with X-linked *amelogenesis imperfecta* (AIH1). *Human Genetics*, **90**, 413–416.

Aoba, T. and Moreno, E. C. (1988). *Microenvironment for Amelogenetic Mineralization and Solubility of Forming Enamel Mineral*. New York: Gordon and Breach.

Aoba, T. and Moreno, E. C. (1989). Mechanism of amelogenetic mineralization in minipig secretory enamel. In *Tooth Enamel V*, ed. R. W. Fearnhead, pp. 163–167. Yokahama, Japan: Florence Publishers.

Aoba, T., Moreno, E. C., Kresak, M. and Tanabe, T. (1989). Possible roles of partial sequences at N- and C-termini of amelogenin in protein-enamel mineral interaction. *Journal of Dental Research*, **68**, 1331–1336.

Archibald, D. D. and Mann, S. (1993). Template mineralization of self-assembled anisotropic lipid microstructures. *Nature*, **364**, 430–433.

Astbury, W. T. (1961). The structure of the fibres of the collagen group and related matters twenty-one years after. *Journal of the Society of Leather Trades' Chemists*, **45**, 186–214.

Bai, P. and Warshawsky, H. (1985). Morphological studies on the distribution of enamel matrix proteins using routine electron microscopy and freeze-fracture replicas in the rat incisor. *Anatomical Record*, **212**, 1–16.

Belcourt, B., Fincham, A. G. and Termine, J. D. (1983). Bovine high molecular weight amelogenin proteins. *Calcified Tissue International*, **35**, 111–114.

Bonar, L. C., Glimcher, M. J. and Mechanic, G. L. (1965). Molecular structure of the neutral soluble proteins of embryonic bovine enamel in the solid state. *Journal of Ultrastructure Research*, **13**, 308–317.

Bonass, W. A., Robinson, P. A., Kirkham, J., Shore, R. C. and Robinson, C. (1994). Molecular cloning and DNA sequence of rat amelogenin and a comparative analysis of mammalian amelogenin protein sequence divergence. *Biochemical and Biophysical Research Communications*, **198**, 755–763.

Boyde, A. (1996). Microstructure of enamel. In *Dental Enamel*. Ciba Foundation Symposium Series, vol. 205, eds. H. C. Slavkin, D. J. Chadwick, G. Cardew and G. B. Winter. New York: Wiley.

Carter, J., Smillie, A. C. and Shephard, M. G. (1984). Proteolytic enzyme in developing porcine enamel. In *Tooth Enamel IV*, eds. R. W. Fearnhead and S. Suga, pp. 229–233. Amsterdam: Elsevier.

Carter, J., Smillie, A. C. and Shepherd, M. G. (1989). Purification and properties of a protease from developing porcine dental enamel. *Archives of Oral Biology*, 34, 195–202.

Cerny, R., Slaby, I., Hammarstrom, L. and Wurtz, T. (1996). A novel gene expressed in rat ameloblasts codes for proteins with cell binding domains. *Journal of Bone and Mineral Research*, 11, 883–891.

Chapman, V. M., Keitz, B. T., Disteche, C. M., Lau, E. C. and Snead, M. L. (1991). Linkage of amelogenin (AMEL) to the distal portion of the mouse X chromosome. *Genomics*, 10, 23–28.

Chen, C.-T., Bartel, P. L., Sternglanz, R. and Fields, S. (1991). The two-hybrid system: a method to identify and clone genes for proteins that interact with a protein of interest. *Proceedings of the National Academy of Sciences, USA*, 88, 9578–9582.

Chen, E. H., Piddington, R., Decker, S., Park, J., Yuan, Z. A., Abram, W. R., Rosenbloom, J., Feldman, G. and Gibson, C. W. (1994). Regulation of amelogenin gene expression during tooth development. *Developmental Dynamics*, 199, 189–198.

Chen, W. Y. and Smith, C. E. (1994). Partial biochemical characterization of sulfated enamel proteins. *Journal of Dental Research*, 73, 146.

Chen, W. Y., Nanci, A. and Smith, C. E. (1995). Immunobloting studies on artifactual contamination of enamel homogenates by albumin and other proteins. *Calcified Tissue International*, 57, 145–151.

Chevray, P. M. and Nathans, D. (1992). Protein interaction cloning in yeast: Identification of mammalian proteins that react with the leucine zipper of Jun. *Proceedings of the National Academy of Sciences, USA*, 89, 5789–5793.

Couwenhoven, R. I. and Snead, M. L. (1994). Early determination and permissive expression of amelogenin transcription during mouse mandibular first molar development. *Developmental Biology*, 164, 290–299.

Couwenhoven, R. I., Davis, C. and Snead, M. L. (1992). Mouse ameloblasts do not transcribe the albumin gene. *Calcified Tissue International*, 45, 367–371.

Crenshaw, M. A. and Bawden, J. W. (1984). Proteolytic activity in embryonic bovine secretory enamel. In *Tooth Enamel IV*, eds. R. W. Fearnhead and S. Suga, pp. 109–113. Amsterdam: Elsevier.

Daculsi, G., Menanteau, J., Kerebel, L. M. and Mitre, D. (1984). Enamel crystals: size, shape, length and growing process; high resolution TEM and biochemical study. In *Tooth Enamel IV*, eds. R. W. Fearnhead and S. Suga, pp. 229–233. Amsterdam: Elsevier.

Den Besten, P. K. and Heffernan L. M. (1989). Separation by polyacrylamide gel electrophoresis of multiple proteases in rat and bovine enamel. *Archives of Oral Biology*, 34, 399–404.

Deutsch, D. (1989). Structure and function of enamel proteins. *Anatomical Record*, 224, 189–210.

Deutsch, D., Palmon, A., Catalano-Sherman, J. and Laskov, R. (1987). Production of monoclonal antibodies against enamelin and against amelogenin proteins of developing enamel matrix. *Advances in Dental Research*, 1, 282–288.

Deutsch, D., Palmon, A., Fisher, L. W., Kolodny, A., Termine, J. D. and Young, M. F. (1991). Sequencing of bovine enamelin ('tuftelin') a novel acidic enamel protein. *Journal of Biological Chemistry*, 266, 16021–16028.

Deutsch, D., Palmon, A., Young, M. F., Selig, S., Kearns, W. G. and Fisher, L. W. (1994). Mapping of the human tuftelin (Tuft1) gene to chromosome-1 by fluorescence in-situ hybridization. *Mammalian Genome*, 5, 461–462.

Deutsch, D., Catalano-Sherman, J., Dafni, L., David, S. and Palmon, A. (1995a). Enamel matrix proteins and ameloblast biology. *Connective Tissue Research*, 32, 97–107.

Deutsch, D., Palmon, A., Dafni, L., Catalano-Sherman, J. and Young, M. F. (1995b). The enamelin (tuftelin) gene. *International Journal of Developmental Biolpgy*, 39, 135–143.

Diekwisch, T. G. H., David, S., Bringas, P., Santos, V. and Slavkin, H. C. (1993). Antisense inhibition of amelogenin translation demonstrates supramolecular controls for enamel HAP crystal-growth during embryonic mouse molar development. *Development*, 117, 471–482.

Diekwisch, T. G. H., Berman, B. J., Genter, S. and Slavkin, H. C. (1994). Initial enamel crystals are not spatially associated with mineralized dentin. *Cell and Tissue Research*, 279, 146–167.

Dodds, A., Simmons, D., Gu, T. T., Zeichner-David, M. and Mac-Dougall, M. (1996). Identification of murine tuftelin cDNA. *Journal of Dental Research*, 75, 26.

Doi, Y., Eanes, E. D., Shimokawa, H. and Termine, J. D. (1984). Inhibition of seeded growth of enamel apatite crystals by amelogenin and enamelin proteins in vitro. *Journal of Dental Research*, 63, 98–105.

Eastoe, J. E. (1965). The chemical composition of bone and tooth. *Advances in Fluorine Research and Dental Caries Prevention*, 3, 5–16.

Fearnhead, R. W., ed. (1989). *Tooth Enamel V, Proceedings of the Fifth International Symposium on the Chemistry, Properties and Fundamental Structure Of Tooth Enamel And Related Tissues*. Yokahama, Japan: Florence Publishers.

Fields, S. and Song, O. (1989). A novel genetic system to detect protein–protein interactions. *Nature*, 340, 245–247.

Fincham, A. G. and Moradian-Oldak, J. (1996). A comparative mass spectrometric analysis of enamel matrix proteins from five different species suggest a common pathway of post-secretory proteolytic processing. *Connective Tissue Research*, 35, 151–156.

Fincham, A. G., Graham, G. N. and Pautard, F. G. E. (1965). The matrix of enamel and related mineralized keratins. In *Tooth Enamel*, ed. R. W. Fearnhead, pp. 117–123. Bristol: Wright.

Fincham, A. G., Belcourt, A. B., Termine, J. D., Butler, W. T. and Cothran, W. C. (1981). Dental enamel matrix: sequences of two amelogenin polypeptides. *Bioscience Reports*, 1, 771–778.

Fincham, A. G., Belcourt, A. B., Lyaruu, D. M. and Termine, J. D. (1982a). Comparative protein biochemistry of developing dental enamel matrix from five mammalian species. *Calcified Tissue International*, 34, 182–189.

Fincham, A. G., Belcourt, A. B. and Termine, J. D. (1982b). Changing patterns of enamel matrix proteins in the developing bovine tooth. *Caries Research*, 16, 64–71.

Fincham, A. G., Bessem, C. C., Lau, E. C., Pavlova, Z., Shuler, C., Slavkin, H. C. and Snead, M. L. (1991a). Human developing enamel proteins exhibit a sex-linked dimorphism. *Calcified Tissue International*, **48**, 288–290.

Fincham, A. G., Hu, Y., Lau, E. C., Slavkin, H. C. and Snead, M. L. (1991b). Amelogenin post-secretory processing in the postnatal mouse molar tooth. *Archives of Oral Biology*, **36**, 305–317.

Fincham, A. G., Moradian-Oldak, J., Simmer, J. P., Sarte, P. E. and Lau, E. C. (1994). Self-assembly of a recombinant amelogenin protein generates supramolecular structures. *Journal of Structural Biology*, **112**, 103–109.

Fincham, A. G., Moradian-Oldak, J. and Sarte, P. E. (1995a). Mass-spectrographic analysis of a porcine amelogenin identifies a single phosphorylated locus. *Calcified Tissue International*, **55**, 398–400.

Fincham, A. G., Moradian-Oldak, J., Diekwisch, T. G. H., Layaruu, D. M., Wright, J. T., Bringas, P. and Slavkin, H. C. (1995b). Evidence for amelogenin 'nanospheres' as functional components of secretory-stage enamel matrix. *Journal of Structural Biology*, **115**, 50–59.

Fong, C. D., Slaby, I. and Hammarstrom, L. (1996). Amelin, an enamel-related protein, transcribed in the cells of epithelial root sheath. *Journal of Bone and Mineral Research*, **11**, 892–898.

Forsman, K, Lind, L., Backman, B., Westermark, E. and Holmgren, G. (1994). Localization of a gene for autosomal dominant amelogenesis imperfecta (ADAI) to chromosome 4q. *Human Molecular Genetics*, **3**, 1621–1625.

Fukae, M. and Tanabe, T. (1987a). Nonamelogenin components of porcine enamel in the protein fraction free from the enamel crystals. *Calcified Tissue International*, **40**, 286–283.

Fukae, M. and Tanabe, T. (1987b). 45Ca-labeled protein found in porcine developing dental enamel at an early stage of development. *Advances in Dental Research*, **1**, 261–266.

Fukae, M., Tanabe, T., Ijiri, H. and Shimizu, M. (1980). Studies on procine enamel proteins: a possible original enamel protein. *Tsurumi University Dental Journal*, **6**, 87–94.

Fukae, M., Tanabe, T., Murakami, C. and Tohi, N. (1996). Primary structure of porcine 89 kDa Enamelin. *Advances in Dental Research*, **10**, 111–118.

Gibson, C. W., Golub, E., Ding, W., Shimokawa, H., Young, M., Termine, J. D. and Rosenbloom, J. (1991). Identification of the leucine-rich amelogenin peptide (LRAP) as the translation product of an alternatively spliced transcript. *Biochemical and Biophysical Research Communications*, **174**, 1306–1312.

Gibson, C., Golub, E., Abrams, W., Shen, G., Ding, W. and Rosenbloom, J. (1992). Bovine amelogenin message heterogeneity: Alternative splicing and Y-chromosomal gene transcription. *Biochemistry*, **31**, 8384–8388.

Goldberg, M. and Septier, D. (1986). Visualization of proteoglycans and membrane-associated components in rat incisor enamel organ using ruthenium hexamine trichloride. *Journal de Biologie Buccale*, **15**, 59–66.

Heuer, A. H., Fink, D. J., Laraia, V. J., Arias, J. L., Calvert, P. D., Kendall, K., Messing, G. L., Blackwell, J., Rieke, P. C., Thompson, D. H., Wheeler, A. P., Veis, A. and Caplan, A. I. (1992). Innovative material processing strategies: a biomimetic approach. *Science*, **255**, 1098–1105.

Heussen, C. and Dowdle, E. B. (1980). Electrophoretic analysis of plasminogen activators in polyacrylamide gels containing sodium dodecyl sulfate and copolymerized substrates. *Analytical Biochemistry*, **102**, 196–202.

Hogan, B., Costantini, F. and Lacey, E. (1986). *Manipulating the Mouse Embryo*. New York: Cold Spring Harbor Laboratory.

Hu, C.-C., Fukae, M., Uchida, T., Qian, Q., Zhang, C. H., Ryu, O. H., Tanabe, T., Yamakoshi, Y., Murakami, C., Dohi, N., Shimizu, M. and Simmer, J. P. (1997). Sheathlin: Cloning, cDNA/ polypeptide sequences, and immunolocalization of porcine enamel sheath proteins. *Journal of Dental Research*, **76**, 648–657.

Ignelzi, M. A., Liu, Y. H., Maxson, R. E. and Snead, M. L. (1995). Genetically engineered mice: tools to understand craniofacial development. *Critical Reviews in Oral Biology*, **6**, 181–201.

Kallenbach, E. (1971). Electron microscopy of the differentiating rat incisor ameloblast. *Journal of Ultrastructure Research*, **35**, 508–531.

Katz, E. P., Mechanic, G. L. and Glimcher, M. J. (1965). The ultracentrifugal and free zone electrophoretic characterization of the neutral soluble proteins of embryonic bovine enamel. *Biochimica et Biophysica Acta*, **107**, 471–484.

Kogaya, Y. and Furuhashi, K. (1988). Sulfated glycoconjugates in rat incisor secretory ameloblasts and developing enamel matrix. *Calcified Tissue International*, **43**, 307–318.

Krebsbach, P. H., Lee, S. K., Matsuki, Y., Kozak, C. A. and Yamada, R. M. (1996). Full-length sequence, localization and chromosomal mapping of ameloblastin: a novel tooth-specific gene. *Journal of Biological Chemistry*, **271**, 4431–4435.

Lagerström, M., Niklas-Dahl, L. M., Iselius, L., Bäckman, B. and Pettersson, U. (1990). Mapping of the gene for X-linked amelogenesis imperfecta by linkage analysis. *American Journal of Human Genetics*, **46**, 120–125.

Lagerström, M., Dahl, N., Nakahori, Y., Nakagome, Y., Backmän, B., Landegren, U. and Pettersson, U. (1991). A deletion in the amelogenin gene in a family with X-linked amelogenesis imperfecta. *Genomics*, **10**, 971–975.

Lagerström-Fermer, M., Nilsson, M., Backmän, B., Salido, E., Shapiro, L., Pettersson, U. and Landegren, U. (1995). Amelogenin signal peptide mutation: correlation between mutations in the amelogenin gene (AMGX) and manifestations of X-linked amelogenesis imperfecta. *Genomics*, **26**, 159–162.

Lau, E. C., Mohandas, T., Shapiro, L. J., Slavkin, H. C. and Snead, M. L. (1989). Human and mouse amelogenin gene loci are on the sex chromosomes. *Genomics*, **4**, 162–168.

Lau, E. C., Simmer, J. P., Bringas, P., Hsu, D., Hu, C.-C., Zeichner-David, M., Thiemann, F., Snead, M. L., Slavkin, H. C. and Fincham, A. G. (1992). Alternative splicing of the mouse amelogenin primary transcript contributes to amelogenin heterogeneity. *Biochemical and Biophysical Research Communications*, **188**, 1253–1260.

Lesot, H., Smith, A. J., Matthews, J. B. and Ruch, J. V. (1988). An extracellular matrix protein of dentin, enamel, and bone shares common antigenic determinants with keratins. *Calcified Tissue International*, **42**, 53–57.

Limeback, H. (1987). Isolation and characterization of porcine enamelins. *Biochemical Journal*, **243**, 385–390.

Limeback, H. and Simic, A. (1989). Porcine high molecular weight enamel proteins are primarily stable amelogenin aggregates and serum albumin-derived proteins. In *Tooth Enamel V*, ed. R. W. Fearnhead, pp. 269–277. Yokahama, Japan: Florence Publishers.

Limeback, H. and Simic, A. (1990). Biochemical characterization of stable high molecular-weight aggregates of amelogenins formed during porcine enamel development. *Archives of Oral Biology*, 35, 459–468.

Limeback, H., Sakarya, H., Chu, W. and MacKinnon, M. (1989). Serum albumin and its acid hydrolysis peptides dominate preparations of mineral-bound enamel proteins. *Journal of Bone and Mineral Research*, 4, 235–241.

Lowenstam, H. A. and Weiner, S. (1989). *On Biomineralization*. Oxford: Oxford University Press.

Lyngstadaas, S. P., Rinses, S., Sproat, B. S., Thrane, P. S. and Prydz, H. P. (1995). A synthetic, chemically modified ribozyme eliminates amelogenin, the major translation product in developing mouse enamel *in vivo*. *EMBO Journal*, 14, 5224–5229.

Maas, R. and Bei, M. (1997). The genetic control of early tooth formation. *Critical Reviews in Oral Biology and Medicine*, 8, 4–39.

Mann, S. (1993). Molecular tectonics in biomineralization and biomimetic materials chemistry. *Nature*, 365, 499–505.

Matsuki, Y., Nakashima, M., Amizuka, N., Warshawsky, H., Goltzman, D., Yamada, K. M. and Yamada, Y. (1995). A compilation of partial sequences of randomly selected cDNA clones from the rat incisor *Journal of Dental Research*, 74, 307–312.

Mechanic, G. L. (1971). The multicomponent re-equilibrating protein system of bovine embryonic enamelin (dental enamel protein): chromatography in deaggregating solvents. In *Tooth Enamel II*, eds. R. W. Fearnhead and M. V. Stack, pp. 88–94. Bristol: Wright.

Menanteau, J., Mitre, D. and Raher, S. (1986). An in-vitro study of enamel protein degradation in developing bovine enamel. *Archives of Oral Biology*, 31, 807–810.

Menanteau, J., Gregoire, M., Dalcusi, G. and Jans, I. (1987). In vitro albumin binding on apatite crystals from developing enamel. *Journal of Dental Research*, 62, 100–104.

Menanteau, J., Meflah, K. and Strecker, G. (1988). The carbohydrate moiety of mineral-bound proteins from fetal enamel: a basis for enamelins heterogeneity. *Calcified Tissue International*, 42, 196–200.

Moradian-Oldak, J., Simmer, P. J., Lau, E. C., Sarte, P. E. and Slavkin, H. C. (1994a). Detection of monodisperse aggregates of a recombinant amelogenin by dynamic light scattering. *Biopolymers*, 34, 1339–1347.

Moradian-Oldak, J., Simmer, P. J., Sarte, P. E., Zeichner-David, M. and Fincham, A. G. (1994b). Specific cleavage of a recombinant murine amelogenin by a protease fraction isolated from bovine tooth enamel. *Archives of Oral Biology*, 39, 647–656.

Moradian-Oldak, J., Sarte, P. E. and Fincham, A. G. (1996a). Description of two classes of proteinases from enamel extracellular matrix cleaving a recombinant amelogenin. *Connective Tissue Research*, 35, 231–238.

Moradian-Oldak, J., Leung, W., Simmer, J. P., Zeichner-David, M. and Fincham, A. G. (1996b). Identification of a novel proteinase (ameloprotease-I) responsible for amelogenin degradation during enamel maturation. *Biochemical Journal*, 318, 1015–1021.

Mosher, D. F., Sottile, J., Wu, C. and McDonald, J. A. (1992). Assembly of extracellular matrix. *Current Opinion in Cell Biology*, 4, 810–818.

Mosteller, D., Han, J. and Broek, D. (1994). Identification of residues of the H-ras protein critical for functional interaction with guanine nucleotide exchange factors. *Molecular and Cellular Biology*, 14, 1104–1112.

Mosteller, R. D., Park, W. and Broek, D. (1995). Analysis of interaction between Ras and CDC25 guanine nucleotide exchange factor using yeast GAL4 two-hybrid system. *Methods in Enzymology*, 255, 135–148.

Munder, T. and Furst, P. (1992). The *Saccharomyces cerevisiae* CDC25 gene product binds specifically to catalytically inactive ras proteins in vivo. *Molecular and Cellular Biology*, 12, 2091–2099.

Nikiforuk, G. and Simmons, N. S. (1965). Purification and properties of protein from embryonic bovine enamel. *Journal of Dental Research*, 44, 1119–1122.

Ogata, Y., Shimokawa, H. and Sasaki, S. (1988). Purification, characterization and biosynthesis of bovine enamelins. *Calcified Tissue International*, 43, 389–399.

Okamura, K. (1983). Localization of serum albumin in dentin and enamel. *Journal of Dental Research*, 62, 100–104.

Overall, C. M. and Limeback, H. (1988). Identification and characterization of enamel proteinases isolated from developing enamel. Amelogeninolytic serine proteinases are associated with enamel maturation in pig. *Biochemical Journal*, 256, 965–972.

Paine, M. L. and Snead, M. L. (1997). Protein interactions during assembly of the enamel organic extracellular matrix. *Journal of Bone and Mineral Reasearch*, 12, 221–227.

Paine, M. L., Deutsch, D. and Snead, M. L. (1996). Carboxyl-region of tuftelin mediates self-assembly. *Connective Tissue Research*, 35, 157–161.

Park, J., Collier, P., Chen, E. and Gibson, C. W. (1994). A beta-galactosidase expression vector for promoter analysis. *DNA and Cell Biology*, 13, 1147–1149.

Robinson, C., Lowe, N. R. and Weatherell, J. A. (1975). Amino acid composition, distribution and origin of 'tuft' protein in human and bovine dental enamel. *Archives of Oral Biology*, 20, 29–42.

Robinson, C., Briggs, H. D., Atkinson, P. J. and Weatherell, J. A. (1981a). Chemical changes during formation and maturation of human deciduous enamel. *Archives of Oral Biology*, 26, 1027–1033.

Robinson, C., Fuchs, P. and Weatherell, J. A. (1981b). The appearance of developing rat incisor enamel using a freeze fracturing technique. *Journal of Crystal Growth*, 53, 160–165.

Robinson, C., Kirkham, J., Briggs, H. D. and Atkinson, P. J. (1982). Enamel proteins from secretion to maturation. *Journal of Dental Research*, 61, 1490–1495.

Robinson, C., Briggs, H. D., Kirkham, J. and Atkinson, P. J. (1983). Changes in the protein components of rat incisor enamel during tooth development. *Archives of Oral Biology*, 28, 993–1000.

Robinson, C., Kirkham, J. and Shore, R. C. (1989a). Extracellular processing of enamel matrix and origin and function of tuft protein. In *Tooth Enamel V*, ed. R. W. Fearnhead, pp. 59–68. Yokahama, Japan: Florence Publishers.

Robinson, C., Kirkham, J. and Fincham, A. (1989b). The enamelin/non-amelogenin problem. A brief review. *Connective Tissue Research*, 22, 93–100.

Robinson, C., Kirkham, J. and Shore, R. (eds.) (1995). *Dental Enamel, Formation to Destruction*. Boca Raton: CRC Press.

Rosenbloom, J., Lally, E., Dixon, M., Spencer, A. and Herold, R.

(1986). Production of a monoclonal antibody to enamelins which does not cross-react with amelogenins. *Calcified Tissue International*, **39**, 412–415.

Salido, E. C., Yen, P. H., Koprivnikar, K., Yu, L. C. and Shapiro, L. J. (1992). The human enamel protein gene amelogenin is expressed from both the X and Y chromosome. *American Journal of Human Genetics*, **50**, 303–316.

Sasaki, S., Takagi, T. and Suzuki, M. (1991a). Amelogenin degradation by an enzyme having acidic pH optimum and the presence of acidic zones in developing bovine enamel. In *Mechanisms and Phylogeny of Mineralization in Biological Systems*, ed. S. Suga, pp. 79–81. New York: Springer.

Sasaki, T., Takagi, T. and Suzuki, M. (1991b). Cyclical changes in pH in bovine developing enamel sequential bands. *Archives of Oral Biology*, **36**, 227–231.

Shimokawa, H., Wassmer, P., Sobel, M. E. and Termine, J. D. (1984). Characterization of cell-free translation products of mRNA from bovine ameloblasts by monoclonal and polyclonal antibodies. In *Tooth Enamel IV*, eds. R. W. Fearnhead and S. Suga, pp. 161–166. Amsterdam: Elsevier.

Shimokawa, H., Sobel, M. E., Sasaki, M., Termine, J. D. and Young, M. F. (1987). Heterogeneity of amelogenin mRNA in bovine tooth germ. *Journal of Biological Chemistry*, **262**, 4042–4047.

Simmer, J. P. and Fincham, A. G. (1995). Molecular mechanisms of dental enamel formation. *Critical Reviews in Oral Biology*, **6**, 84–108.

Simmer, J. P. and Snead, M. L. (1995). Molecular biology of the amelogenin gene. In *Dental Enamel, Formation to Destruction*, eds. C. Robinson, J. Kirkham and R. Shore, pp. 59–84. Boca Raton: CRC Press.

Simmer, J. P., Lau, E. C., Hu, C. C., Bringas, P., Santos, V., Aoba, T., Lacey, M., Nelson, D., Zeichner-David, M., Snead, M. L., Slavkin, H. C. and Fincham, A. G. (1994). Isolation and characterization of a mouse amelogenin expressed in *Escherichia coli*. *Calcified Tissue International*, **54**, 312–319.

Simmons, D., Gu, T. T., Leach, R., Krebsbach, P. H. and MacDougall, M. (1996). Identification and temporal expression of cDNA for murine ameloblastin. *Journal of Dental Research*, **75**, 27.

Skobe, Z., Stern, D. N. and Prostak, K. S. (1995). The cell biology of amelogenesis. In *Dental Enamel; Formation to Destruction*, eds. C. Robinson, J. Kirkham and R. Shore, pp. 23–57. Boca Raton: CRC Press.

Slavkin, H. C. (1976). Towards a cellular and molecular understanding of periodontics: cementogenesis revisited. *Journal of Periodontology*, **47**, 249–255.

Slavkin, H. and Price, P. (eds.) (1992). *Chemistry and Biology of Mineralized Tissues*. Amsterdam: Excerpta Medica.

Slavkin, H. C., Bessem, C., Bringas, P., Zeichner-David, M., Nanci, A. and Snead, M. L. (1988). Sequential expression and differential function of multiple enamel proteins during fetal, neonatal and early postnatal stages of mouse molar organogenesis. *Differentiation*, **37**, 26–39.

Slavkin, H. C., Chadwick, D. J., Cardew, G. and Winter, G. B. (eds.) (1996). *Dental Enamel*. Ciba Foundation Symposium Series, vol. 205. New York: Wiley.

Smales, F. C. (1975). Structural sub-units in prisms of immature rat enamel. *Nature*, **258**, 772–774.

Smith, C. E., Borenstein, S., Fazel, A. and Nanci, A. (1989a). *In vitro* studies of the proteinases which degrade amelogenins in devel-

oping rat incisor enamel. In *Tooth Enamel V*, ed. R. W. Fearnhead, pp. 286–290. Yokahama, Japan: Florence Publishers.

Smith, C. E., Pompura. J. R., Borenstein, S., Fazel, A. and Nanci, A. (1989b). Degradation and loss of matrix proteins from developing enamel. *Anatomical Record*, **224**, 292–316.

Smith, C. E., Chen, W. Y., Issid, M. and Fazel, A. (1995a). Enamel matrix protein-turnover during amelogenesis. Basic biochemical properties of short-lived sulfated enamel proteins. *Calcified Tissue International*. **57**, 133–144.

Smith, C. E., Chen, W. Y., Issid, M. and Fazel, A. (1995b). Matrix protein turnover during amelogenesis: the ultra short-lived sulfated enamel proteins. *Connective Tissue Research*, **32**, 147.

Snead, M. L. (1996). Enamel biology logodaedaly: getting to the root of the problem, or 'Who's on first ...' *Journal of Bone and Mineral Research*, **11**, 899–904.

Snead, M. L., Zeichner-David, M., Chandra, T., Robson, K. J., Woo, S. L. and Slavkin, H. C. (1983). Construction and identification of mouse amelogenin cDNA clones. *Proceedings of the National Academy of Sciences, USA*, **80**, 7254–7258.

Snead, M. L., Lau, E. C., Zeichner-David, M., Fincham, A. G. and Woo, S. L. (1985). DNA sequence for cloned cDNA for murine amelogenin reveal the amino acid sequence for enamel-specific protein. *Biochemical and Biophysical Research Communications*, **129**, 812–818.

Snead, M. L., Luo, W., Lau, E. C. and Slavkin, H. C. (1988). Spatial- and temporal-restricted pattern for amelogenin gene expression during mouse molar tooth organogenesis. *Development*, **104**, 77–85.

Snead, M. L., Lau, E. C., Fincham, A. G., Zeichner-David, M., Davis, C. and Slavkin, H. C. (1989). Of mice and men: Anatomy of the amelogenin gene. *Connective Tissue Research*, **22**, 101–109.

Snead, M. L., Paine, M. L., Chen, L. S., Yoshida, B., Luo, W., Zhou, D., Lei, Y. P., Liu, Y. H. and Maxson, R. E. Jr (1996). The murine amelogenin promoter: transient and permanent transgenic animals. *Connective Tissue Research*, **35**, 41–47.

Staatz, W. D., Fok, K. F., Zutter, M. M., Adams, S. P., Rodriguez, B. A. and Santoro, S. A. (1991). Identification of a tetrapeptide recognition sequence for the alpha2 beta1 integrin in collagen. *Journal of Biological Chemistry*, **266**, 7363–7367.

Strawich, E. and Glimcher, M. J. (1990). Tooth 'enamelin' identified mainly as serum proteins. Major 'enamelin' is albumin. *European Journal of Biochemistry*, **191**, 47–56.

Suga, S. (1970). Histochemical observation of proteolytic enzyme activity in the developing dental hard tissues of the rat. *Archives of Oral Biology*, **15**, 555–558.

Takagi, T., Suzuki, M., Baba, T., Minegishi, K. and Sasaki, S. (1984). Complete amino acid sequence of amelogenin in developing bovine enamel. *Biochemical and Biophysical Research Communications*, **121**, 592–597.

Tanabe, T., Aoba, T., Moreno, E. C., Fukae, M. and Shimizu, M. (1990). Properties of phosphorylated 32 kDa nonamelogenin proteins isolated from porcine secretory enamel. *Calcified Tissue International*, **46**, 205–215.

Tanabe, T., Fukae, M., Uchida, T. and Shimizu. M. (1992). The localization and characterization of proteinases for the initial cleavage of porcine amelogenin. *Calcified Tissue International*, **51**, 213–217.

Termine, J. D. and Torchia, D. A. (1980). 13C-1H Magnetic double-resonance study of fetal enamel matrix proteins. *Biopolymers*, **19**, 741–750.

Termine, J. D., Belcourt, A. B., Christner, P. J., Conn, K. M. and Nylen, M. U. (1980). Properties of dissociatively extracted fetal tooth matrix proteins. I. Principal molecular species in developing bovine enamel. *Journal of Biological Chemistry*, **255**, 9760–9768.

Thesleff, I., Vaahtokari, A. and Partanen, A. M. (1995). Regulation of organogenesis. Common molecular mechanisms regulating the development of teeth and other organs. *International Journal of Developmental Biology*, **39**, 35–50.

Traub, W., Jodaikin, A. and Weiner, S. (1985). Diffraction studies of enamel protein–mineral structural relations. In: *The Chemistry And Biology Of Mineralized Tissues*, ed. W. T. Butler, pp. 221–225. Birmingham, AL: EBSCO.

Uchida, T., Tanabe, T., Fukae, M. and Shimizu, M. (1991). Immunocytochemical and immunochemical detection of a 32 kDa nonamelogenin and related proteins in porcine tooth germs. *Archives of Histology and Cytology*, **54**, 527–538.

Uchida, T., Fukae, M., Tanabe, T., Yamakoshi, Y., Satoda, Y., Murakami, C., Takahashi, O. and Shimizu, M. (1995). Immunochemical and immunocytochemical study of a 15 kDa nonamelogenin and related proteins in the porcine immature enamel: Proposal of a new group of enamel proteins 'sheath proteins'. *Biomedical Research*, **16**, 131–140.

Weiss, K. M., Bollekens, J., Ruddle, F. H. and Takashita, K. (1994). Distal-less and other homeobox genes in the development of the dentition. *Journal of Experimental Zoology*, **270**, 273–284.

Williams, G. C. and Nesse, R. M. (1991). The dawn of Darwinian medicine. *Quarterly Review of Biology*, **66**, 1–22.

Wright, J. T., Robinson, C. and Shore, R. (1992). Enamel ultrastructure in pigmented hypomaturation *amelogenesis imperfecta*, *Journal of Oral Pathology and Medicine*, **21**, 390–394.

Yamada, Y. and Kleinman, H. K. (1992). Functional domains of cell adhesion molecules. *Current Opinions in Cell Biology*, **4**, 819–823.

Yamakoshi, Y., Tanabe, T., Fukae, M. and Shimizu, M. (1989). Amino acid sequence of porcine 25 kDa amelogenin. In *Tooth Enamel V*, ed. R. W. Fearnhead, pp. 314–318. Yokahama, Japan: Florence Publishers.

Zeichner-David, M., MacDougall, M., Vides, J., Snead, M. L., Slavkin, H. C., Turkel, S. B. and Pavlova, Z. (1987). Immunochemical and biochemical studies of human enamel proteins during neonatal development. *Journal of Dental Research*, **66**, 50–56.

Zeichner-David, M., Vides, J., MacDougall, M., Fincham, A., Snead, M. L., Bessem, C. and Slavkin, H. C. (1988). Biosynthesis and characterization of rabbit tooth enamel extracellular matrix proteins. *Biochemical Journal*, **251**, 631–641.

Zeichner-David, M., MacDougall, M., Davis, A., Vides, J., Snead, M. L. and Slavkin, H. C. (1989). Enamelins and amelogenins share the same amino-terminal sequence. *Connective Tissue Research*, **20**, 749–755.

Zeichner-David, M., Thiemann, F. and MacDougall, M. (1993). Tuftelin mRNA expression during mouse tooth organ development. *Journal of Dental Research*, **72**, 203.

Zeichner-David, M., Hsiu, P., Berman, B. and Diekwisch, T. (1994). Immunolocalization of tuftelin during mouse tooth development. *Journal of Dental Research*, **73**, 112.

Zeichner-David, M., Diekwisch, T., Fincham, A., Lau, E., MacDougall, M., Moradian-Oldak, J., Simmer, J., Snead, M. L. and Slavkin, H. C. (1995). Control of ameloblast cell differentiation. *International Journal of Developmental Biology*, **39**, 69–92.

Zhou, P., Byrne, C., Jacobs, J. and Fuchs, E. (1995). Lymphoid enhancer factor 1 directs hair follicle patterning and epithelial cell fate. *Genes and Development*, **9**, 570–583.

Tooth tissues: development and evolution

Part two

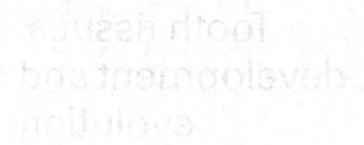

5 Evolutionary origins of dentine in the fossil record of early vertebrates: diversity, development and function

M. M. Smith and I. J. Sansom

5.1 Introduction

Dentine has a central role in the support and function of the tooth. It is formed at the interface of the dental epithelium and dental mesenchyme (see Chapter 4), and is the vital link or selective barrier between the sensory and vascular supply of the pulp and the functional surface of the tooth, usually enamel of the crown (see Chapter 6). How do the formative cells mediate between the environment of the tooth surface and the tissue they maintain, either to allow a sensory function, or to effect repair of the dentine to maintain a supportive but wear-resistant material but still with vital properties? Curiously these are questions that still remain largely unsolved.

As Lumsden (1987) emphasised, the initial appearance of dentine in the exoskeleton, largely as a protective cover to the skin, probably allowed the Ordovician agnathans to detect tactile, temperature and osmotic changes. This view that dentine allowed the earliest vertebrates to engage in electroreception, was advanced by Northcutt and Gans (1983), Gans and Northcutt (1983) and Gans (1987) (see section 5.5.1). Neural crest-derived cells, initially with a sensory function, became one population of skeletogenic cells, possibly providing a sensory function for dentine as one of the first vertebrate hard tissues. Although Tarlo (1965) drew attention to the striking similarity between human orthodentine and the dentine of the dermal armour of an early jawless (agnathan) heterostracan fish, little insight has been gained into how this histological similarity might allow a comparable sensory mechanism. In the heterostracan example the dentine tubercles are not covered by enamel.

It has been recognised since the early studies of Pander (1857) that the outermost parts of the dermal armour in the earliest vertebrates are made of the tissue dentine. However, despite many thorough accounts of the palaeohistology of the enormous range of vertebrate dentinous tissues, much controversy still exists concerning the evolutionary primacy of dentine over bone, and linked to this, whether it is cell containing first (dentine with odontocytes) leading to tubular dentine (pulpal odontoblasts), or a different order of change.

The tuberculate ornament of the thick dermal armour and the skin denticles of microsquamous (flexible armour) vertebrates are considered as homologous structures with teeth, with the possibility that either arose independently of the other (see Smith and Coates, 1998 and Chapter 10). Descriptions of the dentine microstructure in these odontode-derived tissues are frequently compared with dentine of teeth, the latter for a long time being considered as a subsequent evolutionary derivative, perhaps with a modified function. Recent studies have increased our knowledge of early vertebrate hard tissues dramatically, and in this chapter we will attempt to synthesize the early fossil record of dentinous tissues, with special emphasis on possible functional scenarios for their evolutionary development.

5.2. Dentine terminology

5.2.1. Historical
There is considerable terminological confusion in the literature describing exoskeletal tissues, and almost as much in describing the histological variation found in the teeth of living and fossil fishes. Much of this stems from the concept that dentine should be compared with

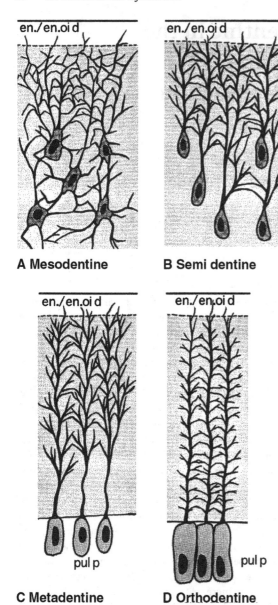

Figure 5.1. **Schematic of four patterns in dentines of early vertebrate dermal armour. Variations of cell arrangement and their interconnecting cell processes were proposed as an imagined evolutionary progression from cellular bone (A–D); the odontoblast cell first retained in the dentine, then evolving to be in a central dental pulp tissue at the formative front, with increasing polarity of the central cell process. (Modified from Smith and Hall, 1993, figure 5.)**

ported. The terms mesodentine, semidentine, metadentine, and orthodentine are relicts from such an approach (Figure 5.1). Developmental differences, represented by stages in a presumed evolutionary series, were explained as a progressive process of retreat of odontoblasts in a centripetal direction towards the pulp cavity. In early types cell bodies were trapped in the matrix (odontocytes) and in later ones free within the pulp cavity (odontoblasts). This was accompanied by an increasing polarity of the odontoblasts, reduction to a single cell process, and the tendency to form a continuous single cell layer at the pulp surface, or separate clusters of cells in groups retained in separate, small pulp canals. The single cell layer would allow specific contacts between odontoblasts as in an epithelium, gap junctions and a presumed isoionic coupling at the pulpo-dentinal surface, as in human orthodentine. The regulation of intradentinal fluid, with both composition and movement independent from that of the pulpal interstitial fluid, may be important in sensory mechanisms. As we find orthodentine early in vertebrate evolution, this sensory function is one predicted for the dermal skeleton of some agnathan fish (see section 5.5.1).

Whilst the above scenario may describe the developmental differences of each evolutionary stage, it can with no certainty be interpreted as a phylogenetic progression. Rather, the apparently unrestricted range of dentine types can be explained as great phenotypic plasticity, or a random pattern of histogenetic diversity based on relatively loose developmental controls. As Raff (1994, p 496) has suggested for the great explosion of body forms in the Cambrian, diversity became stabilised later in evolution when closely integrated developmental controls allowed less experimentation. The four patterns in primitive dentines have taken on an accepted reality as a transformation series (Figure 5.1) largely because they have been figured many times in accounts of dentine diversity (Ørvig, 1967, figures 15, 18–22, 24–28; Moss, 1968, figure 5; Baume, 1980, figure 4.1; Reif, 1982, figure 2; Smith and Hall, 1993, figure 5; Janvier 1996, figure 6.13).

5.2.2.   Terminology

No account of dentine would be complete without reference to two major reviews by Ørvig (1951, 1967) in which he proposed schemes for rationalising the types and variation found to occur in lower vertebrates. In his early paper, the older literature is comprehensively reviewed (see also Peyer, 1968; Baume, 1980). Apart from the four patterns based on distribution of cells and processes in the dentine matrix, Ørvig (1951, figure 1) compares three common types of arrangement (dentine and pulpal tissues) in the tooth structure of fishes (Figure 5.2), based

bone, both at the macroscopic level of tissue arrangement and at the level of the cell and matrix composition.

Ascribing a hierarchical order to dentine diversity had been on the early assumptions that dentine evolved from bone and that some of the early types in the fossil record represent a transformation series (see Ørvig, 1967 for review). Neither of these assumptions can now be sup-

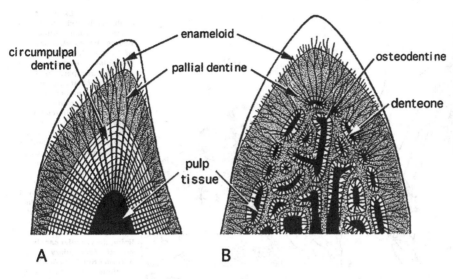

circumpulpal dentine

enameloid

pallial dentine

osteodentine

denteone

pulp tissue

A   B

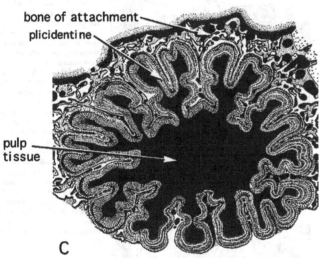

bone of attachment

plicidentine

pulp tissue

C

Figure 5.2A–C. **Examples of three major variations in the arrangement of dentine in the teeth of fish. *A.* An undivided pulp cavity and orthodentine. *B.* A pulp cavity divided by dentine trabeculae and completed by infilling denteones formed around each vascular canal. (modified from Ørvig, 1951, fig. 1C, D). *C.* Folded dentine at the base of a conical crossopterygian tooth (*Eusthenopteron fordii*, from Schultze, 1969, figure 10).**

on some of the drawings from Tomes's (1904) classic textbook of dental anatomy, he also figures several possible configurations of odontoblast cells in relation to the matrixes produced by them (Ørvig, 1967, figures 15–17). The difficulty of Ørvig's scheme is that it is too complex based as it is on different levels of tissue description. What should be emphasised are the levels of complexity in which dentine and the associated soft tissues of the pulp can be arranged, and after that the relationship of cells and their processes to the mineralised matrix and to the collagen fibres contained within. It is implicit that all these features can also be interpreted in fossil tissues, because of the essential imprint of cells and their processes, and the retention of unmineralised collagen fibre spaces in the matrix.

The classification scheme produced by Ørvig (1967, p. 101) is comprehensive in that it considers the three principal types of dentine and all possible combinations of arrangement, but it has proved too unwieldly for common use. The summary by Ørvig (1951) has the advantage of simplicity and is more useful as it is based partly on locational terms. This distinguishes *pallial dentine* (mantle) from *circumpulpal dentine*, distributed around an open pulp cavity (Figure 5.2A). Both pallial and circumpulpal dentine exhibit parallel but branching tubules, and are together called *orthodentine*. Variations of orthodentine include *plicidentine* (Figure 5.2C), a folded arrangement at the base of large conical teeth, and *vasodentine* where capillaries are included in the circumpulpal dentine. In orthodentine the matrix has coarse fibres radial to the outer dentine border in the mantle layer (pallial dentine), and fine fibres parallel to the forming surface in the circumpulpal dentine. Where the pulp cavity is filled with dentine incorporating the

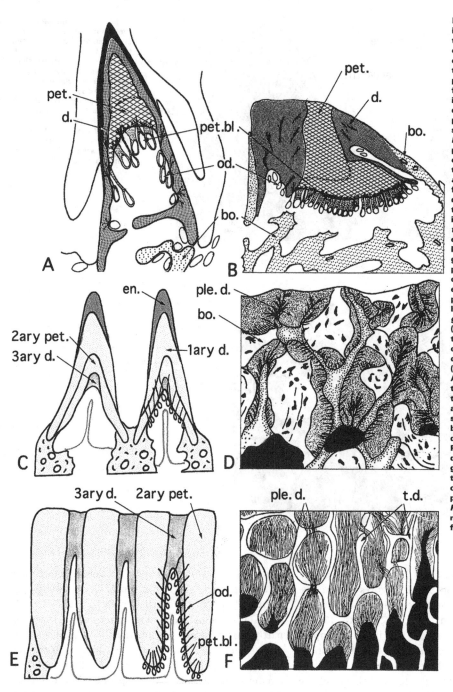

Figure 5.3EA–F. **Examples of infilling (pleromic) dentine in toothplates and dentine covered bones.** *A.* **New tooth of the larval lungfish toothplate, where secondary petrodentine (pet.) first forms from petroblasts (pet. bl.) inside the primary pallial dentine, (d.) dentine production continues at the sides from odontoblasts (od.).** *B.* **In the worn juvenile tooth plate growth of petrodentine continues into the spaces in the bone (bo.)** (*Protopterus aethiopicus* from Smith, 1984, figs. 1, 12). *C.* **Two primary dentine (1ary d.) teeth joined to the bone of the tooth plate in the lungfish** *Neoceratodus forsteri,* **tertiary dentine (3ary d.) forms from odontoblasts lining the vascular canals, as growth of secondary petrodentine (2ary pet.) continues more basally. (en, enamol)** *D.* **Surface of worn palatal bone in the fossil lungfish** *Holodipterus gogoensis,* **pleromic dentine (ple. d.) has grown into the soft tissue space of the bone to compensate for wear and lack of denticle/tooth replacement (from Smith, 1977, fig. 11).** *E.* **Adult** *Neoceratodus* **toothplate where the primary tooth structure has worn away, leaving columns of secondary petrodentine lined by tertiary circumvascular dentine around the canals.** *F* **Pleromo-aspidin, dentine (ple. d.) infills soft tissue spaces by growth between the aspidin trabeculae in the dermal bone of a heterostracan branchial plate (t. d., trabecular dentine;** *Psammolepis paradoxa,* **modified from Ørvig, 1976, figure 5).**

pulpal vascular supply it is called *osteodentine* (Figure 5.2B). This first develops as trabecular dentine and later it becomes compact, as dentinal osteones (*denteones*) fill in the soft tissue spaces around the blood vessels (Figure 5.2B; see also Ørvig, 1967, figures 16 – 17). This circumvascular tissue is fine-fibred and tubular and compares with circumpulpal dentine in its fine structure.

A special arrangement of osteodentine is found in flat crushing teeth and those adapted to resist wear, and in these the regular, vertical arrangement of the vascular canals gives the dentine a tubate appearance, often mis-leadingly called tubular dentine. In *columnar* or *tubate dentine* (tubate was a suggestion by Ørvig to avoid ambi-guity with all tubular forms of dentine) each canal is

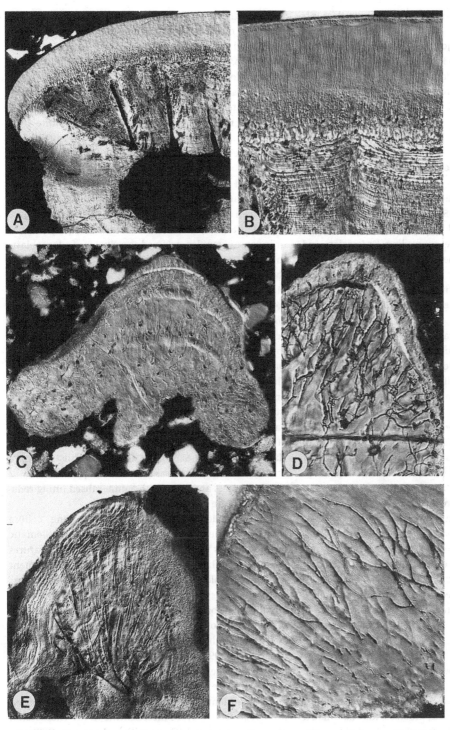

Figure 5.4A–F.
Photomicrographs taken in Nomarski DIC optics of sections of agnathan dermal odontodes, to show dentine types in these tubercles/ denticles. All sections are from the Harding Sandstone of Colorado, USA (Caradoc, Ordovician). A, B. *Astraspis desiderata* large tubercle with enameloid cap and several vascular canals (A), dentine with very fine tubules, these broaden at the dentine-enameloid junction (B) and continue as finer tubules (v. c., vascular canals; specimen no. BU 2268; A, × 75; B, × 200; from Smith *et al.*, 1996, figures 1 A,B).). C, D. *Skiichthys halsteadi* dermal denticle, cap of enameloid over mesodentine with odontocytes, and basal flanges of osteocytic bone (specimen no. BU 2272, × 100; from Smith and Sansom, 1997, Plate 1, figure 1). D. coronal ridge with interconnecting network of tubules and terminal branches into enameloid (specimen no. BU 2610, × 600; from Smith and Sansom 1997, Plate 1, figure 3). E. *Eriptychius americanus* dentine ridge with wide calibre tubules, few branches, and thickened walls similar to peritubular dentine, many close spaced growth lines (specimen no. BU 2253, × 200; from Smith *et al.*, 1996, figure 2C). F. Earliest known thelodont denticle, longitudinal section of coronal dentine, fine diameter tubules with few interconnections (specimen No. BU 2591, × 600; from Sansom *et al.*, 1996, figure 4C).

surrounded by circumvascular dentine, but, unlike more irregular osteodentine it is built onto a framework of very specialised hypermineralised dentine, called *petrodentine* (see Smith, 1984 for a survey of this in lungfish toothplates, and Figure 5.3).

Many atypical dentinous tissues are encountered in the fossil record (Figure 5.4). Notable amongst these are the atubular dentines described, from Lower Silurian shark-like scales, as *lamellin* by Karatajùte-Talimaa *et al.* (1990) and Karatajùte-Talimaa and Novitskaya (1992), and the atubular basal bodies of certain conodont elements (Sansom, 1996).

It is apparent that a major revision and redefinition of dentine terminology and types may be required, although here we have adopted the pre-existing terminology and applied it more specifically to the material discussed below.

5.3.   Earliest dentines

The fossil record of dentine now extends back to the Late Cambrian, approximately 510 million years ago (Smith, M.P. et al., 1996), followed by a rapid diversification of dentinous tissues into the succeeding Ordovician and Lower Silurian periods. The variety of dentinous tissues encountered in this time interval appears to greatly exceed the diversity seen in the subsequent 450 million year history of vertebrate development.

5.3.1.   Fossil tissue characters
Interpretation of fossil tissues requires an understanding of the biological features left behind in the mineralised part, often stained naturally by post-mortem diagenetic processes in the geological environment. Taphonomic artifacts (such as bioerosion and recrystallisation of phosphate) must be disentangled from the characters of biomineralisation in the fossil specimens. It is possible to characterise fossil hard tissues on the basis of a number of features, including: differences in the level of mineralisation of the organic matrix, these differences forming a record of the biological processes of growth (incremental lines); patterns of mineralisation (calcospheritic growth, interglobular spaces); included cell bodies (lacunae); included cell processes (tubules, lateral and terminal branching); progressive hypermineralisation of tubule space (peritubular dentine); collagen fibre bundles with unmineralised cores appearing as tubules (Sharpey's fibre tubules), and vascular spaces (capillaries in vasodentine), and central canals in osteodentine (denteonal vascular canal).

5.3.2.   Early dentine types: cell and tubule arrangement
Moss (1968, pp. 49–50), proposed that 'the induction of ectomesenchymal cells into odontoblasts may have had a range of expressivity' which he explained as due to variations in the inductive stimulus or the competency to respond. It is likely that dentinous tissues in early vertebrates are evidence of developmental lability early in the history of vertebrate dermal skeletal biomineralisation. We have attempted to simplify descriptions of dentine

found in the dermal armour of the earliest vertebrates; focussing on the variation in position of the cell body, the branching pattern of the cell processes within the dentine matrix, and the arrangement of the pulp canals assumed to house the odontoblast cell bodies.

Most early dentines do not have the odontoblast cell body located in a separate lacuna within the matrix. Also, the majority of these primitive vertebrate hard tissue complexes do not have cells included in the bone, rather the tissue is described as aspidin (coarse-fibred acellular bone). Nearly all show a combination of fine and broad tubules extending through the tissue to the surface, varying in the degree of branching. The location of the cell body (because it is not preserved in the hard tissues as a lacuna) is assumed to be at the pulp surface, or in some examples, grouped as a cluster of odontoblasts in the wide, separate pulp canals from which the tubules radiate. Dentines in the Cambrian fish *Anatolepis* (Smith, M.P. et al., 1996), the Ordovician fishes *Astraspis* (Figure 5.4A,B) and *Eriptychius* (Figure 5.4E), and the conodonts *Chirognathus* (Figure 5.5B) and *Oulodus* (Figure 5.5C) are acellular but all with tubules, although they are difficult to categorize as any transitional dentine (metadentine of Ørvig, 1967, p 102), nor are they a form of orthodentine. They all show great diversity in the pattern of tubule branching, possibly related to cell phenotypic variability (see section 5.4.1), and the relationship of dentine to papillary soft tissues. The dentine tubules in *Astraspis* are of notably fine calibre, of the order of 0.1 micron (see Halstead 1987 and Sansom et al., 1997), whilst in *Eriptychius* they are considerably coarser, approximately 2 microns across and often with a mineralised lining reducing the calibre of the tubule.

Atubular dentines, and indeed the fine calibre tubular dentine of *Astraspis*, are particularly problematic when it comes to identifying them as dentine. Features such as globular calcification, crystallite arrangement and their topological relationship to other components of the hard tissue complex are particularly useful. As already mentioned, atubular dentines are encountered in the crowns of scales of the shark-like mongolepid fish (Karatajùte-Talimaa et al., 1990; Karatajùte-Talimaa and Novitskaya, 1992; Karatajùte-Talimaa, 1995), as well as the basal bodies of many conodonts, including *Pseudooneotodus* (Figure 5.5D; Sansom, 1996).

Cellular dentines also appear during this early period in vertebrate history, as mesodentine in the exoskeleton of *Skiichthys* (Figure 5.4C, D; Smith and Sansom, 1997) and similar tissues in the basal bodies of elements of the conodont genus *Neocoleodus* (Figure 5.5A; Sansom et al., 1994).

In specimens which develop as a result of superimpositional growth it is particularly difficult to distinguish

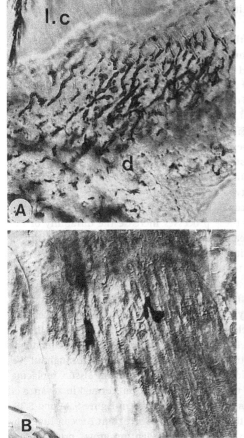

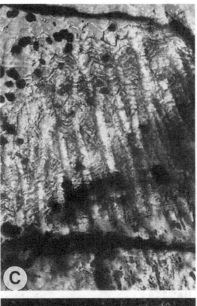

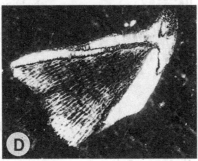

Figure 5.5A–D.
Photomicrographs of conodont element tissues, to show dentine types in the basal body. A–C are from the Harding Sandstone of Colorado, USA (Caradoc, Ordovician); D is from the Clarita formation (Wenlock, Silurian) of Oklahoma, USA. A. *Neocoleodus* **sp., iron oxide infilled network of tubules (l.c., lamellar crown; d., dentine; specimen no. BU 2256, × 250; from Smith et al., 1996, figure 3F). B.** *Chirognathus duodactylus,* **broad calibre tubules and calcospheretic incremental pattern (specimen no. BU 2251, × 250; from Smith et al., 1996, figure 3B). C.** *Oulodus serratus,* **regular, close-spaced tubules and calcosperetic growth pattern (specimen no. BU 2276, × 250; from Smith et al., 1996, figure 4C). D.** *Pseudooneotodus tricornis,* **closed tubules formed from either cell processes or attachment fibre bundles, taken in polarised light (specimen no. BU 2277, × 90; from Sansom, 1996, figure 3C).**

bone from mesodentine, and from semidentine, in those mixed component structures where dentine ridges are fused onto bone. Bone and dentine become intimately mixed tissues when continuous growth of dentine into the bone occurs as a pleromic phenomenon (see section 5.5.3).

In summary, there is an assumed widespread developmental plasticity in dentine formed in the dermal armour of early vertebrates and in the papillary vascular tissues, both derived in development from a presumed ectomesenchymal papilla (see Smith, 1991), located beneath the epithelium. This has resulted in great diversity of tissue types amongst which there is a lack of consensus on the pattern of evolutionary change.

5.4. Developmental aspects of dentines in lower vertebrates

With the benefit of recent work on the structural development of dentines in extant lower vertebrates, it is possible

to offer some suggestions about the growth, patterning and phenotypic expression amongst fossil tissues.

5.4.1. Odontoblast phenotypes

In an attempt to explain the developmental basis of some of this diversity in early dentines, with regard to a potential difference in the odontoblast cells, we have taken an example of new information on odontogenesis in the shark tooth to illustrate a later diversification. Also, some specialisations contributing to wear-resistant tissues in non-replacing teeth found in fossil and extant forms.

These recent ultrastructural studies of extant tissues have provided more than one example of two different derivative cells of the odontoblast stem population lining the pulp surface. For example, odontoblasts function in the production of both shark enameloid (the first tissue to form) and the circumpulpal dentine (Figure 5.6A; Sasagawa, 1994), but as yet no molecular data is available. The organic matrix of the enameloid is formed first from the cells of the papilla, and they continue to participate in its progressive mineralisation as the matrix of the dentine is formed beneath it from a second

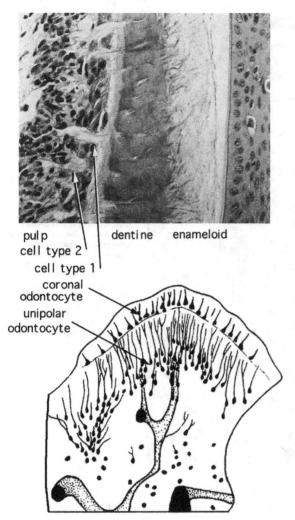

pulp   dentine   enameloid
cell type 2
cell type 1
coronal
odontocyte
unipolar
odontocyte

Figure 5.6. **A. Developing tooth section at the stage of dentine mineralisation of the angel shark (*Squatina japonica*). This shows two types of odontoblast, dark cells (type 1) and light cells (type 2), both concerned with the production and progressive mineralisation of enameloid and dentine. (Photo micrograph by courtesy of Ichiro Sasagawa, stain H and E). B. Section through a dentine tubercle of a placoderm (*Phlyctaenaspis acadica*, from Gross, 1957, figure 9A) in which two cell types are indicated, both are odontocytes in the dentine but one is situated superficially and described as a special coronal cell.**

cell population. Cell type one (Figure 5.6A) retains a process in the mineralizing enameloid, the other appears to be confined to the dentine beneath. This is an area of research dealt with in the enameloid literature, but is currently underexplored and would benefit from innovative molecular studies. The second example is the differentiation of petroblasts as a distinct cell type from odontodes in the continuous growth of hypermineralised dentines (see discussion below).

Many examples can be found amongst fossil dentines where varieties of dentine seem to suggest two different cell types involved in the production of the mineralised tissues, either in tandem, or sequentially. In placoderm dentine, the suggestion is that coronal odontocytes produce the coronal tissue (Figure 5.6B), and others the semidentine of the papillary dentine. The other examples are found in osteodentine, trabecules of dentine added to by circumvascular dentine filling in around the blood vessels as denteones (Figure 5.2B). In lungfish toothplates, petrodentine without tubules is deposited first from the papilla (Figure 5.3A, C, D) followed by normal dentine lining the vascular canals. This process also occurs in tubate osteodentine of many crushing teeth in chondrichthyans and actinopterygians, with an enameloid/acrodin type as the first formed tissue penetrated by vascular canals lined by normal circumvascular dentine.

All of these processes increase the durability of the teeth. Wear-resistant petrodentine forming the composite dentition in lungfish (Lison, 1941, Smith, 1984) is essential to the continuous growth process of their non-shedding tooth plates. The cells in petrodentine (petroblasts) differentiate from the odontoblast layer but have the ability to produce a form of dentine which is as hard as enamel, only it is produced from the pulpal surface, and therefore, capable of replacement in the functional tooth. Petrodentine forms continuously throughout life and provides a wear resistant, growing tissue, replacing the worn tissue at the surface, but formed in advance of need. This process is illustrated in Figure 5.3A and 5.3B, showing the larval teeth of the extant African lungfish, *Protopterus*, in which petrodentine grows progressively and expansively into the bone spaces below (Smith, 1985). In the Australian lungfish, *Neoceratodus*, the process is similar (Figure 5.3C and 5.3E), these diagrams illustrate the sequential production of $1°$ pallial dentine, $2°$ petrodentine, and $3°$ circumvascular dentine in the teeth. The dentine lining each of the vascular canals (tertiary dentine) contributes to columns penetrating the petrodentine of the adult tooth plate (Figure 5.3E). Petrodentine does not have tubules in the mature tissue, and during its production the cell processes are so fine that they can only be demonstrated with the electron microscope (Smith, personal observations). These cells (petroblasts) are a unique specialisation of dipnoan toothplates, although convergently found also in holocephalan tooth plates. Alongside the petroblasts are the normal odontoblasts producing the tertiary circumvascular dentine. Both are examples of two phenotypes of odontoblast forming a layer at the dentino-pulpal junction and maintaining two different tissue types.

5.4.2.   Matrix fibre pattern

The interpretation of grouping and major directional differences of collagen fibres within the dentine matrix of fossil tissues relies on the property of the biological crystals of hydroxyapatite to remain in the orientation in which they form in life, parallel to the c-axis of the collagen fibre, both within the intermolecular space and around the fibrils. Interpretation in fossil material depends almost entirely upon the use of polarised light, as the birefringence pattern determines the direction of the crystal orientation and hence the orientation of the original fibres. This topic is covered in a major and comprehensive account of polarised light observations on all dentine and enamel tissues by Schmidt and Keil (1967). In extant tissues, collagen fibre direction and arrangement can be determined from the intrinsic birefingence of collagen in polarised light, and by stained preparations of the organic matrix, or direct visualisation in ultrathin sections. The outer mantle dentine has fibres of von Korff incorporated in the matrix together with the surface parallel fine fibres, the latter are the only type in circumpulpal dentine. Fine fibres parallel to the forming surface are the basis of the circumvascular dentine of the denteones of osteodentine, and a different coarse-fibred, woven arrangement in the primary trabeculae.

Much controversy has existed in the dental literature concerning the cellular origin of the fibres of von Korff in the mantle predentine matrix and also their continued existence between the mature odontoblasts lining the pulp cavity, as so-called corkscrew fibres of von Korff (Lester and Boyde, 1968). The possible existence of two phenotypes, an odontoblast secreting the soluble matrix precursors with reciprocal organisation of the cells and orientation of the radial fibres, and the other forming a joined layer of cells secreting the fine-fibred circumpulpal matrix and becoming a pulpo-dentinal membrane controlling calcium ion flow (Linde, 1995), has never really been tested with modern molecular methods for gene expression patterns. The existence of two cell phenotypes at the pulpal surface in the teeth of the lemon shark (as mentioned in section 4.5.1) has been proposed on ultrastructural evidence, and the suggestion made that this concerns the production and maintenance of both enameloid (a product of both odontoblasts and dental epithelium) and circumpulpal dentine from this papillary layer of cells (Figure 5.6A, cell types 1 and 2; Sasagawa, 1994). It is interesting to reflect that the production of osteodentine as already outlined may involve a second cell phenotype making the intrapulpal trabecular dentine with one type of fibre arrangement, as the basis of osteodentine, and reactivation of odontoblasts to deposit the parallel-fibred circumvascular dentine of the denteones (see section 5.5.3).

In some of the early dentine types in the dermal skeleton, there are intriguing examples where two cell types, both variant end-cells of the preodontoblast, have been mentioned as a possible explanation of histology. One is the placoderm example, *Phlyctaenaspis acadica*, where Gross (1957, figure 9) has described and drawn special cells in the tubercle crown as distinct from the cells in the semidentine of the 'body' of the tubercle (Figure 5.6B, coronal odontocytes, unipolar odontocytes). Ørvig (1967, figure 36; 1980, figures 16, 17) has also suggested that the cap tissue of *Astraspis* has a row of cell lacunae above the dentine of the tubercle, although Sansom *et al.* (1997) have noted that the cell processes identified by Ørvig originate in the dentine (Figure 5.4A,B).

5.4.3.   Calcification pattern

The calcification of the organic matrix of dentine may take place in two different ways, described by Schmidt and Keil (1967, pp. 68–82) as inotropic and spheritic mineralisation. The former is when the hydroxyapatite crystals are associated with the collagen fibres so that they lie alongside, or within the molecular structure of the fibrils, the c-axes of the crystals lying parallel to the long axes of the fibres (Boyde, 1974; Höhling *et al.*, 1974). Inotropic mineralisation predominates in mantle dentine where bundles of collagen are arranged radially and larger than those of circumpulpal dentine. In this region mineralisation occurs in advance of the mineralisation front along the von Korff bundles (Boyde and Reith, 1969), the presence of which may be the reason for mantle dentine being less calcified that of the inner dentine (Mjör, 1966). Spheritic calcification is initiated from nucleating centres with a radial arrangement of crystallites, forming calcospherites. Nucleation is focused on either matrix vesicles or other matrix components, such as phosphoproteins, or a more hydrated, soluble form of calcium phosphate. The currently favoured view is that calcification of dentine is entirely through the initiation and growth of calcospherites, as far as the intertubular dentine is concerned.

The direction of the biological mineralisation is retained in fossil dentine and can be determined with polarised light microscopy (see Schmidt and Keil, 1967 for a full explanation). It is unknown as to which is the predominant method of calcification in fossil dentines, but in the earliest types, conodonts and Cambro-Ordovician fish, spheretic calcification is widespread.

Examples in human dentine of a large scale globular dentine, in which large interglobular spaces are retained as relatively unmineralised areas, are known to be associated with hypophosphataemia. It is known that the size of calcospherites increases from small bodies in the outermost dentine, to the largest, about one third towards the pulp (see Schmidt and Keil, 1967, pp. 73–81). Within each calcospherite the two patterns of mineralisation coexist, radial and collagen-orientated (Shellis, 1983). Within the calcospherites, peritubular dentine will also form and extend beyond the borders of the globule (Shellis, 1983, figures 15–16).

Peritubular dentine is a third type of mineralisation of dentine, occurring on the walls of the intertubular dentine, and formed in the ground substance of the matrix by exclusion of the organic material to become very highly mineralised. The higher mineral content of peritubular dentine is shown by secondary ion microscopy (Lefèvre et al., 1976, figures 10, 14). This technique also highlights the absence of peritubular dentine in the first calcification of predentine (Lefèvre et al., 1976, figures 2–5), and its absence from the interglobular spaces (Lefèvre et al., 1976, figures 6–9). Occasionally, in the larger interglobular regions, peritubular dentine is present in human dentine. Shellis (1983, figures 9, 16) demonstrated that peritubular dentine may exist independently of the calcospheretic area.

These patterns of mineralisation are a largely unexplored area of research in early fossil examples. We believe that peritubular dentine may also occur in the Ordovician vertebrate *Eriptychius*, where there is clearly a differentiated sheath of mineralisation surrounding the coarse dentine tubules (Figure 5.4E). Peritubular mineralisation may be the process through which some other fossil dentines appear atubular, by complete closure of the tubule, although this is somewhat speculative at present.

5.5. Functional considerations

Here we draw attention to some functional advantages of this extreme plasticity of the papillary-derived ectomesenchyme cell, producing the range of exoskeletal hard tissues in vertebrate teeth and tubercles. Two properties of the odontoblast cell should be emphasised: first, the essential ability to move away from the secreted product and remain at the pulpal surface but able to interact through numerous fine cell processes and respond to change at the outer surface; second, that of producing a hypermineralised tissue simply as a peritubular dentine around each process, but also as the more specialised petrodentine and an infilling process termed pleromic dentine (section 5.5.3). These two properties are essential features determining the process of reactionary dentine.

5.5.1. Sensory

Most textbooks on dental anatomy will state that it is difficult to explain why the dentine–pulp complex should be so sensitive, and although the dominant sensation is perceived as pain, this almost certainly was not the case early in evolution of the tissue (Halstead 1987; Tarlo, 1965). However, the mechanism responsible for this sensitivity might be of evolutionary significance for a functional role of dentine in a peripheral transduction system at the interface of the fish with its environment.

New data on mammalian dentine may provide an explanation of an ancient mechanism, with a link between the fluid flow through dentinal tubules and discharge of intradentinal nerves. Our descriptions of very fine diameter tubules of one of the early vertebrate dentines as in the Ordovician fish *Astraspis*, together with extensions into the enameloid cap (Sansom et al., 1997) may be part of a simple sensory system involving fluid movement. Also the molecular sieve properties of enamel (Chapter 17), where high osmotic pressures remove water from the pores, may occur also in enameloid and be transmitted by the enameloid tubules in continuity with dentine tubules, and registered as water efflux at the surface.

It is known from the work on cat dentine (Vonsgaven and Mathews, 1991, 1992a, b), that all stimuli producing a sensory input produce movement of the tubule contents, but new data has shown that intradentinal nerves are much more sensitive to outward than to inward flow through the dentine (Mathews and Vonsgaven, 1994). This observation forms the basis of the suggestion that velocity of fluid flow around the odontoblast process could be very great, and sufficient to cause movement of the cell body and nerve ending into the tubule, damage causing excitation of the nerve. This simple transduction mechanism may well have been in operation in the earliest forms of vertebrate dermal armour, and not a more elaborate mechanism, such as one involving stretch-sensitive ion channels, or other exquisitely sensitive receptor mechanisms (Mathews and Vonsgaven, 1994).

If sensory functions of the integument were derived from placode cells (Gans and Northcutt, 1983), then the neural crest-derived dentine, with cells derived from a primitive neural network, would provide a third sensory input from the environment, as also suggested by Northcutt and Gans (1983, p. 8). It is generally accepted that dentine has mechano-receptive properties, although the mechanism for transduction is not understood. The three main theories have been recently reviewed by Hildebrand et al. (1995). Gans (1989) suggested that odontoblasts, as early skeletogenic cells producing hydroxyapatite in the dermal armour, may have contributed to enhanced sensory capacity, through the dielectric properties of the crystallites. This allowed other electroreceptors in the armour to detect the difference between distance and nearby signals. However, mineralised tissues other than dentine, could have served this function, and may have been present in the earliest dermal armour; these possibilities are discussed in two recent papers (Smith *et al.*, 1996; Young *et al.*, 1996).

One can speculate that the original sensory role of dentine was to provide a hydrodynamic system for ion concentration detection, or osmotic pressure regulation (as suggested previously by Lumsden and Cronshaw, 1979). Alternatively, cell processes in narrow tubules may simply have been able to respond to persistent, mechanical sources of irritation and trigger the reparative response to fill in the spaces with more dentine, as discussed in section 5.5.3. It may also have been a property that any mineralised tissue shared and could respond through a piezoelectric effect to trigger further production of new, replacement units, as is known to happen in bone remodelling at all sites in the body.

Admittedly the functional advantages of the early types of dentine will remain elusive, based as it is on the fossil record. Only a few groups of extant fish retain dentine as part of their dermal skeleton, for example the chondrichthyans, coelacanth, and armoured catfishes, and there has been a general decrease in the extent of biomineralisation of the dermal armour through geological time. However, only by building scenarios based on information from the fossil record and the constraints this places on development, can we attempt to understand the evolution of function of dentine, which we assume is as diverse as the morphological types present in early vertebrates.

5.5.2.   Support

A rather more basic role of dentine could simply be for support or attachment of the whole denticle especially when free in the epithelium, as in sharks, thelodonts and conodonts, and not attached to dermal bone. Our view is that the dentine of conodonts could have had little function other than a flexible base to the lamellar crown of enamel, possibly providing attachment to the soft tissues of the oral cavity, as suggested by Smith *et al.*, (1987). The basal body of conodont elements is usually covered by a comparatively thick crown of enamel, which would presumably form a barrier to external stimuli. It is notable that the majority of post-Ordovician conodonts appear to lack tubules within their basal body, possibly suggesting that fluid flow and the transduction of movement along tubules, had become a redundant feature of this tissue. Instead, many of these tissue types have an atubular dentine incorporating what might be interpreted as attachment fibres (Smith *et al.*, 1996). Again the assumption is that the many varieties of conodont basal body are a reflection of great phenotypic plasticity of the mesench-

ymal cell type, with unconstrained developmental strategies, in the vertebrate pattern of development, arising from the novel neural crest as part of an epithelial mesenchymal, interactive, developmental tissue (Smith, 1991; Smith and Hall, 1993).

5.5.3.   Reactionary pleromic dentine

It is interesting to note that one of the evolutionarily early properties of odontoblasts is the ability to provide reactionary pleromic dentine in the soft tissue spaces in response to, or in advance of wear of the dermal armour. This function, to provide the possibility of a continuous supply of wear-resistant tissue, is also seen in human dentine with the formation of tertiary dentine (see Chapter 6).

We could speculate that there is an early molecular mechanism for dentine repair of the type that fills in soft tissue spaces beneath a wearing surface. As further abrasion removed the dentine tubercles, this reactionary dentine continued to grow invasively into all vascular spaces, to provide a minimum thickness of sclerotic dentine ahead of abrasive wear (Figure 5.7). In this way odontoblasts provide a prophylactic functional response, proposed as an activity of these cells early in the fossil record (Gross, 1935, plate II, figure 2; Ørvig, 1976, figures 10–13).

In many fossil dentines, a translucent, pleromic form grows in the soft tissue spaces between the trabeculae of spongy bone, and fills in from the functional surface inwards, to make a wear resistant tissue. Ørvig (1967) in his major review of early vertebrate hard tissues has given a comprehensive account of these pleromic hard tissues (those infilling spaces) in teeth and in the dermal armour of agnathan vertebrates (Figure 5.3F). The process of formation was analysed as early as 1930 by Gross (1930, 1935) and given the descriptive term 'massive spongiosa'. The process of prophylactic production of pleromic dentine, its development and growth, is illustrated in the dermal armour of psammosteid bone (Figure 5.7; Ørvig 1976, figures 10–13), where the process once initiated at the surface continues into the available spaces beyond the stage when the dentine tubercles are worn down (Figure 5.7C, D).

In a discussion of secondary dentine in the dermal armour of psammosteids Halstead (1969) noted that Kiaer (1915) had been the first to illustrate the response to serious abrasion. In these fossils, where the vascular spaces were open at the surface they had become filled by plugs of dentine. Halstead (1969) and Tarlo (1965) provided many examples of this tissue in the dermal armour of psammosteiform agnathans. He described one in which secondary dentine was produced in the pulp cavity and continued to invade across into the spongy bone

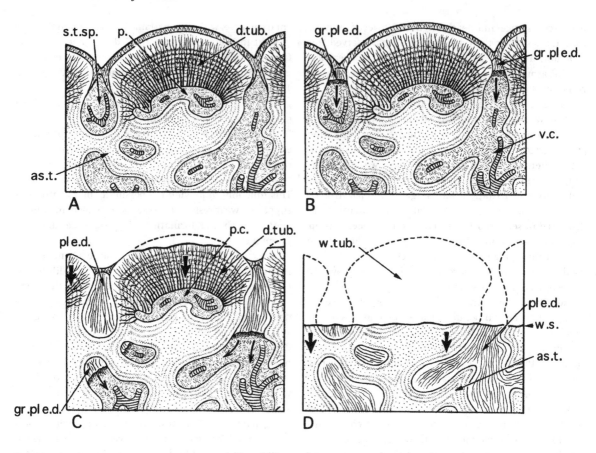

Figure 5.7A–D. **Four stages of progressive wear and pleromic infill of dentine in psammosteid dermal bone. A. Young, unworn dermal tubercles (d. tub.) above aspidin trabeculae (as. t.); open pulp cavity (p) and soft tissue spaces (s. t. sp.); B. Early stage of growth of pleromic dentine (gr. ple. d.) into vascular soft tissue (v. c.); C. Wear of tubercle and invasive growth of dentine to fill the superficial spaces; D. Complete loss of tubercle dentine (w. tub.), worn pleromic dentine and flattened bone surface (w. s.), large arrows show direction of wear (after Ørvig, 1976, figs. 10–13.).**

(Halstead, 1969, figure 16C), and compared this with the response in human teeth. It was suggested that stem cells surviving in the vascular, connective tissue at the base of the dentine tubercles were induced to become odontoblasts and deposit dentine in the available cancellar spaces of the spongy bone (Figure 5.7A, B). Schmidt and Keil (1971, p. 197, figure 148) refer to secondary dentine in *Psammolepis* and suggest that osteoblasts were transformed into odontoblasts filling the spongiosa with newly secreted dentine.

From the above summary of the early fossil literature it can be seen that the provision of a wear-resistant tissue by pleromic dentine formation is a widespread phenomenon of many fish. We cannot determine which cell type is capable of this activity in the fossil examples, but similar phenomena occur in teeth of extant forms in which cellular production of the tissues has been studied (see Smith, 1985). Although the examples given so far are in the dermal armour, there are many others of pleromic

dentine in both fossil and extant lungfish tooth plates, because these are never shed but retained throughout life. In these, pleromic dentine contributes to the functional wear-resistant surface by invasively spreading into all soft tissue spaces, deep to the functional surface. Denticle-covered bones of one of the Devonian lungfish (*Holodipterus*, Figure 5.3D; Smith, 1977) show the production of pleromic dentine in response to wear. All the soft tissue spaces in the trabecular bone are filled with this type of dentine, as the denticles have worn down.

In the light of findings on molecular mechanisms involved in pulpal repair of dentine in response to damage but not resorption (Chapter 6), it could be suggested that TGFß, sequestered in the matrix, is released as a bioactive molecule and triggers odontoblast differentiation from a stem cell population of cells (subodontoblasts). It is possible that a back-up population of cells, comparable with subodontoblasts (in that they have received the initial signal) could respond to 'stress' stimuli

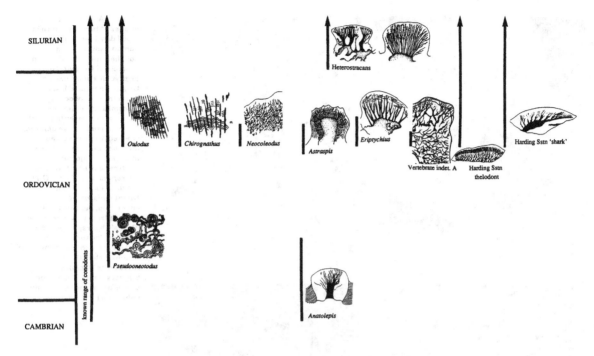

Figure 5.8. **Stratigraphic range chart of earliest vertebrates with vignettes of dentine as described from sections of dermal tubercles, denticles and conodont elements.**

to produce reactionary pleromic dentine in all these fossil examples. These precursor cells, of the same lineage as odontoblasts, are always in the pulp space and can respond to stimuli, by the conserved molecular mechanisms for odontoblast differentiation, to one required for repair, as suggested for human dentine (Chapter 6). Also, the invasive, migratory property of the stem mesenchyme cells (neural crest-derived) is implied from the many types of dermal armour in which dentine grows between the dentine tubercles and fills the subepithelial vascular spaces to reinforce the dentine surface with pleromic dentine.

From the suggestions above, it can be seen that repair and a prophylactic property are ancient and adaptive properties of odontoblasts, cells derived in embryogenesis from the neural ectoderm, which after migration to many subepithelial sites differentiate to produce a dentine phenotype.

5.6.   Conclusions and future directions

Amongst the earliest occuring vertebrates, dentine tissue is found in the dermal armour of fish and in the feeding elements of conodonts, this seems to predate bone of both acellular and cellular types. The earliest certain

dermal armour recorded is that of *Anatolepis* (Smith M. P. et al., 1996), in the Late Cambrian. Another possible, but enigmatic, Late Cambrian form may represent vertebrate dermal armour (Young et al., 1996) but its structure is problematic. The fossil record illustrates that dentine with tubules is the first recognised type of tissue (the acellular supporting tissue of *Anatolepis* is of unknown affinity at present), and that it grows from a papillary tissue (pulp) retaining process of the cell within the mineralised cell product (*Anatolepis*, Figure 5.8). A very large range of tissue types is found early on in the history of vertebrates, as an example of great diversity. This we have assumed is due to great developmental lability and conserved control mechanisms of the assumed neural crest-derived mesenchyme cells (Smith, 1991). However, it is not possible to determine the primacy of tissue types at this early time interval, because great phenotypic plasticity is apparent (see these placed within their stratigraphic context in Figure 5.8).

This great diversity of dentine in the fossil record poses a problem for meaningful discussion of the polarity of dentine characters in cladistic analysis of early vertebrate relationships. In the future a revised terminology will also be essential to avoid the ambiguity evident from the many studies of fossil dentine and to allow reliable comparison throughout the vertebrate fossil record. Corroboration of the primitive condition will depend upon

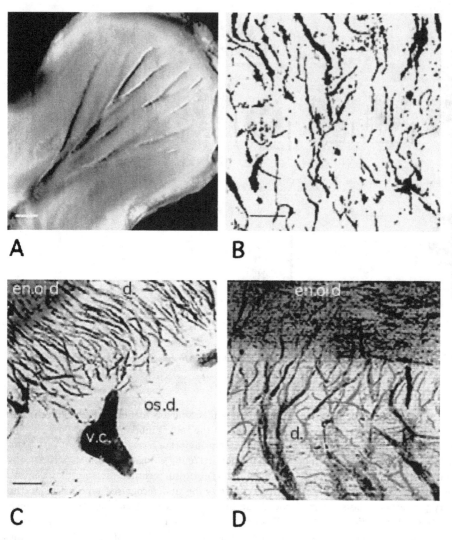

A   **B**

C   **D**

Figure 5.9A–D. **Four confocal images of tubule pattern in fossil dentines and tertiary human dentine, from cut polished surfaces. *A.* Dermal tubercle of Ordovician *Eriptychius americanus*: Odyssey confocal laser scanning microscope, joint transmitted/fluorescent image (scale bar represents 25 μm). *B.* Human reparative dentine: stacked brightest video image, focal planes at 2 μm intervals: image enhanced by contrast and blur filter (scale bar represents 20 μm). *C, D.* Tooth of Tertiary shark *Odontaspis*, osteodentine and enameloid, dentine tubules emerge from vascular canals (C, scale bar represents 50 μm); dentine tubules branch and give rise to finer tubules in the enameloid (D, scale bar represents 20 μm): stacked brightest video image, focal planes at 2 μm intervals: image enhanced by sharp filter.**

the results of ongoing cladistic analysis, including non-histological characters, of the phylogeny of early vertebrates. Only then will it be possible to propose the order of tissue evolution and to discard the false concept of a transition series, the latter influencing the names such as, mesodentine, semidentine and metadentine.

Continuing research into image analysis is providing ways of making comparisons between fossil and extant dentines in their tubule size, density and pattern, including that of irregular tertiary human dentine (Figure 5.9B). This type of dentine, often also formed within a pulp stone, is assumed to be from a triggered response by stem cells not previously engaged in dentine production. Some of these possible mechanisms are discussed in Chapter 6. The simplest type of tubule may seem to be that of the Ordovician fish *Eriptychius*, but this does not imply phylogenetic primacy (Figure 5.9A). Examples of

dentine tubules continuing beyond the dentine across the enamel dentine junction and into the most coronal layer are many, and present in the other Ordovician form *Astraspis* (Figure 5.4A, B). Those of a Triassic shark form sprays of wide tubules arising from vascular canals in the osteodentine (Figure 5.9C), these divide and branch into finer ones crossing into the enameloid (Figure 5.9D). In all cases a mineral ion supply function is assumed, but a sensory function cannot be excluded with the present state of knowledge (section 5.5.1).

We do not have a firm idea of the function of these early dentine tissues, except that in the case of dermal armour, it appears to contribute to a wear-resistant reactionary dentine, by continuous growth in advance of the loss of dentine in response to initial attritional wear. One unexplored area is to make real comparisons of tissue histology and growth between dentine of oral denticles and teeth, and that in the dermal armour of fossil verte-brates to elucidate the enigmatic question of the sensory role of this tissue.

Summary

We have attempted to place the earliest types of dentine from the fossil record within postulated scenarios of the functions of this tissue in the dermal armour of the earliest known vertebrates. The assumption is that the fundamental properties of dentine, those conserved through evolution, may be proposed by comparing fossil tissues with known structure of dentine in mammalian teeth. These properties include sensitivity; ability to become hypermineralised in response to use (reactionary dentine); continuous growth; reparative ability; and cells acting as a calcium ion modulating layer.

Descriptions of fossil dentines rely predominantly on the interpretations of the spaces left behind within the biomineralised tissue. These record biological features of the cells, cell processes, vascular tissue, fibre bundles, and pauses in growth of the mineral phase within the organic matrix. Historically, many descriptions have used terminology, either based solely on these interpreta-tions, or on comparison with bone, based on the assump-tion that an evolutionary series exists as a fossil record of the transformation of bone into dentine. Terms such as mesodentine, semidentine and metadentine have been used unrealistically in phylogenies with this assumption, and also with the concept that they are confined to specific taxa. More recent proposals of a modular developmental mechanism for the dermal skeleton and teeth in the jaws (see Smith and Coates, 1998; Chapter 10), in which in phylogeny each structure derived from developmental modules (i.e. dentine odontodes, dermal bone, cartilage) can be selectively present or absent, and differentiate independently, allow a different interpret-ation of the fossil record and clarification of terminology.

We have demonstrated that, in Cambro-Ordovician vertebrates, there is great diversity of dentine tissues, including some which are atubular. Amongst recent dis-coveries are the earliest fish with dentine preceding cellular bone, the Late Cambrian *Anatolepis*, varieties of dentine in conodont elements, together with the earliest recorded occurence of dentine in sharks and thelodonts in the Ordovician.

It is suggested that production in early psammosteid heterostracan dermal armour of a wear-resistant tissue, pleromic dentine, which grows invasively into all spaces in the spongy bone from the pulp spaces of the dentine tubercles, is an early example of reactionary dentine.

This could be produced by the same molecular mechan-isms that operate in extant mammalian dentine.

Acknowledgements

We are indebted to Valya Karatajùte-Talimaa for many discussions on the variety of dentines in early fossil vertebrates and the opportunity to examine her collec-tions of microvertebrates, and to Philippe Janvier, Daniel Goujet and Paul Smith for ongoing discussions over many years on fossil tissues in fish dermal armour. Thanks are due to Peter Pilecki and Tim Watson (Resora-tive Dentistry, KCL Guy's campus) for collaboration with the confocal microscopy and image analysis. We are especially grateful to Martha Richter and Sheila Jones for comments made on the first draft, and to my co-editor Mark Teaford for many constructive suggestions. MMS acknowledges grants from the Royal Society for two exchange visits to Lithuania, and to the Dental school at UMDS, and NERC for grants to pursue this research (GR/38543; GR3/10272, jointly held with Dr M. P. Smith at Earth Sciences, Birmingham University, and Anatomy and Cell Biology, UMDS, London University).

References

Baume, L. J. (1980). The biology of pulp and dentine, a historic, terminologic-taxonomic, histologic-biochemical, embryonic and clinical survey. In *Monographs in Oral Science*, ed. H. M. Myers, pp. 1–246. Basel: Karger.

Boyde, A. (1974). Transmission electron microscopy of ion beam thinned dentine. *Cell Tissue Research*, **152**, 543–550.

Boyde, A., and Reith, E. J. (1969). The pattern of mineralisation of rat molar dentine. *Zeittschrift fur Zellforschung Mikroscopic Anatomie*, **94**, 479–486.

Gans, C. (1987). The neural crest: a spectacular invention. In *Developmental and Evolutionary Aspects of the Neural Crest*, ed. P. F. A. Maderson, pp. 361–379. New York: Wiley.

Gans, C. (1989). Stages in the origin of vertebrates: analysis by means of scenarios. *Biological Reviews*, **64**, 221–265.

Gans, C., and Northcutt. R. G. (1983). Neural crest and the origin of the vertebrates: a new head. *Science*, **220**, 268–274.

Gross, W. (1930). Die Fische des mittleren Old Red Süd-Livlands. *Geologica et Paläeontologica*, **18**, 121–156.

Gross, W. (1935). Histologische Studien am Aussenskelett fossiler Agnathan und Fische. *Paleontographica*, **83**, 1–60.

Gross, W. (1957). Mundzähne und Hautzähne der Acanthodier und Arthrodiran. *Paleontographica*, **109A**, 1–40.

Halstead, L. B. (1969). Calcified tissues in the earliest vertebrates. *Calcified Tissue Research*, **3**, 107–124.

Halstead, L. B. (1987). Evolutionary aspects of neural crest-derived skeletogenic cells in the earliest vertebrates. In *Developmental and Evolutionary Aspects of the Neural Crest.*, ed. P. F. A. Maderson, pp. 339–358. New York: Wiley.

Höhling, H. J., Ashton, B. A. and Köster, H. D. (1974). Quantitiative electron microscopic investigations of mineral nucleation in collagen. *Cell Tissue Research*, **148**, 11–26.

Hildebrand, C., Fried, K., Tuisku, F. and Johansson, C. S. (1995). Teeth and tooth nerves. *Progress in Neurobiology*, **45**, 165–222.

Janvier, P. (1996). *Early vertebrates.* Oxford monographs in geology and geophysics, 33. New York: Oxford University Press.

Karatajùte-Talimaa, V. N. (1995). The Mongolepidida: scale structure and systematic position. *Geobios* **19**, 35–37.

Karatajùte-Talimaa, V. N. and Novitskaya, L. I. (1992). *Teslepis*, a new mongolepid elasmobranchian fish from the Lower Silurian of Mongolia. *Paleontologicheskii Zhurnal*, **4**, 36–47.

Karatajùte-Talimaa, V. N., Novitskaya, L. I., Rozman, Kh. S., and Sodov, Zh. (1990). *Mongolepis*, a new elasmobranch genus from the Lower Silurian of Mongolia. *Paleontologicheskii Zhurnal*, **1**, 76–86.

Kiaer, J. (1915). Upper Devonian fish remains from Ellesmere Land, with remarks on *Drepanaspis*. *Second Report Norwegian expedition 'Fram' 1899–1902*, **33**, 1–72.

Lefèvre, R., Frank, R. and Voegel, J. C. (1976). The study of human dentine with secondary ion microscopy and electron diffraction. *Calcified Tissue Research*, **19**, 251–261.

Lester, K. S. and Boyde, A. (1968). The question of von Korff fibres in mammalian dentine. *Calcified Tissue Research*, **1**, 273–287.

Linde, A. (1995). Dentine mineralisation and the role of odontoblasts in calcium transport. *Connective Tissue Research*, **33**, 163–170.

Lison, L. (1941). Recherches sur la structure et l' histogenèse des dents des Poissons Dipneustes. *Archives du Biologie, Paris*, **52**, 279–320.

Lumsden, A. G. S. (1987). The neural crest contribution to tooth development in the mammalian embryo. In *Developmental and Evolutionary Aspects of the Neural Crest*, ed. P. F. A. Maderson, pp. 261–300. New York: Wiley.

Lumsden, A. G. S. and Cronshaw, M. A. (1979). Dentine sensitivity. *Guy's Hospital Gazette*, 29 September, 374–378.

Mathews, B. and Vonsgaven, N. (1994). Interactions between neural and hydrodynamic mechanisms in dentine and pulp. *Archives of Oral Biology*, **39**, 875–955.

Mjor, I. A. (1966). Microradiography of human cornal dentine. *Archives of Oral Biology*, **11**, 225–234.

Moss, M. L. (1968). The origin of vertebrate calcified tissues. In *Current Problems of Lower Vertebrate Phylogeny*, ed. T. Ørvig, pp. 359–371. Stockholm: Almquist & Wiskell.

Northcutt, R. G., and Gans, C. (1983). The genesis of neural crest and epidermal placodes: a reinterpretation of vertebrate origins. *Quarterly Review of Biology*, **58**, 1–28.

Ørvig, T. (1951). Histologic studies of Placoderms and fossil Elasmobranchs. I. The endoskeleton, with remarks on the hard tissues of lower vertebrates in general. *Arkiv för Zoologie*, **2**, 321–454.

Ørvig, T. (1967). Phylogeny of tooth tissues: evolution of some calcified tissues in early vertebrates. In *Structural and Chemical Organization of Teeth*, ed. A. E. W. Miles, vol. 1, pp. 45–110. London: Academic Press.

Ørvig, T. (1976). Paleohistological notes. 3. The interpretation of pleromin (pleromic hard tissue) in the dermal skeleton of psammosteid heterostracans. *Zoologica Scripta*, **5**, 35–47.

Pander, C. H. (1857). *Über die Placodermen des Devonischen (1857) Systems*, pp. 1–106. St Petersburg.

Peyer, B. (1968). *Comparative Odontology*, pp. 1–347; Plates 1–88 and I–VIII. Chicago: University of Chicago Press.

Raff, R. A. (1994). Developmental mechanisms in the evolution of animal form: origins and evolvability of body plans. In *Early Life on Earth*, Nobel Symposium No. 84, ed. S. Bengtson, pp. 489–500. New York: Columbia University Press.

Reif, W.-E. (1982). Evolution of dermal skeleton and dentition in vertebrates: the odontode-regulation theory. *Evolutionary Biology*, **15**, 287–368.

Sansom, I. J. (1996). *Pseudooneotodus*: a histological study of an Ordovician to Devonian vertebrate lineage. *Zoological Journal of the Linnean Society*, **118**, 47–57.

Sansom, I. J., Smith, M. P. and Smith, M. M. (1994). Dentine in conodonts. *Nature*, **368**, 591.

Sansom, I. J., Smith, M. M. and Smith, M. P. (1996). Scales of thelodont and shark-like fishes from the Ordovician of Colorado. *Nature*, **379**, 628–630.

Sansom, I. J., Smith, M. P., Smith M. M. and Turner, P. (1997). *Astraspis* : the anatomy and histology of an Ordovician fish. *Palaeontology*, **40**, 625–643.

Sasagawa, I. (1994). The fine structure of the dentine matrix and mineralization during dentinogenesis in sting rays, elasmobranchs. *Bulletin de Institute Océanaographique, Monaco*, **14**, 167–174.

Schmidt, W. J. and Keil, A. (1967). *Polarizing Microscopy of Dental Tissues*. Oxford: Pergamon Press.

Schultze, H.-P. (1969). Die Faltenzähne der rhipidistiiden Crossopterygier, der Tetrapoden un der Actinopterygier-Gattung *Lepisosteus*, nebst einer Beschreibung der Zahnstruktur von *Onychodus* (struniiformer Crossopterygier). *Palaeontographica Italiana*, **65**, 63–136.

Shellis, R. P. (1983). Structural organisation of calcospherites in normal and rachitic human dentine. *Archives of Oral Biology*, **28**, 85–95.

Smith, M.M. (1977). The microstructure of the dentition and dermal ornament of three dipnoans from the Devonian of Western Australia. *Philosophical Transactions of the Royal Society of London Series B*, **281**, 29–72.

Smith, M. M. (1984). Petrodentine in extant and fossil dipnoan dentitions: microstructure, histogenesis and growth. *Proceedings of the Linnean Society NSW.*, **107**, 367–407.

Smith, M. M. (1985). The pattern of histogenesis and growth of tooth plates in larval stages of extant lungfish. *Journal of Anatomy*, **140**, 627–643.

Smith, M. M. (1991). Putative skeletal neural crest cells in early Late Ordovician vertebrates from Colorado. *Science*, **251**, 301–303.

Smith, M. M. and Coates, M. I. (1998). Evolutionary origins of the vertebrate dentition: phylogenetic patterns and developmental evolution. *European Journal of Oral Sciences*, **106**, 482–500.

Smith, M. M. and Hall, B. K. (1993). A developmental model for evolution of the vertebrate skeleton and teeth: the role of cranial and trunk neural crest. *Evolutionary Biology*, **27**, 387–448.

Smith, M. M. and Sansom, I. J. (1997). Exoskeletal remains of an Ordovician fish from the Harding Sandstone, Colorado. *Palaeontology*, **40**, 645–658.

Smith, M. M., Sansom, I. J. and Smith, M. P. (1996). Teeth before armour: the earliest vertebrate mineralised tissues. *Modern Geology*, **20**, 303–319.

Smith, M. P., Briggs, D. E. G. and Aldridge, R. J. (1987). A conodont animal from the Lower Silurian of Wisconsin, U.S.A., and the apparatus architecture of panderodontid conodonts. In *Palaeobiology of Conodonts*, ed. R. J. Aldridge, pp. 91–104, British Micropalaeontology Society Series. Chichester: Ellis Horwood.

Smith, M. P., Sansom, I. J. and Repetski, J. E. (1996). Histology of the first fish. *Nature*, **380**, 702–704.

Tarlo. B. J. (1965). The origin of sensitivity in dentine. *Royal Dental Hospital Magazine, London, N.S.*, **1**, 11–13.

Tomes, C. S. (1904). *A Manual of Dental Anatomy*, sixth edition. London: Churchill.

Vonsgaven, N. and Mathews, B. (1991). The permeability of cat dentine *in vivo* and *in vitro*. *Archives of Oral Biology*, **36**, 641–646.

Vonsgaven, N. and Mathews, B. (1992a). Fluid flow through cat dentine *in vivo*. *Archives of Oral Biology*, **37**, 175–185.

Vonsgaven, N. and Mathews, B. (1992b). Changes in pulpal flow and in fluid flow through dentine produced by autonomic and sensory nerve stimulation in the cat. *Proceedings of the Finnish Dental Society*, **88** (Suppl. 1), 491–497.

Young, G. C., Karatajùte-Talimaa, V. and Smith, M. M. (1996). A possible Late Cambrian vertebrate from Australia. *Nature*, **383**, 810–812.

6 Pulpo-dentinal interactions in development and repair of dentine

A. J. Smith

6.1. Introduction

Dentine and pulp are developmentally and functionally related to such an extent that they should be considered together as a complex. This interrelationship becomes of critical importance both during physiological development and maintenance of the tissues and also, in tissue repair after injury. As our understanding of the molecular and cellular events during tissue formation and repair expands, it is becoming increasingly apparent that many of the events of tissue formation may be mirrored during repair after injury of the tissue (Lesot et al., 1993, 1994). Thus, elucidation of these events becomes a prerequisite for development of new, effective treatment modalities to effect tissue repair in the dentine–pulp complex after injury.

6.2. Pulp–dentine complex: from development to repair

The early stages of tooth initiation give rise to the enamel organ and dental papilla of the developing tooth germ. As development proceeds, epithelial-mesenchymal interactions between tissue compartments control tooth morphogenesis and cytodifferentiation (Kollar, 1983; Ruch, 1987; Slavkin, 1990; Thesleff et al., 1991). A number of aspects of these processes during the earlier stages of tooth development are dealt with in the first section of this volume (e.g. Chapter 3). These processes lead to a basement membrane being interposed between the inner dental epithelium and mesenchymal cells of the dental papilla that will give rise to the odontoblasts responsible for secretion of dentine. This basement mem-

brane has been hypothesized to act as a mediator of the epithelial-mesenchymal interactions which control odontoblast differentiation (Thesleff, 1978; Ruch et al., 1982; Ruch, 1987; Thesleff and Vaahtokari, 1992). It is interesting to suggest that the unmineralised matrix at the pulp–dentine interface might fulfill a similiar role post-developmentally as equivalent to a basement membrane, or pulpo-dentinal membrane, for both the maintenance of the odontoblast phenotype and mediation of tissue repair. Our present knowledge of odontoblast differentiation suggests control by a functional network consisting of matrix molecules and growth factors. Epithelially derived growth factor molecules, sequestered by basement membrane-associated components may interact with dental papilla cells and regulate the transcription of genes encoding for homeoproteins such as msx2 and proteins belonging to the TGFβ family (Bègue-Kirn et al., 1994). The former may in turn modulate the expression of genes involved in cytoskeletal assembly, whilst the latter may regulate matrix production by the cell.

During primary dentinogenesis, much of the circumpulpal dentine of the tooth will be laid down. Concomitant with this will be the development of the vasculature and neural network of the pulp–dentine complex, the majority of which is confined to the pulpal compartment. Whilst much attention has focussed on the differentiation and subsequent behaviour of the odontoblasts, little is known of the processes responsible for development of the vasculature and nerves of the pulp–dentine complex. Morphological studies, however, have demonstrated that a rich capillary plexus and nerve plexus exists in the subodontoblast/odontoblast 'odontogenic' zone of the active dentine–pulp complex. Certainly, there appears to be a relationship between dentinogenic activity and the pulpal vasculature and nerves, since ageing leads to a reduction in vascular supply (Bernick, 1967a) and both a loss and a degeneration of myelinated and unmyelinated axons (Bernick, 1967b).

There also exists a relationship between the morphological appearance of the odontoblast and its synthetic and secretory activity. The odontoblast is a post-mitotic cell and, unless damaged as a result of tissue injury, will exist for the functional life of the tooth. However, its morphological appearance can vary widely throughout its life cycle (Takuma and Nagai, 1971; Fox and Heeley, 1980; Couve, 1986; Romagnoli et al., 1990) and under physiological conditions can be related to the secretory activity of the cell. A high secretory activity for the odontoblast is observed during primary dentinogenesis, but becomes much slower during physiological secondary dentine formation (Baume, 1980) after completion of root formation.

Tertiary dentine represents a matrix laid down at specific foci of the pulp–dentine interface in response to environmental stimuli. Whilst secretion of a tertiary dentine matrix can occur at quite high rates (Baume, 1980), there appears to be far less of a relationship between odontoblast morphology and secretory behaviour which is perhaps a reflection of the pathological conditions pertaining. However, this may in part be a consequence of the broad spectrum of circumstances which gives rise to tertiary dentinogenesis. To clarify our understanding of tertiary dentinogenesis, we have proposed the following definitions (Lesot et al., 1993; Smith et al., 1994):

> Reactionary Dentine: a tertiary dentine matrix
> secreted by surviving post-mitotic odontoblasts
> in response to an appropriate stimulus
>
> Reparative Dentine: a tertiary dentine matrix
> secreted by a new generation of odontoblast-like
> cells in response to an appropriate stimulus, after
> the death of the original post-mitotic
> odontoblasts responsible for primary and
> physiological secondary dentine secretion.

The fundamental difference between these two variants of tertiary dentine is in the cells responsible for their synthesis, which will reflect the intensity of the stimulus responsible for their production (Figure 6.1). With strong stimuli, odontoblast death occurs and tertiary dentinogenesis (reparative) requires recruitment and differentiation of a new generation of odontoblast-like cells from pulpal precursor cells before tissue repair can take place. Reactionary dentinogenesis, however, represents modulation of activity of the existing odontoblasts and thus, perhaps might be considered to lie somewhere between the physiological and pathological ends

TERTIARY DENTINOGENESIS

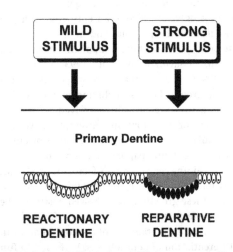

Figure 6.1. **Schematic illustration of differing responses of tertiary dentinogenesis. Reactionary and reparative dentines secreted by surviving post-miotic odontoblasts (white) and new odontoblast-like cells (black) respectively.**

of the spectrum of the pulp–dentine response. Such a view might concur with its formation in response to attrition, which could be regarded as a consequence of normal function.

6.3.   Development and maintenance of the pulp-dentine complex

In discussing the development and maintenance of the pulp–dentine complex, emphasis will focus predominantly on the odontoblasts which are responsible for the secretion of dentine. However, the importance of other components of the complex must be recognised, especially the vasculature in providing nutrition for dentinogenesis and the fibroblasts and other cells which help to support the activity of odontoblasts.

6.3.1.   Epigenetic signalling
It is now recognised that specific epigenetic signals trigger odontoblast differentiation. Isolation of the dental papilla from such epigenetic signals precludes the opportunity for odontoblast differentiation (Kollar and Baird, 1970; Ruch and Karcher-Djuricic, 1971a). The importance of a stage-space-specific inner dental epithelium and basement membrane in the epithelial-

mesenchymal interactions controlling odontoblast differentiation and the possible role of this basement membrane in sequestering growth factor molecules has been discussed above. However, it is interesting that traditional views of the main basal lamina components, e.g., type IV collagen, laminin, nidogen and heparan sulphate proteoglycans, having an epithelial origin are being modified. Both epithelial and mesenchymal cells appear to be able to contribute components of embryonic basal laminae (Marinkovich et al., 1993; Thomas and Dziadek, 1993). A functional network of these components and growth factors, including members of the TGFβ superfamily, are probably involved in terminal differentiation of odontoblasts. Attempts have been made to simulate these developmental events by culture of isolated mouse dental papillae in a semi-solid agarose medium containing active growth factors (TGFβ1, BMP2 and IGF1) immobilised with heparin (Bègue-Kirn et al., 1992, 1994). These growth factors showed positive, albeit differential, effects on odontoblast cytodifferentiation. Whilst TGFβ1 induced gradients of cytological and functional differentiation of odontoblasts, BMP2 led to functional differentiation in restricted areas and IGF1 allowed cytological differentiation in the absence of apical matrix secretion. Comparison of gene expression during this in vitro odontoblast differentiation with that seen in vivo indicated that TGFβ1 and BMP2 induced similiar patterns of expression (Bègue-Kirn et al., 1994). These studies suggested that upregulation of msx2 transcription together with upregulation of members of the TGFβ superfamily are important events in the cytological and functional differentiation of odontoblasts. Upregulation of members of the TGFβ superfamily may be very pertinent to subsequent pulpo-dentinal interactions since the dentine matrix would essentially sequester this pool of growth factors and provide a possible reservoir of these molecules that could participate in cell–matrix interactions.

6.3.2.   Cell competence

The events during tooth development which lead to odontoblast differentiation are dependent on the epigenetic signalling discussed above, but also require the presence of cells of the dental papillae which are capable of responding to the appropriate signals. Studies of ^{3}H-thymidine incorporation (Ruch and Karcher-Djuricic, 1971a, b; Ruch et al., 1976) suggest that preodontoblasts must undergo a minimum number of cell cycles before they are competent to respond to specific epigenetic signals leading to odontoblast differentiation. It has been hypothesised (Ruch, 1990, 1995) that cell kinetics may determine the progressive emergence of competency amongst the preodontoblast cell population. Such competency on the part of precursor cells will also be critical for the pulp–dentine complex in a reparative situation. Under such circumstances, induction of differentiation of new odontoblast-like cells will occur from the pulpal cell population. In younger teeth, a population of undifferentiated mesenchymal cells (Höhl's cells) derived from the same cells as those which go on to become odontoblasts, can be found in the subodontoblast area of the pulp. Presumably, such cells would have the necessary competence to respond to inductive stimuli leading to differentiation of odontoblast-like cells. Whilst anecdotal, these cells are considered to reduce in number in older teeth and any precursor cells for odontoblast differentiation during repair would have to be drawn from elsewhere in the pulpal cell population.

6.3.3.   Maintenance of odontoblast phenotype

Following odontoblast differentiation, maintenance of the phenotypic morphology of the cell requires contact between the odontoblasts and their extracellular matrix. During culture of odontoblasts, their phenotypic morphology will only be preserved as long as contact is maintained between the cells and the dentine matrix (Munksgaard et al., 1978; Heywoord and Appleton, 1984). The requirement for intact dentine matrix tissue, however, can be fulfilled by isolated dentine extracellular matrix components (Lesot et al., 1986). Dissociated dental papillae cultured on Millipore filters coated with these isolated dentine extracellular matrix components showed maintenance of odontoblast differentiation over the 4-day period. Whilst these observations imply that cell–matrix interactions between the odontoblasts and dentine matrix are required for maintenance of the odontoblast phenotype, the nature of the interactions remains unclear. It might be reasonable to speculate, however, that the upregulation of expression of various genes including those for the TGFβ superfamily upon odontoblast differentiation might lead to accumulation of growth factors and other molecules within the dentine extracellular matrix, which may be associated with these cell–matrix interactions. Thus, pulpo-dentinal interactions might play key roles both in maintenance of the odontoblast phenotype and also, tissue repair leading to differentiation of new odontoblast-like cells.

6.4.   Repair of the pulp–dentine complex

6.4.1.   Matrix-bound bioactive molecules as epigenetic signals

The importance of specific epigenetic signals in triggering odontoblast differentiation has been identified above and has highlighted the important role that growth factors can play. Homologies between these growth factors and those involved in signalling during development of the *Drosophila* embryo indicate that there may be considerable evolutionary conservation in molecular mechanisms of morphogenetic signalling (Thüringer and Bienz, 1993). Conservation in molecular mechanisms involved in odontoblast differentiation during development and repair would also, therefore, appear to be a possibility.

Finkelman *et al.* (1990) have reported the presence of several growth factors in human dentine extracellular matrix, including TGFβ, and have hypothesized that their release from dentine during demineralisation by bacterial acids might invoke the response of odontoblasts to dental caries. Several approaches have implicated members of the TGFβ superfamily in odontoblast differentiation during development. Isolated rabbit dentine extracellular matrix components were able to initiate functional differentiation of odontoblasts in dissociated mouse dental papillae cultured in a semi-solid agarose medium, which was blocked by incubation of the active matrix component fractions with a neutralising anti-TGFβ antibody (Bègue-Kirn *et al.*, 1992). Recombinant TGFβ1 could substitute for the isolated dentine matrix components in the initiation of odontoblast differentiation as long as it was immobilised by heparin or other inactive matrix components (Bègue-Kirn *et al.*, 1992) implying a functional network of growth factors and matrix molecules during odontoblast differentiation. Both immunolocalization and *in situ* hybridisation have also implicated the TGFβ superfamily in odontoblast differentiation: TGFβ1, 2 and 3 (Cam *et al.*, 1990; D'Souza *et al.*, 1990; Pelton *et al.*, 1990, 1991; Millan *et al.*, 1991; Wise and Fan, 1991; Thesleff and Vaahtokari, 1992; Heikinheimo *et al.*, 1993) and BMP2, 4 and 6 (Lyons *et al.*, 1990; Vainio *et al.*, 1993; Heikinheimo, 1994). Upregulation of expression of some of the members of the TGFβ superfamily by odontoblasts after differentiation (Bègue-Kirn *et al.*, 1994) would allow these molecules to become sequestered within the dentine extracellular matrix. Thus, the concept of dentine matrix sequestering and acting as a reservoir of growth factors, especially of the TGFβ superfamily, and their subsequent release during tissue injury leading to induction of repair in the pulp–dentine complex is well founded.

Such a reservoir of growth factors within the matrix would explain the inductive effect of native dentine, demineralised dentine and dentine extracellular matrix components in dental tissue repair. As early as 1957, Seltzer and Bender reported that when dentine chips, produced as a result of operative procedures, were accidentally displaced into the pulp tissue, they served as stimuli for the induction of new matrix deposition on the surface of the chips. Such fragments of dentine can often be observed to be surrounded by odontoblast-like cells that secrete a tubular dentine matrix (Héritier *et al.*, 1990; Mjör *et al.*, 1991; Tziafas *et al.*, 1993). As a result of experiments involving implantation of trypsin-isolated dental papillae into empty pulp chambers of crown molars and subsequently transplanting the teeth into subcutaneous tissue of the ear in new born mice, Héritier *et al.*, (1990) proposed that a factor with analogous properties to those in the dental basement membrane was present in dentine.

The bio-active nature of dentine matrix was investigated further by Smith *et al.* (1990) when lyophilised preparations of isolated dentine matrix components were implanted into exposed cavities prepared in ferret canine teeth. *In vivo* morphogenetic activity was demonstrated for both EDTA-soluble matrix components and also, non-collagenous components released from the insoluble dentine matrix after collagenase digestion. Differentiation of a new generation of columnar, polarized odontoblast-like cells was induced in direct association with the existing insoluble matrix with secretion of a tubular reparative dentine matrix. This response was one of reparative dentinogenesis since the pulpal exposures had resulted in odontoblast loss at the injury sites. Whilst there was some evidence that the preparations of implanted dentine matrix components were able to diffuse around the injury site, in all cases tissue repair was seen to occur in association with insoluble dentine matrix, either the residual dentine beneath the cavity or displaced dentine fragments.

However, the role of the insoluble matrix appears to be more than one of simple immobilisation. When similiar implantation experiments were performed but with implantation of the dentine matrix components immobilised on cellulose membranes, their biological effects on the induction of reparative dentinogenesis were blocked (Smith *et al.*, 1995b). Thus, the existing insoluble dentine matrix may provide a substrate for immobilisation and also potentiation of bioactive molecules in the matrix. Synergistic interactions between the extracellular matrix and growth factors have been suggested (Schubert, 1992). Both heparin (McCaffrey *et al.*, 1992) and proteoglycans (Ruoslahti and Yamaguchi, 1991) have been reported to potentiate and modulate the activity of TGFβ and other growth factors. Such interactions may well play a role in the induction of odontoblast-like cell differentiation. In examining the association between TGFβ1 and dentine extracellular matrix, Smith *et al.* (1998) have shown that whilst some of the growth factor is present in active form, free within the matrix, a greater proportion is in association with molecules such as beta-glycan (Type III receptor) and latency-associated peptide (TGFβ LAP). Thus, the association between TGFβ1 and the extracellular matrix may be significant in both the presentation and regulation of the activity of the growth factor.

The studies of Finkelman et al. (1990) on growth factors in dentine examined only the soluble tissue compartment of the dentine and used a bioassay that could not distinguish the different isoforms of TGFβ. Subsequently, both the soluble and insoluble tissue compartments of both rabbit and human dentine matrices have been analysed for TGFβ1, 2 and 3 using an ELISA (Cassidy et al., 1997). This has shown that some species differences appear to exist in terms of expression of TGFβ isoforms, although TGFβ1 was the predominant isoform in both species examined. Importantly, TGFβ1 was present in both the soluble and insoluble tissue compartments of the dentine matrix in similar amounts. This may be pertinent in consideration of tissue repair within the pulp–dentine complex since immobilisation of the TGFβ in the insoluble matrix may facilitate its presentation to cells involved in reparative processes at the pulp–dentine interface.

Attempts have been made to induce reparative dentinogenesis in pulp-capping situations using BMPs, which are members of the TGFβ superfamily of growth factors. Implantation of either crude BMP (Nakashima, 1990; Lianjia et al., 1993), recombinant BMP2 and 4 (Nakashima, 1994), or recombinant osteogenic protein-1 (BMP7; Rutherford et al., 1993), at sites of dental tissue injury led to repair, albeit with an initially non-physiological dentinogenic response. Generally, the initial response was one of secretion of an osteo-dentine, or fibro-dentine matrix, which was then followed by secretion of a tubular dentine matrix. This initial response of fibro-dentine or osteo-dentine has also been observed frequently in pulp-capping situations with calcium hydroxide (Baume, 1980; Yamamura, 1985) and dentine preparations (Tziafas et al., 1992). The atubular fibro- or osteo-dentine reparative matrix may provide a temporary sealing of the pulp to allow healthy conditions to prevail and as such, may represent a non-dentinogenic response. However, the atubular fibro-dentine matrix might also provide a substratum on which secondary induction of true reparative dentinogenesis could occur leading to differentiation of a new generation of odontoblast-like cells and secretion of tubular reparative dentine (Baume, 1980). Ruch (1985) has hypothesized a dynamic role for the fibro-dentine matrix on pulpal cells, analogous to that of the basement membrane during tooth development. This concurs with the view postulated by Baume (1980) that the fibro-dentine exerted an epigenetic influence over the differentiating cells lining the fibro-dentine. It is still unclear as to why true reparative dentinogenesis with secretion of a tubular matrix can occur directly in some situations of tissue repair, but proceeds indirectly through the intermediary secretion of a fibro-dentine matrix in others. It could be speculated that initial fibro-dentine matrix secretion occurs when the exisiting dentine matrix has become modified in some way such as to lose its bioactive properties and acts as an insoluble substrate which can sequester appropriate epigenetic factors to influence subsequent induction of true reparative dentinogenesis. It may also represent a non-dentinogenic response being secreted by cells not of odontoblastic lineage and thus of different composition.

6.4.2. Pulpal precursor cells and reparative dentinogenesis

Reparative dentinogenesis involves the differentiation of a new generation of odontoblast-like cells from pulpal precursor cells to replace those lost as a result of tissue injury. Whilst it has been possible to identify possible epigenetic signals which might trigger this process, it is also necessary to have cells present in the vicinity of the tissue injury which are competent to respond to these epigenetic signals. The role of cell kinetics in allowing cells to achieve competency for odontoblast differentiation to take place during development (Ruch, 1990, 1995) has already been discussed. However, in the mature pulp the cell population is far more heterogeneous than in the dental papilla during development. Thus, it is necessary to understand which particular type(s) of cells in the pulpal population can give rise to a new generation of odontoblast-like cells after injury and also, which cells are competent to respond to an appropriate epigenetic signal.

Undifferentiated mesenchymal cells in the subodontoblastic cell-rich zone (Höhl's cells) may be derived from the dividing pre-odontoblasts before their terminal differentiation into odontoblasts. Some of these cells will have been subjected to the same experience as the original post-mitotic odontoblasts of the pulp–dentine complex except for the epigenetic trigger to initiate odontoblast differentiation. Thus, if an appropriate epigenetic signal is provided by the dentine matrix after tissue injury, these cells would be well placed to respond and differentiate into odontoblast-like cells. It has been suggested that these cells may be involved in odontoblast-like cell differentiation when they could be stimulated by specific molecular signals without any replication of their DNA (Cotton, 1968; Takuma and Nagai, 1971; Slavkin, 1974; Torneck and Wagner, 1980). If these cells are precursors of odontoblast-like cells, then their ability to do so will depend on their vitality. Whilst

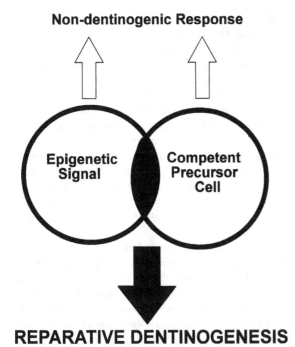

Non-dentinogenic Response

Epigenetic Signal

Competent Precursor Cell

REPARATIVE DENTINOGENESIS

Figure 6.2. **Schematic illustration of the importance of both an appropriate epigenetic signal and a competent precursor cell to induce a specific response of reparative dentinogenesis.**

odontoblasts are recognised as post-mitotic cells, which are maintained for the life of the tooth, little is known of the life span of the undifferentiated mesenchymal cells in the Höhl's cell-rich zone and the influence of processes such as apoptosis. However, a reduction in the number of cells is seen in the pulp with ageing (Symons, 1967) and this would correlate with the anecdotal clinical reports that dental tissue repair is more difficult to effect in older teeth.

An origin for odontoblast-like cells from other cells in the core of the pulp, including undifferentiated mesenchymal cells, fibroblasts or vascular cells, is feasible where competency might be achieved after cell division (Sveen and Hawes, 1968; Feit et al., 1970; Fitzgerald, 1979; Yamamura et al., 1980; Yamamura, 1985). Whilst no definitive data exists on the precise origin of odontoblast-like cells, it seems probable that many of these different cells may be able to participate in the overall reparative processes after tissue injury. However, the specificity of the dentinogenic response may well be determined by the nature of the cell responding as well as the epigenetic signal. This would help to explain the diversity of reparative matrices, e.g. fibro- and osteo-dentines, often observed in different studies during initial stages of repair of the pulp–dentine complex. It could be speculated that a true dentinogenic response may only be seen when

odontoblast-like cells are derived from the undifferentiated mesenchymal cells of Höhl's cell-rich zone, whilst other cells participating in the reparative process give rise to non-specfic matrices with atubular or poorly tubular structure. In fact, it is possible that contributions from a heterogeneous mixture of the cells may be responsible for secretion of a reparative matrix. Thus, the induction of true reparative dentinogenesis may depend both on the availability of appropriate epigenetic signals and precursor cells of specific lineage (Figure 6.2).

6.5. Physiological versus pathological responses of the dentine–pulp complex

The pulp–dentine complex can respond to environmental stimuli by secretion of a tertiary dentine matrix beneath sites of stimuli. Whilst reparative dentinogenesis clearly represents a pathological response because of odontoblast loss and differentiation of new odontoblast-like cells, reactionary dentinogenesis perhaps lies between this and the normal physiological behaviour of odontoblasts. As defined earlier, reactionary dentine is secreted by surviving post-mitotic odontoblasts in response to an appropriate stimulus after injury and thus represents modulation of activity of the existing odontoblasts. Whether a tertiary dentine matrix can be defined as reactionary or reparative in nature depends on knowledge of the chronological events within the tissue after injury. Whilst many studies on the reaction of the pulp–dentine complex to dental materials have concluded that tertiary dentine formation occurs as a reaction to these materials, the absence of information on its formation precludes identification of the processes involved. However, time-course studies with animal models can provide valuable information on the processes taking place during reactionary dentinogenesis.

Smith et al. (1994) implanted lyophilised preparations of isolated dentine matrix components (the same fractions used for study of reparative dentinogenesis described above) in unexposed cavities in ferret teeth, which gave rise to stimulation of odontoblast synthetic and secretory activity. Significant deposition of reactionary dentine at the pulp–dentine interface could be observed as early as 14 days after implantation and was restricted to those areas of the odontoblast layer where the cells were in direct contact with the cavity through their dentinal tubules. The absence of odontoblast cell death was verified by examination of early periods of implantation. It, therefore, appears that reactionary dentinogenesis can result from interaction between existing

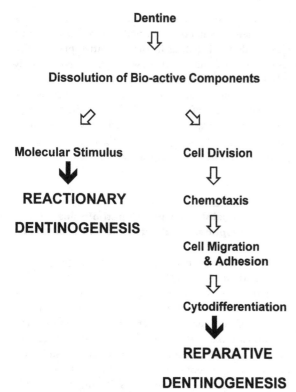

Figure 6.3. Schematic illustration of the biological events underlying reactionary and reparative dentinogenesis.

odontoblasts and appropriate molecular stimuli leading to an increase in odontoblast synthetic and secretory activity. To investigate the nature of the molecular stimulus, the active preparation of isolated dentine matrix components was further purified by affinity chromatography on heparin-agarose. The bound material, eluting with a similiar affinity to members of the TGFβ family, was capable of stimulating a strong response of reactionary dentinogenesis in the unexposed ferret tooth cavity model (Smith *et al.*, 1995*a*). Analysis of the purified dentine matrix fraction showed it to contain approximately 100 pg TGFβ1 per implantation site.

This response of the odontoblasts to a preparation containing TGFβ1 is not dissimilar to that seen when mouse dental papillae were cultured in the presence of TGFβ1 in the absence of potentiating molecules, such as heparin, and stimulation of extracellular matrix secretion by these cells was observed (Bègue-Kirn *et al.*, 1992). TGFβ and related members of this family of growth factors are known to influence matrix synthesis and have been shown to modulate the synthesis of collagens (Ignotz and Massagué, 1986; Takuwa *et al.*, 1991; Yu *et al.*, 1991) and proteoglycans (Rapraeger, 1989; Yu *et al.*, 1991). Whilst it appears that members of the TGFβ family may

be able to stimulate both reactionary and reparative dentinogenesis, the nature of the processes involved may differ markedly. During reactionary dentinogenesis, there appeared to be direct stimulation, or modulation of the odontoblasts, where the growth factor could diffuse down the dentinal tubules and interact with the cell. With reparative dentinogenesis, there will be a cascade of events involving cell division, chemotaxis, cell migration, cell adhesion and cytodifferentiation and a synergistic association, or functional network, of growth factor and extracellular matrix components seems necessary to initiate this cascade. However, in both cases dissolution of growth factors and matrix components from dentine after injury might initiate subsequent processes (Figure 6.3).

A response of reactionary dentinogenesis can be observed in human teeth subject to attrition, which might be regarded as a consequence of normal tissue function. Worn rodent teeth have been described as containing osteo- and secondary dentine as long ago as 1850 (Tomes, 1850) and in a comparative light and electron microscopic study, Takuma *et al.*, (1968) have described the tip of the erupting rat incisor as comprising fibro- or osteodentine. However, it is contentious as to whether this should be regarded as a reactionary dentine since it appears to arise initially as a result of impaired devlopment of the enamel organ in this area of the tooth. With time, functional wear of the occlusal tip of the rat incisor will result in loss of tissue at this site and the fibrodentine matrix deposited is likely to be the result of reparative rather than reactionary dentinogenesis. It is interesting to consider, however, what processes might give rise to reactionary dentinogenesis in human teeth after attrition. In osteogenic tissues, forces of pressure can induce resorption mediated by multi-nucleated giant cells of osteoclastic lineage, but such cells are not a feature of attrition involving the pulp–dentine complex. Fibroblasts in the periodontal ligament appear to be responsible for both synthetic and degradative events during tissue remodelling. No information appears to exist on the ability of cells of the pulp–dentine complex to mediate remodelling of dentine, but if they did exhibit such a capacity this could lead to dissolution of bioactive dentine components, thereby inducing a response of reactionary dentinogenesis. The observation that both hydroxyproline and non-collagenous protein levels of dentine decline towards the occlusal tip of the rabbit incisor (Smith and Leaver, 1981) would support this hypothesis. However, considerable experimentation is still required to elucidate the processes taking place in the pulp–dentine complex during attrition.

Summary

Many of the molecular and cellular events of tissue formation in the pulp–dentine complex appear to be mirrored during repair after injury of the tissue. Epigenetic signalling, involving functional networks of growth factors and extracellular matrix molecules, and cell competence appear to be critical for development, maintenance and repair of the pulp–dentine complex. Bioactive components in the dentine matrix may participate in pulpo-dentinal interactions which could play key roles both in the maintenance of the odontoblast phenotype and in the induction of tissue repair. The specificity of dentinogenic response during repair might well be determined by the nature of the cells responding as well as the epigenetic signal.

References

Baume, L. J. (1980). The biology of pulp and dentine. In *Monographs in Oral Science*, ed. H. M. Myers. Basel: Karger.

Bègue-Kirn, C., Smith, A. J., Ruch, J. V., Wozney, J. M., Purchio, A., Hartmann, D. and Lesot, H. (1992). Effect of dentin proteins, transforming growth factor β1 (TGFβ1) and bone morphogenetic protein 2 (BMP2) on the differentiation of odontoblasts *in vitro*. *International Journal of Developmental Biology*, 36, 491–503.

Bègue-Kirn, C., Smith, A. J., Loriot, M., Kupferle, C., Ruch, J. V. and Lesot, H. (1994). Comparative analysis of TGFβs, BMPs, IGF, msxs, fibronectin, osteonectin and bone sialoprotein gene expressions during normal and *in vitro* induced odontoblast differentiation. *International Journal of Developmental Biology*, 38, 405–420.

Bernick, S. (1967a). Age changes in the blood supply to human teeth. *Journal of Dental Research*, 46, 544–550.

Bernick, S. (1967b). Effect of ageing on the nerve supply to human teeth. *Journal of Dental Research*, 46, 694–699.

Cam, Y., Neumann, M. R. and Ruch, J. V. (1990). Immunolocalization of TGFβ1 and EGF-receptor epitopes in mouse incisors and molars. Demonstration of in vitro production of transforming activity. *Archives of Oral Biology*, 35, 813–822.

Cassidy N, Fahey M, Prime S. S. and Smith A. J. (1997). Comparative analysis of transforming growth factor-β isoforms 1–3 in human and rabbit dentine matrices. *Archives of Oral Biology*, 42, 219–223.

Cotton, W. R. (1968). Pulp response to cavity preparation as studied by the method of thymidine 3H autoradiography. In *Biology of the Dental Pulp Organ*, ed. S. B. Finn, pp. 60–102. Tuscaloos: University of Alabama Press.

Couve, E. (1986). Ultrastructural changes during the life cycle of human odontoblasts. *Archives of Oral Biology*, 31, 643–651.

D'Souza, R. N., Happonen, R. P., Ritter, N. M. and Butler, W. T. (1990). Temporal and spatial patterns of transforming growth factor β1 expression in developing rat molars. *Archives of Oral Biology*, 35, 957–965.

Feit, J., Metelova, M. and Sindelka, Z. (1970). Incorporation of 3H-thymidine into damaged pulp. *Journal of Dental Research*, 49, 783–785.

Finkelman, R. D., Mohan, S., Jennings, J. C., Taylor, A. K., Jepsen, S. and Baylink, D. J. (1990). Quantitation of growth factors IGF-1, SGF/IGF-II and TGFβ in human dentin. *Journal of Bone Mineral Research*, 5, 717–723.

Fitzgerald, M. (1979). Cellular mechanisms of dentinal bridge repair using 3H-thymidine. *Journal of Dental Research*, 58, 2198–2206.

Fox, A. G. and Heeley, J. D. (1980). Histological study of pulps of human primary teeth. *Archives of Oral Biology*, 25, 103–110.

Heikinheimo, K. (1994). Stage specific expression of decapentaplegic-Vg-related genes 2, 4 and 6 (bone morphogenetic proteins 2, 4 and 6) during human tooth morphogenesis. *Journal of Dental Research*, 73, 530–597.

Heikinheimo, K., Happonen, R. P., Miettinen, P. J. and Ritvos, O. (1993). Transforming growth factor β2 in epithelial differentiation of developing teeth and odontogenic tumours. *Journal of Clinical Investigation*, 91, 1019–1027.

Héritier, M., Dangleterre, M. and Bailliez, Y. (1990). Differentiation of odontoblasts in mouse dental papillae recombined with normal or chemically treated dentinal matrices. *Archives of Oral Biology*, 35, 917–924.

Heywood, B. R. and Appleton, J. (1984). The ultrastructure of the rat incisor odontoblast in culture. *Archives of Oral Biology*, 29, 327–329.

Ignotz, R. A. and Massagué, J. (1986). Transforming growth factor-β stimulates the expression of fibronectin and collagen and their incorporation into the extracellular matrix. *Journal of Biological Chemistry*, 261, 4337–4345.

Kollar, E. J. (1983). Epithelial-mesenchymal interactions in the mammalian integument: tooth development as a model for instructive interaction. In *Epithelial-Mesenchymal Interactions in Development*, eds. R. M. Sawyer and J. F. Fallon, pp. 87–102. New York: Praeger.

Kollar, E. J. and Baird, G. (1970). Tissue interactions in embryonic mouse tooth germs. II. The induction role of the dental papilla. *Journal of Embryology and Experimental Morphology*, 24, 173–184.

Lesot, H., Smith, A. J., Meyer, J. M., Staubli, A. and Ruch, J. V. (1986). Cell-matrix interactions: influence of non-collagenous proteins from dentin on cultured dental cells. *Journal of Embryology and Experimental Morphology*, 96, 195–209.

Lesot, H., Bègue-Kirn, C., Kubler, M. D., Meyer, J. M., Smith, A. J., Cassidy, N. and Ruch, J. V. (1993). Experimental induction of odontoblast differentiation. *Cells and Materials*, 3, 201–217.

Lesot, H., Smith, A. J., Tziafas, D., Bègue-Kirn, C., Cassidy, N. and Ruch, J. V. (1994). Biologically active molecules and dental tissue repair: a comparative review of reactionary and reparative dentinogenesis with the induction of odontoblast differentiation in vitro. *Cells and Materials*, 4, 199–218.

Lianjia, Y., Yuhao, G. and White, F. H. (1993). Bovine bone morphogenetic protein-induced dentinogenesis. *Clinica Orthopedica*, 295, 305–312.

Lyons, K. M., Jones, C. M. and Hogan, B. L. M. (1990). The TGFβ-related DVR gene family in mammalian development. In *Postimplantation Development in the Mouse*. CIBA Foundation Symposium 165, pp. 219–234. Chichester: Wiley.

Marinkovich, M. P., Keene, D. R., Rimberg, C. S. and Burgeson, R. E. (1993). Cellular origin of the dermal-epidermal basement membrane. *Developmental Dynamics*, **197**, 255–267.

McCaffrey, T. A., Falcone, D. J. and Du, B. (1992). Transforming growth factor-β1 is a heparin-binding protein: identification of putative heparin-binding regions and isolation of heparins with varying affinitiy for TGF-β1. *Journal of Cell Physiology*, **152**, 430–440.

Millan, F. A., Denhez, K., Kondaian, P. and Akhurst, R. K. (1991). Embryonic gene expression patterns of TGFβ1, β2 and β3 suggest different developmental functions *in vivo*. *Development*, **111**, 131–144.

Mjör, I. A., Dahl, E. and Cox, C. F. (1991). Healing of pulp exposures: an ultrastructural study. *Journal of Oral Pathology and Medicine*, **20**, 496–501.

Munksgaard, E. C., Richardson, W. S. and Butler, W. T. (1978). Biosynthesis of phosphoprotein by rat incisor odontoblasts in *in vitro* culture. *Archives of Oral Biology*, **23**, 583–585.

Nakashima, M. (1990). The induction of reparative dentine in the amputated pulp of the dog by bone morphogenetic protein. *Archives of Oral Biology*, **35**, 493–497.

Nakashima, M. (1994). Induction of dentin formation on canine amputated pulp by recombinant human bone morphogenetic proteins (BMP) -2 and -4. *Journal of Dental Research*, **73**, 1515–1522.

Pelton, R. W., Dickinson, M. E., Moses, H. L. and Hogan, B. L. M. (1990). *In situ* hybridization analysis of TGFβ3 RNA expression during mouse development: comparative studies with TGFβ1 and β2. *Development*, **110**, 609–620.

Pelton, R. W., Saxén, B., Jones, M., Moses, H. L. and Gold, J. I. (1991). Immunohistochemical localization of TGFβ1, TGFβ2 and TGFβ3 in the mouse embryo: expression patterns suggest multiple roles during embryonic development. *Journal of Cell Biology*, **115**, 1091–1105.

Rapraeger, A. (1989). Transforming growth factor (type β) promotes the addition of chondroitin sulfate chains to the cell surface proteoglycan (syndecan) of mouse mammary epithelia. *Journal of Cell Biology*, **109**, 2509–2518.

Romagnoli, P., Mancini, G., Galeotti, F., Franchi, E. and Piereoni, P. (1990). The crown odontoblasts of rat molars from primary dentinogenesis to complete eruption. *Journal of Dental Research*, **69**, 1857–1862.

Ruch, J. V. (1985). Development, form and function of odontoblasts. *Journal of Dental Research*, **64**, 489–498.

Ruch, J. V. (1987). Determinisms of odontogenesis. *Cell Biology Reviews*, **14**, 1–112.

Ruch, J. V. (1990). Patterned distribution of differentiating dental cells: facts and hypotheses. *Journal de Biologie Buccale*, **18**, 91–98.

Ruch, J. V. (1995). Tooth crown morphogenesis and cytodifferentiations: candid questions and critical comments. *Connective Tissue Research*, **31**, 1–8.

Ruch, J. V. and Karcher-Djuricic, V. (1971a). Action de la 5-fluorodeoxyuridine sur la differenciation in vitro de molaires d'embryons de souris. *Archives de Biologie*, **82**, 115–129.

Ruch, J. V. and Karcher-Djuricic, V. (1971b). Mise en évidence d'un rôle spécifique de l'epithélium adamantin dans la différenciation et le maintien des odontoblastes. *Annales d'Embryologie et de Morphogenese*, **4**, 359–366.

Ruch, J. V., Karcher-Djuricic, V. and Thiebold, J. (1976). Cell division and cytodifferentiation of odontoblasts. *Differentiation*, **5**, 165–169.

Ruch, J. V., Lesot, H., Karcher-Djuricic, V., Meyer, J. M. and Olive, M. (1982). Facts and hypotheses concerning the control of odontoblast differentiation. *Differentiation*, **21**, 7–12.

Ruoslahti, E. and Yamaguchi, Y. (1991). Proteoglycans as modulators of growth factor activities. *Cell*, **64**, 867–869.

Rutherford, R. B., Wahle, J., Tucker, M., Rueger, D. and Charette, M. (1993). Induction of reparative dentine formation in monkeys by recombinant human osteogenic protein-1. *Archives of Oral Biology*, **38**, 571–576.

Schubert, D. (1992). Collaborative interactions between growth factors and the extracellular matrix. *Trends in Cell Biology*, **2**, 63–66.

Seltzer, S. and Bender, I. B. (1957). Some influences affecting repair of the exposed pulps of dogs teeth. *Journal of Dental Research*, **37**, 678–687.

Slavkin, H. C. (1974). Tooth formation: a tool in developmental biology. *Oral Science Reviews*, **4**, 1–136.

Slavkin, H. C. (1990). Molecular determinants of tooth development: a review. *Critical Reviews in Oral Biology and Medicine*, **1**, 1–16.

Smith, A. J. and Leaver, A. G. (1981). Distribution of the EDTA-soluble non-collagenous organic matrix components of rabbit incisor dentine. *Archives of Oral Biology*, **26**, 643–649.

Smith, A. J., Matthews, J. B. and Hall R. C. (1998). Transforming growth factor B1 (TGF-B1) in dentine matrix: ligand activation and receptor expression. *European Journal of Oral Sciences*, **106**, 179–184.

Smith, A. J., Tobias, R. S., Plant, C. G., Browne, R. M., Lesot, H. and Ruch, J. V. (1990). *In vivo* morphogenetic activity of dentine matrix proteins. *Journal de Biologie Buccale*, **18**, 123–129.

Smith, A. J., Tobias, R. S., Cassidy, N., Plant, C. G., Browne, R. M., Bègue-Kirn, C., Ruch, J. V. and Lesot, H. (1994). Odontoblast stimulation in ferrets by dentine matrix components. *Archives of Oral Biology*, **39**, 13–22.

Smith, A. J., Cassidy, N., Perry, H., Bègue-Kirn, C., Ruch, J. V. and Lesot, H. (1995a). Reactionary dentinogenesis. *International Journal of Developmental Biology*, **39**, 273–280.

Smith, A. J., Tobias, R. S., Cassidy, N., Bègue-Kirn, C., Ruch, J. V. and Lesot, H. (1995b). Influence of substrate nature and immobilisation of implanted dentin matrix components during induction of reparative dentinogenesis. *Connective Tissue Research*, **32**, 291–296.

Sveen, O. B. and Hawes, R. R. (1968). Differentiation of new odontoblasts and dentin bridge formation on rat molar teeth after tooth grinding. *Archives of Oral Biology*, **13**, 1399–1412.

Symons, N. B. B. (1967). The microanatomy and histochemistry of dentinogenesis. In *Structural and Chemical Organization of Teeth*, vol. 1, ed. A. E. W. Miles, pp. 317–318. New York: Academic Press.

Takuma, S. and Nagai, N. (1971). Ultrastructure of rat odontoblasts in various stages of their development and maturation. *Archives of Oral Biology*, **16**, 993–1011.

Takuma, S., Kurahashi, Y. and Tsuboi, Y. (1968). Electron microscopy of the osteodentine in rat incisors. In *Dentine and Pulp: Their Structure and Reactions*, ed. N. B. B. Symons. Dundee: University of Dundee Press.

Takuwa, Y., Ohse, C., Wang, E. A., Wozney, J. M. and Yamahita, K.

(1991). Bone morphogenetic protein-2 stimulates alkaline phosphatase activity and collagen synthesis in cultured odontoblastic cells MC3T3. *Biochemical and Biophysical Research Communications*, **174**, 96–101.

Thesleff, I. (1978). Role of the basement membrane in odontoblast differentiation. *Journal de Biologie Buccale*, **6**, 241–249.

Thesleff, I. and Vaahtokari, A. (1992). The role of growth factors in determination and differentiation of the odontoblastic cell lineage. *Proceedings of the Finnish Dental Society*, **88** (Suppl.1), 357–368.

Thesleff, I., Partanen, A. M. and Vainio, S. (1991). Epithelial-mesenchymal interactions in tooth morphogenesis: the roles of extracellular matrix, growth factors and cell surface receptors. *Journal of Craniofacial Genetics and Developmental Biology*, **11**, 229–237.

Thomas, P. and Dziadek, M. (1993). Genes coding for basement membrane glycoproteins Laminin, Nidogen and Collagen IV are differentially expressed in the nervous system and by epithelial, endothelial and mesenchymal cells of the mouse embryo. *Experimental Cell Research*, **208**, 54–67.

Thüringer, F. and Bienz, M. (1993). Indirect autoregulation of a homeotic *Drosophila* gene mediated by extracellular signalling. *Proceedings of the National Academy of Sciences, USA*, **90**, 3899–3903.

Tomes, J. (1850). On the structure of the dental tissues in the order Rodentia. *Philosophical Transactions of the Royal Society of London*, **140**, 529–567.

Torneck, C. D. and Wagner, D. (1980). The effect of a calcium hydroxide cavity liner on early cell division in the pulp subsequent to cavity preparation and restoration. *Journal of Endodontics*, **6**, 719–723.

Tziafas, D., Kolokuris, A., Alvanou, A. and Kaidoglu, K. (1992). Short-term dentinogenic response of dog dental pulp tissue after its induction by demineralised or native dentine or predentine. *Archives of Oral Biology*, **37**, 119–128.

Tziafas, D., Lambrianidis, T. and Panagioltis, B. (1993). Inductive effect of native dentin on the dentinogenic potential of adult dog teeth. *Journal of Endodontics*, **19**, 116–122.

Vainio, S., Karavanova, I., Jowett, A. and Thesleff, I. (1993). Identification of BMP-4 as a signal-mediating secondary induction between epithelial and mesenchymal tissues during early tooth development. *Cell*, **75**, 45–58.

Wise, G. E. and Fan, W. (1991). Immunolocalization of transforming growth factor beta in rat incisors. *Journal of Oral Pathology and Medicine*, **20**, 74–80.

Yamamura, T. (1985). Differentiation of pulpal cells and inductive influences of various matrices with reference to pulpal wound healing. *Journal of Dental Research*, **64**, 530–540.

Yamamura, T., Shimono, M., Koike, H. and Terao, M. (1980). Differentiation and induction of undifferentiated mesenchymal cells in tooth and periodontal tissue wound healing and regeneration. *Bulletin of the Tokyo University Dental College*, **21**, 181–222.

Yu, Y. M., Becvar, R., Yamada, Y. and Reddi, A. H. (1991). Changes in the gene expression of collagens, fibronectin, integrin and proteoglycans during matrix-induced bone morphogenesis. *Biochemical and Biophysical Research Communications*, **177**, 427–432.

7 Prismless enamel in amniotes: terminology, function, and evolution

P. M. Sander

7.1. Introduction

Amniota, the group consisting of reptiles, birds, and mammals, is characterized by monotypic enamel, a hypermineralized tissue secreted exclusively by ectodermal cells, the ameloblasts. While the enamel of Tertiary, Quaternary, and Recent mammals has received much study (Koenigswald and Sander, 1997a; Chapter 8), that of reptiles, toothed birds, and mammal-like reptiles is much less well known. One reason is that most mammals have prismatic enamel, which is easily studied in thin sections with polarized light, whereas the enamel of most non-mammalian amniotes (summarily referred to as reptiles hereafter) lacks prisms. This prismless enamel can only adequately be studied with the scanning electron microscope (SEM) because the light microscope cannot resolve individual crystallites. Enamel types differing in crystallite arrangement thus can look the same in light micrographs. Published SEM surveys of reptile enamel do not predate the last decade (Carlson, 1990a, b; Dauphin, 1987a, b, 1988; Dauphin et al., 1988; Sahni, 1987). There are two additional reasons for the low level of interest in reptilian enamel structure. First, the main impetus for studying reptile enamel in the past was not interest in the matter per se but in the origin of mammalian prismatic enamel. Second, paleontologists and systematists working with mammals traditionally focus on the teeth because mammalian teeth are rich in characters compared with the postcranial skeleton. The reverse is true for non-mammalian amniotes.

The purpose of this chapter is to provide an overview of the structural complexity and diversity of reptilian enamel and to explore the evolutionary constraints on its structure, be they functional or phylogenetic. The enamel structure of numerous investigated taxa clearly indicates that amelogenesis in prismless enamel is, at the cellular level, fundamentally different from that of prismatic mammalian enamel. These putative differences in amelogenesis are the foundation for a new model for the origin of prismatic enamel.

The great structural diversity of reptilian enamel, which was first recognized by Carlson and Bartels (1986), can only be adequately described and interpreted with a complex, hierarchical terminology (Figure 7.1) because existing terminologies for prismless enamel are too simple and misleading, and those for prismatic enamel are not applicable.

This review primarily draws on my own work on prismless amniote enamel (Sander, 1991, 1992, 1997a, b, 1999), which is based on a wide range of diadectomorphs (1 taxon), synapsids (26 taxa), anapsids (3 taxa), and diapsids (39 taxa). It covers all major groups as well as all major feeding modes and tooth morphologies. However, few comparisons were made *within* families or genera, leaving the field wide open for future work. Only Dauphin (1988) and Carlson (1990a, b) have also published a broad range of scanning electron micrographs of reptilian enamel. These, as well as their accompanying descriptions, are of limited value, however, because the planes of section were too few in number and poorly constrained, and the descriptive terminology was inadequate. Enamel must be studied in three planes of section to understand its structure in three dimensions. Nevertheless, taking these limitations into account, the results of these authors are directly comparable to mine, i.e. different specimens of taxa studied by me as well as by Dauphin (1988) or Carlson (1990a, b) show the same enamel structure.

Methods of preparation are discussed in detail elsewhere (Sander, 1999). In short, teeth or tooth fragments

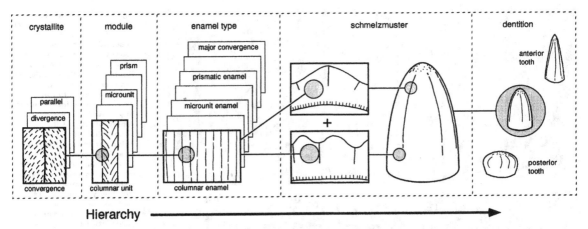

Figure 7.1. **Hierarchical approach to the study of reptilian enamel microstructure similar to that of Koenigswald and Clemens (1992) for mammals. The terminology is structured by levels of complexity; see text for details. Only the more important terms are incorporated into this figure.**

are embedded in resin to facilitate sectioning with a rock saw. After grinding and polishing, the plane of section is etched for 5–10 seconds with 10% HCL, dried, coated with gold, and studied with the SEM.

7.2. The evolutionary origin of monotypic enamel

Amniote enamel is undoubtedly of purely ectodermal origin as shown by numerous lines of evidence, discussed by Smith (1989) and Slavkin *et al.* (1984). In fact, there is evidence that the genetic basis of enamel formation has changed little throughout the history of the vertebrates (Slavkin *et al.*, 1984) and that ectodermal enamel may even occur in conodonts, thus extending the record of this tissue back into the Late Cambrian (Sansom *et al.*, 1992; see also Chapter 10). Ectodermal enamel is solely a product of cells of ectodermal origin, i.e. ameloblasts, which is why it was called monotypic enamel by Smith (1989). Evidence for its monotypic origin is primarily appositional growth as documented by incremental lines and crystallites perpendicular to the tooth surface (Sansom *et al.*, 1992). Bitypic enamel, on the other hand, is also a hypermineralized tissue on the surface of teeth and scales, but is produced by the cooperation of neural crest-derived cells, i.e. odontoblasts, with the ameloblasts (Smith, 1989). Bitypic enamel is also called enameloid and is best known in chondrichthyians.

Monotypic enamel that is structurally more complex than just crystallites perpendicular to the tooth surface is first seen in sarcopterygian fishes (Smith, 1989) and well documented in their descendants, the amniotes. The structures observed in sarcopterygian enamel by Smith

(1989) are very similar to the kind observed in reptiles and discussed in detail below. Little is known about the nature and structure of the hypermineralized covering on the teeth of the other large group of tetrapods derived from sarcopterygian fishes, the amphibians. While amphibians are traditionally believed to have enameloid (Poole, 1967), at least some fossil forms have monotypic enamel (Smith, 1992) and others may have both bitypic and monotypic enamel (Kawasaki and Fearnhead, 1983).

7.3. History of terminology

Important reviews of previous terminologies for reptilian enamel structure can be found in Smith (1989), Lester and Koenigswald (1989), Carlson (1990a) and Koenigswald and Clemens (1992). As the incentive for these, and most of the other previous studies proposing terminologies for reptilian enamel, was to trace the evolutionary origin of the mammalian enamel prism, most of the terminologies aim at distinguishing enamel with prisms from that without prisms. Considerable confusion has been created with inconsistent usage of terms (Sander, 1999), but only the most pertinent points can be reviewed here.

Previous workers distinguished two important structural grades in enamel without prisms: enamel in which all crystallites are parallel to each other and perpendicular to the outer enamel surface, and enamel in which the crystallites are not parallel but are organized into some kind of structural unit that is not a prism. This subdivision is similar to the distinction of Moss (1969) between 'continuous' enamel and 'discontinuous' enamel. Enamel at the second structural level ('discontinuous') was variously called 'pseudoprismatic' (Poole,

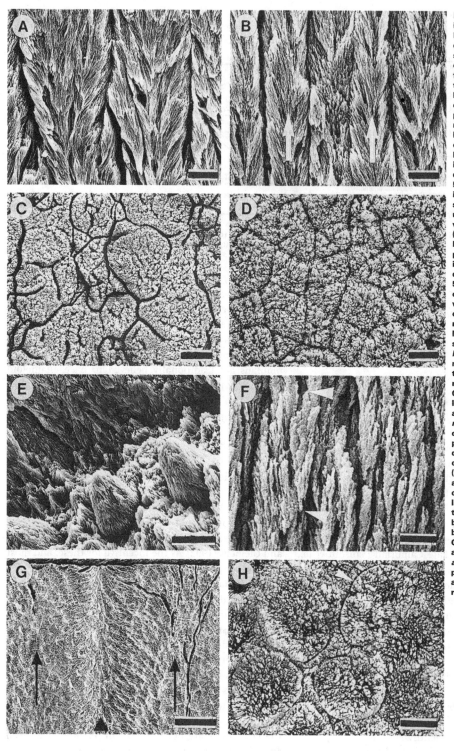

Figure 7.2A–H. **Terminology illustrated at the module level. For definitions and descriptions of these features, see text. Note that the planes of section discussed in this figure refer to the module, not to the entire tooth. A. Longitudinal section of columnar units, indeterminate phytosaur (scale bar represents 10 μm). B. Longitudinal section of columnar units with central line of divergence (arrows), indeterminate tyrannosaurid dinosaur (scale bar represents 5 μm). C. Columnar units with irregularly rounded cross sections, indeterminate rauisuchid. Note the great variability in unit diameter (scale bar represents 10 μm). D. Columnar units with polygonal cross-sections, indeterminate tyrannosaurid dinosaur (scale bar represents 5 μm). E. Columnar convergence units in oblique view, indeterminate ceratopsid dinosaur (scale bar represents 10 μm). F. Microunits in longitudinal section in the nothosaur _Nothosaurus_ sp. A single microunit extends between the two arrowheads (scale bar represents 2.5 μm). G. Compound units in longitudinal section in the alligatorid crocodile _Alligator mississippiensis_. The compound units consist of polygonal columnar units that diverge evenly from the centre of the compound units (arrows). The compound units are delimited by planes of columnar unit convergence like the one in the centre of the micrograph (from the black triangle upwards; scale bar represents 50 μm). H. Cross-sections of prisms in the agamid lizard _Uromastyx aegyptia_. The most complete prism sheath is marked by an arrowhead (scale bar represents 3 μm).**

1956; Lester and Koenigswald, 1989), 'preprismatic' (Carlson, 1990a; Koenigswald and Clemens, 1992), and 'protoprismatic' (Smith, 1989). These terms, however, should no longer be used because of their evolutionary connotations. Many taxa described this way are completely unrelated to mammals or not part of the direct ancestry of mammals (Sander, 1997a, 1999). In addition, the terms are inadequate to describe the great variety of structures found in reptilian enamel.

'Nonprismatic', 'aprismatic', and 'prismless' were applied to enamel of the first structural grade (Lester and Koenigswald, 1989; Carlson, 1990a; Koenigswald and Clemens, 1992). However, the terms 'aprismatic' and 'nonprismatic' (e.g. Carlson, 1990a; Koenigswald and Clemens, 1992) are problematic because they imply that all other enamels have prisms. Recently a consensus has emerged (Clemens, 1997; Sander, 1997a, Wood and Stern, 1997) that the term 'prismless enamel' should be retained and used outside formal terminologies to denote all kinds of enamel, at all levels of the hierarchy, that lack prisms.

The shortcomings of existing terminology, particularly in describing the great structural variety of prismless enamel, led Sander (1999, see also Koenigswald and Sander, 1997b) to develop a hierarchical terminology for prismless enamel based solely on structure and not on any supposed phylogenetic relationships with prismatic enamel.

7.4.   Hierarchical terminology

Many natural systems are hierarchically ordered and best described that way (Lakes, 1993; Francillon-Vieillot et al., 1990), and the hierarchical approach is well suited for describing the structural complexity and diversity of amniote enamel. Based on Koenigswald and Clemens (1992), I recognize five levels (Figure 7.1) starting with the lowest: the crystallite level, the module level, the enamel type level, the schmelzmuster level, and the dentition level (Sander, 1999; Koenigswald and Sander, 1997b). The structural units of one level are the building blocks of the structural units at the next higher level (Figure 7.1). The terminology discussed below is explained and illustrated more fully elsewhere (Sander, 1995; Koenigswald and Sander, 1997b). The terminology is designed to be also applicable to the prismless monotypic enamel of other vertebrate groups such as sarcopterygians and amphibians.

At the crystallite level, the orientation of the individual crystallites relative to the enamel-dentine junction (EDJ) is described. The EDJ generally serves as the major reference plane in enamel descriptions because this is where amelogenesis starts and enamel as a tissue forms as crystallites in phases of appositional growth. Crystallites either can be normal to the EDJ or at a different, though generally high, angle. If this is the case, crystallites will diverge and converge looking outward from the EDJ. Crystallites may diverge or converge continuously in a *zone of changing crystallite orientation* or with an abrupt change of direction along a plane or a line. Such a plane or line is called a *crystallite discontinuity*.

At the next level, *modules* are repeatable volumes of enamel that are delimited by crystallite discontinuities and/or by zones of changing crystallite orientation (Figure 7.2). The *enamel prism* in mammalian enamel is probably the most familiar module but only one of the possibilities. A feature of many modules not expressed in prisms is that there may be several orders of modules, i.e. smaller modules can themselves be combined into a volume of enamel bounded by planar crystallite discontinuities. Most enamel units are elongate and developed as *columnar divergence units* (in short, *columnar units*), consisting of a linear zone of crystallite divergence surrounded by planes of crystallite convergence (Figure 7.2A–D). The cross-section of these columnar units is usually polygonal but triangular and diamond shapes may occur as well (Figure 7.2D). Columnar units may vary in length: some are almost continuous extending nearly through the entire enamel layer (Figure 7.2B) while others are only a few times longer than wide and then are replaced by other, slightly offset columnar units. Other distinctive enamel units are *columnar convergence units* in which the crystallites converge on the unit centre (Figure 7.2E), and *microunits*, in which bundles of diverging crystallites are only a few crystallites across and a few crystallites long (Figure 7.2F). *Compound units* are groupings of columnar units or microunits into higher order units (Figures 7.2G, 7.3, 7.4E).

As opposed to most columnar units, *enamel prisms* (Figure 7.2H) extend without interruption, splitting, or merging from the EDJ close to the *outer enamel surface* (OES). This is due to the fact that each individual prism is secreted by a single ameloblast. Prisms are bundles of crystallites surrounded by a curved plane of crystallite convergence (the prism border or prism sheath) which may be open or closed (Figure 7.2H). Prisms are able to decussate, which means that adjacent prisms can cross each other. Decussation may occur in layers, called Hunter–Schreger bands, or in an irregular fashion (Koenigswald and Sander, 1997a); decussation of any kind of module that is not a prism has not been observed.

At the enamel type level, *enamel types* and *major discontinuities* are the important descriptive categories (Figures 7.3, 7.4). Enamel types are large volumes of enamel that

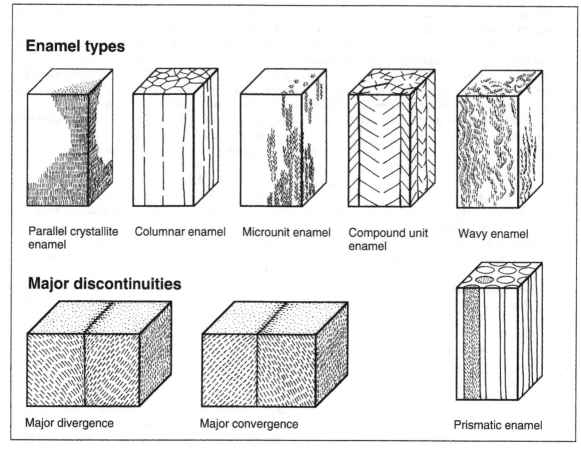

Enamel types

Parallel crystallite enamel   Columnar enamel   Microunit enamel   Compound unit enamel   Wavy enamel

Major discontinuities

Major divergence   Major convergence   Prismatic enamel

Figure 7.3. **Schematic representation of selected features at the enamel type level. The more important reptilian enamel types and major discontinuities are illustrated and drawn to the same scale. For details see text.**

consist of crystallites or modules (enamel units, prisms) each of the same kind and same orientation (Figure 7.3). Major discontinuities are planes of crystallite and module divergence (*major divergence*) or convergence (*major convergence*) that extend through much if not all of the enamel layer (Figures 7.3, 7.4D, 7.5E, H). Major discontinuities are common in prismless enamel but rare in prismatic enamel.

Important enamel types in prismless enamel are *parallel crystallite enamel*, *columnar enamel* and *microunit enamel*. Parallel crystallite enamel is the simplest enamel type and consists of crystallites that are all parallel to each other ('nonprismatic' of Carlson, 1990a). In addition, the crystallites are usually, but not always, normal to the EDJ (Figures 7.3, 7.4A). Parallel crystallite enamel thicker than about 20 μm always has well-developed incremental lines. Columnar enamel consists of columnar units of the same orientation (Figures 7.3, 7.4B, C, G, 7.5C). A specific and very widespread kind of columnar enamel is the *basal unit layer* (BUL), a thin layer of polygonal columnar units adjacent to the EDJ. Microunit enamel consists

entirely of microunits of the same orientation (Figure 7.4D). It should be remembered that both microunits and columnar units can be combined to form *compound unit enamel* (Figures 7.3, 7.4E).

The spatial distribution of enamel types and major discontinuities in the enamel cap is described as the *schmelzmuster* of a tooth (Figure 7.1, 7.5). Also entering the schmelzmuster description are microscopic features of the enamel surface. The most simple schmelzmuster is a single layer of parallel crystallite enamel (Figure 7.4A). However, most schmelzmusters in prismless amniote enamel are more complex.

Schmelzmusters can vary from tooth to tooth in a single dentition. This variability at the *dentition level* is virtually unexplored in prismless enamel. In mammalian enamel, it seems to be correlated with great morphological differences between tooth types (Koenigswald, 1997), and the same appears likely for reptiles.

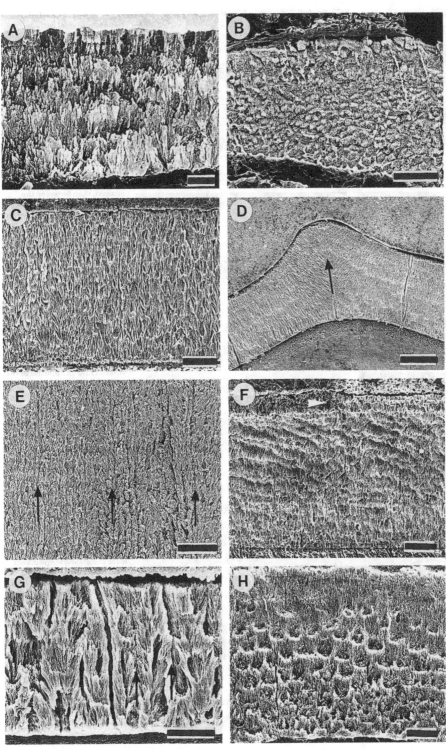

Figure 7.4A–H. **Typical examples of enamel microstructure in non-mammalian amniotes. In all micrographs except E, the enamel–dentine junction (EDJ) is visible at the bottom and the outer enamel surface (OES) at the top. A. The varanid lizard** *Varanus salvator.* **Longitudinal section, the enamel layer consists of a poorly developed basal unit layer and of parallel crystallite enamel (scale bar represents 2.5 μm). B. The procolophonid** *Procolophon trigoniceps.* **This oblique section ground from the EDJ towards the outside reveals that the entire enamel layer consists of columnar enamel (scale bar represents 30 μm). C. An indeterminate tyrannosaurid dinosaur. Longitudinal section, the enamel layer consists of columnar enamel only. Some columnar units extend through the entire enamel thickness (scale bar represents 30 μm). D. The nothosaur** *Nothosaurus* **sp., longitudinal section. The only enamel type in the schmelzmuster is microunit enamel. Note the major divergence (arrow) bisecting the enamel ridge (scale bar represents 50 μm). E. The durophagous ichthyosaur** *Omphalosaurus nisseri.* **Cross-section, the EDJ is beyond the top of the image. The enamel consists of compound unit enamel. Three compound units are intersected in this micrograph; their approximate centres are marked by arrows (scale bar represents 50 μm). F. The iguanodontid dinosaur** *Iguanodon* **sp., longitudinal section. The schmelzmuster consists of coarse wavy enamel covered by a thin layer of fine wavy enamel (arrowhead; scale bar represents 30 μm). G. The pelycosaur** *Dimetrodon,* **cross-section. The enamel consists of typical synapsid columnar enamel. One columnar unit is delimited by arrows (scale bar represents 10 μm). H. The tritheledontid cynodont** *Pachygenelus.* **Cross-section, note that the enamel is clearly prismatic (scale bar represents 10 μm).**

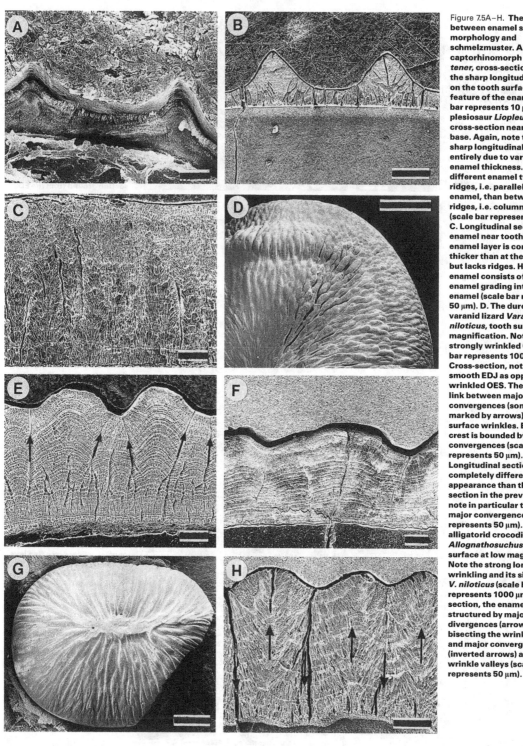

Figure 7.5A–H. **The relationship between enamel surface morphology and schmelzmuster. A. The captorhinomorph *Dictyobolus tener*, cross-section. Note that the sharp longitudinal ridges on the tooth surface are a feature of the enamel (scale bar represents 10 μm). B. The plesiosaur *Liopleurodon ferox*, cross-section near crown base. Again, note that the sharp longitudinal ridges are entirely due to variations in enamel thickness. There is a different enamel type in the ridges, i.e. parallel crystallite enamel, than between the ridges, i.e. columnar enamel (scale bar represents 100 μm). C. Longitudinal section of enamel near tooth tip. The enamel layer is considerably thicker than at the crown base but lacks ridges. Here the enamel consists of columnar enamel grading into microunit enamel (scale bar represents 50 μm). D. The durophagous varanid lizard *Varanus niloticus*, tooth surface at low magnification. Note the strongly wrinkled OES (scale bar represents 1000 μm). E. Cross-section, note the smooth EDJ as opposed to the wrinkled OES. There is a close link between major convergences (some are marked by arrows) and enamel surface wrinkles. Each wrinkle crest is bounded by major convergences (scale bar represents 50 μm). F. Longitudinal section with a completely different appearance than the cross-section in the previous figure, note in particular the rarity of major convergences (scale bar represents 50 μm). G. The alligatorid crocodile *Allognathosuchus* sp., tooth surface at low magnification. Note the strong longitudinal wrinkling and its similarity to *V. niloticus* (scale bar represents 1000 μm). H. Cross-section, the enamel is structured by major divergences (arrows) bisecting the wrinkle crests and major convergences (inverted arrows) at the wrinkle valleys (scale bar represents 50 μm).**

7.5. Phylogenetic constraints on schmelzmuster

One important aspect of research on enamel structure in reptiles is, of course, its application to phylogenetic questions. Based on the currently available sample, which was studied with a standardized methodology (Sander, 1999), synapomorphic enamel structures are not very widespread in higher taxa because the plesiomorphic condition, as observed in sarcopterygian fishes (Smith, 1989), and homoplasies are more common. The use of enamel characters in detailed analyses of the interrelationships of genera and species is largely unexplored except for the genus *Varanus* and clearly invites further study.

The largest clade to be characterized by a specific enamel structure is the synapsids which, with very few exceptions, have the same enamel type dominating their schmelzmuster (Sander, 1997a). This type, synapsid columnar enamel, consists of short, often discontinuous columnar units with irregular, polygonal cross-sections (Figures 7.4G, 7.6C). This schmelzmuster varies (between synapsid species) mainly in the thickness of the enamel layer. This uniformity in schmelzmuster is rather surprising in comparison with the variety encountered in other higher taxa such as dinosaurs (Sander, 1997b, 1999) or squamates (Sander, 1999), but it is significant for prism evolution in advanced synapsids and mammals (see below, Sander, 1997a).

The only other higher reptile taxon to show a possible enamel synapomorphy is the ichthyosaurs. Of the five taxa studied by me (representing four families), only one did not have microunits at the module level. Microunits are also clearly visible in the scanning electron micrographs of *Ophthalmosaurus* published by Dauphin (1988, plate 4, figure 7), confirming their occurrence in a fifth family. Dauphin (1988, plate 4, figures 8–11) also pictured the enamel of *Platypterygius* but the quality of the micrographs is insufficient for detecting microunits. Apart from the presence of microunits, the schmelzmusters of these ichthyosaur taxa have little else in common. The microunits may occur as microunit enamel (*Tholodus, Ichthyosaurus,* cf. *Stenopterygius*), may be combined into compound units (*Omphalosaurus,* Figure 7.4E, *Ichthyosaurus,* cf. *Stenopterygius*), and may be part of a schmelzmuster with parallel crystallite enamel (*Tholodus*). Some ichthyosaur schmelzmusters are structured by major discontinuities (*Tholodus, Ichthyosaurus*) while others are not (*Omphalosaurus,* cf. *Stenopterygius*). The only other occurrence of microunits is in some sauropterygia (in *Nothosaurus,* Figures 7.2F, 7.4D, and *Liopleurodon,* Figure 7.5C).

Among Dinosauria, at present only the clade of advanced ornithopods including *Iguanodon* and the Hadrosauridae is characterized by a distinctive enamel synapomorphy, wavy enamel. *Wavy enamel* (Figures 7.3, 7.4F) is not built up by enamel units but consists of individual crystallites in a staggered and helical arrangement that has a very distinctive appearance in SEM micrographs in every plane of section. Wavy enamel is notable for its lack of incremental lines.

7.6. Functional constraints on schmelzmuster

As phylogenetic constraints are probably of little importance in reptilian schmelzmusters, functional constraints must have explanatory power if reptilian enamel structural diversity is more than purely accidental. The most obvious starting point is diet. However, Dauphin (1988) found no good link between enamel structure and diet, partially because of the rather broad categories employed. A more refined approach, using eight different ecomorphotypes (Sander, 1999), was primarily based on tooth shapes but also on other indicators of dietary preference. Nonetheless, a good correlation with schmelzmuster could only be found for some ecomorphotypes (Sander, 1999)

In defining the ecomorphotypes and linking them to schmelzmusters, it became apparent that the geometry of the finished enamel cap, and in particular its surface morphology, need to be considered in addition to its internal structure. This is opposed to the situation in mammals in which the mechanical properties of the enamel layer are of major importance (Koenigswald, 1980; Pfretzschner, 1994; Rensberger, 1997). Noticeably, the differences between reptilian and mammalian enamel structure are correlated with the differing patterns of tooth replacement in the two groups, i.e. continuous in most reptiles (see Chapter 13) but only once in mammals. The ramifications of this will be discussed more fully below.

The consideration of functional constraints on reptilian enamel structure will proceed along two lines: the importance of internal structure *per se* and the importance of enamel cap geometry and surface morphology.

7.6.1. Importance of internal structure

Two functions are relevant for the adaptive value of enamel microstructure: resistance to abrasion and mechanical stability of the tooth. The importance of both functions is insufficiently explored in reptiles and the remarks about adaptive value of internal structure must be considered preliminary.

Enamel contributes to tooth stability primarily by preventing deformation and fracture (Rensberger, 1997;

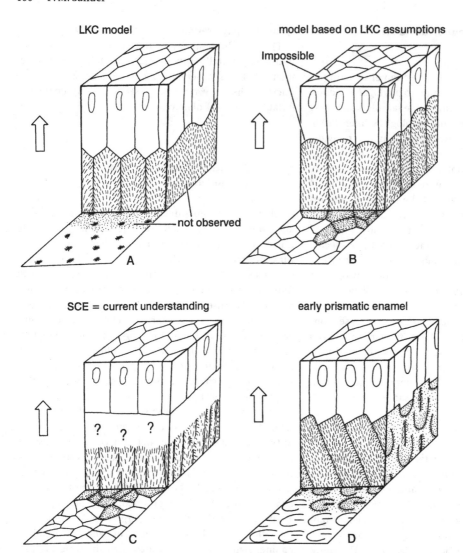

Figure 7.6A–D. **Three-dimensional diagrams of the relationship between ameloblasts and enamel structure. These diagrams illustrate the falsification of the LKC (after the authors: Lester and Koenigswald, 1989; Carlson, 1990a) model of prism origin and the revised model of prism origin. A. Secretion of preprismatic enamel as envisaged by the LKC model. A conical Tomes' process certainly is possible but the depicted enamel structure with lines of crystallite convergence was not found in any synapsid nor in any other reptile enamel. B. Hypothetical relationship between synapsid columnar enamel and its secreting ameloblasts. This diagram is based on the assumption of the LKC model that enamel crystallites are always oriented perpendicular to the secretory surface. If this were true, the ameloblasts would have to be irregularly polygonal in cross-section, mirroring the shape of the columnar units in synapsid columnar enamel (SCE), and have concave secretory surface. Neither conditions have been observed and are unlikely to occur in any ameloblast epithelium. C. Model of amelogenesis of SCE. Just as in living amniotes, ameloblasts almost certainly were roughly hexagonal in shape in Mesozoic synapsids as well. Thus there is no one-to-one relationship of the columnar units with the ameloblast matrix. As in many other prismless enamel types, the mineralization front in SCE apparently lagged behind the presumably flat secretory surface of the receding ameloblasts. The structure of SCE was determined by a combination of self-organization of crystal growth and extracellular control mechanisms (enamel proteins) of the ameloblasts. D. Amelogenesis in Mesozoic prismatic enamel probably conformed to that observed in prismatic enamel today. The evolutionary steps from SCE to prismatic enamel must have included the mineralization front moving into close proximity of the ameloblast secretory surface, the establishment of perpendicular crystallite orientation, and the differentiation of the secretory surface into the Tomes' process and the 'shoulder'.**

Chapter 18). However, this function requires sufficiently thick enamel, as found in most mammals, and may be fairly unimportant in the often rather thin reptilian enamel. Exceptions are the durophagous forms (feeding on hard-shelled prey) that have thick enamel. These and many other kinds of reptiles, particularly those with thicker enamel, have columnar enamel which is the most common enamel type encountered in reptiles (Figures 7.2A–D, 7.4B, C, G, 7.5B, C). This suggests that columnar enamel is more adaptive than the structurally more simple parallel crystallite enamel. Possible improvements may lie in mechanical strength as well as in abrasion resistance. Abrasion resistance of parallel crystallite enamel may be poor because of its well-developed incremental lines acting as planes of weakness. These assumptions have not been tested experimentally, however. Arguing *against* the biomechanic superiority of columnar enamel over parallel crystallite enamel is the observation that, in some cases, fractures preferentially follow columnar unit borders that consist of planes of crystallite convergence. However, planar crystallite discontinuities in general do not seem to act as planes of weakness. Clearly, more work needs to be done on the potential functional differences between these enamel types.

Another explanation for the preponderance of columnar enamel could possibly lie in its mode of deposition. Columnar enamel may be the structural byproduct of a fast secretion rate. Fast secretion, in turn, is advantageous in the development of a thick enamel layer. Parallel crystallite enamel, on the other hand, may have a much slower rate of secretion. Knowing as little as we do about amelogenesis in prismless enamel from direct observation, this intriguing possibility can only be evaluated incompletely at present. A comparison of several reptile taxa indicates that thick parallel crystallite enamel has more closely spaced and more pronounced incremental lines than columnar enamel. Assuming that the time interval represented by one incremental line is the same in columnar enamel and parallel crystallite enamel, this would indeed indicate faster secretion of columnar enamel.

Only one particular enamel type seems to have evolved for specific functional needs. This is wavy enamel (Figures 7.3, 7.4F), which is exclusively found in advanced ornithopod dinosaurs, animals who fed on tough plant matter. Ornithopod teeth show high rates of wear, which are compensated for by rapid tooth replacement. As tooth wear is largely controlled by the abrasion resistance of the enamel, selective pressure may have favoured the evolution of a more wear-resistant enamel type. The primary function of wavy enamel may thus be to reduce abrasion.

Abrasion presumably proceeds by the dislocation of individual crystallites. Thus the irregular arrangement of the crystallites and the lack of incremental lines in wavy enamel would prevent wholesale loss of crystallites because adjacent crystallites are not affected to the same degree by the same force vector. Another possible function of wavy enamel could be as a simple crack-stopping mechanism analogous to that seen in enamel prism decussation in mammals (Chapter 18). The staggered and helical arrangement of the crystallites leads to an interlocking of neighbouring crystallites not seen in the other prismless enamel types, possibly preventing the growth of cracks.

Possibly also of adaptive value are the prismatic enamel (Figure 7.2H) of the agamid lizard *Uromastyx* (Cooper and Poole, 1973; Sander, 1999) and the peculiar convergence unit enamel of the sphenodontid *Sphenodon* (as observed in the micrographs of Carlson, 1990b, plate 180). Both taxa are herbivores and are characterized by an acrodont dentition in which the teeth are not replaced. Heavy wear in these dentitions produces a single secondary occlusal surface over the entire tooth row which facilitates a precise cutting action. Lack of tooth replacement combined with wear naturally places great selection pressure on preserving every tooth as long as possible. An internal enamel structure adapted to this need certainly seems a likely evolutionary consequence. The reduced rate of tooth replacement in mammals compared with that in reptiles has also been related to the evolution of prismatic enamel in the synapsid lineage (Grine and Vrba, 1980).

7.6.2.   Importance of enamel surface morphology

Particular tooth shapes have evolved repeatedly in reptiles as adaptations to certain diets and modes of feeding. Features of enamel surface morphology, i.e. carinae, ridges, wrinkles, and bumps are an important component of these tooth shapes and have also evolved repeatedly. Theoretically, such a surface morphology can be achieved by enamel deposition starting at an already sculptured EDJ or by enamel deposition starting at a smooth EDJ and producing the surface sculpture. The latter is usually observed, i.e. most complex surface morphologies are the product of amelogenesis. Although the reason for this is presently not understood, biomechanical as well as amelogenetic causes are imaginable. The enamel cap thus is characterized by a complex geometry with a smooth inner surface and a complex outer surface. The observation that amelogenesis produced enamel surface morphology and cap geometry provides the key to understanding prismless enamel structure in reptiles (Sander, 1999).

The most impressive case in point is the crushing teeth of durophagous animals which, with the exception of the placodonts, all have a bulbous shape with longitudinal wrinkles (e.g. Figure 7.5D, G). Taxa studied include some ichthyosaurs (*Phalarodon, Tholodus, Omphalosaurus*), one mosasaur (*Globidens*), some species of *Varanus* (*V. albigularis, V. exanthematicus, V. niloticus*, Figure 7.5D–F), the teiid lizard *Dracaena*, and a crocodile (*Allognathosuchus*, Figure 7.5G,H). In all taxa, the wrinkles were a feature of the enamel, and all taxa had rather different schmelzmusters (Figure 7.5D–H). The biomechanical explanation for these wrinkles was addressed by Preuschoft *et al.* (1974) and Sander (1990).

Surface features produced by amelogenesis were also observed in conical teeth with ridges, ribs, and carinae

(Figures 7.4D, 7.5A,B). Only serrations along cutting edges are usually a feature preformed at the EDJ. Due to the limited thickness of the enamel layer, there is an upper size limit for structure expressed solely by the enamel layer. Thus, features that define the general shape of a tooth, such as major cutting edges or the large longitudinal ridges of many ornithopod dinosaur teeth, are preformed by the dentin core.

Enamel surface morphology and cap geometry as the major constraints on reptilian schmelzmusters, as opposed to phylogeny and biomechanics of the internal structure on mammalian schmelzmusters, makes much more sense if the differences in tooth replacement are considered. Teeth are replaced only once in mammals but continuously in most reptiles. In mammals, the individual tooth needs to be preserved as it has to last for the lifetime of the animal; not so in most reptiles, as any one tooth will be replaced soon. However, enamel surface morphology, cap geometry, and general tooth shape need to be the same in each replacement tooth. Reptilian enamel production is thus optimized to ensure an appropriate enamel surface morphology and cap geometry, and not a microstructure designed for a long life.

7.7. Amelogenesis in prismless amniote enamel

Amelogenesis is very poorly understood in prismless amniote enamel. First, only a handful of studies about living reptiles exist (Woerderman, 1921; Kvam, 1946, 1958; Poole, 1957; Peyer, 1968; Slavkin et al., 1984; Piesco and Kallenbach, 1985), as opposed to mammals (review in Moss-Salentijn et al., 1997). Second, a high proportion of extinct taxa has schmelzmusters and enamel types not observed in recent reptiles. Nevertheless, the finished product, i.e. the enamel microstructure, permits certain inferences about amelogenesis even in extinct taxa. This is of particular interest because previous enamel classifications and models for the origin of mammalian enamel prisms relied heavily on untested assumptions about reptilian amelogenesis.

7.7.1.   Observations on living reptiles

The published studies on amelogenesis in living reptiles agree on certain observations but appear to disagree on others. The transmission electron microscopy (TEM) studies by Slavkin et al. (1984) and Piesco and Kallenbach (1985) concur that the enamel minerals are secreted together with the enamel proteins so that crystallites form from the start. However, these authors focused on the initial stage of amelogenesis at the EDJ and did not

investigate the main phase. Older light microscopy studies focused instead on the main phase of amelogenesis but could not resolve the initial phase. The older studies showed that a thick layer of proteinaceous enamel precursor is formed without mineral and that this is gradually converted to enamel crystallites (Poole, 1957; Peyer, 1968). A synthesis of these studies would suggest that, in the initial stages of amelogenesis, crystallites mineralize close to the secretory faces of the ameloblasts, as is the case throughout the secretory phase in mammals (Moss-Salentijn et al., 1997), but that during the main phase, mineralization lags well behind the advancing secretory epithelium. Both light microscopy and TEM studies note one very significant fact: the secretory end of the ameloblasts is flat, i.e. a Tomes' process is lacking in reptiles. Tomes' processes are instrumental in the deposition of mammalian enamel prisms (Warshawsky, 1988; Moss-Salentijn et al., 1997).

7.7.2.   Inferences from microstructure

Perhaps the most obvious of the conclusions (about reptile amelogenesis) derived from enamel structure is that the structural units, the modules, as well as larger blocks delimited by major discontinuities, do not reflect the arrangement, shape, and size of the ameloblasts, particularly their secretory ends (Sander, 1997a 1999). This is only the case in prismatic enamel, where one ameloblast secretes one prism (Carlson, 1990a; Lester and Koenigswald, 1989; see also Sander, 1997a). The structural units in reptilian enamel are so variable in size and shape, even in a single enamel type, that it is inconceivable that the ameloblasts that deposited them were equally variable (Figure 7.6B). In addition, the crystallites in reptilian enamel cannot always be oriented perpendicular to the (presumably flat) secretory surface (Sander, 1995). One reason is that structural units such as columnar units would require a concave secretory surface of the ameloblast (Figure 7.6B), which appears rather unlikely. In the past it has been firmly assumed that crystallites in any kind of enamel are always orientated perpendicular to the secretory surface (Carlson, 1990a; Lester and Koenigswald, 1989).

Certain enamel types, such as columnar enamel (e.g. Figure 7.2A, B, 7.5C) have an uncanny resemblance to inorganic precipitates, eggshells, and prismatic bivalve shells, all of which are not the result of a secretory epithelium exerting close control on a finished structure. As shown by Ubukata (1994) for bivalve shells, this kind of structure is the result of simple geometrical selection on spherulitic growth. If columnar enamel were formed the same way, this would explain its common occurrence in enamel more than about 10 μm thick. Spherulitic growth may require less cellular control than oriented secretion of a protein matrix through a cell membrane. Less cellular control, as compared with the situation in mammals, is precisely what is to be expected when the mineralization front lags behind the matrix secretion front.

7.7.3.   Hypotheses of amelogenesis

Combining the amelogenetic interpretation of the finished enamel structure of fossil and recent reptiles with the scarce observations of amelogenesis in recent reptiles, a fairly coherent hypothesis of amelogenesis in complex prismless reptile enamel types emerges. Enamel deposition consists of an initial phase in which mineralization closely follows secretion of a proteinaceous matrix. The possible expression of this phase is the ubiquitous basal unit layer which, on structural grounds, was hypothesized to be a relatively close reflection of the ameloblast matrix (Sander, 1995). Transmission electron micrographs of this initial phase in *Alligator mississippiensis* published by Slavkin *et al.* (1984, figure 2c) possibly depict a basal unit layer.

For the main phase of amelogenesis, the observation that structural units do not reflect ameloblast arrangement predicts that the mineralization front must lag well behind the secretion front, which is precisely what the work of Poole (1957) and Peyer (1968) shows. The ameloblasts secrete a proteinaceous matrix, the structure of which is difficult to resolve in the light microscope. Only after a few tens of micrometers of this matrix are laid down, mineralization moves on from the early mineralized layer near the EDJ into the thick layer of proteinaceous matrix. At this time, the varied and sometimes complex structure of prismless enamel is established, independently of the arrangement of the ameloblast epithelium. This mode of amelogenesis also explains the weak expression or lack of incremental lines in enamel types such as columnar enamel, microunit enamel, and wavy enamel in which the crystallites are not perpendicular to the secretory surface.

A second, different hypothesis of amelogenesis explains the most simple reptilian enamel type, i.e. parallel crystallite enamel, which is characterized by well-developed incremental lines and crystallites perpendicular to them. In this case, formation of the crystallites must occur adjacent to the secretory surface and nearly simultaneously with the secretion of the proteinaceous matrix. Amelogenesis of parallel crystallite enamel would thus continue the mode of initial enamel formation into the main phase.

The two modes of amelogenesis probably also differ in the rate of enamel deposition: slow in parallel crystallite enamel and fast in complex enamel types. The two modes should not be viewed as mutually exclusive but as the end members of a spectrum. They also may have both been active during the formation of a single, complex schmelzmuster.

7.7.4.   Morphogenesis of enamel cap geometry

The origin of complex surface morphologies from a smooth EDJ can be explained in a number of ways, in accordance with the widely differing internal structure of very similar surface morphologies. In schmelzmusters dominated by parallel crystallite enamel, epithelial folding seems to be most important. The incremental lines in parallel crystallite enamel record successive positions of the ameloblast layer during amelogenesis (Figure 7.5E,F,H). Tracing these lines suggests that complex surface morphologies and the concordant major discontinuities are commonly produced by folding of the epithelium. Most often, folding increases with ongoing amelogenesis (e.g. *Varanus niloticus*, Figure 7.5D–F) but early establishment of a wrinkled epithelium followed by outward growth can also be deduced.

Epithelial folding as a morphogenetic mechanism may also work to some extent in other prismless schmelzmusters. However, other phenomena such as geometric selection on spherulitic growth are important in these as well. Yet another possibility is seen in schmelzmusters with more than one enamel type. Surface features such as sharp ridges (e.g. *Liopleurodon*, Figure 7.5B) can be the product of different rates of growth in different enamel types which can be deduced from the spatial distribution of the enamel types and sometimes from the spacing of the incremental lines (Sander, 1999). A mechanism suggested by Schmidt and Keil (1971), but not documented by them or subsequent workers, is the differential onset of amelogenesis but synchronous termination resulting in differential thickening of the enamel layer.

7.7.5.   Prismless enamel in mammals

Prismless enamel also occurs in the schmelzmuster of many mammals as an outer layer covering the prismatic layer. This prismless outer enamel generally consists of parallel crystallites and is assumed to have been deposited by a flat secretory surface after the reduction of the Tomes' processes towards the end of amelogenesis (Lester and Koenigswald, 1989). Prismless enamel as the sole constituent of the schmelzmuster is only observed among Cenozoic mammals in forms with reduced teeth such as in some whales (Ishiyama, 1987) and in some monotremes (Lester and Archer, 1986).

7.8.   Evolutionary origin of prismatic enamel

The evolutionary origin of mammalian prismatic enamel has aroused much interest over the years (e.g. Poole, 1967; Moss, 1969; Osborn and Hillman, 1979; Grine and Vrba, 1980; Frank et al., 1988; Lester and Koenigswald, 1989; Carlson, 1990a; Stern and Crompton, 1995; Clemens, 1997; Sander, 1997a; Wood and Stern, 1997) and, as already noted, actually was the main incentive for the study of prismless enamel in reptiles. However, the first coherent model of prism origin was only recently proposed in two independent papers (Lester and Koenigswald, 1989; Carlson, 1990a) and is termed the LKC model after its authors (Figure 7.6A,B).

This model explains the evolution of prisms by the shape evolution of the Tomes' process from a flat secretory surface via a hypothetical conical structure also called a Tome's process (Figure 7.6A) to the complex asymmetrical Tomes' process in mammals which secretes prismatic enamel (Figure 7.6D, Lester and Koenigswald, 1989, Carlson, 1990a). According to the model, the one-to-one relationship between ameloblast and structural unit evolved at the stage of the hypothetical conical structure, i.e. before the evolution of prismatic enamel. A similar approach to that of the LKC model was taken by Stern and Crompton (1995) who attempted to trace the shape evolution of the secretory surface from the finished enamel structure as observed with the SEM.

The inferences and observations about reptilian enamel structure and amelogenesis discussed above run counter to the assumptions of the LKC model of prism origin (Sander, 1991). The poor correlation between internal structure and ameloblast epithelium in reptiles in general (Sander, 1991, 1999), and in non-mammalian synapsids in particular, invalidates the LKC model (Sander, 1997a). In addition, synapsid columnar enamel is incompatible with the two other implicit premises of the LKC model, i.e. that prismless synapsid enamel consists of structural units defined by lines of crystallite convergence, the so-called preprisms or pseudoprisms (Figure 7.6A), and that the enamel crystallites are always orientated perpendicular to the secretory surface (Figure 7.6B). The preprism or pseudoprisms were not observed in any synapsid investigated by scanning electron microscopy (Sander, 1997a) nor do they occur in any reptile (Sander, 1999). Their structure was apparently hypothesized based on one plane of section only.

The new model of prism origin (Sander, 1997a) takes into account two important facts: first, the observations of, and inferences about, amelogenesis in prismless enamel in general and synapsid columnar enamel in particular (Figure 7.6C); second, the lack of intermediates between synapsid columnar enamel and prismatic enamel in the fossil record. The new model thus states that prism evolution consists of a rapid sequence of hypothetical steps that evolutionarily transform an enamel organ that secretes synapsid columnar enamel (Figure 7.6C) to one that secretes prismatic enamel (Figure 7.6D). Judging from its structure, synapsid columnar enamel was formed the same way as most other prismless enamel, with the mineralization front lagging well behind the protein matrix secretory surface (Figure 7.6C). The first step in the new model is that the enamel mineralization front moves into close proximity of the ameloblast secretory surface. Next, crystallite formation perpendicular to the secretory surface, i.e. the cell membrane, is established. As the last step, the secretory surface differentiates into the Tomes' process and the 'shoulder' (Figure 7.6D). The 'shoulder' is the flat rim of the secretory surface of the ameloblast from which the Tomes' process arises.

The relative ease and speed of prism evolution is underscored by the very patchy occurrence of prisms in advanced cynodonts and Mesozoic mammals (Clemens, 1997; Sander, 1997a; Wood and Stern, 1997). This pattern can be explained by multiple convergent evolution (Clemens, 1997) as well as by repeated loss (Wood and Stern, 1997), most probably by a combination of both.

7.9.   Summary

Prismless amniote enamel ('reptile' enamel) is characterized by greater structural variety but often lesser complexity than mammalian prismatic enamel. This structural variety has only recently been recognized and necessitated the development of a new descriptive terminology. The new terminology takes the same hierarchical approach as that of Koenigswald (1980) and Koenigswald and Clemens (1992) and recognizes the crystallite level, the module level, the enamel type level, the schmelzmuster level and the dentition level. Common enamel types are columnar enamel, microunit enamel and parallel crystallite enamel. The spatial distribution of enamel types and major discontinuities in an enamel cap makes up the schmelzmuster.

Phylogenetic constraints on enamel microstructure are rare in reptile enamel, and are only seen in the occurrence of microunit enamel and wavy enamel. This particular enamel type of advanced ornithopod dinosaurs seems to reduce abrasion. Apart from wavy enamel, functional constraints on microstructure are difficult to prove.

The best correlation between microstructure and function is documented in adaptive enamel surface morphologies and accompanying cap geometries such as the wrinkles in the bulbous teeth of durophagous reptiles. These taxa document clearly that reptilian amelgenesis is best viewed as enamel surface morphogenesis, and thus tooth surface morphogenesis, in addition to providing a protective cover for the tooth. This is correlated with the continuous tooth replacement of most reptiles which places less emphasis on durability of the enamel layer and more on a consistent tooth surface morphology in each tooth generation.

Amelogenesis in prismless amniote enamel, as documented in the mineralized tissue, differs fundamentally from that of prismatic enamel. The structural units do not reflect individual ameloblasts, and self-organization of growing crystallites may play a significant role in establishing structural patterns. Two types of amelogenesis may be distinguished in prismless enamel: one in which the mineralization front lags behind protein matrix secretion and one in which mineralization closely follows protein matrix secretion. The former applies to most prismless enamel types with complex structures while the latter is confined to parallel crystallite enamel.

Applying these insights to prismless enamel of non-mammalian synapsids leads to the falsification of current models of the evolutionary origin of prismatic enamel. A revised model explains the rapid and convergent evolution of prismatic enamel in Mesozoic mammals.

Acknowledgements

Destructive techniques as employed in enamel research depend heavily on the benevolence of curators and collection managers of other institutions in providing tooth specimens. I am extremely grateful to all those who have supported my research in this way, fortunately they are too numerous to mention.

This paper benefited greatly from the prolonged and intensive exchange of ideas with W. v. Koenigswald (Bonn), H.-U. Pfretzschner (Tübingen), W. A. Clemens (Berkeley), D. N. Stern (Cambridge), and C. B. Wood (Providence). The probing and insightful remarks of two anonymous reviewers and M. M. Smith (London) enabled me to formulate an improved hypothesis of reptilian amelogenesis. I also would like to thank M. M. Smith for the kind invitation to contribute to this volume.

D. Kranz and G. Oleschinski (Bonn) are acknowledged for their highly professional support in preparing the illustrations. This research was made possible by the generous financial support of the Deutsche Forschungsgemeinschaft (project nos. Sa 469/1-1 to Sa 469/1-4 and Ko 627/16-1).

References

Carlson, S. J. (1990*a*). Vertebrate dental structures. In *Skeletal Biomineralization: Patterns, Processes and Evolutionary Trends*, vol. 1, ed. J. G. Carter, pp. 531–556. New York: Van Nostrand Reinhold.

Carlson, S. J. (1990*b*). Part 9. Vertebrate dental structures, plates 176–192. In *Skeletal Biomineralization: Patterns, Processes and Evolutionary Trends*, vol. 2, ed. J. G. Carter, pp. 73–79. New York: Van Nostrand Reinhold.

Carlson, S. J. and Bartels, W. S. (1986). Ultrastructural complexity in reptilian tooth enamel. *Geological Society of America Abstracts with Program*, **18**, 558.

Clemens, W. A. (1997). Characterization of enamel microstructure terminology and applications in systematic analyses. In *Tooth Enamel Microstructure*, eds. W. v. Koenigswald and P. M. Sander, pp. 85–112. Rotterdam: A. A. Balkema.

Cooper, J. S. and Poole, D. F. G. (1973). The dentition and dental tissues of the agamid lizard, *Uromastyx. Journal of Zoology, London*, **169**, 85–100.

Dauphin, Y. (1987*a*). Implications of preparation processes on the interpretation of reptilian enamel structure. *Paläontologische Zeitschrift*, **61**, 331–337.

Dauphin, Y. (1987*b*). Présence de prismes dans l'émail des dents jugales d'*Oligokyphus* (Synapsida, Tritylodontidae): implications phylétiques. *Comptes rendus des Séances de l'Academie des Sciences, Paris*, **304**, 941–944.

Dauphin, Y. (1988). L'email dentaire des reptiles actuels et fossiles: repartition de la structure prismatique, son role, ses implications. *Palaeontographica*, **203**, 171–184.

Dauphin, Y., Jaeger, J. J. and Osmolska, H. (1988). Enamel microstructure of ceratopsian teeth (Reptilia, Archosauria). *Geobios*, **21**, 319–327.

Francillon-Vieillot, H., Buffrenil, V. de, Castanet, J., Géraudie, J., Meunier, F. J., Sire, J. Y., Zylberberg, L. and Ricqles, A. de (1990). Microstructure and mineralization of vertebrate skeletal tissues. In *Skeletal Biomineralization: Patterns, Processes and Evolutionary Trends*, vol. 1, ed. J. G. Carter, pp. 471–530. New York: Van Nostrand Reinhold.

Frank, R. M., Sigogneau-Russell, D. and Hemmerle, J. (1988). Ultrastructure study of triconodont (Prototheria, Mammalia) teeth from the Rhaeto-Liassic. In *Teeth Revisited: Proceedings of the VII International Symposium on Dental Morphology*, eds. D. E. Russell, I.-P. Santoro and D. Sigogneau-Russell, pp. 101–108. Paris: Memories du Muséum National d'Histoire Naturelle, Sciences de la Terre.

Grine, F. E. and Vrba, E. S. (1980). Prismatic enamel: a pre-adaptation for mammalian diphyodonty? *South African Journal of Science*, **76**, 139–141.

Ishiyama, M. (1987). Enamel structure in odontocete whales. *Scanning Microscopy*, **1**, 1071–1079.

Kawasaki, K. and Fearnhead, R. W. (1983). Comparative histology of tooth enamel and enameloid. In *Mechanisms of Tooth Enamel Formation*, ed. S. Suga, pp. 229–238. Tokyo: Quintessence.

Koenigswald, W. v. (1980). Schmelzmuster und Morphologie in Molaren der Arvicolidae (Rodentia). *Abhandlungen der Senckenbergischen Naturforschenden Gesellschaft*, **539**, 1–129.

Koenigswald, W. v. (1997). The variability of enamel at the dentition level. In *Tooth Enamel Microstructure*, eds. W. v.

Koenigswald, and P. M. Sander, pp. 193–202. Rotterdam: A. A. Balkema.

Koenigswald, W. v. and Clemens, W. A. (1992). Levels of complexity in the microstructure of mammalian enamel and their application in studies of systematics. *Scanning Microscopy*, **6**, 195–218.

Koenigswald, W. v. and Sander, P. M., eds. (1997a). *Tooth Enamel Microstructure*. Rotterdam: A.A. Balkema.

Koenigswald, W. v. and Sander, P. M. (1997b). Glossary of terms used for enamel microstructure. In *Tooth Enamel Microstructure*, eds. W. v. Koenigswald and P. M. Sander, pp. 267–280. Rotterdam: A.A. Balkema.

Kvam, T. (1946). Comparative study of the ontogenetic and phylogenetic development of dental enamel. *Norske Tannlaegeforen Tidskrift Supplement*, **56**, 1–198.

Kvam, T. (1958). The teeth of *Alligator mississippensis*. III. Development of enamel. *Journal of Dental Research*, **37**, 540–546.

Lakes, R. (1993). Materials with structural hierarchy. *Nature*, **361**, 511–515.

Lester, K. S. and Archer, M. (1986). A description of the molar enamel of a middle Miocene monotreme (*Obdurodon*, Ornithorhynchidae). *Anatomy and Embryology*, **174**, 145–151.

Lester, K. S. and Koenigswald, W. v. (1989). Crystallite orientation discontinuities and the evolution of mammalian enamel – or, when is a prism? *Scanning Microscopy*, **3**, 645– 663.

Moss, M. L. (1969). Evolution of mammalian dental enamel. *American Museum Novitates*, **2360**, 1–39.

Moss-Salentijn, L., Moss, M. L. and Sheng-tien Yuan, M. (1997). The ontogeny of mammalian enamel. In *Tooth Enamel Microstructure*, eds. W. v. Koenigswald and P. M. Sander, pp. 5–30. Rotterdam: A.A. Balkema.

Osborn, J. W. and Hillman, J. (1979). Enamel structure in some therapsids and Mesozoic mammals. *Calcified Tissue International*, **29**, 47–61.

Peyer, B. (1968). *Comparative Odontology*. Chicago: University of Chicago Press.

Pfretzschner, H. U. (1994). Biomechanik der Schmelzmikrostruktur in den Backenzähnen von Grosssäugern. *Palaeontographica*, A **234**, 1–88.

Piesco, N. P. and Kallenbach, E. (1985). The fine structure of early tooth formation in an iguanid lizard, *Anolis carolinensis*. *Journal of Morphology*, **183**, 165–176.

Poole, D. F. (1956). The structure of the teeth of some mammal-like reptiles. *Quarterly Journal of Microscopical Science*, **97**, 303–312.

Poole, D. F. G. (1957). The formation and properties of the organic matrix of reptilian tooth enamel. *Quarterly Journal of Microscopical Science*, **98**, 349–364.

Poole, D. F. G. (1967). Phylogeny of tooth tissues: enameloid and enamel in recent vertebrates, with a note of the history of cementum. In *Structural and Chemical Organization of Teeth*, ed. A. E. W. Miles, pp. 111–149. New York: Academic Press.

Preuschoft, H., Reif, W. E. and Müller, W. H. (1974). Funktionsanpassungen in Form und Struktur an Haifischzähnen. *Zeitschrift für Anatomie und Entwicklungsgeschichte*, **143**, 315–344.

Rensberger, J. M. (1997). Mechanical adaptation in enamel. In *Tooth Enamel Microstructure*, eds. W. v. Koenigswald and P. M. Sander, pp. 237–257. Rotterdam: A. A. Balkema.

Sahni, A. (1987). Evolutionary aspects of reptilian and mammalian enamel structure. *Scanning Microscopy*, **1**, 1903–1912.

Sander, P. M. (1991). The structure of lower tetrapod tooth enamel and its bearing on the origin of the enamel prism. *Journal of Vertebrate Paleontology*, **11**, 54A.

Sander, P. M. (1992). The ultrastructure of reptilian and non-mammalian synapsid tooth enamel: new descriptive concepts. *Journal of Vertebrate Paleontology*, **13**, 50A–51A.

Sander, P. M. (1997a). Non-mammalian synapsid enamel and the origin of mammalian enamel prisms: the bottom-up perspective. In *Tooth Enamel Mcrostructure*, eds. W. v. Koenigswald and P. M. Sander, pp. 40–62. Rotterdam: A.A. Balkema.

Sander, P. M. (1997b). Jaws and teeth. In *Encyclopedia of Dinosaurs*, eds. P. J. Currie and K. Padian, pp. 717–725. San Diego: Academic Press.

Sander, P. M. (1999). The microstructure of reptilian tooth enamel: terminology, function and phylogeny. *Münchner geowissenschaftliche Abhandlungen, Reihe A*, **38**, 1–102.

Sansom, I. J., Smith, M. P., Armstrong, H. A. and Smith, M. M. (1992). Presence of the earliest vertebrate hard tissues in conodonts. *Science*, **256**, 1308–1311.

Schmidt, W. J. and Keil, A. (1971). *Polarizing Microscopy of Dental Tissue*. Oxford: Pergamon Press.

Slavkin, H., Zeichner-David, M., Snead, M. L., Graham, E. E., Samuel, N. and Ferguson, M. W. J. (1984). Amelogenesis in Reptilia: evolutionary aspects of enamel gene products. *Symposia of the Zoological Society of London*, **52**, 275–304.

Smith, M. M. (1989). Distribution and variation in enamel structure in the oral teeth of sarcopterygians: its significance for the evolution of a protoprismatic enamel. *Historical Biology*, **3**, 97–126.

Smith, M. M. (1992). Microstructure and evolution of enamel amongst osteichthyan fishes and early tetrapods. In *Structure, Function and Evolution of Teeth*, eds. P. Smith and E. Tchernov, pp. 1–19. Tel Aviv: Freund Publishing House.

Stern, N. D. and Crompton, A. W. (1995). A study of enamel organization, from reptiles to mammals. In *Aspects of Dental Biology: Paleontology, Anthropology and Evolution*, ed. J. Moggi-Cecchi, pp. 1–25. Florence: International Institute for the Study of Man.

Ubukata, T. (1994). Architectural constraints on the morphogenesis of prismatic structure in Bivalvia. *Paleobiology*, **37**, 241–261.

Warshawsky, H. (1988). The teeth. In *Cell and Tissue Biology*, ed. L. Weiss, pp. 595–640. Baltimore: Urban & Schwarzenberg.

Woerderman, M. W. (1921). Beiträge zur Entwicklungsgeschichte von Zähnen und Gebiss der Reptilien. IV. Über die Anlage und Entwicklung der Zähne. *Archiv für Mikroskopielle Anatomie*, **95**, 265–395.

Wood, C. B. and Stern, D. N. (1997). The earliest prisms in mammalian and reptilian enamel. In *Tooth Enamel Microstructure*, eds. W. v. Koenigswald, and P. M. Sander, pp. 63–84. Rotterdam: A. A. Balkema.

8 Two different strategies in enamel differentiation: Marsupialia versus Eutheria

W. von Koenigswald

8.1. Introduction

Tooth morphology is one of the most important tools for the identification of fossil mammals. It is not only the outer morphology which bears important information about the phylogeny and biomechanics of ancient animals, but the microstructure of the enamel itself that is also full of information. Using enamel structure only, it is almost impossible to identify taxa at the species level or to reconstruct the specific diet of an animal. Nevertheless, enamel microstructure can be used for more general inferences about biomechanics and phylogeny.

Knowledge of the complex enamel microstructure of mammalian teeth has increased dramatically during recent decades. A hierarchical system of levels of complexity was established (Koenigswald and Clemens, 1992) to facilitate the comparison and correlation of results (Table 8.1). The lowest level of complexity is the crystallite level, followed by the prism level, and the enamel type level. Enamel types are defined by the orientation of enamel prisms. Since several enamel types may form the enamel cap of a tooth, the schmelzmuster level focuses on the three-dimensional arrangement of these enamel types. The highest level of complexity is the dentition level, since different teeth may have different schmelzmusters, especially when they differ distinctly in morphology.

Two stratigraphically separated phases of enamel differentiation can be distinguished in the fossil record (Figure 8.1). During the first phase, prisms became separated from the interprismatic matrix and thus formed the prismatic enamel characteristic of most mammals. This development occurred during the Late Triassic and the Jurassic, and it is most probable that it occurred independently in the various lineages of Mesozoic mammals (Clemens, 1997; Sander, 1997; Wood and Stern, 1997).

Early prismatic enamel is characterized by prisms rising radially in relation to the enamel–dentine junction (EDJ). Therefore radial enamel is regarded as the most primitive type of prismatic enamel. During a second phase, additional enamel types were developed. It is of great significance that these derived enamel types did not replace radial enamel but arose in combination with it forming a complex schmelzmuster. This second phase of enamel differentiation is roughly related to the mammalian radiation during the Tertiary and the Quarternary.

The comparison of marsupial and placental enamel is of great interest since both groups share prismatic (radial) enamel as a plesiomorphic or primitive character. Thus, in all likelihood Marsupialia and Eutheria separated between the two phases of enamel differentiation. Even if the last common ancestor of both subclasses is not yet identified, it probably belongs to the Eupantotheria (Krebs, 1991), and the underived prismatic enamel was probably like that in *Dryolestes* sp. (Lester and Koenigswald, 1989). All further differentiation of enamel microstructure post-dates the dichotomy. In both mammalian subclasses, carnivorous and herbivorous species of small and large body size developed. This placed similar biomechanical demands on the enamel of both groups. These similar conditions should make a comparison of the enamel microstructure in Marsupialia and Eutheria most interesting.

8.2. Methods

The complex terminology used for the description of the enamel microstructure was recently compiled in a glossary by Koenigswald and Sander (1997).

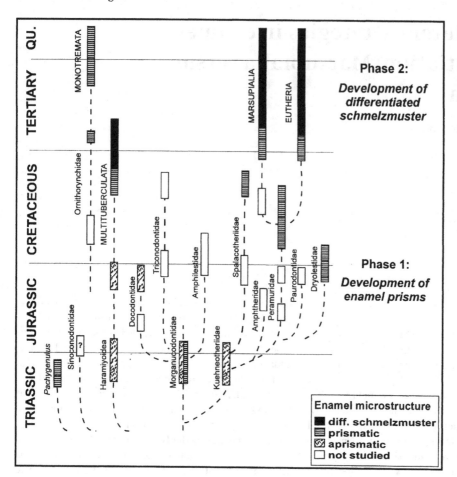

Figure 8.1. **Stratigraphic position of the two phases of differentiation in mammalian enamel. During the Triassic and Jurassic, enamel prisms evolved, most probably in various lineages. During the Upper Cretaceous and Tertiary, Multituberculata, Marsupialia and Eutheria apparently developed differentiated schmelzmuster with various enamel types independently. (Phylogenetic tree after Carroll (1988), data on the enamel partially after Wood and Stern (1997) and Sander (1997).)**

Previous research focused on different aspects of enamel. That made comparisons difficult. The hierarchical system established by Koenigswald and Clemens (1992) allowed easier comparisons at the various levels. However, to identify the enamel types of the schmelzmuster, an understanding of the three-dimensional arrangement of prisms is needed. There are some non-destructive methods to study enamel microstructure, especially the prism cross-sections near the outer surface (Boyde, 1985). Unfortunately, important information about the course of the prisms between the EDJ and the outer enamel surface (OES) cannot be gained by these methods.

Thus, to reconstruct the spatial arrangement of the various enamel types, vertical, transverse, and tangential sections are required. Sections are made from teeth embedded in resin, and the sections are etched to make the prisms visible. Documentation of schmelzmuster and enamel types was mainly done with a scanning electron microscope, but reflected light microscopy, especially in dark field, is also very helpful for an easy and fast overview because of the 'fibre optic' effects of prisms (Koenigs-

wald and Pfretzschner, 1987). Enamel investigation techniques are described in more detail in Koenigswald (1980), Koenigswald and Clemens (1992), and Martin (1992).

Table 8.1. *Levels of complexity in the mammalian enamel (according to Koenigswald and Clemens, 1992)*

Levels	Characterization
Dentition	Variation in schmelzmuster throughout the dentition
Schmelzmuster	Three-dimensional arrangement of enamel types in one tooth
Enamel type	Orientation of prisms relative to the EDJ and to one another (parallel or decussating, and nature of the decussation). Differences in orientation of IPM crystallites relative to prisms are recognized as subtypes
Prism type	Prism type describes the characters of a prism (particularly cross-section, absolute prism size, prism density, crystallite orientation inside prisms, IPM surrounding prisms, and seam)
Crystallite	Orientation of crystallites

Note: EDJ, enamel–dentine junction; IPM, interprismatic matrix.

Table 8.2. *Distribution of structural enamel features in Didelphidae and Australian marsupials*

Taxon	Tooth position	Radial enamel	Simultaneous prism deviation	HSB	Irregular enamel	Zipper enamel	Border line	Inter-row sheets
		Prisms remain parallel		Decussating prisms				
Didelphidae	M	■						
Dasyuridae	M	■			(■)			
	C	■						
Thylacinidae	M	■			■			
	C	■			■			
Peramelidae	M	■						■
Phascolarctidae	M	■					■	
	I	■						
Diprotodontidae	M	■	■			■		■
	I	■			■			
Vombatidae	M	■		■				
	I	■		■				
Thylacoleonidae	P	■	■		(■)			■
	I	■			■			
Phalangeridae	M	■	■				(■)	■
	I	■						
Potoroidae	M	■	■			■		■
	I	■	■					■
Macropodidae	M	■	■		(■)	■		■
	I	■	■			■		■
Pseudocheiridae	M	■	■				(■)	■
	I	■	■					■
Petauridae	M	■	■					■
	I	■	■					■
Acrobatidae	M	■						
	I	■						

Note: I, incisor; C, canine; P, premolar; M, molar; ■ regularly present; (■) occasionally present; HSB, Hunter–Schreger bands.

The enamel of the Eutheria has been studied by a great number of authors (see Koenigswald, 1997a for a survey of the enamel differentiation at the schmelzmuster level). By contrast, marsupial enamel has not been studied as frequently. Fortunately, most authors (e.g. Boyde and Lester, 1967) have used the same terminology in studies of marsupial and placental enamel. The enamel microstructure of fossil and extant Marsupialia from Australia are summarized in Table 8.2 based on recent investigations (Koenigswald, 1994a). Here a basic comparison of the enamel types in both subclasses shall be given.

It is important to note that, at the crystallite or prism level, similarities between marsupials and placentals may be related to common heritage. But when details of derived enamel types, like Hunter–Schreger bands or inter-row sheets are discussed, it should be remembered that these structures were probably developed independently. Most of the enamel types found in placentals can be found in marsupials as well (Koenigswald, 1994a) so that very few additional terms for specific structures need to be defined. However, the *frequency* of the occurrence of the enamel types differs widely between Eutheria and Marsupialia. These differences are so obvious that they can be easily described – even when a calculation in percentages would be extremely difficult.

The comparison here describes the various enamel types and their significance in both groups. In addition, an evolutionary model is given as a possible explanation for the observed differences.

8.3.   Results

8.3.1.   Radial enamel

Radial enamel is characterized by prisms rising apically in a radial direction. It is found in most Mesozoic mammals. Due to their orientation, prisms penetrate the occlusal surface at almost right angles. According to several observations, this condition seems to reduce the effects of abrasion, especially when opposing tooth surfaces are sliding past each other (Rensberger and Koenigswald, 1980; Boyde and Fortelius, 1986). Therefore radial enamel, even if primitive, is very effective in maintaining sharp cutting edges.

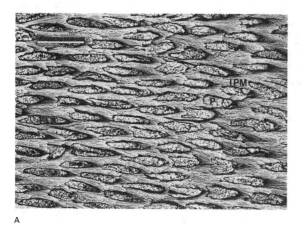
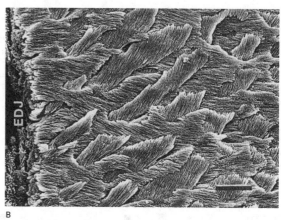

A   B

Figure 8.2A,B. Radial enamel in *Dasyurus viverrinus*. A. In transverse section the rising prisms are truncated. The interprismatic matrix (IPM) is anastomosing (KOE 1718 exp. 33262; scale bar represents 10 μm). B. In the longitudinal section the angle between rising prisms and IPM is visible (KOE 1718, exp. 33257; scale bar represents 10 μm). (The material used and indicated by KOE numbers is stored in the enamel collection of the University Bonn. The scanning electron micrographs, indicated by exp. numbers, were taken by the author at the Palaeontological Department at the University Bonn. EDJ indicates the enamel–dentine junction.)

In Marsupialia this enamel type is dominant. It is the only enamel type in most small dasyurids (Figures 8.2), and the interprismatic matrix (IPM) is usually at a large angle to the prism direction (Figure 8.2B). Even the molars of large marsupials like *Macropus titan* or *Diprotodon optatum* show radial enamel as the dominant enamel type.

In placental mammals of small body size (e.g. most Lipotyphla) radial enamel is common and often forms the only enamel type. With increasing body size, other enamel types are incorporated into the schmelzmuster, mainly Hunter–Schreger bands. Nevertheless, radial enamel plays a significant role in almost every schmelzmuster of placental mammals, even if only as a small layer. An analysis of evergrowing molars in various rodents (Koenigswald *et al.*, 1994) demonstrates that the radial enamel occurs primarily on the pushside of the enamel band (i.e. the side which has first contact to the antagonist during mastication). Only very few mammals of large body size have radial enamel as the dominant enamel type, for instance the Paleocene periptychid *Carsioptychus* (Koenigswald *et al.*, 1987) among terrestrial mammals and the sirenians among marine mammals (Pfretzschner, 1994).

Even if radial enamel is currently present in the teeth of most placental mammals, it has been largely replaced by more derived enamel types during the course of evolution. Moreover, radial enamel itself underwent a progressive change in evolution due to a reorientation of the IPM. In the primitive condition, the IPM varies in direction but remains more or less parallel to the prism direction, as in early primates and in the outer enamel of Carnivora. The prism sheath and the seam are the major discontinuities where crystallites meet at angles (Lester

and Koenigswald, 1989). At a more derived stage, the IPM is more uniformly orientated and at a large angle to the prism direction, as in Lipotyphla. The IPM may anastomose between prisms or form inter-row sheets (Boyde, 1965). Herbivorous mammals with hypsodont molars developed a specific modification of the radial enamel serving as the innermost layer near the EDJ and meeting specific biomechanical demands related to hypsodonty as in bovids or equids (Pfretzschner, 1994). In other words, due to differences in the elasticity of dentine and enamel, severe stress concentrations occur at the EDJ in hypsodont molars. The stress is directed radially and, in the modified radial enamel, the inter-row sheets, as well as the prisms, seem to be oriented optimally to counter these stresses.

8.3.2.   Tangential enamel and the simultaneous prism deviation (SPD)

In tangential enamel, as in radial enamel, all prisms are parallel to each other and do not decussate. In contrast to radial enamel where prisms are directed radially outward from the EDJ, the prisms of tangential enamel show a distinct lateral deviation. In derived stages, the prisms do not raise apically at all but are strictly horizontal, for example, in arvicoline molars parallel to the occlusal surface. Tangential enamel is often combined with radial enamel, from which this enamel type is evolutionarily derived (Koenigswald, 1997b) (also see Figures 8.3). At the transition between these enamel types, the prisms have to change direction. This occurs at a straight line parallel to the EDJ and is defined as 'simultaneous prism deviation' (SPD). In contrast to the various forms of prism decussation, all prisms remain parallel to each other and retain the same organization of neighbouring prisms.

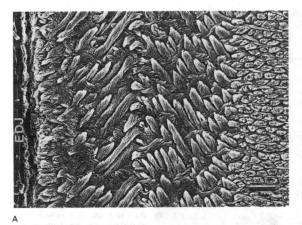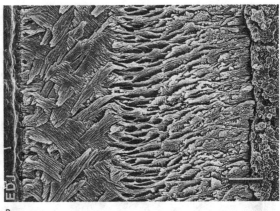

A
B

Figure 8.3A,B. **The simultaneous prism deviation (SPD) is common in Marsupialia but an exception among Eutheria. EDJ indicates the enamel–dentine junction, thus inner is left, outer is right. (Scale bars represent 10 μm. Detail of the cross-section of the incisor enamel of the marsupial *Thylogale thetis*. The position is indicated in Figure 8.5B. Two distinct simultaneous prism deviations separate two layers of tangential enamel in the inner layer and the radial enamel in the outer layer (KOE 1677, exp. 33253. Simultaneous prism deviation in the incisor of the placental soricid *Blarina brevicauda*. An inner layer of tangential enamel is overlain by radial enamel (KOE 1435, exp. 33251).**

In Marsupialia, simultaneous prism deviation is a very frequent structural element. Often it occurs several times in a sequence (Figure 8.3A). In the elongated incisors of kangaroos, three layers of enamel with different lateral inclination are superimposed and separated by simultaneous prism deviations. If one accepts the longitudinal margin as the occlusal surface of the procumbent incisors, the two inner layers may be regarded as tangential enamel (Figures 8.3). In some areas, macropodid incisors show an even higher number of simultaneous prism deviations. In molars of diprotodontid marsupials such multiple simultaneous prism deviations are found regularly (see Figure 8.5B).

The enamel of the fossil South American marsupials is almost unknown, but the incisors of the fossil rodent-like marsupial *Groeberia* displayed simultaneous prism deviations too (Koenigswald and Pascual, 1990).

In Eutheria, tangential enamel, as well as simultaneous prism deviation, is very rare, despite the fact that tangential enamel was discovered and defined in the molars of arvicoline rodents. It is limited to the molars of a few cricetine rodents (Loenigswald, 1980; Fahlbusch, 1987). Tangential enamel also occurs in the incisors and molars of some soricines (Figure 8.3B; Koenigswald, 1997b). Even if this enamel type might be detected in some additional placental mammals, it remains an exception for this group, while it is very common in Marsupialia.

It may be added here that simultaneous prism deviation occurs as well in the incisor schmelzmuster of ptilodontoid Multituberculata (Sahni, 1979). Radial enamel forming an inner layer continues with a simultaneous prism deviation into tangential enamel. This is the only enamel differentiation at the schmelzmuster level known from Mesozoic mammals.

8.3.3.   Prism decussation

The decussation of enamel prisms is a common feature in placental mammals (Boyde, 1969), but as discussed elsewhere (Koenigswald, 1997a), several types of prism decussation have to be differentiated. These have been designated Hunter–Schreger bands, irregular enamel, and Zipper enamel.

Hunter–Schreger bands (HSB)

Hunter–Schreger bands were originally described as light and dark bands in enamel viewed by light microscopy (Hunter, 1778; Schreger, 1800). Koenigswald and Pfretzschner (1987) demonstrated that the correlation of prism orientation and light and dark banding was due to a 'fibre optic' effect. HSB are layers of decussating prisms which normally are oriented transversely to the growing axis of a tooth (Koenigswald and Sander, 1997). Prisms change from one band to the next forming transitional zones in which the prisms change direction. One of the characteristic features of HSB is their regular bifurcation in a very specific pattern (Figure 8.4B). Each set of bands only bifurcates in one direction.

In Eutheria, HSB are the characteristic enamel type, at least in most forms with the body size of a hare or larger. In Primates and other groups, HSB become more distinct with increasing body size. In Lipotyphla, HSB have only been found in *Erinaceus* and the giant hedgehog *Deinogalerix* (Koenigswald, 1997a). In rodents, well-developed HSB are found at least in their incisors. This indicates that HSB developed several times among Eutheria (Koenigs-

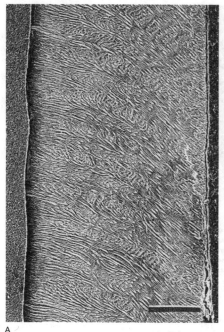

A

B

Figure 8.4A,B. **Hunter–Schreger bands (HSB) common in Eutheria occur in Marsupialia only in Vombatidae. Here the incisor of the extant marsupial *Vombatus ursinus* is illustrated. A. In the longitudinal section of the incisor. An inner zone of radial enamel present near the EDJ (right). The HSB are visible as decussating layers of prisms. They extend almost to the outer surface and are well identifiable (B) from the outside under lateral illumination (KOE 265, exp. 22135, scale bar represents 100 μm). B. Due to the light guide effect, the different prism orientations within the HSB results in light and dark bands. Dark and light areas reflect (thick) prism layers of different orientation. In light bands prisms point towards the source of light (to the left). The light bands bifurcate towards the right, dark ones to the left. The regular bifurcation is characteristic for HSB.**

wald, 1997*a*). HSB also show a great variation in orientation and thickness among the various groups of Eutheria (Kawai, 1955). In some groups, the lateral orientation of the HSB – as seen in tangential sections or from the outside – is strongly modified. Vertical HSB are typical for Rhinocerotidae (Rensberger and Koenigswald, 1980) and U-shaped HSB occur in Chalicotheriidae and Brontotheriidae (Koenigswald, 1994*b*). In some rodent incisors, various kinds of reorientation were observed (Wahlert and Koenigswald, 1985; Bruijn and Koenigswald, 1994).

The occurrence of HSB, predominantly in animals of larger body size, has traditionally been interpreted biomechanically, as a protection against the increased tension forces generated in mastication (Koenigswald *et al.*, 1987). The net effect is that the decussating layers of prisms probably function as crack-stopping mechanisms (Pfretzschner, 1988; Chapter 18).

In Marsupialia, HSB have only been found to date in the continuously growing incisors and molars of Vombatidae (Figure 8.4). The enamel of *Vombatus ursinus* lacks enamel tubules characteristic of most marsupial enamels. One might assume that the presence of HSB with strongly bent prisms prohibit the existence of tubuli. Lester et al. (1988) demonstrated that *Tarsipes* and *Yalkaparidon* lack tubules as well, but have no HSB. Gilkeson (1997) demonstrated that tubules follow prisms through strong bends passing a simultaneous prism deviation. Therefore a correlation of HSB with the lack of tubules is not convincing.[1] In comparison with placentals, the lack of HSB in larger marsupials is astonishing. Special emphasis was placed on the

study of teeth of large marsupials like *Thylacinus, Diprotodon, Thylacoleo* and giant kangaroos, but surprisingly nowhere could HSB be identified (Koenigswald, 1994*a*). Therefore HSB remain a rare exception among Marsupialia, despite the fact that this structure which is regarded to have such a great biomechanical significance in placental teeth. Why marsupials could survive without HSB will be discussed below.

Irregular enamels

In contrast to the HSB, the 'irregular' enamels form a diverse group of structures. Their common character is the decussation of prisms in irregular bundles or as individual prisms (Figure 8.5A). They do not form regular layers.

In Eutheria, irregular enamel occurs in a few groups. The most significant occurrence is the '3-D enamel' of Proboscidea (Pfretzschner, 1994). In lagomorph molars, irregular enamel (Figure 8.5B) is combined with radial enamel (Koenigswald and Clemens, 1992). In rodent molars, irregular enamel occurs in a few caviomorphs and arvicolines (Koenigswald *et al.*, 1994). In all of these cases, it is most plausible that the irregular enamel derived from various modifications of HSB.

In Marsupialia, irregular enamel has also been found in several groups, as in *Thylacinus*, in *Diprotodon* incisors, *Thylacoleo* and in a few macropodids (Young *et al.*, 1990;

---

[1] Beier (1983) supposedly noted HSB from several other Marsupialia, but they turned out to represent other enamel types.

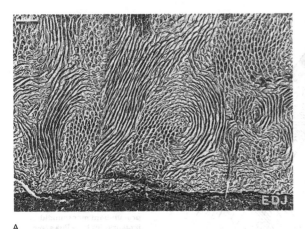
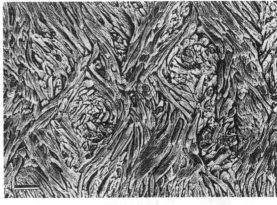

A   B

Figure 8.5A,B. **Irregular enamels in Marsupialia and Eutheria. A.** *Thylacinus cynocephalus,* **longitudinal section of the enamel of canine with irregular enamel. From tangential sections it is obvious that prisms decussate in bundles and not in layers (KOE 1721, exp. 26524–6; scale bar represents 30 µm). B.** *Hypolagus brachygnathus,* **enamel of a premolar in a transverse section. The irregular enamel forms the inner layer (radial enamel from the outer layer is just visible at the bottom) (KOE 1062, exp. 11451; scale bar represents 10 µm).**

Koenigswald, 1994*a*; Stefen, 1995). However, this enamel type normally is restricted to specific areas of specific teeth and occurs together with radial enamel. Thus, in contrast to the occurrence in Eutheria, no relation to HSB has been found.

Zipper enamel and border zone

Two structural elements were found in marsupials not occurring in placentals. Both occur where areas of radial enamel from the two lateral sides of a tooth meet in a crest. The first, found in sharp edges between two fields of radial enamel, has been termed zipper enamel. In a very narrow strip along the crest, prisms decussate very regularly in thin layers at about right angles. This structure differs from HSB mainly in its lack of lateral extension (Koenigswald, 1994*a*), in other words, it is strictly limited to the edge of the tooth and doesn't extend further laterally. Zipper enamel was found in *Diprotodon* molars, in the cutting edges of the premolar of *Thylacoleo*, and in the crests of molars and incisors of Macropodidae (Figure 8.6).

The second structural element is a distinct zone of changing prism direction. It was found in *Phascolarctus* by Young *et al.* (1990) and occurs in a similar fashion in some Phalangeridae and Pseudocheiridae (Koenigswald, 1994*a*). It is not yet possible to reconstruct the prism course in this narrow zone since the structure is not nearly as regular as in the zipper enamel. This feature has been termed the 'border zone' (Koenigswald, 1994*a*). However, placental mammals show no such feature in this topographical situation.

8.4.   Conclusions

The enamel of Eutheria and Marsupialia is formed by the same enamel types. Except for the radial enamel inherited by both groups from their last common ancestor, the more derived enamel types were apparently developed independently in each group. The biomechanical benefits of the various enamel types reflect general mechanical concerns that transcend traditional dietary categories. For instance, radial enamel is regarded as resistant to abrasion while HSB prevent large-scale fracture of enamel, yet both occur in herbivorous and carnivorous mammals. Therefore the preference for a specific diet can generally not be deduced solely from enamel structure. For such a purpose, the combination of dental morphology (Kay, 1975; Maier, 1984), dental microwear (Teaford, 1988, 1993) and enamel structure analysis may be particularly useful (Teaford *et al.*, 1996).

Another important point to be drawn from this study is that marsupial and placental enamel can *not* be structurally distinguished in most cases. Only very few structural elements like the zipper enamel type and border zone occur exclusively in Marsupialia, but these structures are limited to a few specific taxa. Tubules cannot be used either. Although they are common in marsupial enamel, they are missing in Vombatidae and a few more taxa (Tomes, 1849; Boyde and Lester, 1967; Lester *et al.*, 1988; Gilkeson and Lester, 1989). On the other hand, short tubules are found in many placental mammals like Lipotyphla and Primates. Only if tooth morphology narrows down the taxonomic affinities, can enamel be used to distinguish between groups (as shown for *Zygomaturus* by Rich *et al.* 1987). Again, if the same derived

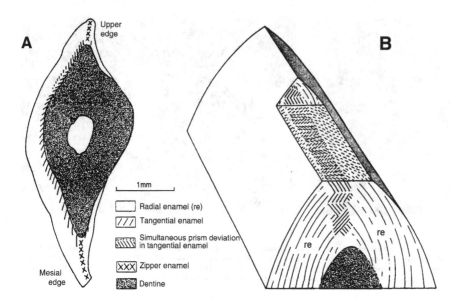

Figure 8.6A. Diagrammatic vertical cross-section of a procumbent macropodid incisor indicating the position of layers of tangential enamel and the zipper enamel in the upper and mesial edges. A detail of this cross-section is shown as a scanning electron micrograph in Figure 8.3A. B. Diagrammatic aspect of a molar loph with the zipper enamel in the cutting edge between fields of radial enamel (re). Even if the prisms decussate, zipper enamel differs from HSB due to the strict lateral limitation. The zipper enamel in the upper and mesial edges of the procumbent macropodid incisors similarly links two fields of radial enamel.

enamel types occur in Eutheria and Marsupialia they probably developed convergently.

Due to the great variety of tooth morphology among placentals and marsupials, one might expect a much greater diversity of enamel types in these two groups. Even if the details of amelogenesis remain unknown, this probably reflects limitations in the basic setup of enamel-secreting ameloblasts allowing only a very limited number of possible prism orientations and the IPM angles. If we follow the traditional view (e.g. Boyde, 1965; Schroeder, 1987) that the prisms are the mineralized pathway of one secreting ameloblast and of its Tomes' process; ameloblast mobility in the epithelium has to be investigated in order to understand these constraints imposed by amelogenesis (Seilacher, 1970; Reif, 1975).

Even if the same enamel types occur in Marsupialia and in Eutheria, great differences can be observed in the *frequency* of their occurrence. This is summarized in Table 8.3. Marsupialia generally exhibit radial enamel with simultaneous prism deviations, while in Eutheria, HSB are most common. Besides the documentation of differences between placental and marsupial enamel, a basic underlying question is why the Marsupialia retained radial enamel as the dominant enamel type while Eutheria developed HSB?[2]

The traditional argument is that HSB give the tooth a series of decussating structural elements, much like those in plywood, which can stop crack propagation. But is this the only way to achieve this goal? In the underived

state of mammalian enamel (e.g., dryolestids), prisms and IPM are not well separated and the angle between prisms and IPM is low and varying (Sander, 1997). By contrast, Marsupialia from the Cretaceous show an increasing angle between the IPM and prism direction (Wood, 1991, 1992). In *Didelphis virginiana* the enamel is already derived and shows a very distinct and regular angle between prisms and IPM (Figure 8.7).

Thus it seems that there might be two different strategies to acquire a reinforcing enamel structure: (1) prisms can be crossed with IPM, or (2) prisms can be crossed with prisms.

Marsupials seem to have chosen the strategy of crossing prisms with IPM at an early stage (at least in the upper Cretaceous; Wood and Stern, 1997). Perhaps the strategy was so successful that they did not have to change it even when they grew larger. The benefit of this strategy is that the enamel-secreting ameloblasts do not have to change their positions with regard to neighbouring ameloblasts. The development of inter-row-sheets, or Boyde's type 2 enamel, is linked with an IPM orientated at a high and regular angle to the prisms. Simultaneous prism deviation is the logical and subsequent step in enamel differentiation. Only in a *further* step, could true prism decussation, as in HSB, zipper enamel or irregular enamel, follow.

The Eutheria followed the second strategy. The IPM was kept parallel to the prisms for a long time. In order to achieve a decussating structure, prisms were crossed with prisms forming HSB. The structure of prisms decussating in HSB, with the IPM parallel to the prisms, is still present in various orders. In Carnivora, the IPM is mainly parallel to the prisms, in radial enamel as well as in HSB

---

[2] The idea that Marsupialia retained the more primitive enamel because their ameloblasts cannot construct HSB is falsified by the Vombatidae.

Table 8.3. *Comparison of the occurrence of structural enamel features in Marsupialia and Eutheria*

Enamel features	Marsupialia	Eutheria
Radial enamel	Dominant	Always present, but mostly subordinated
Interprismatic matrix (a) IPM at low but varying angle to prisms (b) IPM at an angle (c) Inter-row sheets (d) Modified radial enamel	(a) Not in Australian marsupials (b) Regularly (c) Regularly (d) E.g. in *Diprotodon*	(a) Plesiomorphic condition (b) Increasingly during evolution (c) E.g. in *Lipotyphla* (d) In hypsodont molars
Simultaneous prism deviation	Regularly	Only in molars of some Cricetinae and in Arvicolinae
Hunter–Schreger bands	Exclusively in Vombatidae	Regularly in larger mammals
Irregular enamel	In several cases, *Not* derived from HSB	In several cases, derived from HSB
Zipper enamel	Diprotodontia, Thylacoleonidae, Macropodidiae	*Not* present
Border zone	Phascolarctidae, Phalangeroidea	*Not* present

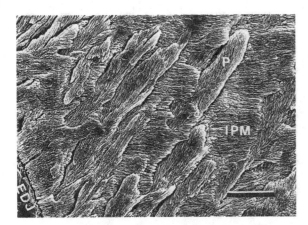

Figure 8.7. ***Didelphis virginiana*, longitudinal section of a lower premolar. The crystallites of the interprismatic matrix (IPM) are oriented at a distinct angle towards the prism (P) direction (KOE 2733 exp. 33472).**

8.5.   Summary

Eutheria and Marsupialia share prismatic enamel as common heritage. Further differentiation of enamel types and complex schmelzmusters occurred independently. While the same basic enamel types were developed in both groups, the frequency with which these enamel types occur differs widely. In Marsupialia, radial enamel with simultaneous prism deviation is dominant. Eutheria generally exhibit Hunter–Schreger bands. This difference can be interpreted biomechanically. At an early stage of evolution, the Marsupialia crossed prisms and interprismatic matrix to achieve a reinforced, plywood-like structure. Eutheria kept the interprismatic matrix parallel to the prisms but crossed prisms with prisms in Hunter–Schreger bands. So both groups reached a similar plywood structure in different ways. This may be a classic case of parallel evolution.

Note added in proof: While this paper was in press, the author had the opportunity to investigate many marsupials from South America (Koenigswald and Goin, 2000). Basically, they seem to follow the same strategy as the Australian marsupials. Both groups share the first step of enamel differentiation, the decussation of prisms and IPM. As a consequence, both groups evolved synchronous prism deviation as the primary way to strengthen the enamel. According to the systematic and stratigraphic distribution patterns of synchronous prism deviation, this feature must have arisen several times independently. This strengthens the conclusions drawn from the Australian marsupials.

In the further differentiation of the enamel of South American marsupials, differences and similarities with Australian marsupials were observed. Neither zipper enamel, nor the border line or irregular enamel was found in the South American marsupials. HSB, however, occur in a very restricted group of Borhyaenidae. It is

(Figure 8.8A). Only at a later stage of differentiation is the IPM reorientated to include an angle with the prisms as in most derived ungulates as well as in the prosimian primate *Daubentonia* (Figure 8.8B). Even in the highly derived uniserial HSB in some rodent incisors (Sciuridae and some Myoxidae) the IPM is parallel to the prisms (Figure 8.8C). In some myoxids and most Myomorpha (Figure 8.8D) the IPM is reorientated to form a right angle with both sets of prism layers thus strengthening the third direction in an orthogonal system. Martin (1992) could distinguish two groups of caviomorph rodents by the orientation of the IPM in relation to the prism.

If this model can be confirmed by further investigations, the differentiation of the enamel in Eutheria and Marsupialia can be used as a classical example of parallel evolution, where, from a common starting point, different strategies were followed. From this perspective, comparable enamel types were developed due to the limited potential of the enamel-forming cells.

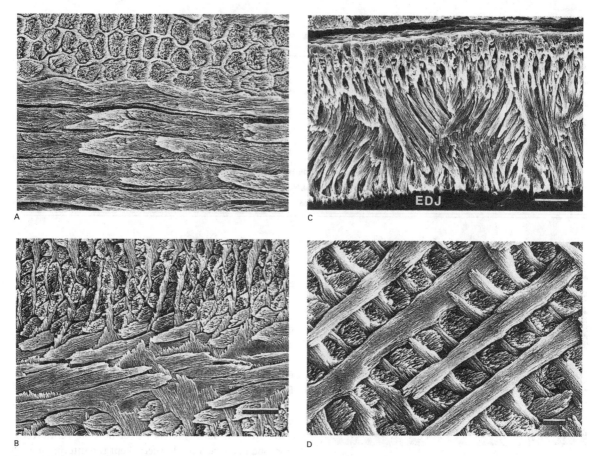

Figure 8.8A–D. **Orientation of IPM in placental enamels. A. Thick HSB in placental *Smilodon fatalis (= californicus)* with the IPM oriented parallel to the prisms as common in Carnivora (KOE 358 exp. 33286; scale bar represents 10 μm). B. *Daubentonia madagascariensis*, longitudinal section through lower incisor with IPM at a high angle to the prisms in HSB (KOE 275 exp. 33295; scale bar represents 10 μm). C. Uniserial HSB in the lower incisor of *Myoxus glis* with the IPM oriented parallel to the prisms (KOE 35 exp. 33279; scale bar represents 10 μm). C. Uniserial HSB in the lower incisor of *Dicrostonyx torquatus* IPM at right angles to both sets of prisms (KOE 1003 exp. 99068; scale bar represents 3 μm).**

noteworthy that this group, like all Borhyaenidae, was carnivorous, while the Vombatidae, the only group of Australian marsupials to develop HSB, feed only on plant material. Thus, enamel types are not related to specific diets. Enamel differentiation more probably reflects the tendency to strengthen the enamel in general with increasing masticatory forces (Koenigswald *et al.*, 1987).

Acknowledgements

This study was supported by the Deutsche Forschungsgemeinschaft (KOE 627–18/1). For technical help in preparing the illustrations I am indebted to G. Oleschinski and D. Kranz from the Department of Palaeontology in Bonn. Special thanks are due to Mark Teaford for his comments on this paper and his help in improving the English.

References

Beier, K. (1983). Hunter–Schreger-Bänder im Zahnschmelz von Känguruhs (Macropodinae, Marsupialia). *Zoologischer Anzeiger*, **210**, 315–332.

Boyde, A. (1965). The structure of developing mammalian dental enamel. In *Tooth Enamel*, ed. M. G. Stark, and R. W. Fearnhead, pp. 163–194. Bristol: Wright.

Boyde, A. (1969). Electron microscope observation relating to the nature and development of prism decussation in mammalian dental enamel. *Bulletin du Groupement International pour la Recherche Scientifique en Stomatologie*, **12**, 151–207.

Boyde, A. (1985). The tandem scanning reflected light microscope. Part 2. Pre-micro '84 applications at UCL. *Proceedings of the Royal Microscopical Society*, **20/3**, 131–139.

Boyde, A. and Fortelius, M. (1986). Development, structure and function of rhinoceros enamel. *Zoological Journal of the Linnean Society*, **87**, 181–214.

Boyde, A. and Lester, K. S. (1967). The structure and development of marsupial enamel tubules. *Zeitschrift für Zellforschung*, **82**, 558–576.

Bruijn, H. de and Koenigswald, W. v. (1994). Early Miocene rodent faunas from the eastern Mediterranean area. Part V. The genus *Enginia* (Muroidea) with a discussion of the structure of the incisor enamel. *Proceedings of the Koninklijke Nederlandse Akademie van Wetenschappen*, **97**, 381–405.

Carroll, R. L. (1988). *Vertebrate Paleontology and Evolution*. New York: W. H. Freeman.

Clemens, W. A. (1997). Characterization of enamel microstructure terminology and applications in systematic analyses. In *Tooth Enamel Microstructure*, eds. W. v. Koenigswald and P. M. Sander, pp. 85–112. Rotterdam: A. A. Balkema.

Fahlbusch, V. (1987). The Neogene mammalian faunas of Ertemte and Harr Obo in Inner Mongolia (Nei Mongol), China. 5. The genus *Microtoscoptes* (Rodentia: Cricetidae). *Senckenbergiana Lethaea*, **67**, 345–373.

Gilkeson, C. (1997). Enamel tubules in Australian marsupials. In *Tooth Enamel Microstructure*, eds. W. v. Koenigswald and P. M. Sander, pp. 113–121. Rotterdam: A. A. Balkema.

Gilkeson, C. F. and Lester, K. S. (1989). Ultrastructural variation in the enamel of Australian marsupials. *Scanning Microscopy*, **3**, 177–191.

Hunter, J. (1778). *The Natural History of the Human Teeth: Explaining their Structure, Use, Formation, Growth and Diseases*, second edution. London: J. Johnson.

Kawai, N. (1955). Comparative anatomy of bands of Schreger. *Okajimas Folia Anatomica Japonica*, **27**, 115–131.

Kay, R. F. (1975). Functional adaptations of primate molar teeth. *American Journal of Physical Anthropology*, **43**, 195–216.

Koenigswald, W. v. (1980). Schmelzmuster und Morphologie in den Molaren der Arvicolidae (Rodentia). *Abhandlungen der Senckenbergischen Naturforschenden Gesellschaft*, **539**, 1–129.

Koenigswald, W. v. (1994a). Differenzierungen im Zahnschmelz der Marsupialia im Vergleich zu den Verhältnissen bei den Placentalia (Mammalia). *Berliner Geowissenschaftliche Abhandlungen*, **E13**, 45–81.

Koenigswald, W. v. (1994b). U-shaped orientation of Hunter–Schreger bands in the enamel of *Moropus* (Chalicotheriidae, Mammalia) in comparison to some other Perissodactyla. *Annals of the Carnegie Museum*, **63**, 49–65.

Koenigswald, W. v. (1997a). Brief survey of enamel diversity at the schmelzmuster level in Cenozoic placental mammals. In *Tooth Enamel Microstructure*, eds. W. v. Koenigswald and P. M. Sander, pp. 137–161. Rotterdam: A. A. Balkema.

Koenigswald, W. v. (1997b). Evolutionary trends in the differentiation of mammalian enamel ultrastructure. In *Tooth Enamel Microstructure*, eds. W. v. Koenigswald and P. M. Sander, pp. 203–235. Rotterdam: A. A. Balkema.

Koenigswald, W. v. and Clemens, W. A. (1992). Levels of complexity in the microstructure of mammalian enamel and their application in studies of systematics. *Scanning Microscopy*, **6**, 195–218.

Koenigswald, W. v. and Goin, F. (2000). Enamel differentiation in South American marsupials and a comparison of placental and marsupial enamel. *Paleontographica*, **A256** (in press).

Koenigswald, W. v. and Pascual, R. (1990). The Schmelzmuster of the Paleogene South American rodentlike marsupials *Groeberia* and *Patagonia* compared to rodents and other Marsupialia. *Paläontologische Zeitschrift*, **64**, 345–358.

Koenigswald, W. v. and Pfretzschner, H. U. (1987). Hunter-

Schreger-Bänder im Zahnschmelz von Säugetieren: Anordnung und Prismenverlauf. *Zoomorphology*, **106**, 329–338.

Koenigswald W. v. and Sander, P. M. (1997). Glossary for terms used for enamel microstructure. In *Tooth Enamel Microstructure*, eds. W. v. Koenigswald and P. M. Sander, pp. 267–280. Rotterdam: A. A. Balkema.

Koenigswald, W. v., Rensberger, J. M. and Pfretzschner, H. U. (1987). Changes in the tooth enamel of early Paleocene mammals allowing increased diet diversity. *Nature*, **328**, 150–152.

Koenigswald, W. v., Sander, M., Leite, M., Mörs, T. and Santel, W. (1994). Functional symmetries in the schmelzmuster and morphology in rootless rodent molars. *Zoological Journal of the Linnean Society*, **110**, 141–179.

Krebs, B. (1991). Das Skelett von *Henkelotherium guimarotae* gen et sp. nov. (Eupantotheria, Mammalia) aus dem Oberen Jura von Portugal. *Berliner Geowissenschaftliche Abhandlungen*, **A133**, 1–110.

Lester, K. S. and Koenigswald, W. v. (1989). Crystallite orientation discontinuities and the evolution of mammalian enamel – or, when is a prism? *Scanning Microscopy*, **3**, 645–663.

Lester, K. S., Archer, M., Gilkeson, C. F. and Rich, T. (1988). Enamel of *Yalkparidon coheni*: representative of a distinct order of Tertiary zalambdodont marsupial. *Scanning Microscopy*, **2**, 1491–1501.

Maier, W. (1984). Tooth morphology and dietary specialization. In *Food Acquisition and Processing in Primates*, eds. D. J. Chivers, B. A. Wood and A. Bilsborough, pp. 303–330. New York: Plenum.

Martin, T. (1992). Schmelzmikrostruktur in den Inzisiven alt- und neuweltlicher hystricognather Nagetiere. *Palaeovertebrata*, *Mémoire Extraordinaire*, 1992, 1–168.

Pfretzschner, H. U. (1988). Structural reinforcement and crack propagation in enamel. In *Teeth Revisited. Proceedings of the VIIth International Symposium on Dental Morphology*. ed. D. E. Russell, J.-P. Santoro and D. Sigogneau-Russell, pp. 133–143. Mémoires Museum National de Histoire Naturelle (sér. C), 53.

Pfretzschner, H. U. (1994). Biomechanik der Schmelzmikrostruktur in den Backenzähnen von Grossäugern. *Palaeontographica*, **A 234**, 1–88.

Reif, W.-E. (1975). Lenkende und limitierende Faktoren in der Evolution. *Acta Biotheoretica*, **24**(3–4), 136–162.

Rensberger, J. M. and Koenigswald, W. v. (1980). Functional and phylogenetic interpretation of enamel microstructure in rhinoceroses. *Paleobiology*, **6**, 477–495.

Rich, T. H., Fortelius, M., Rich, P. V. and Hooijer, D. A. (1987). The supposed *Zygomaturus* from New Caledonia is a rhinoceros: a second solution to an enigma and its paleogeographic consequences. In *Possums and Opossums: Studies in Evolution*. ed. M. Archer, pp. 769–778. Sidney: Surrey Beatty & Son and Royal Zoological Society of New South Wales.

Sander, M. (1997). Non-mammalian synapsid enamel and the origin of mammalian enamel prisms: The bottom-up perspective. In *Tooth Enamel Microstructure*, eds. W. v. Koenigswald and P. M. Sander, pp. 40–62. Rotterdam: A. A. Balkema.

Sahni, A. (1979). Enamel ultrastructure of certain North American cretaceous mammals. *Palaeontographica*, **166**, 37–49.

Schreger, D. (1800). Beitrag zur Geschichte der Zähne. *Beiträge für die Zergliederkunst*, **1**, 1–7.

Schroeder, D. (1987). *Orale Strukturbiologie.* Stuttgart: Thieme.

Seilacher, A. (1970). Arbeitskonzept zur Konstruktionsmorphologie. *Lethaia*, **3**, 393–396.

Stefen, C. (1995). Zahnschmelzdifferenzierungen bei Raubtieren. [Carnivora, im Vergleich zu Vertretern der Oreodonta, Arctocyonidae, Mesonychidae, Entelodontidae (Placentalia), Thylacoleodontidae, Dasyuridae und Thylacinidae (Marsupialia)]. Unpublished dissertation, University of Bonn.

Stern, D., Crompton, A. W. and Skobe, Z. (1989). Enamel ultrastructure and masticatory function in molars of the American opossum, *Didelphis virginiana. Zoological Journal of the Linnean Society*, **95**, 311–334.

Teaford, M. F. (1988). A review of dental microwear and diet in modern mammals. *Scanning Microscopy*, **2**, 1149–1166.

Teaford, M. F. (1993). Dental microwear and diet in extant and extinct *Theropithecus*: preliminary analyses. In Theropithecus: *The life and death of a primate genus*, ed. N. D. Jablonski, pp. 331–349. Cambridge: Cambridge University Press.

Teaford, M. F., Maas, M. C. and Simons, E. L. (1996). Dental microwear and microstructure in early Oligocene Fayum primates: implications for diet. *American Journal of Physical Anthropology*, **101**, 527–543.

Tomes, J. (1849). On the structure of dental tissues of marsupial animals, and more especially of the enamel. *Philosophical Transactions of the Royal Society London*, **139**, 403–412.

Wahlert, J. H. and Koenigswald, W. v. (1985). Specialized enamel in incisors of eomyid rodents. *Novitates, American Museum of Natural History*, **2832**, 1–12.

Wood, C. B. (1991). Enamel structure and the phylogeny of extant and fossil new world opossums. *Journal of Vertebrate Paleontology*, **11**(3), abstract 216.

Wood, C. B. (1992). Comparative studies of enamel and functional morphology in selected mammals with tribosphenic molar teeth: Phylogenetic applications. Unpublished PhD thesis, Harvard University.

Wood, C. B. and Stern, D. (1997). The earliest prisms in mammalian and reptilian enamel. In *Tooth Enamel Microstructure*, eds. W. v. Koenigswald and P. M. Sander, pp. 63–83. Rotterdam: A. A. Balkema.

Young, W. G., Stevens, M. and Jupp, R. (1990). Tooth wear and enamel structure in the mandibular incisors of six species of kangaroo (Marsupialia: Macropodinae). *Memoirs of the Queensland Museum*, **28**, 337–347.

9 Incremental markings in enamel and dentine: what they can tell us about the way teeth grow

M. C. Dean

9.1. Introduction

Those who study human evolution do so for one, or more, of several reasons. For many, the primary goal is to define phylogenetic relationships between species of fossil hominids. For others, the human fossil record provides unique insights into primate biology that help us understand living primates in a wider context. Some view the fossil record differently. It represents one way in which we can begin to understand the mechanisms that underlie morphological change during evolution. In fact, the hominoid fossil record preserves a few details that allow us to determine the ways in which tooth tissues grow after ameloblasts and odontoblasts have become fully differentiated. In recent years, developmental biologists have come much closer to understanding the mechanisms of certain key embryological processes. Evolutionary biologists have also come closer to explaining how morphological change comes about during evolution. Identifying the common processes in ontogeny that regulate and control morphological change during phylogeny is once again one of the primary goals in evolutionary biology.

With regard to teeth and jaws, there has been an explosion of knowledge about their embryonic development and about morphogenesis of different tooth types (see Zhao, Weiss and Stock, chapter 11) as well as about the control of dental patterns at the molecular level (see Weiss, 1993; Chapter 5.1, 2). Studies in the embryo of individual rhombomeric crest populations are beginning to establish ultimate homologies between cranial elements and to elucidate which molecular pathways have changed during vertebrate evolution (Koentges and Lumsden, 1996). For the most part, studies that are specifically on tooth development centre on events

before hard tissues are formed and build on earlier observations of how tooth germs develop (Butler, 1956). Some recent studies have come tantalisingly close to linking older theories about fields of dental development for specific tooth shapes with spatially restricted domains of specific gene expression (Sharpe, 1995). A lot less is known about the nature and control of growth processes that, for example, determine enamel thickness, or which regulate the rates of proliferation of newly differentiated ameloblasts and odontoblasts in teeth of different shapes and sizes in different species. Obsorn (1993) has attempted to address aspects of this issue with a computer simulation of tooth morphogenesis that specifically concentrates on the physical interaction of dividing cells at the inner enamel epithelium with each other and with the basement membrane and stellate reticulum. However, something more can be learned from studying incremental markings in teeth. That is, how cells move during secretion, the rate at which they secrete enamel and dentine matrix, and the time during which they are active in their secretory phase. Since incremental markings are preserved in fossil teeth they offer a way of studying evolutionary processes at the cellular level as well. They can provide a timescale for dental development even, for example, in long extinct dinosaurs (Erikson, 1996a).

9.2. Using incremental markings in enamel and dentine to study tooth growth

Enamel, dentine and cementum each grow incrementally. Alternating periods of slower and faster growth during development are evident from incremental mark-

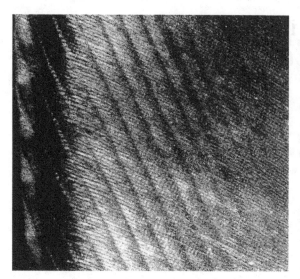

Figure 9.1. **Laser confocal photomicrograph (reflectance mode) of striae of Retzius in the lateral enamel of a modern human molar. Fieldwidth 400 microns (adjacent striae are approximately 45 microns apart).**

Figure 9.2. **Laser confocal photomicrograph (reflectance mode) of daily cross-striations between adjacent striae of Retzius in enamel. There are nine cross-striations between the striae in this individual. Fieldwidth approximately 160 microns.**

ings detected in each tissue by histological or ultrastructural examination. Very slow growing cementum often contains seasonal bands that may sometimes be about a year apart (Lieberman, 1993). Odontoblasts and ameloblasts secrete matrix at a greater rate but over a shorter period of time than do cementoblasts. Enamel and dentine therefore, contain daily increments of growth (and occasionally even subdaily infradian or ultradian increments of growth).

The various markings visible in human enamel and dentine were first described by Owen (1840–1845), Andresen (1898), von Ebner (1902, 1906) and Retzius (1837). Originally, many of these markings were described in the context of a debate at the turn of the century about whether enamel and dentine were the secretory products of the ameloblasts and odontoblasts, or whether, parts of ameloblasts and odontoblasts actually mineralised themselves (petrified processes) and therefore 'turned into' enamel and dentine (Mummery, 1924). Some of them (Owen's lines and irregular striae of Retzius) we now recognise as non-incremental (Boyde, 1989). They do not represent regular secretions occuring throughout the growth period, but rather are accentuated lines associated with upsets and disturbances during tooth formation. They are, nevertheless, crucial to histological studies of tooth growth since they mark surfaces of enamel and dentine forming at one time period (Dean, 1995a). The remaining lines or markings in enamel and dentine can be grouped into two sets which are visible using polarized light and confocal microscopy and occasionally in scanning electron microscopy. So called short

period lines are daily markings (or ultradian markings). These are now considered to include daily von Ebner's lines in dentine, and daily cross-striations or alternating daily varicosities and constrictions along enamel prisms (Figures 9.1, 9.2). Long period lines are typically coarser markings in teeth and can be traced through enamel and dentine over considerable distances. Long period lines (Figure 9.3) in humans and great apes are spaced between 6 and 10 days apart (but most commonly 7, 8 or 9 days) in enamel and dentine (Fukuhara, 1959; Beynon, 1992; Dean et al., 1993b; Dean and Scandrett, 1995, 1996). In the extinct baboon (Theropithecus brumpti) Swindler and Beynon (1992) have also observed a periodicity that resembles modern humans and great apes. In other monkeys, long period lines are reported to be 4 or 5 days apart (Okada, 1943; Bowman, 1991). This all seems to suggest some broad link between line periodicity and body size. However, the number of cross-striations between striae of Retzius across a range of primates correlates to some extent with body size and other life history parameters (M. C. Dean, in preparation).

Andresen lines in human dentine and regular striae of Retzius in enamel seem to be homologous long period markings (possibly with a common underlying cause) and are equal in number when they can be counted between accentuated markings in enamel and dentine of the same individual (Dean, 1995a). Early studies by Gysi (1931) and Asper (1916) suggested that cross-striations in enamel were daily markings and these authors (together

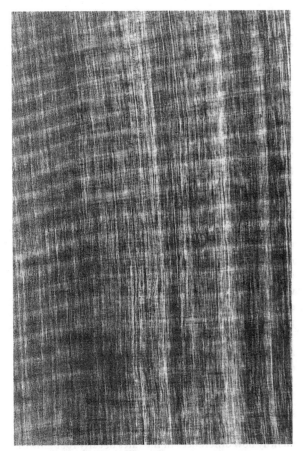

Figure 9.3. **Polarised transmitted light image of long period (Andresen) lines in modern human dentine. Between the long period lines it is possible to see daily incremental lines (von Ebner's lines). In this individual there are eight short period daily lines between adjacent pairs of long period lines. Fieldwidth approximately 350 microns.**

mented more experimental evidence for the daily periodicity of markings in enamel and dentine in a range of other animals including primates. In particular, Professor Hisashi Shinoda has continued the work of Okada and colleagues in Japan and has documented crucial changes in blood chemistry that occur as incremental structural changes are formed (Shinoda, 1984) as well as some of the mechanisms that underlie the formation of certain incremental markings.

Shellis (1984) has applied this knowledge of circadian incremental markings in enamel to a study of the way enamel extends and spreads over the dentine cap. Shellis (1984) made use of both the short period daily cross-striations in enamel and the slope (angle D) of the long period lines (regular striae of Retzius) to the enamel dentine junction, as well as the direction of growth of ameloblasts (angle I). Shellis used these known variables to calculate the rate of proliferation of newly differentiated ameloblasts at the growing tooth apex. Shellis defined this daily rate ('c') as the 'enamel extension rate' ($c=\{(\sin I/\tan D) - (\cos I)\}$). Shellis derived and used this equation to calculate changes in the enamel extension rate in human teeth which slowed from about 20 microns per day in the cuspal region to about 4 microns per day in the cervical region of human permanent teeth. He also commented that this equation was appropriate for determining dentine extension rates in growing tooth roots when each variable of the equation was known or measurable.

9.3.   Using incremental markings in studies of fossil tooth growth

with Boyde, 1963) were the first to use them in studies of tooth growth. More substantial evidence for the periodicity of markings in human and primate tooth tissues followed with the work of Isaac Schour and colleagues and of Okada and Mumura. Initially, with experimental data from one child and then subsequently from another two children, Schour and Poncher (1937) and Massler and Schour (1946) were able to calculate daily rates of enamel and dentine formation and the active secretory life in days of the odontoblasts and ameloblasts in human deciduous and some permanent teeth. Massler and Schour (1946) also counted cross-striations between their experimental labels in enamel and tied these counts in with the calculated daily rates in human teeth. Further experimental studies by Melsen et al. (1977) on dentine provided more data for humans. Okada (1943), Yilmaz et al. (1977), Bromage (1991), Erikson (1996a, 1996b) and Ohtsuka and Shinoda (1995) have subsequently docu-

Recently, and against this established background of information about incremental markings in human and primate tooth tissues, there has been great progress in understanding from histological studies how human, fossil hominid and other primate teeth grow (Bromage and Dean, 1985; Beynon and Wood, 1986; 1987; Beynon and Dean, 1988; Dean et al., 1993a). We have a much clearer idea of how great ape teeth grow from histological studies (Dean and Wood, 1981; Beynon et al., 1991a,b; Dean and Beynon, 1991). We have also begun to understand more about human and hominoid tooth root growth (Dean et al., 1992, 1993a,b) and about rates of dentine formation in primates (Dean, 1993; Dean and Scandrett, 1995, 1996). There is an established background to these histological studies of tooth growth and one can now confidently predict that more could be learned about the different mechanisms of enamel and dentine growth in primates using these approaches.

9.4.   **Problems of hominoid tooth growth**

The fast growing teeth of, for example a dog or a cat, early in development appear on a x-ray in outline and then seem to 'fill in' as enamel and dentine are secreted over the period the tooth takes to form completely. Incremental markings in these teeth follow the contours of the growing tooth and look not unlike tree rings, although they are tooth shaped in outline. Hominoid teeth begin to mineralize at the cusp tip and grow in length until the root apex is completed (Dean, 1989). Incremental markings in these teeth look quite different since the contours of the tooth at different stages of growth are open towards the apex.

The way a tooth grows, in this gross sense at least, is a reflection of the time available for it to grow. It is independent, however, of the size of a tooth which reflects only numbers of cells active at one time. The rate at which a hominoid tooth grows in length is influenced by three things. Firstly, by the time available in the growth period to grow the tooth: secondly, by the rate at which new ameloblasts and odontoblasts differentiate apically: thirdly, by the eventual length of a tooth. At any given rate of increase in tooth length, a long tooth will take more time to form than a short tooth. Tooth length is one of many things determined by sexual or natural selection, since differing tooth lengths are either functionally or socially advantageous in the adult. The enamel of the tooth crown is one component of total tooth length.

Among living and extinct hominoids the time it takes to form tooth crowns of a similar tooth type is broadly similar (Dean, 1995b). This is despite there being differences in crown height and differences in the length of time between birth and adulthood in hominoids (Dean, 1995b). It is the rates at which roots form that differ most significantly between humans and apes for example (Dean and Wood, 1981). The crowns of permanent teeth in hominoids are very different in size and shape and may have thick or thin enamel. By examining the incremental growth lines in enamel and dentine it is possible to say something about how tooth crowns, as variable as this, might form in similar time periods. The first aim of the present study is to show how there may be different ways of growing cuspal enamel of different thicknesses in primates. This can be achieved by counting and measuring incremental markings in cuspal enamel.

For a number of reasons the first permanent molar tooth is regarded as a key tooth developmentally in primates (Smith, 1986, 1989, 1994; Smith and Tompkins, 1995). Its emergence into the mouth marks the time many primates are weaned (Smith, 1991). Its emergence also marks the time that brain growth (in volume at least) is 90% complete in all primates (Ashton and Spence, 1958). Because of the high infant mortality among primates around the time of weaning, there are a number of fossil hominids with first permanent molars just in occlusion in the fossil record. For these reasons and others it has been studied often in humans, fossil hominids and in many living primates. It is a useful tooth to study in order to discover more about rates of root growth. If crown formation times in first permanent molar teeth are indeed broadly comparable in hominoids, as suggested, then because the age at which teeth emerge into occlusion is different among hominoids, the rate at which the first part of the root forms is likely to reflect this (Macho and Wood, 1995). By this argument, if the first half of the first permanent molar roots from two different taxa were the same length, then the teeth of one that erupt at 3 or 4 years, will have faster growing roots than the teeth of the other, that erupt at 6 or 7 years of age. It is then, the first part of root growth that occurs before the emergence of the tooth that is likely to be different between humans and apes. This conclusion has been drawn on more than one occasion (Dean and Wood, 1981; Dean, 1985; Macho and Wood, 1995). Root growth after this time may continue for a very long time in both humans and apes and show few differences.

A second aim of the present study is to attempt to identify how differences in the rates of early root growth might occur. One way of finding out more about the mechanisms that control the way tooth roots grow in length is to study incremental markings in the dentine of both living and fossil teeth. Ideally, some estimate of the time period between the end of crown formation and tooth emergence into functional occlusion would be most desirable.

The data presented here are in three sections and address the aims reviewed above. In the first section the role that the daily secretory rate of ameloblasts plays in the control of cuspal enamel thickness during growth of the crowns of teeth will be considered. Some data will be presented that shed light on the different ways in which cuspal enamel grows in primates. In the second section, the role that daily secretory rates of odontoblasts may or may not play in the early part of root growth in human and ape first permanent molars will be considered. Data that describe the daily rates of dentine formation in these teeth will be presented. In the third section, some preliminary data from an ongoing study that give an indication of the rate of root extension of the first half of root formation in *Homo habilis* will be presented and discussed briefly in the light of similar data for humans and apes.

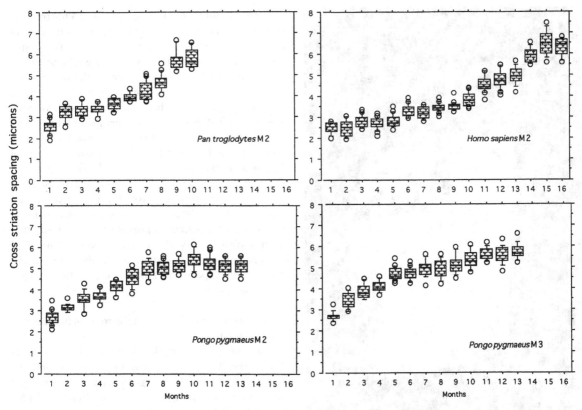

Figure 9.4. **Cross-striation spacings (in microns) for consecutive monthly zones of enamel in the cusps of M2's of** *Homo sapiens* **and** *Pan* **troglodytes and of an M2 and an M3 of** *Pongo pygmaeus***. Each box plot for each month has a central infilled region with a line representing the median value of the cross-striation spacing. The 25th and 75th percentiles are represented by the upper and lower borders of the infilled box, and the 10th and 90th percentiles by the whiskers and outliers beyond these are indicated by open circles. The plots are described in the text.**

9.4.1. How cuspal enamel may grow differently in primate teeth

In this part of the study enamel cross-striations were tracked through the cuspal enamel of two orang-utan molars (a second permanent molar and a third permanent molar), a chimpanzee second permanent molar, and a human second permanent molar just lateral to the maximally decussating (so-called gnarled) enamel under the first formed mesiobuccal cusp. This was done from photomontages of the whole cuspal region of each tooth taken in polarised light at × 250. Teeth for this study were chosen for the clarity and number of cross-striations which could be easily measured in the cuspal enamel. Cross-striations along each prism were marked on the montages and followed through the entire thickness of the enamel. Naturally, prisms weave in and out of the plane of section but many can be followed for long distances in two dimensions. Zones of enamel were identified that represented approximate near-monthly bands of growth by counting 30 cross-striations from the enamel-dentine junction (EDJ) to 'month one' and so on

to the surface of the cuspal enamel. In this way the whole of the cuspal enamel was divided into near-monthly zones. Within each near-monthly zone, and some distance either side of the central track, the distance between adjacent cross-striations was measured. This was done by taking a measurement across six cross-striations in series and then dividing the total length by five to obtain the average for 1 day's secretion in this field of view. This procedure was repeated 15 to 20 times in each zone such that the average values for 75 to 100 cross-striations is represented in the results for each zone on the photomontages.

The median values for all of these measurements, along with the 10th, 25th, 75th and 90th percentiles and the outliers for each monthly zone, are presented as box plots for the orangutan molars and for the chimpanzee and human second permanent molar teeth (Figure 9.4). What is clear from the box plots in Figure 9.4 is that the ameloblasts that secrete enamel matrix in the human and chimpanzee molar tooth cusps do so at a slow rate for many months before increasing in rate towards the

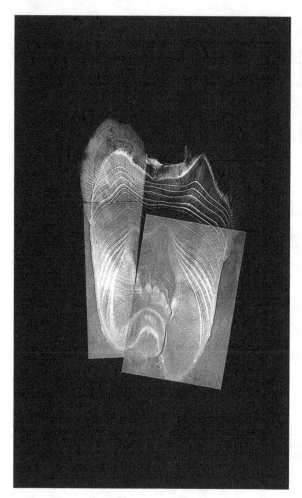

Figure 9.5. **Montage of a ground section of a modern human M1 seen in incident ultraviolet light. Rates of dentine formation in the crown and root dentine were determined from this tooth. There are approximately 30 fluorescing tetracycline labels in the dentine (some are reflected in the enamel as well) of this tooth.**

more rapidly but then remain around 4 or 5 microns per day for 8 or 9 months. These data suggest that the control of enamel secretion at a cellular level in the cuspal regions of hominoid teeth may be different between one species and another and even from tooth type to tooth type. It may even vary from one cusp to another in the same tooth. However, the data presented here also, suggest a way of looking for similarities and differences in the way cuspal enamel grows between different hominoid taxa that directly reflects the pattern of activity of the ameloblasts through developmental time. This kind of approach may help reveal developmental mechanisms that may reflect processes through which evolution can operate, and may determine significant different thicknesses of cuspal enamel in primate teeth in a way that linear measurements of enamel thickness cannot. In short, ontogenetic information can provide some insight into the way cuspal enamel has come to be thick or thin in some primates.

9.4.2.   How dentine grows in the roots of hominoid teeth

The daily incremental markings in dentine can be made clearer than they often appear in ground sections by rendering tooth sections anorganic, or by demineralising teeth and block staining with silver (see Dean, 1995a for review). Fossil dentine is by its very nature anorganic and paradoxically, incremental markings often show more clearly in fossil dentine than in well-fixed modern human teeth. Dentine can also be labelled easily and this is one reason why the experimental evidence for there being daily lines in dentine is rather better than that for enamel. A combination of approaches that make use of labels and of incremental markings in dentine can reveal a great deal about the way dentine in the crowns and roots of teeth grows.

The question addressed here is whether there are any differences in the daily rates of dentine formation between humans and apes that might underlie the initially faster rates of root elongation (extension rates) in ape permanent molars. In this study, human first permanent molar teeth with a large number (approximately 30 in total) of tetracycline antibiotic labels in the dentine and enamel were used (Figure 9.5). These labels fluoresce in incident ultraviolet light and are easily seen in fluorescence microscopy of tooth tissues. Also visible in the same field of view as the fluorescing lines (but in polarised light) were long-period incremental markings which could be counted between the fluorescing lines by superimposing the two images (Figure 9.6). First, enamel cross-striations were used to calibrate the time interval between the label lines that appeared in the enamel and dentine forming at the same time (Dean *et al.*, 1993a;

outer cuspal enamel. For 10 months in human cuspal enamel the median rate of enamel secretion is below 3.5 microns per day. Only after 10 months in this human tooth cusp does the rate increase. In the chimpanzee molar some 5 months of enamel formation is below the median rate of 3.5 microns per day. Subsequently, median rates of enamel formation rise to 5 or 6 microns per day.

In contrast, the ameloblasts in the cuspal enamel of the orang-utan take less time than the human molar (13 months rather than 16 months) to form the slightly thinner cuspal enamel but take more time than do those of the chimpanzee molar (13 months rather than 10 months). However, they follow a different pattern of secretory activity to those in the human and chimpanzee second molar. Median rates of enamel secretion increase

Figure 9.6. **The micrograph on the left of the pair is taken in incident ultraviolet light and successive fluorescing tetracycline lines are numbered 13 to 23. The micrograph on the right is of exactly the same field of view but viewed in transmitted polarised light. In this image a long series of long period incremental markings (Andresen lines) can be seen. By superimposing the two images, counts of long period lines between fluorescing lines were made. In addition (not shown) the equivalent, fluorescing lines in the enamel can be calibrated by counting daily cross-striations between adjacent lines. Each long period line in enamel and dentine is 8 days apart in this individual.**

Dean and Scandrett, 1995, 1996). Second, the distance along dentine tubules was measured between label lines of known time interval from the EDJ pulpwards, i.e. along the direction of dentine formation. This procedure was repeated for several 'tracks' in the crown, at the cervix and in the root dentine. This procedure is a way of looking for gross similarities or differences in the way odontoblasts secrete dentine matrix throughout tooth formation (Dean and Scandrett, 1995).

The first permanent molar of an orang-utan was also used in this study for comparison with the human molar tooth. In this tooth there were four or five tetracycline labels in the dentine that could be calibrated using the enamel forming at the same time. In addition, long-period markings in the dentine of the orang-utan were also examined. These long-period markings (Dean, 1995a) have the same periodicity as striae of Retzius in the enamel of the same individual (in this animal they were 9 days apart; Figure 9.2 is a confocal image of cross-striations between adjacent striae in the enamel of the M1 used in this study). The long-period lines and the fluorescent lines were used in this study to calculate the period of time between successive positions of odontoblasts as they formed dentine matrix. Several 'tracks' were again identified along dentine tubules in the crown, at the cervix and in the root of the orang-utan molar tooth. The distance between successive lines was then measured along the direction of the dentine tubules. Figures 9.7A,B

illustrates the label lines and the tracks used in the human and orang-utan permanent molars. Figures 9.8 shows plots of the cumulative distance travelled by odontoblasts against time in the human and orang-utan teeth.

There is a striking similarity between the graphs of human and the ape teeth. In fact, rates of dentine formation calculated from the slopes of the lines suggest in both teeth, dentine formation begins slowly at between 1 and 2 microns per day. It then rises to between 4 and 5 microns per day and falls off again as the pulp is approached. These are typical 'S-shaped' growth curves and reflect the secretory activity of the odontoblasts during dentine formation in the crown and root of both a human and ape molar tooth. The conclusion to be noted from this study is that, unlike cuspal enamel formation where we can identify differences in the rates at which ameloblasts secrete enamel over time, there appear to be no obvious differences in the rates at which dentine formation occurs between the molar teeth of humans and apes over the time period that could account for the way the first portion of ape tooth roots extends at a faster rate than in humans.

These data are the first that provide estimates of the rates of dentine formation from the beginning of dentine matrix secretion to the point at which secondary dentine begins to form. Even though the data presented here relate only to two teeth, they do confirm the findings of other experimental studies (Schour and Hoffman, 1939;

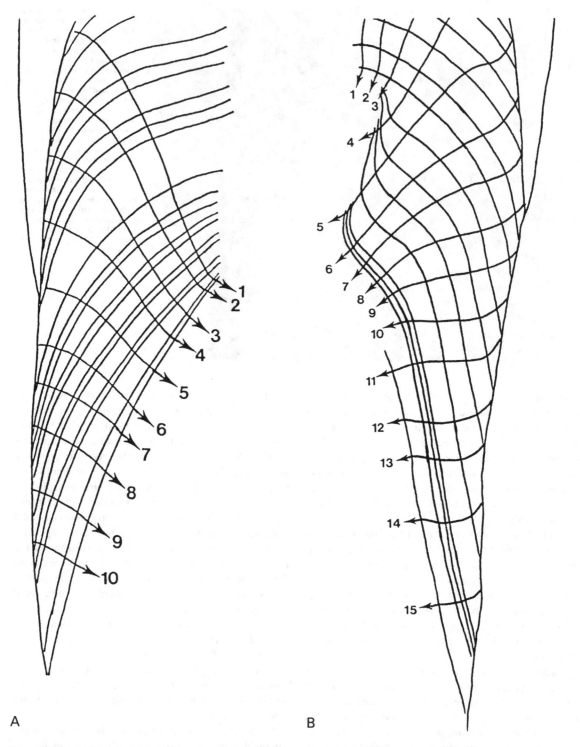

A   B

Figure 9.7A. **The dentine and enamel of the human M1 used in this study. Tetracycline lines in the dentine are drawn in. Across these, ten 'tracks' follow dentine tubules from the first formed dentine towards the pulp chamber. Cumulative measurements were made between successive tetracycline lines along each 'track'. B. The orang-utan M1 used in this study. Both tetracycline lines and long period lines used to calibrate dentine formation in this study are drawn in. 'Tracks 4 to 15' only were used for comparison with the human data in this study. These are indicated in the drawing.**

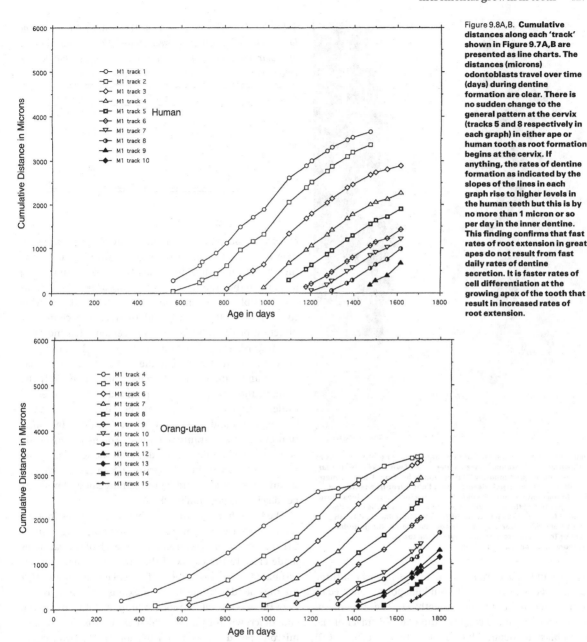

Figure 9.8A,B. **Cumulative distances along each 'track' shown in Figure 9.7A,B are presented as line charts. The distances (microns) odontoblasts travel over time (days) during dentine formation are clear. There is no sudden change to the general pattern at the cervix (tracks 5 and 8 respectively in each graph) in either ape or human tooth as root formation begins at the cervix. If anything, the rates of dentine formation as indicated by the slopes of the lines in each graph rise to higher levels in the human teeth but this is by no more than 1 micron or so per day in the inner dentine. This finding confirms that fast rates of root extension in great apes do not result from fast daily rates of dentine secretion. It is faster rates of cell differentiation at the growing apex of the tooth that result in increased rates of root extension.**

Molnar *et al.*, 1981; Bowman, 1991; Bromage, 1991) that suggest what we know of rates of dentine formation in non-human primates appears to match those known for humans. Differences in the daily rate of dentine matrix secretion (or mineralisation) cannot then be held responsible for differences in the rates at which tooth crowns and roots extend in length. The only other factor that can account for faster root extension rates in the first part of great ape and other non-human primate tooth roots is the rate at which numbers of cells differentiate at the open growing apex of the root.

9.4.3.   How to estimate the extension rate of tooth roots

Three things need to be known to estimate the rate at which teeth grow in length: (1) the daily rate at which cells produce matrix, (2) the direction of cell movement, and (3) the number of mature secretory cells active at any one time (their rate of differentiation). As outlined above, Shellis (1984) has expressed the 'extension rate' of teeth at the EDJ in the crown or at the cement-dentine junction (CDJ) in the root in the equation $c=d\{(\sin I/\tan D)-\cos I\}$. In order to calculate the rate of extension of tooth

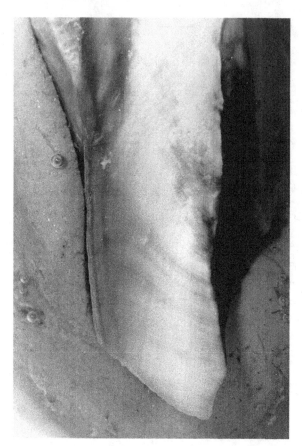

Figure 9.9. **A sieving fragment of a permanent molar tooth OH 16 from Olduvai Gorge, Tanzania, viewed under alcohol with a Wild binocular microscope at approximately 80 times magnification. The specimen was illuminated with polarised incident light. Strong accentuated lines in the dentine show the direction of the incremental lines. Note the cementum at the root surface and the position of the granular layer of Tomes just deep to this which looks like a dark band in this micrograph. Measurements were made just deep to the granular layer of Tomes and close to the cervix of the tooth. These measurements appear in Table 9.1.**

Table 9.1. *Data for 'Angle I', 'Angle D', the daily rate of dentine formation, 'd' (measured between 50 and 100 microns from the granular layer of Tomes in humans and apes) and 'c' the extension rate calculated from the formula described in the text*

Angle I	Angle D	Daily rate d	Extension rate c
115	11	2.5	12.7
115	11	2.5	12.7
130	10	2.5	12.5
109	9	2.5	15.7
117	11	2.5	12.6
122	13.5	2.5	10.2
133	14.5	2.5	9.6

crowns and roots from living or fossil species three things need to be measured from high-power reflected or transmitted light images of teeth: (1) The amount of tissue secreted in a day (2) the direction of travel of the ameloblast or odontoblast relative to the EDJ or CDJ (which can be inferred from the alignment of an enamel prism or dentine tubule), and (3) the angle that the active cell sheet subtends to the EDJ (which is a reflection of the number of active secretory cells).

Figure 9.9 is a photomicrograph of a naturally fractured crown and root belonging to Olduvai Hominid 16 (OH 16). The tooth is a fragment of a permanent molar fractured axially and here it has been photographed under ethyl alcohol and in incident polarised light to eliminate unwanted reflectance from the unpolished tooth surface. From direct visualisation of this sieving fragment and others under alcohol using a Wild bino-

cular microscope and goniometer eyepiece at 80 times magnification it was possible to measure the angle the incremental markings in the dentine make with the CDJ (as close to the granular layer of Tomes as possible). It was also possible to see and measure the angle the dentine tubules make to the EDJ and to track the change in direction of the tubules from the cement–dentine junction towards the pulp chamber. It is likely the daily rates of dentine secretion, close to the root surface, are similar in apes and humans (and therefore more than likely the same in this and other fossil hominids). In fact there is evidence that in gibbons, great apes, humans (and even in cave bears and pigs) the rates of dentine mineralisation close to the granular layer of Tomes is constantly between 2 and 3 microns per day even though it may rise to very different values thereafter (Dean, 1998). We can surmise then that would be in the range of 2–3 microns per day in this position in OH 16.

We have then all three variables needed to estimate a range of root extension rates in this tooth fragment of OH 16. A range of measurements for each of the variables made close to the cervix of this tooth fragment give an estimated extension rate of between 9.6 and 15.7 microns per day (Table 9.1). This extension rate is greater than those reported for modern human teeth close to the end of crown formation which is approximately 3–6 microns per day (Dean and Beynon, 1991; Dean *et al.*, 1993*a*; Liversidge *et al.*, 1993). It also fits with other observations on this specimen, which suggest a comparatively fast rate of tooth root extension in *Homo habilis* at the root cervix (Dean, 1995*a*). Given more data of this kind it is likely one could say more confidently whether tooth development in general in *Homo habilis* simply resembled that in *Australopithecus* and *Paranthropus* as well as in extant great apes where the extension rate has been estimated at around 13 or more microns per day (Bromage and Dean, 1985; Beynon *et al.*, 1991*b*; Dean *et al.*, 1993*b*; Dean, 1995*b*) or whether it had shifted to reflect the prolonged period of growth and development we associate with modern *Homo sapiens*. At present, in the light of the data presented here and elsewhere (Dean, 1995a) this seems highly unlikely.

Summary

Teeth preserve a record of the way they grow in the form of incremental markings in enamel, dentine and cementum. They make it possible for us to reconstruct some of the developmental processes that operated in the teeth of our fossil ancestors and they provide a way of exploring the mechanisms that underlie morphological change during evolution. New discoveries in developmental biology and in evolutionary biology will eventually mean that we can ask more focused questions about the nature of the relationship between ontogeny and phylogeny.

Acknowledgements

The research reported here was supported by grants from The Leakey Foundation, The Royal Society and The Leverhulme Trust. I am grateful to the Government of Tanzania for allowing me permission to work with fossil hominid material from Olduvai Gorge. I am grateful to the trustees of The National Museum of Kenya and to Dr Meave Leakey for permission to work on material housed in the Palaeontology Department of the Kenya National Museum, Nairobi. I am also grateful for the opportunity to contribute to this volume and thank its editors as well as those who have provided comments on an earlier draft of this paper.

References

Andresen, V. (1898). Die Querstreifung des Dentins. *Deutsche Monatsschrift fur Zahnheilkunde.* Sechzehnter Jahrgang (Leipzig), **38**:386–389.

Asper, von H. (1916). Uber die 'Braune Retzinusgsche Parallelstreifung' im Schmelz der menschlichen Zahne. *Schweizerische Vierteljahrsschrift fur Zahnheilkunde.* **26**:275–214.

Ashton, E. H. and Spence, T. F. (1958). Age changes in the cranial capacity and foramen magnum of hominoids. *Proceedings of the Zoological Society of London*, **130**, 169–181.

Beynon A. D. (1992). Circaseptan rhythms in enamel development in modern humans and Plio-Pleistocene hominids. In *Structure, Function and Evolution of Teeth*, eds. P. Smith and E. Tchernov, pp.295–309. London and Tel Aviv: Freund Publishing House.

Beynon A. D., and Dean M. C. (1988). Distinct dental development patterns in early fossil hominids. *Nature*, **335**, 509–514.

Beynon, A. D. and Wood, B. A. (1986). Variations in enamel thickness and structure in East African hominids. *American Journal of Physical Anthropology*, **70**, 177–193.

Beynon, A. D. and Wood, B. A. (1987). Patterns and rates of enamel growth in molar teeth of early hominids. *Nature*, **326**, 493–496.

Beynon, A. D., Dean, M. C., and Reid, D. J. (1991a). Histological study on the chronology of the developing dentition in gorilla and orang-utan. *American Journal of Physical Anthropology*, **86**, 189–203.

Beynon, A. D., Dean, M. C., and Reid, D. J. (1991b). On thick and thin enamel in hominoids. *American Journal of Physical Anthropology*, **86**, 295–309.

Bowman J. E. (1991). Life history, growth and dental development in young primates: a study using captive rhesus macaques. D.Phil. thesis, University of Cambridge.

Boyde, A. (1963). Estimation of age at death from young human skeletal remains from incremental lines in dental enamel. In *Third International Meeting in Forensic Immunology, Medicine, Pathology and Toxicology*, London, April 1963, Plenary session 11A.

Boyde, A. (1989). Enamel. In *Handbook of Microscopic Anatomy*, vol. V6, *Teeth*, pp. 309–473. Berlin: Springer.

Bromage, T. G. (1991). Enamel incremental periodicity in the pigtailed macaque: a polychrome fluorescent labelling study of dental hard tissues. *American Journal of Physical Anthropology*, **86**, 205–214.

Bromage T. G. and Dean M. C. (1985). Re-evaluation of the age at death of immature fossil hominids. *Nature*, **317**, 525–527.

Butler, P. M. (1956). The ontogeny of molar pattern. *Biological Reviews*, **31**, 30–70.

Dean, M. C. (1985). Variation in the developing root cone angle of the permanent mandibular teeth of modern man and certain fossil hominids. *American Journal of Physical Anthropology*, **68**, 233–238.

Dean, M. C. (1987). Growth layers and incremental markings in hard tissues: a review of the literature and some preliminary observations about enamel structure in *Paranthropus boisei*. *Journal of Human Evolution*, **16**, 157–172.

Dean, M. C. (1989). The developing dentition and tooth structure in primates. *Folia Primatologica*, **53**, 160–177.

Dean, M.C. (1993). Daily rates of dentine formation in macaque tooth roots. *International Journal of Osteoarchaeology*, **3**, 199–206.

Dean, M. C. (1995a). The nature and periodicity of incremental lines in primate dentine and their relationship to periradicular bands in OH 16 (*Homo habilis*). In *Aspects of Dental Biology; Paleontology, Anthropology and Evolution*, ed. J. Moggi-Cecchi, pp. 239–265. Florence: Angelo Pontecorboli.

Dean, M. C. (1995b). Developmental sequences and rates of growth in tooth length in hominoids. In *Proceedings of the 10th International Symposium on Dental Morphology*, ed. R. J. Radlanski and H. Renz, pp. 308–313. Berlin: 'M' Marketing Services.

Dean, M. C. (1998). Comparative observations on the spacing of short-period (von Ebner's) lines in dentine. *Archives of Oral Biology*, **43**, 1009–1021.

Dean, M. C. and Beynon, A. D. (1991). Tooth crown heights, tooth wear, sexual dimorphism and jaw growth in hominoids. *Zeitschrift für Morphologie und Anthropologie*, **78**, 425–440.

Dean, M. C. and Scandrett, A. E. (1995). Rates of dentine mineralization in permanent human teeth. *International Journal of Osteoarchaeology*, **5**, 349–358.

Dean, M. C. and Scandrett, A. E. (1996). The relation between enamel cross-striations and long-period incremental markings in dentine in human teeth. *Archives of Oral Biology*, **41**, 233–241.

Dean, M. C. and Wood, B. A. (1981). Developing pongid dentition and its use for ageing individual crania in comparative cross-sectional growth studies. *Folia Primatologica*, **36**, 111–127.

Dean, M. C., Beynon, A. D. and Reid, D. J. (1992). Microanatomical estimates of rates of root extension in a modern human child from Spitalfields, London. In *Structure, Function and Evolution of Teeth*, eds. P. Smith and E. Tchernov, pp. 311–333. London and Tel Aviv: Freund Publishing House.

Dean, M. C., Beynon, A. D., Thackeray, J. F. and Macho, G. A. (1993a). Histological reconstruction of dental development and age at death of a juvenile *Paranthropus robustus* specimen, SK 63, from Swartkrans, South Africa. *American Journal of Physical Anthropology*, **91**, 401–419.

Dean, M. C., Beynon, A. D., Reid, D. J. and Whittaker, D. K. (1993b). A longitudinal study of tooth growth in a single individual based on long and short period incremental markings in dentine and enamel. *International Journal of Osteoarchaeology*, **3**, 249–264.

Ebner V. von (1902). Histologie der Zahne mit Einschluss der Histogenese. In *Handbuch der Zahnheilkunde*, ed. J. Scheff, pp. 243–299. Vienna: A. Holder.

Ebner V. von (1906). Uber die Entwicklung der leimgebenden Fibrillen, insbesondere im Zahnbein. *Sitzungsberichte der Mathematisch – Naturwissenschaftlichen Klasse der Kaiserlichen Akademie der Wissenschaften in Wien*, vol. 115, Abteilung, **111**, 281–347.

Erickson, G. M. (1996a). Incremental lines of von Ebner in dinosaurs and the assessment of tooth replacement rates using growth line counts. *Proceedings of the National Academy of Sciences, USA*, **93**, 14623–14627.

Erikson, G. M. (1996b). Daily deposition of dentine in juvenile *Alligator* and assessment of tooth replacement rates using incremental line counts. *Journal of Morphology*, **228**, 189–194.

Fukuhara, T. (1959). Comparative anatomical studies of the growth lines in the enamel of mammalian teeth. *Acta Anatomica Nipponica*, **34**, 322–332.

Gysi, A. (1931). Metabolism in adult enamel. *Dental Digest*, **37**, 661–668.

Koentges, G. and Lumsden, A. (1996). Rhombencephalic neural crest segmentation is preserved throughout craniofacial ontogeny. *Development*, **122**, 3229–3242.

Lieberman, D. E. (1993). Life history variables preserved in dental cementum, microstructure. *Science*, **261**, 1162–1164.

Liversidge, H. M., Dean, M. C. and Molleson, T. I. (1993). Increasing human tooth length between birth and 5.4 years. *American Journal of Physical Anthropology*, **90**, 307–313.

Macho, G. A. and Wood, B. A. (1995). The role of time and timing in hominoid dental evolution. *Evolutionary Anthropology*, **4**, 17–31.

Massler, M. and Schour, I. (1946). The appositional life span of the enamel and dentin-forming cells. *Journal of Dental Research*, **25**, 145–156.

Melsen, B., Melsen, F. and Rolling, I. (1977). Dentine formation rate in human teeth. *Calcified Tissue Research*, 23:R16, abstract 62.

Molnar, S., Przybeck, T. R., Gantt, D. G., Elizondo, R. S. and

Wilkerson, J. E. (1981). Dentin apposition rates as markers of primate growth. *American Journal of Physical Anthropology*, **55**, 443–453.

Mummery, J. H. (1924). *The Microscopic and General Anatomy of the Teeth: Human and Comparative*. Oxford: Oxford University Press.

Ohtsuka, M. and Shinoda, H. (1995). Ontogeny of circadian dentinogenesis in the rat incisor. *Archives of Oral Biology*, **40**, 481–485.

Okada, M. (1943). Hard tissues of animal body. Highly interesting details of Nippon studies in periodic patterns of hard tissues are described. *Shanghai Evening Post*, medical edition of September 1943, pp. 15–31.

Osborn, J. W. (1993). A model simulating tooth morphogenesis without morphogens. *Journal of Theoretical Biology*, **165**, 429–445.

Owen R. (1840–1845). *Odontography: or a Treatise on the Comparative Anatomy of the Teeth; Their Physiological Relations, Mode of Development and Microscopic Structure in the Vertebrate Animals*, vol. 1, Text and vol. 2, plates. London: Baillière.

Retzius, A. (1837). Bemerkungen uber den innern Bau der Zahne, mit besonderer Ruchsicht auf den im Zahnknochen vorkommenden Rohrenbae. *Müllers Archiv fur Anatomie und Physiologie*, pp. 486.

Schour, I. and Hoffman, M. M. (1939). Studies in tooth development. II. The rate of apposition of enamel and dentine in man and other animals. *Journal of Dental Research*, **18**, 91–102.

Schour I. and Poncher H. G. (1937). Rate of apposition of enamel and dentin, measured by the effect of acute fluorosis. *American Journal of Diseases of Children*, **54**, 757–776.

Sharpe, P. T. (1995). Homeobox genes and orofacial development. *Connective Tissue Research*, **32**, 17–25.

Shellis, R. P. (1984). Variations in growth of the enamel crown in human teeth and a possible relationship between growth and enamel structure. *Archives of Oral Biology*, **29**, 697–705.

Shinoda, H. (1984). Faithful records of biological rhythms in dental hard tissues. *Chemistry Today*, **162**, 43–40 (in Japanese).

Smith, B. H. (1986). Dental development in *Australopithecus* and *Homo*. *Nature*, **323**, 327–330.

Smith, B. H. (1989). Dental development as a measure of life history in primates. *Evolution*, **43**, 683–688.

Smith, B. H. (1991). Age of weaning approximates age of emergence of the first permanent molar in non-human primates. *American Journal of Physical Anthropology*, (Suppl.), **12**, 163–164.

Smith, B. H. (1994). Patterns of dental development in *Homo*, *Australopithecus*, *Pan* and *Gorilla*. *American Journal of Physical Anthropology*, **94**, 307–325.

Smith, B. H. and Tompkins, R. L. (1995). Towards a life history of the Hominidae. *Annual Review of Anthropology*, **24**, 257–279.

Swindler, D. R. and Beynon A. D. (1992). The development and microstructure of the dentition of *Theropithecus*. In *Theropithecus. The Life and Death of a Primate Genus*, ed. N. G. Jablonski, pp. 351–381. Cambridge: Cambridge University Press.

Weiss, K. M. (1993). A tooth, a toe and a vertebra: the genetic dimensions of complex morphological traits. *Evolutionary Anthropology*, **2**, 121–134.

Yilmaz, S., Newman, H. N. and Poole D. F. G. (1977). Diurnal periodicity of von Ebner growth lines in pig dentine. *Archives of Oral Biology*, **22**, 511–513.

Evolution of tooth shape and dentition

Part three

10 Evolutionary origins of teeth and jaws: developmental models and phylogenetic patterns

M. M. Smith and M. I. Coates

10.1. Introduction

The long-established scenario of teeth evolving from dermal denticles in direct association with the evolutionary origin of jaws, is based on the observation that shark (chondrichthyan) dermal denticles and oral denticles are homologous and contiguous with teeth. This theory, however, no longer accounts for the diversity of new data emerging from the fossil record. Certain denticle-covered fossil agnathan fishes (thelodonts) are now known to have well-developed oropharyngeal denticles. Moreover, the conodonts, a large and geologically extensive fossil group with a naked body but phosphatic mineralized feeding apparatus, are now thought to belong within the vertebrate clade. This projects the evolutionary origin of teeth, or at least a specialized suite of oral denticles, back to a point some 50 million years earlier than previously thought. These data suggest that 'teeth', in a broad sense of the term, may precede jaw evolution, and possibly precede the evolution of dermal armour in some primitive vertebrates. Alternatively, other new evidence shows that toothless armoured fish probably existed contemporaneously with conodonts. These include taxa with either dentine plus acellular bone-like basal tissue (*Anatolepis*, M. P. Smith *et al.*, 1996), or acellular skeletal tissue, but no dentine (Young *et al.*, 1996). Such discoveries challenge previously held views on the relative primitiveness of different tissues, and predict important stages in the evolution of signalling molecules for regulatory mechanisms during craniate mineralized skeletogenesis. Furthermore, the precise geological dates associated with these data, when placed in the context of recent phylogenies of early vertebrates, require hypotheses of significant 'ghost lineages' (Norell, 1992: i.e. those lineages that are implied by emergent gaps in phyloge-

netic trees plotted against a time axis) in order to account for a diversity of vertebrate types emergent before the Cambrian–Ordovician boundary (see Figure 10.3). The whole concept of teeth linked to jaws as an integrated functional unit, constant throughout evolution and development, needs to be reassessed.

The theory that teeth only evolved with the imperative of biting jaws, states that marginal oral skin denticles became enlarged to function as teeth once the first branchial arches became modified to close tightly with a dorsoventral opposing action. Such enlarged marginal denticles were specialized to form in 'advance of need' from a subepithelial dental lamina, instead of 'on demand' as space became available in the skin. Furthermore, these denticles were endowed uniquely with morphological and pattern information, specific for the precise functioning of the dentition. This information could concern spatial morphological variation (heterodonty as in the Port Jackson shark: see Figure 10.4C), tooth location in particular regions of the oral cavity, and timing of formation linked to shedding of functional teeth and eruption. Control of tooth size is very precise, with dimensional increases for new teeth in an exact gradation relative to body size and morphological change (see section 10.6). The survival value of an efficient dentition throughout an animal's life history is extremely high, and likely to be selected for and conserved in evolution, with similarly conserved essential developmental controls and regulatory mechanisms.

One of the key innovations at the origin of craniates was the developmental potential of a fourth germ layer, neural crest-derived mesenchyme, to make skeletal tissues in the head, in particular teeth (Gans and Northcutt, 1983). This population of cells, in migrating ventrally from a dorsal position in the neural ectoderm to new anterior–posterior (A/P) sites, enabled alternative positional information to be carried to interactive sites with an epithelium of different embryonic position. One section of the cranial neural crest alone is responsible for making the dentition (see Graveson et al., 1997), its intrinsic odontogenic potential is such that from cells prior to emigration, teeth are formed in explants with an appropriate interactive epithelium, and without these crest cells no teeth are formed. New research using cell lineage tracing and molecular markers for the location of gene activation, indicates that skeletogenic neural crest-derived mesenchyme, may confer positional information on the branchial arches by its intrinsic Hox gene code. The combinatorial Hox code present in hindbrain neural crest is iterated in the crest-derived mesenchyme of the gill arches (which in certain teleost fishes are the only tooth-bearing components of the visceral skeleton). The sequence and source of signalling molecules in this patterning mechanism has yet to be discovered. It is known that in the vast majority of vertebrates tested so far, odontogenesis depends upon cranial neural crest-derived mesenchyme reaching the appropriate oral epithelium (see Chapter 1). Here there may be a real difference (and a subject for future research), between gill arch dental patterning in non-tetrapod jawed vertebrates, and patterning of the mandibular arch and dentition, which in extant tetrapods is the only tooth-bearing component of the visceral skeleton. Moreover, the mandibular arch originates in part from midbrain-derived crest, which is subject to the influence of a series of more derived homeobox-containing regulatory genes, instead of the (perhaps more primitive) hindbrain, hyoid and gill arch Hox codes. In these developmental differences may reside the mechanism for evolutionary preservation of dentition patterning independently of jaw morphology and function, perhaps so essential for the survival value of efficient jaws and teeth.

10.2.   Dermal denticles: homology with teeth

Williamson (1849), Hertwig (1879), and Goodrich (1907) each assumed that teeth and skin denticles were in some sense similar structures, but it was Schaeffer's (1977) review of fish dermal skeletons that first clearly linked this similarity with shared morphogenetic mechanisms. Schaeffer's hypothesis suggested that 'the dermal skeleton develops from a single, modifiable morphogenetic system'. Moreover, regulation of morphogenetic change through such an interactive system would have significant implications for dermal skeletal diversity and evolution. First, Reif (1982) and then Smith and Hall (1993)

coupled this concept with Ørvig's (1967) on the constancy of the odontode as a fundamental dermal skeletal unit, and developed it further. Each of these discussions identifies the fine tuning of these highly conserved developmental, epithelial-mesenchymal mechanisms as responsible for the diversity within ranges of contributory tissues, replacement and development patterns, and denticle and tooth morphologies (see section 10.6).

Ørvig (1967) first proposed the term odontode to include only those dermal skeletal units which shared developmental and structural properties in common with teeth, and to replace the general terms 'skin teeth' or 'denticles'. Odontode characteristics include the following (among other functional and morphological criteria): (a) formation within a single undivided dental papilla bounded by an epithelial dental organ; (b) formation superficially in the dermis and not from a sub-epithelial dental lamina; and (c) replacement from below, but may be replaced laterally or remain and form odonto-complexes by lateral and superimpositional growth (Ørvig, 1977). Precise distinctions (apart from positional criteria) between jaw teeth and odontodes were not established, and Ørvig (1977) recognized further difficulties with distinguishing odontodes from ornamental tubercles elsewhere in the dermal skeleton. Reif (1982) resolved the issue, in part, by expanding the odontode concept to include all dermal denticles, with teeth as a distinct sub-set. Reif defined teeth as developing (a) in advance of their requirement, and (b) exclusively from a dental lamina of either a permanent type, or a new infolding for every new replacement tooth (as described for actinopterygian and urodele teeth: Kerr 1960).

All three tissue types (enamel/enameloid, dentine/pulp, bone/cementum) are integral parts of an odontode, but any one could have evolved earlier than the others, and occur independently within the fossil record. Developmentally, this could result from suppression of part of the odontode differentiation programme (e.g. through a heterochronic shift in enamel development, as discussed in Smith, 1995), or if one cell type in the odontode system had not acquired competence.

Variation in timing of signalling molecule expression, ligand-binding site expression, or transcription factor production, could produce different odontode products in different combinations from a single morphogenetic system. This would provide a developmental basis for evolutionary change within the dermal skeleton (including teeth), and produce the various odontode and bone combinations illustrated by Reif (1982). Odontogenic cell fate and cell potential are topics discussed in section 10.6 on developmental models. Data for this are provided by heterotopic transplantation and by heterochronic, heterotypic recombinations in explant experiments. Current research concerns mostly higher vertebrate tooth-forming tissues, but similar experiments on fish teeth and (dermal) denticles will need to be considered for a more informed assessment of similarity between the two odontogenic systems.

How do we decide on the basis for homology of denticles and teeth? Constancy of developmental process has often been cited as a key criterion for recognition of homology (e.g. Bolker and Raff, 1996). However, as Hall (1995) states, summarizing many workers' views on homology: 'commonality of development at some level is pivotal if structures are to be recognised as homologous … while many do share developmental pathways, mechanisms, or constraints, equally many do not'. Provisional hypotheses about levels of developmental homology between morphogenetic systems in different taxa can be tested against phylogenetic analyses. Patterns of developmental evolution may therefore emerge from phylogenetic analyses, instead of prior assumptions about developmental conservation and homology (in effect, evolutionary scenarios) dictating the reconstruction of phylogenetic interrelationships.

10.3.   Denticles and teeth in the oropharynx

Distinctions between two types of odontode, teeth and skin denticle, have been given in the previous section, but here we consider the position in many fish where the lining of the mouth and pharynx has a covering of denticles as well as teeth (*sensu* Reif, 1982). Both of these contribute to the functional dentition, but retain distinctive developmental differences. Ørvig (1977) attempted to summarize this by concluding that 'such a distinction is possible by considering position and function in combination with mode and place of formation, size and structure'. Smith (1988) used lungfish dentitions to predict such developmental differences, arguing that teeth could be distinguished from denticles by their position, consistent shape, incrementally increased size, sequential times of development, and different tissue composition. Tissue composition difference would seem to be the most significant character on which to argue for a real, non-homologous, distinction between teeth and denticles. This difference is exemplified by the presence of petrodentine (an extremely highly mineralized form of dentine, Smith, 1984) as the main tissue component of the teeth in *Andreyevichthys epitomus* whereas it is absent from adjacent denticles, despite their use as part of the dentition (M. M. Smith personal observations).

Ørvig's (1978) description of the mandibular dermal skeletons of the fossil actinopterygian fish, *Birgeria groenlandica*, *Boreosomus piveteaui* and *Colobodus* cf. *bassani*, provides an alternative example. Like most actinopterygians (discussed in section 10.5), these fish have teeth with an enameloid cap (acrodin) and basal to this an enamel collar (ganoine), whereas the odontodes have only a thin, continuous, monolayer enamel covering (the squamation of certain catfish provides an interesting extant exception to this rule: Sire and Huysseune, 1996). However, in these fossil taxa, a field of 'dental units' or 'small teeth' (Ørvig, 1978) lies close to the marginal teeth, and each has an apical enameloid 'wart' and a diminutive tooth-like morphology. Ørvig believed that the 'development of odontodes with apical acrodin was a specialised condition … imitating the jaw-teeth in the vicinity'. Other denticles (tubercles), situated more distantly from the toothed margin, are shaped more irregularly and lack any enameloid capping tissue (Figure 10.1).

Such differences between adjacent denticles signify major differences at the level of histodifferentiation and morphogenesis. 'Odontodes' with apical enameloid are restricted to the vicinity of the dentition on dermal bones of the jaws, thus exemplifying a 'dental field'. These odontodes have the potential to make teeth of varying size but with consistent shape and composition, as distinct from the nearby dermal denticles (tubercles). Furthermore, Smith (1995) concludes that enameloid is a more recent phylogenetic development than enamel. The difference between these tissues results from a heterochronic shift in timing of epithelial cell differentiation relative to odontoblasts, enamel production (with a later time of ameloblast differentiation) being the primitive condition. A review of these tissue types in osteichthyan fishes can be found in Smith (1992), and enamel (ganoine) distribution in actinopterygian fishes in Richter and Smith (1995).

10.4.   Origins of teeth before jaws

The assumption that exoskeletal denticles evolved before ornamented bone tesserae and teeth (Reif, 1982) implies that neither teeth nor pharyngeal denticles were present in naked or armoured agnathan vertebrates. As noted in the introductory section, such hypotheses that teeth evolved from skin denticles as jaws evolved from the branchial arch series, need to be re-evaluated in the light of important new findings. The first is van der Bruggen and Janvier's observation (1993) that oropharyngeal denticles resembling those of jawed vertebrates occur in the thelodont agnathan, *Loganellia*, and the second is the

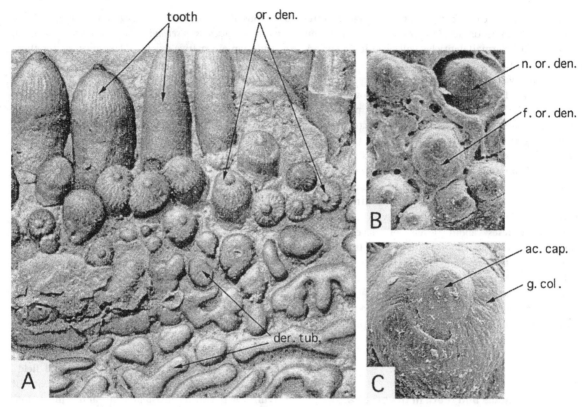

Figure 10.1A–C. **Representation of the dermal bones in the jaws of basal actinopterygian fishes to show the apical wart of acrodin (ac. cap) on the teeth and oral 'denticles' (or. den.; odontodes) of the dental unit. By contrast the adjacent, irregular shape dermal tubercles (der. tub.) at the labial margin of the tooth field do not have acrodin but only a continuous cover of enamel (after Ørvig, 1978, figures 7–9). A. Lower jaw of** *Colobodus* **cf.** *bassanii* **Alessandri, Triassic of Monte San Giorgio, Southern Switzerland. B, C. Palatal dermal bone with all denticles formed of apical acrodin (ac. cap) and ganoine of the collar (g. col.), showing one newly erupted oral denticle (n. or. den.) and one worn functional oral denticle (f. or. den.).** *Boreosomus piveteaui* **Nielsen, Eotriassic of Cap Stosch, East Greenland.**

reinterpretation of conodont elements as a basal verte-brate tooth-like apparatus (Figure 10.2) (Aldridge *et al.*, 1993).

Van der Brugghen and Janvier's (1993) discovery of specialized oropharyngeal denticles in the thelodont, *Loganellia*, constitutes significant and independent evi-dence that such dentitions may evolve independently of jaws. These denticles fall into two categories: (a) 'pharyn-geal denticles or tooth whorls associated with the gill bars, or within the extrabranchial ducts', compared with those of sharks, and (b) a second population occuring within the snout region and centrally in the oropharyn-geal cavity. These latter denticles are not associated with pharyngeal skeletal supports, thus lending some support for the model suggested (section 10.6), in which the developmental evolution of jaws and teeth are un-coupled. It is fascinating to conclude that some oropha-ryngeal denticles could have functioned as 'teeth' or 'tooth plates' before jaws had evolved.

Although Pander (1856) first considered conodont ele-ments as isolated fish teeth, they have been dismissed as

such by most workers. Recently, almost 80 years later, conodont assemblages in bedding planes have been dis-covered. These show only a bilateral organization (rather than dorsoventrally opposing rows of tooth-like cusps), so for this reason Pander's fish-tooth interpretation has also been rejected. It should, however, be noted that pharyn-geal teeth of cyprinid fish are also organized bilaterally as two sets of teeth functioning together in a ventro-dorsal direction, but biting against a basioccipital kerati-nous pad. It is also significant that the rostrocaudally graded series of divergent conodont morphologies consti-tutes a further dentition-like characteristic (Figure 10.2). At least three lines of evidence now suggest that cono-dont elements deserve reevaluation as most closely related to vertebrate oropharyngeal denticles, and perhaps teeth: (a) the discovery of conodont soft tissue anatomy (notochord; muscle segments: Aldridge *et al.*, 1993); (b) the identification of vertebrate hard tissues within conodont histology (enamel; dentine: Sansom *et al.*, 1992, 1994); and (c) functional models indicating that the conodont apparatus served as a feeding structure

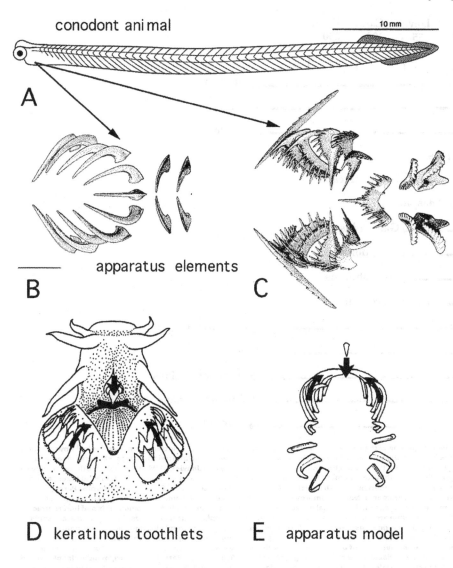

conodont animal

10 mm

A

apparatus elements

B   C

D keratinous toothlets   E apparatus model

Figure 10.2A–E. **Diagram of the conodont animal and reconstructed apparatuses to show their diversity of form and putative arrangements (E based on the hagfish model). A.** *Clydagnathus windsorensis*, **Carboniferous of Scotland (from M. Purnell, 1995, with permission) to show the relative size of the eyes and caudal fin, and position of the apparatus (origin of arrows). Below are enlargements of two types of apparatus in ventral and dorsal view, situated** *in vivo* **ventrocaudal to the eyes. B. Reconstruction of apparatus of** *Drepanodus robustus* **Hadding, 1933, based on isolated elements, with five pairs of bilaterally symmetrical ones, one median, and two posterior pairs (from Dzik, 1991, figure 3B). C. Reconstruction of the spatial arrangement of elements in the apparatus of** *Manticolepis subrecta* **Miller and Youngquist 1947, based on clusters and natural assemblages (from Dzik, 1991, figure 15). This shows the same arrangement as in A, but with a more complex morphology, arranged as five pairs in the rostral basket, an unpaired medial symmetrical element, and two pairs in the caudal platform complex. D. Keratinous toothlets of the extant agnathan,** *Myxine*, **in antero-ventral view, a suggested analogue to the conodont apparatus, arrows show direction of bite (from Krejsa** *et al.*, **1990a, figure 3B). E. Reconstruction of elements of** *Panderodus* **into a model with the direction of bite shown by the arrows. Antero-ventral view comparable with the hagfish (with permission, I. J. Sansom 1992).**

(Figure 10.2; Purnell and von Bitter, 1992; Purnell, 1993). Krejsa *et al.*, (1990a) suggests that the cyclostome feeding apparatus provides an interpretive model for functional analysis of the conodont apparatus (Figure 10.2D) and Sansom (1992) presented an apparatus model with the paired elements located ventrally in this way (Figure 10.2E).

Such evidence of early agnathans with oropharyngeal denticles would appear to be consistent with the suggestion that hagfish and lamprey keratin toothlets are antecedents of the enamel–dentine complexes of modern teeth (Slavkin *et al.*, 1991; Slavkin and Diekwisch, 1996). This assumes that the embryological development of keratin toothlets in hagfish and the apatitic enamel–dentine odontodes of vertebrates is homologous, and

that keratin biomineralization represents the primitive state for teeth (Krejsa *et al.*, 1990a, b). Smith and Hall (1990), however, addressing much the same question, reach somewhat different conclusions.

Lamprey keratinous toothlets appear to be similarly divergent from vertebrate teeth, and Smith and Hall (1990), followed by M. M. Smith *et al.*, (1996), concluded that neither hagfish nor lamprey dentitions are homologous in any obvious way with primitive odontodes; furthermore, a number of objections need to be overcome before any such hypothesis can be accepted (*contra* Krejsa *et al.*, 1990b).

Conodonts are, therefore, thought to represent an alternative vertebrate experiment with functional 'teeth', and perhaps one which evolved prior to any

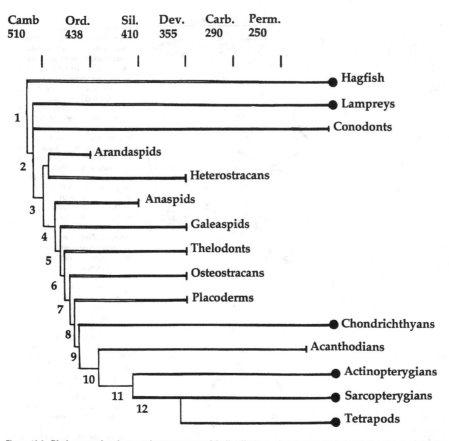

Camb	Ord.	Sil.	Dev.	Carb.	Perm.
510	438	410	355	290	250

Figure 10.3. **Phylogeny of major vertebrate groups with distribution of major dentoskeletal structures plotted against geological time axis (not drawn to scale). Interrelationships scheme mostly from Janvier, 1996 (for gnathostome crown group interrelationships, see references in text; geological dates (MYBP) from Cowie and Bassett, 1990; taxon time ranges taken from Forey and Janvier, 1994 unless stated otherwise in text). Double lines indicate ghost lineages; thick lines show taxonomic time ranges. Line termination as a solid circle indicates living taxon; as T bar indicates extinct taxon. Abbreviations: Camb., Cambrian; Ord., Ordovician; Sil., Silurian; Dev., Devonian; Carb., Carboniferous; Perm., Permian.** *Node 1*: hagfish plus more advanced craniates. *Node 2*: lampreys, conodonts, plus more advanced craniates. Possible origin of phosphatic feeding apparatus, enamel + dentine; external branchial basket derived from neural crest. *Node 3*: arandaspids and heterostracans, plus more advanced stem-gnathostomes. Dermal armour with basal bone, no evidence of oropharyngeal denticles. *Nodes 4 and 5*: anaspids, followed by galeaspids plus more advanced stem-gnathostomes. *Node 6*: thelodonts (paraphyletic group) plus more advanced stem-gnathostomes. Oropharyngeal denticles present, showing degree of patterning; short whorl-like compound denticles associated with pharyngeal pouches. *Node 7*: osteostracans (no evidence of oropharyngeal denticles), plus more advanced stem-gnathostomes. Possible presence of jaw adductor musculature. *Node 8*: Placoderms plus crown-gnathostomes. Articulated internal gill arches; mandibular plus hyoid arch present; oropharyngeal denticles present. *Node 9*: Gnathostomes: chondrichthyans plus osteichthyans. Tooth-whorl compound denticles possibly associated with jaws and gill arches in all primitive taxa. Chondrichthyan stem: retention of external component of branchial skeleton; tooth whorl replacement series on mandibular arch; dental lamina present; enameloid present on teeth; basal bone loss from dermal denticles. *Node 10*: stem-osteichthyans: loss of external branchial arches. Acanthodians paraphyletic relative to more advanced osteichthyans; later acanthodians show reduction of denticles associated with mandibular arch; no evidence of true teeth or dental lamina. *Node 11*: osteichthyans: actinopterygians plus sarcopterygians (paraphyletic relative to tetrapods). Primitive members retain symphysial tooth whorls; dental lamina with tooth replacement. Actinopterygian stem: acrodin enameloid caps evolve on all oropharyngeal denticles plus teeth; teeth develop on gill arches. *Node 12*: tetrapod stem. Post-cranial dermal skeleton reduced; oropharyngeal denticles restricted increasingly to mandibular arch and dermal bones lining buccal cavity.

mineralized exoskeleton (M. M. Smith *et al.*, 1996). Similarly, Janvier (1996) thought that 'the structure of conodont elements is much derived relative to that of the classic dentinous tissues of vertebrates' (see Chapter 5). However, discoveries of earlier vertebrate dermal armour from the Late Cambrian in both North America (Repetski, 1978; M. P. Smith *et al.*, 1996) and Australia (Young *et al.*, 1996), require even this assumption to be modified. As noted in the introduction to this chapter, these new finds suggest that early vertebrates were very diverse and already armoured and naked lines were contemporaneous in the Late Cambrian. In summary, while conodonts are naked but have a mineralized feeding apparatus, a further two taxa are now known to have had dermal armour, of which one has dentine odontodes and the other has a pore-canal system. Note that such diverse discoveries are quite consistant with the phylogenetic hypothesis shown in Figure 10.3.

10.5.   Bones with teeth

Pharyngeal denticles have long been known to develop in relation to the whole branchial arch series in gnathostome fishes, and especially those of the Osteichthyes (Nelson, 1970, Lauder and Liem, 1983). If evolutionary sense is to be made of denticle distributions throughout extant and fossil taxa (and the underlying odontogenic systems), then these data must be placed within a phylogenetic framework (e.g. Figure 10.3). Furthermore, in order to construct developmental–evolutionary hypotheses, primitive (plesiomorphic) conditions need to be identified for each of the constituent phylogenetic subgroups. The fossil record of gnathostome dentitions is fortunately extensive, and provides a broad outline of large-scale evolutionary changes. Early details of these events, however, are obscured by uncertainty about the interrelationships of the major basal gnathostome clades (Figure 10.3). In addition to the major living groups (chondrichthyans and osteichthyans), there are two significant extinct taxa, the placoderms and acanthodians. Of these, the acanthodians have closest affinities with basal osteichthyans, whereas the taxonomic position of placoderms is problematic. Several authors have attempted to align placoderms with chondrichthyans (e.g. Goujet, 1982; Jarvik, 1980; see Patterson, 1992 for a recent refutation of this hypothesis), while others have suggested that they are stem-group osteichthyans (see Gardiner, 1990 and references therein). In the present work (following Janvier, 1996), placoderms are considered the most primitive gnathostomes.

As noted earlier, evidence of oropharyngeal denticles in thelodonts (van der Brugghen and Janvier, 1993) indicates a degree of tooth-like specialization occuring before the origin of jaws. The phylogenetic position of thelodonts is rather uncertain; they appear to be a paraphyletic assemblage, including the ancestry of osteostracans (i.e. jawless stem-group gnathostomes) plus all more advanced near-jawed and jawed vertebrates (Figure 10.3; Janvier, 1996). It may therefore be significant that certain features of the loganellid thelodont 'dentition' recur in basal gnathostomes, suggesting that these are indeed genuinely primitive character states. In particular, the whorl-like compound pharyngeal denticles resemble similar features of basal gnathostome dentitions; especially those denticles associated with early chondrichthyan pharyngeal mucous membranes and gill arches (e.g. Cobelodus, Zangerl and Case, 1976). Whorl-like series of replacement teeth are generally assumed to be a chondrichthyan feature, occuring in all known articulated specimens. Holocephalans (=chimaeroids or bradyodonts) are the outstanding exception to this arrangement, having plate-like crushing dentitions. Yet even in these, the toothplates are derived from a conventional, lingual to labial pattern of dental growth and replacement (Patterson, 1992; contra Reif, 1982).

All chondrichthyans possess a dentition consisting of, or derived from, discrete, whorl-like families of replacement teeth, distributed evenly around upper and lower jaw margins (Figure 10.4C; Zangerl, 1981). In modern sharks, such teeth tend to be: (1) narrow-based (enabling a single member of each tooth replacement family to flip into the functional jaw margin); (2) replaced rapidly (shed teeth show little signs of wear; functional teeth therefore maintain optimum condition); and (3) mono-cusped for gouging and cutting (which enables a small predator to remove chunks from a much larger prey species; thus increasing vastly the range of potential food sources). In contrast, primitive chondrichthyan teeth are usually 'cladodont', i.e. (1) broad-based (indicating that more than one member of each tooth replacement family contributes to the functional jaw margin); (2) often show signs of wear (suggesting a slower tooth replacement cycle); and (3) have a tall central cusp flanked by smaller accessory cusps (functionally suited for piercing and grasping). These early chondrichthyan teeth therefore appear to be rather less specialized than those of modern predatory sharks, and adapted more generally for (simple prey capture and retention (Williams, 1990).

Fossil chondrichthyan pharyngeal or mucous membrane denticles are rare; nevertheless, the combined presence of pharyngeal denticles in early fossil taxa, and some recent true sharks (e.g. Cetorhinus, Parker and Boeseman, 1954; Cappetta, 1987), suggests strongly that they are a primitive chondrichthyan characteristic.

Placoderms (Figure 10.4A) the problematica of basal gnathostome phylogeny, have well developed jaws but no recognisable teeth (Denison, 1978). Rhenanids seem to be the most informative group with respect to the primitive condition of the placoderm dentition, with minimal presence of oropharyngeal denticles occurring on the palatoquadrate and meckelian cartilages (Denison, 1978). Oral denticles are otherwise known on certain placoderm parasphenoids (e.g. buchanosteid arthrodires), which are thought to be a further primitive characteristic.

Isolated acanthodian scales date back to the Lower Silurian, and acanthodians may be expected (subject to the provisional phylogenetic hypothesis discussed earlier, Figure 10.3) to originate at the divergence of osteichthyan from chondrichthyan ancestry. Acanthodian dentitions are poorly developed relative to those of subsequent osteichthyan groups, and many taxa are entirely toothless (Moy-Thomas and Miles, 1971; Denison, 1979). Single teeth may be ankylosed to the jaws, and short tooth whorls may occur, both at the mandibular symphysis and elsewhere along the jaw margin (Figures

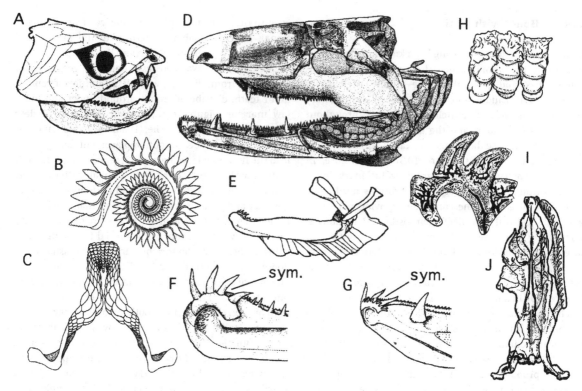

Figure 10.4A–J. **Varieties of jawed vertebrate dentitions (asterisk indicates extinct taxon). A. *Hadrosteus*,* an arthrodire placoderm: lateral view of skull showing toothless jaws with shearing gnathal plates (from Denison, 1978, figure 69). B. *Helicoprion** a eugeneodontid basal chondrichthyan: the symphyseal tooth whorl (from Carroll, 1988, figure 5.8). C. *Heterodontus*, a neoselachian chondrichthyan, the lower jaw with replacement tooth families showing a gripping (anterior) and grinding (posterior) dentition (from Cappetta, 1987, figure 24). D. *Eusthenopteron*,* a fish-like stem tetrapod: neurocranium, parasphenoid, palate and mesial surface of mandible, hyoid and gill arches with dental plates (arrows; reconstructed from Jarvik, 1980, figures 107, 110, 125,). E. *Howqualepis*,*a primitive actinopterygian: mandible with symphyseal tooth whorl (sym; from Long, 1988, figure 21). F. *Onychodus*,* an onychodont primitive sarcopterygian: symphyseal tooth whorl (from Moy-Thomas and Miles, 1971, figure 6.22). G. *Holoptychius*,* a porolepiform primitive sarcopterygian: symphyseal tooth (from Moy-Thomas and Miles, 1971, figure 6.22). H. *Ptomacanthus*,* a climatiid acanthodian (stem-group osteichthyan): tooth whorls from upper jaw (from Carroll, 1988, figure 6.5). I. *Nostolepis*,* a climatiid acanthodian (stem-group osteichthyan): sectioned tooth whorl, lateral view (from Ørvig, 1973, figure 5). J. *Naja*, a colubroid snake: palatal view of cranium showing presence of teeth on palatine and pterygoid bones (arrows, right side palatal and marginal bones removed; from Romer, 1956, figure 67). A and B, anterior to right of figure; C and J, anterior to top of figure; D to G and I, anterior to left of figure; H mesial surface, dorsal to top of figure.**

10.4H, I; Denison, 1979). Reif (1982) notes that such tooth whorls exist in two forms: (a) short, bow-shaped series of four teeth with fused bases; and (b) complete spirals consisting of up to 16 teeth with fused bases. The short tooth series were probably shed periodically and not added to *in situ*, while the complete whorls may have been retained throughout life with replacement added at one margin.

Recent analyses of acanthodian interrelationships indicate that the toothed groups, ischnacanthids and climatiids, are the most primitive (Long, 1986). The combination of such dental and dermal skeletal features, suggests that they may be closely related to the evolutionary root of extant osteichthyan lineages. Living osteichthyans are divided into two major clades: actinopterygian (ray-finned) and sarcopterygian (lobe-finned) fishes (Figure 10.3). *Cheirolepis*, the most primitive actinopterygian known in detail (Pearson and Westoll, 1979; Gardiner and Schaeffer, 1989), has a dentition consisting of tooth rows on the dermal jaw bones, and denticle-fields on the dermal bones lining the oral cavity. Densely denticulate, branchial tooth plates are known in related primitive taxa (e.g. *Mimia* and *Moythomasia*, Gardiner, 1984), and *Polypterus* (Jollie, 1984, Figure 17), the living sister-group of all other actinopterygians, indicating that these are a primitive characteristic.

Tooth whorl morphologies are usually assumed to be absent in actinopterygians, but Long (1988) describes a short symphyseal whorl including six teeth in *Howqualepis* (Figure 10.4E) and *Tegeolepis* (Dunkle and Schaeffer, 1973) displays a similar whorl-like symphyseal tooth series (M.I.C. personal observations, Cleveland Museum of Natural History, Ohio, CMNH 5518). Both of these Devonian taxa occupy a basal position in actinopterygian phylogeny.

The diversity of recent actinopterygian dentitions is

well illustrated in reviews such as that by Lauder and Liem (1983). Pharyngeal jaws usually consist of enlarged gill arches 4 and 5, with associated pharyngeal bones bearing teeth. In certain 'pharyngognaths', the absence of teeth on the mandibular arch means that these jaws, situated at the rear of the pharynx, provide the only effective biting apparatus. The zebrafish (Danio rerio), a cyprinid teleost and a widely used model for genetic, molecular and developmental studies, exemplifies this condition, with teeth occurring only on the lower element (ceratobranchial) of the fifth gill arch. Dental morphotypes differing in size, shape, and function can exist in the same species (Francillion-Vieillot et al., 1994, Huysseune, 1995). Significantly, in all of these examples of dental variation, the replacement teeth, even if deep within the bone, are developed from a dental lamina; sometimes with loss of connection to the epithelium, just as in mammalian teeth.

Like basal actinopterygians, early sarcopterygian dentitions include toothed marginal jaw bones, palatal bones bearing denticle fields, and numerous tooth plates associated with the gill skeleton, branchial pouches, and the roof of the oropharyngeal cavity (Figure 10.4D). Current analyses of sarcopterygian interrelationships place coelacanths and lungfishes as successive outgroups relative to the Tetrapoda (e.g. Janvier, 1986; Cloutier and Ahlberg, 1995). Early sarcopterygian groups include the coelacanths, porolepiforms and onychodonts (Moy-Thomas and Miles, 1971; Schultze, 1973); porolepiforms are probably allied to the lungfish lineage, while onychodont affinities are currently unresolved. Significantly, porolepiforms are among the earliest sarcopterygian body-fossils (Jessen, 1980), and, like the onychodonts, these primitive taxa bear mandibular symphyseal tooth-whorls (Figure 10.4F,G).

The origin of tetrapods is most closely allied with the osteolepiform sarcopterygians, of which the Upper Devonian species Eusthenopteron foordi is known in exceptional detail (Figure 10.4D; Jarvik, 1980). In addition to well-developed marginal tooth series and palatal denticle fields, the palatines and vomers bear large fangs, with complementary fangs present on the mesial surface of the lower jaw (Figure 10.4D). It is these, and similar fangs in porolepiforms, which most clearly display the characteristic infolded dentine that is retained in early tetrapod dentitions, but lost in lungfishes; hence the term 'labyrinthodont' applied to various (paraphyletic) groups of early tetrapods.

Jarvik's (1980, and reference therein) description of Eusthenopteron includes a comprehenive account of the branchial and other oropharyngeal toothplates (Figure 10.4D). Unfortunately, little evidence is available concerning the phylogenetic pattern of oropharyngeal

dental reduction occurring at the emergence of limbed, and eventually terrestrial tetrapods. It appears that branchial toothplates were present in at least one basal clade of amphibian-like tetrapods, the colosteids (Hook 1983; Coates 1996). Intriguingly, the persistence (?) of such toothplates in the pharynx correlates with the appearence of a denticle patch on the newly expanded posterior portion of the parasphenoid in these early amphibians (absent in basal amniotes), suggesting retention, or evolution, of a palatal bite (cf. extant teleosts, as discussed earlier). Basal tetrapod dentitions consist otherwise of a mixture of fish-like denticle fields and fangs (Clack 1994). These are usually considered lost or modified with the emergence of modern amphibians and amniotes, but it is worth noting that additional rows of palatal teeth persist in several non-mammalian tetrapods. Examples include caecilian amphibians (palatine-vomerine tooth row; Carroll 1988), and among amniotes Sphenodon (palatine tooth row), and snakes (e.g. colubrids) with arrays of non-marginal teeth on the pterygoids and palatines (Figure 10.4J; Romer 1956).

10.6.   Developmental models

The evolution of vertebrate dentitions is linked inseparably to the evolving role of neural crest cells during vertebrate embryogenesis. The emergence and migration of a distinct, skeletogenic, neural crest cell population is almost certainly a vertebrate/craniate synapomorphy (Moss, 1968a,b; Gans and Northcutt, 1983). These neural crest cells contribute three distinct tissue complexes to the dermis, visceral arches and anterior neurocranium: dentine plus bone of attachment, dermal spongy bone, and cartilage plus fibrous musculo skeletal attachment (Couly et al., 1993; Koentges and Lumsden, 1996). This ability to make a phosphatic dermal skeleton, dependent on interaction between epithelium and ectomesenchyme, may be linked to a new developmental role, after gene duplication of the transcripts of the homeobox-containing Msx gene family.

Extensive diversification of the Msx gene family probably occurred (by duplication) concurrently with the origin of vertebrates (Holland, 1991). Transcripts of two of these genes (of perhaps three) in amniotes, Msx-1 and Msx-2, are localized in premigratory and migratory neural crest, and in ectomesenchymal regions in mandibular, hyoid and other branchial arches (Takahashi et al., 1991). They are expressed concomitantly in the adjacent ectoderm, and are involved in epithelial-mesenchymal signalling in tissue interactions considered to be essential for development of the dermal and visceral skeleton, including teeth (Mackenzie *et al.*, 1992; Satokata and Mass, 1994; Vastardis *et al.*, 1996). Studies of *Msx-1/2* in the mouse present evidence from double knockouts that these genes are involved in initiation of the entire dentition, as well as for tooth bud position (see Chapter 1). From this work linking neural crest derivatives and *Msx* expression, comparative questions have been posed and answers sought in the zebrafish dentition (Hickman and Smith, 1994). It has already been shown that the dermal component of fin development involves the participation of trunk neural crest mesenchyme (Smith *et al.*, 1994), and at least four of five related *Msx* genes (Akimenko *et al.*, 1995). This suggests further comparative questions about possible relations between the more elaborate dermo visceral skeletal products in teleosts (relative to tetrapods) and the significantly more diverse array of these genes in zebrafish.

The origin of the lamprey branchial cartilages from neural crest chondroblasts illustrates clearly the central importance of neural crest-derived tissues to early craniate evolution (Newth, 1956). Although these branchial basket cartilages are not topographically and structurally equivalent to the internal arches of the gnathostome pharynx (Mallatt, 1996; see section 10.8), both structures may be considered homologous at the more inclusive level of cartilaginous branchial supports. Schaeffer (1975) suggested that it was simply an alternative developmental strategy; the result of differential streaming of neural crest relative to the paraxial mesoderm. From a more recent perspective, the positional information carried by the migrating cells would still retain the combinatorial Hox code specific for each A/P level suggested as the patterning mechanism for the skeletal derivatives of these cells (for a general review see Holland and Garcia-Fernandez, 1996). However, the possible neural crest origin of the internal and external lamprey velar bar cartilages, or the ventral longitudinal piston cartilage (structures associated with the 'old mouth': Mallatt, 1996), remains unknown. Langille and Hall (1988) commented that there was at present no evidence for a neural crest origin of the mucocartilage in this region.

Researchers are now beginning to decipher this complicated process and although Newth (1956) supported a concept of necessary inductive contact between branchial ectomesenchyme and endoderm for visceral cartilage development, he nevertheless concluded that premigratory crest is not determined to produce cartilage but is competent on receiving an inductive signal. The role of the *chn* gene in zebra fish, however, which autonomously specifies neural crest pharyngeal chondroblasts at the beginning, rather than end, of crest cell migration (Schilling *et al.*, 1996), challenges this persistent hypothesis. Schilling and colleagues' (1996) results support earlier suggestions that arch precursor fates appear to be specified *before* any major morphogenetic movements (Schilling and Kimmel, 1994). Lumsden (1987) also maintained that both murine and bird cranial ectomesenchyme is specified to form cartilage before migration but after contact with dorsal cranial ectoderm, as previously suggested (Bee and Thorogood, 1980), but the nature of the signal is unknown. Nevertheless, transplant and explant experiments in urodele amphibians indicate that teeth (ultimately) remain dependent upon an endodermal signal to form from crest-derived mesenchyme (Sellman, 1946; Graveson *et al.* 1997). However, recombination explants in mice (Lumsden, 1987) showed that itinerant contact with the endoderm was not an absolute requirement for tooth development from cranial ectomesenchyme. However, the latest work by Imai *et al.* (1998) is totally pertinent to this question; they show that mouse oral ectoderm is specified to make teeth by contact with the foregut endoderm. This may be an indication of evolutionary developmental change in the inductive system and is perhaps the explanation of the seemingly different results from Lumsden (1987), in that the timing of contact of endoderm with potential odontogenic ectomesenchyme is critical in the mouse. We have also commented (Smith and Coates, 1998) that 'Phylogenetic data showing the primitive presence of oropharyngeal denticles in prejawed fishes as a preadaptation to the gnathostome dentition, strongly suggests that the endoderm is the essential requisite ... in dental development.'

Janvier (1981) in discussing the embryology of the pregnathostome anatomical pattern accepted the then current hypothesis that ectoderm must have invaded the oropharyngeal cavity to provide the dermal elements of the branchial arch skeleton. This explanation is no longer necessary, given the experimental evidence that teeth and cartilage may have a dependence on inductive endodermal signals. Like cartilage, dermal bone formation may also be prespecified in cranial ectomesenchyme and require an epithelial post-migration site for a permissive inductive signal before phenotypic expression. Tooth formation differs from skeletogenesis in that the permissive signal is rectricted topographically to the mandibular arch oral epithelium (at least in the mouse: see Chapter 1 for details of a model for mammalian dentition), although, as elsewhere, the cranial crest cells may be pre-specified. This experimental evidence of separation of odontogenic from osteo- and chondrogenic neural crest potential, provides the basis for Smith and Hall's (1993) model, explained below as modular development (Figure 10.5).

The migrating neural crest population in the cranial

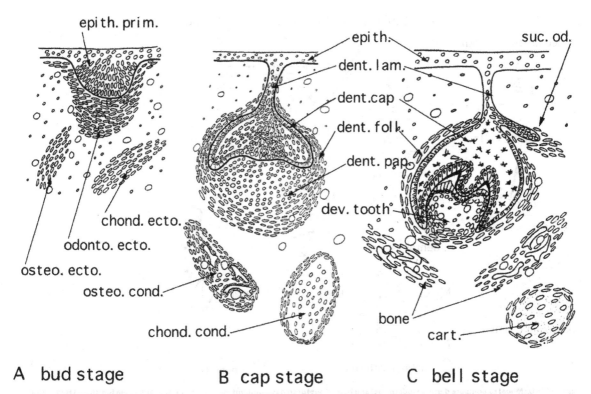

epith. prim.

epith.

suc. od.

dent. lam.

dent. cap

dent. folk.

dent. pap.

chond. ecto.

odonto. ecto.

dev. tooth

osteo. ecto.

osteo. cond.

chond. cond.

bone

cart.

A bud stage   B cap stage   C bell stage

Figure 10.5. **Modular representation of three stages of morphodifferentiation in tooth development in a jaw in which cartilage and bone are also developing in topographic relation to the tooth germ. Three condensations of ectomesenchyme represent osteogenic (osteo. ecto.), chondrogenic (chond. ecto.) and odontogenic (odonto. ect.) developmental modules, the last adjacent to a thickening or epithelial band (epith. prim.), together these form the odontogenic primordium at the tooth bud stage. At the cap stage the odontogenic mesenchyme has formed into two morphogenetic groups, dental papilla (dent. pap.) and encapsulating dental follicle (dent. foll.), the bone condensation (osteo. cond.) with a periosteal group, and the cartilage condensation (chond. cond.) with a perichondral group. The boundary groups in all three modules can form the fibrous connective tissue, the stem cells for each main tissue type and can transform to make perichondral bone in membrane outside the cartilage, and secondary cartilage beneath the fibrous periosteum. At the bell stage, tooth shape has been determined and histogenesis of dentine and enamel has begun. The follicle will give rise to the attachment tissues (cementum, periodontal ligament and alveolar bone of mammals). A primordium for successional teeth (suc. od.) forms a lingual extension from the dental epithelium and is the odontogenic developmental clone of one tooth family in the model. Dermal basal bone of the jaw and primary cartilage (cart.) of the visceroskeleton has also differentiated. (Modified from Smith and Hall, 1993. figure 1.)**

and branchial regions has apparent organizing properties, and may carry positional information relating to dorsoventral and rostrocaudal axes (see Thorogood, 1994; Schilling and Kimmel, 1994; Schilling et al., 1996). Hindbrain crest cells (rhombomeric ectomesenchyme) are patterned by a segmentally derived combination of Hox genes. Thus segmental identity in the head periphery is transmitted from the neurectoderm by means of the Hox code in migrating neural crest, so that each arch expresses a different and 'arch-specific' Hox gene combinatorial identity (Hunt et al., 1991; corroborated by Rijli et al., 1993). However, while it is assumed that 'this axial level-specific code is thought to be imposed on an initially naive ectoderm and any mesoderm' (Thorogood, 1994), the precise role of a prepattern in the ectoderm and endoderm remains unclear, and may differ between the three different tissue types associated with the branchial arches. New research in the chick provides evidence of an independent patterning mechanism for the branchial

arches (Veitch et al., 1999), peripheral to (ectoderm and endoderm) and integrated with that of neural crest.

Once the crest cells have migrated into the branchial arches, the ectomesenchyme forms interactive morphogenetic units with the epithelium (Schaeffer, 1977), where each modular unit may be (finally) committed to form either bone, cartilage, denticles (odontodes), or teeth (Hall, 1975; Smith and Hall, 1993, Figure 10.1; Graveson et al. 1997). Such a model allows for dermal odontodes (teeth and denticles) to develop and evolve independently of dermal bone or branchial cartilage (i.e. some degree of separation of odontogenic from osteo- and chondrogenic neural crest potential: Smith and Hall, 1993), but it cannot predict which condition is primitive for vertebrates. Neither can this model predict whether dermal denticles evolved first in the oropharyngeal cavity, or as a body covering. Only the fossil record can provide these data, because no extant agnathan possesses mineralized exoskeletal denticles, oropharyngeal denticles, or teeth.

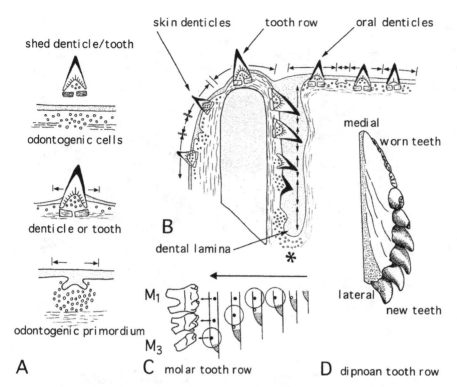

Figure 10.6A–D. **Model to compare denticle (odontode) and homodont tooth development adapted from Reif, 1982 (see Smith and Hall, 1993, figure 2) and proposals for a general heterodont vertebrate model. A. The odontogenic primordium is postulated to generate a zone of inhibition around it (arrows), maintained around the denticle/tooth, a new primordium can only form when a denticle is shed. B. By comparison with oral and skin denticles, teeth from each row, form at the end of an epithelial invagination (dental lamina) when they are out of the zones of inhibition (asterisk). This point is analogous to a clonal primordium, a growing margin at which odontodes are formed when the cell mass has reached a critical level. C. The clonal model for tooth development (mouse molars, after Lumsden, 1979). The three lower jaw mouse molars (M_1–M_3) are generated successively from a pre-specified molar clone (shaded epithelium), first one, then two, and three tooth buds develop (arrow). The large circles represent the zones of inhibition, the white dots tissues at the critical mass, and black dots the initiation of tooth primordia. D. Alongside the shark tooth row is one tooth row of a fossil dipterid (lungfish) tooth plate, to show where new teeth are added at the labial margin of palatal and pre-articular dermal bones of the jaws. Teeth are not shed in this form of dentition and become progressively worn towards the medial margin (after Smith, 1988, figure 2B).**

Smith and Hall (1993) suggest that the basis of the odontode, the conjectured primitive exoskeletal building block, consists of interactive dental cap epithelium plus dental papillary and follicular ectomesenchyme (Figure 10.5). This 'morphogenetic system' is associated with two regulatory processes: neural crest migration, and inductive interaction between ectoderm/endoderm and mesenchyme. The prevailing model of tooth evolution (Reif, 1982) also requires the presence of an inhibitory field around each non-growing odontode. When free of the inhibitory influence, only body growth allows the insertion of new single odontodes in intervening spaces. This model currently lacks experimental or molecular data, such as agonistic and antagonistic signals, but is an area in which new techniques will almost certainly furnish important facts towards an understanding of the control process (see Osborn, 1978, for an alternative model, and further discussion of Reif's earlier research).

Patterning the 'morphogenetic system' in terms of control of tooth/denticle shape and consistent increase

in size is another dimension to the model as outlined above. In the past, two opposing theories have been considered, and both are discussed by Lumsden (1979) as the 'field gradient theory' with an external morphogen source acting on identical primordia (preodontodes), and the 'developmental clone' with intrinsic controls on lineage-related primordia descended from a prespecified clone, with a 'limited repertoire' of ultimate differentiation. Lumsden's studies on the development of mouse molars favoured the latter model, and an extension of this to basal vertebrates would complete the model proposed by Smith and Hall (1993; Figures 10.5, 10.6C). This could predict that the undifferentiated mass of ectomesenchyme at the base of the dental lamina in the shark model (asterisk, Figure 10.6B) would produce teeth of one predetermined shape for each replacement family with a growth-related increase in size. Spacing of individual teeth, and size, would depend upon the rate of cell division in the primordial clonal ectomesenchyme and its distance from the inhibitory influence (arrows and asterisk, Figure 10.6B). Similarly the tooth row on a dipnoan tooth plate (Figure 10.6D) would be generated by a 'developmental clone' at the lateral margin specified for that shape and conforming in size to a consistent logarithmic increase. Each dentition would be patterned by initiation of a number of clones dependent on the degree of heterodonty illustrated as three tooth classes in mammals (Osborn, 1978, figure 7). Recent understanding of how differentiated cell types arise in a uniform population of epithelial cells comes from a study of molecular markers of sensory cell differentiation in *Drosophila*. It seems that a mechanism for autologous signalling operates between Notch and Delta genes by lateral inhibition of Delta expression in surrounding cells to those generating high levels of the gene (the proteins encoded are transmembrane signalling molecules as receptor and ligand and all cells can make both molecules). Delta is focally expressed at the future sites of neuromasts. Recently, *Delta 1, 2* and *3*, mammalian homologues of the *Drosophila* gene have been demonstrated in rodent tooth morphogenesis (Mitsiadis *et al.*, 1995). In this example Notch is downregulated by a mesenchymal signal in the immediately adjacent dental epithelium and may be a prerequisite for odontogenic fate. Either of these mechanisms could provide a developmental control for spacing out the primordial clones by autogenetic controls. The possible molecular controls for initiating the clonal number are discussed in Chapter 1.

Assuming that either, or both, the skin and the oral epithelium had the (primitive) ability to produce fields of replacing odontodes, the question remains as to how tooth development evolved from this system. Reif (1982) suggests that a dental lamina developed as an infolding of the epithelium, thus producing an inhibition-free zone, away from the superficial denticles. The clonal primordium pre-specified for tooth shape, size and tissue composition would be spatially distinct from that producing the oral denticles. As a distinct primordium, tissue differences could arise and be maintained within the clone. Within this fold (linear epithelial primordium), teeth could develop in advance of functional tooth loss, ready to serve as biting elements (Figure 10.6), and as tooth families spaced along the jaw. Whether this fold evolved once, or convergently, in several lineages is uncertain, but a dental lamina would have had many functional advantages, including the provision of a protected developmental site, the possibility of rapid increase in size within each tooth family (see Chapter 13), and a faster replacement cycle. In summary, our understanding of how a dentition can evolve from a morphogenetic system producing non-growing, replaceable oral and skin denticles, is limited. And the principal cause of this limitation is a profound lack of information concerning the regulatory events which produce a dental lamina and serial tooth families (See chapters 1, 12).

10.7.   Jaw evolution and function

In contrast to established evolutionary scenarios, it appears likely that the origins of jaws and teeth are relatively discrete events. This uncoupling suggests different functions in the earliest jaws and teeth, with jaws originating for suspension or suction feeding, and teeth for food apprehension or sampling evolving either earlier or later.

Suspension and suction feeding are the two most common mechanisms of aquatic feeding, and may therefore have important implications for the analysis of evolutionary patterns of feeding mechanisms (Lauder, 1985). Teeth contribute to such systems, but may not be required at all. Thus Mallatt's (1996) proposal, that jaws evolved to shut the mouth during exhalation in the oropharyngeal ventilation cycle, and only later acquired a prey-grasping function, appears credible. Mallatt (1996) concludes that 'changes leading to the grasping jaws of gnathostomes were adaptations for improved ventilation, rather than feeding adaptations'.

Perhaps the most serious problem with Mallatt's hypothesis is that it considers the mandibular arch to be a modified anteriormost agnathan gill arch, followed initially by an unmodified gill-bearing hyoid arch. However, as Forey and Janvier (1993) have stated, 'the way in which jaws have developed (*in evolution*) remains unknown', and the mandibular arch may never have been gill bearing. Similarly, every attempted identification of an unmodified, gill-bearing hyoid arch in primitive gnathostomes (Watson's (1937) aphetohyoidean theory), has been effectively refuted (summary in Maisey, 1986 and further discussion in Maisey, 1989). Janvier (1993) emphasized that jaws were never part of a respiratory system but always associated with feeding mechanisms, and that the lamprey velar skeleton is a closer mandibular homologue, with both medial and lateral elements. Furthermore, the lamprey velar apparatus and gnathostome mandibular arch tissues are each supplied by the trigeminal nerve. The question of when jaws evolved may therefore shift to one which asks 'when did teeth or tooth-bearing dermal bones become associated with the mandibular cartilages and transform it into a biting apparatus?'.

There is an intriguing link between these evolutionary ideas and the molecular coding differences that exist between those for the mandibular arch compared with the hyoid and branchial arches, as mentioned in the introduction. Direct evidence is now available showing that midbrain crest cells contribute to the mandibular arch dentition (Imai *et al.*, 1996). Moreover, none of the neural crest-derived mesenchyme that populates this first arch can be prespecified by *Hox* gene expression (Sharpe, 1995; Chapter 1). Crest cells derive from the anterior hindbrain (rhombomeres 1 and 2) and the midbrain (Lumsden *et al.*, 1991) and although *Hoxa-2* is expressed up to the r1/r2 boundary there is a lack of *Hox* gene expression in the r2 crest because it is independently downregulated as the cells emerge. As Koentges and Lumsden (1996) point out, the *Hoxa-2* knockout phenotype, therefore, only displays a partial complement of first arch skeletal elements despite a complete absence of second arch structures, those of solely midbrain crest are present and also those of r1/r2 origin, the mouse homologues of the chick jaw articulation (articular, pterygoquadrate) and the dermal squamosal bone. We would predict that, whereas the pattern and shape of these elements is specified by downregulation of *Hoxa-2* gene expression, that of the more distal skeleton (midbrain crest-derived) is non-Hox gene coded but dependent on others that may be more derived homeobox-containing genes, such as *Dlx-2* (Qui *et al.* 1995) and *Otx-2*. Indeed, in Otx-2 heterozygous mutants the cranial phenotype is deficient in solely midbrain crest-derived elements (Matsuo *et al.*, 1995), i.e. those mapped by Koenteges and Lumsden in the chick (1996) as distal Meckel's cartilage, dentary, maxilla and palatine. It appears that the mandibular arch skeleton is specified by a precise, topographic, combinatorial overlap of crest derived from midbrain and rhombomeres 1 and 2, a complex patterning mechanism, and indeed those bones bearing teeth in 'higher' vertebrates (i.e. amniote tetrapods) are those derived from midbrain crest. Perhaps for this reason mice deficient in *Hoxa-2* gene function did not have any additional teeth on the transformed hyoid arch. At least this demonstrates an uncoupling of the developmental mechanism for specifying the jaw articulation from that part of the mandibular arch providing the teeth and bony support; as we have suggested, this is worth exploring with new experimental data.

Consequently, if we adopt the suggestion that the skeletal elements of jaws were always used for feeding and not initially just for respiration, and that the homologue is present in the velar skeleton of the lamprey, then a study of the expression pattern of *Otx-2*, *Hoxa-2* and *Hoxb-2* in this agnathan fish could give considerable insight into the developmental evolution of the jaws. If the development of teeth and jaws is modular, and either one can be developed or suppressed independently, then the keratinous toothlets of living agnathans, which we suggest are non-homologous with apatitic teeth, and the mineralized conodont feeding structures, represent alternative feeding strategies, each with separate developmental controls.

10.8.   Phylogenetic conclusions

Many of the diverse aspects of tooth morphology and development discussed in this chapter were examined in Reif's (1982) influential work on the odontode regulation theory. In this, the odontode represents the archetypical dermal skeletal unit. As such, it represents an assembly of morphological and developmental characters, which are likely to become less coherent as more primitive dermal skeletal and tooth-like structures are discovered. Recognition of conodonts as vertebrates may represent the beginning of this process. If they are the sister-group of subsequent armoured vertebrates, then it appears that (a) odontogenic and chondrogenic components of the neural crest preceded the development of osteogenic potential, and (b) that cranial crest mineralized tissues preceded those derived from the trunk crest. The odontode (*sensu* Reif, 1982 and Smith and Hall, 1993) therefore seems to have evolved within pre- Ordovician stem-gnathostomes, and between nodes 2 and 3 according to the phylogenetic hypothesis in Figure 10.3. Perhaps further investigation of keratinous toothlets in extant agnathans (and certain gnathostomes) may inform questions about necessary preconditions for odontode evolution.

As suggested in the introduction, the evolutionary precedence of scales before teeth seems less likely, and, as Schaeffer (1977) commented on the relation between scales and fin rays, it may be better to regard both as alternative manifestations of a common underlying morphogenetic system. Similarly, jaw evolution can no longer be tied in a simple way to the origin of teeth, even if teeth are defined strictly as products of a dental lamina. As might be expected, there are real difficulties in identifying dental lamina generated teeth early in the fossil record. Tooth whorl replacement families associated with jaws in early chondrichthyans resemble smaller versions with fused bases found in the branchial region, and these in turn resemble the short, whorl-like denticle series found in the branchial pouches of loganellid thelodonts. Similar short tooth whorls in basal osteichthyans (acanthodians) indicate strongly that such specialized denticles associated with visceral arches originated in the gnathostome stem (*contra* Reif, 1982). But if these short whorls were shed as a whole, then they do not record the early evolution of serial tooth replacement or dental laminae. Instead, the whorl-like morphology may document a feature of denticle growth associated with branchial arches, which was subsequently modified into a tooth replacement system. It would be especially interesting if these oropharyngeal denticle growth patterns could be linked exclusively to the presence of internal gill supports.

Thelodonts are probably a paraphyletic group with respect to more advanced pre-gnathostomes and gnathostomes (Janvier, 1996). Thelodonts may also include sub-adult osteostracans (of which juveniles are unknown). This could provide an explanation of the apparent absence of teeth or specialized oropharyngeal denticles in the closest sister-group to jawed vertebrates (osteostracans). Clearly, there is a problem here, with (secondary?) absence of tooth-like denticles in those groups which may include incipient stages in early jaw evolution. Mallatt's (1996) evolutionary scenario could be tailored to fit this stretch of vertebrate phylogeny, but it seems equally likely that thelodonts may have already grasped and retained prey within their denticled, pre-jawed, oro-branchial apparatus, as indeed did conodonts much earlier in evolution and as pharyngonath recent teleosts currently do. This is consistent with the likely function of the earliest known chondrichthyan replacement teeth (Williams, 1990), and removes the need for a teleological explanation of teeth as a necessary improvement on edentate, clamping anterior gill bars (Mallatt, 1996).

The precise definition of teeth, relative to denticles, seems to depend upon an inferred ontogenetic and phylogenetic hierarchy. Ørvig's (1977) morphological and histological criteria for teeth are less restrictive than Reif's (1982): formation within a dental lamina in advance of requirement. The more general conditions may represent more primitive properties of oropharyngeal denticles (Ørvig's definition of teeth might originate at node 2, and Reif's at node 9 in Figure 10.3), but this kind of inference should be treated with caution. Again, the thelodont data is important because it suggests that oropharyngeal denticle patterning and regionalization preceded dental lamina and jaw evolution. It also appears that individual oropharyngeal denticle morphology is more precisely regulated than that of skin denticles (cf. Ørvig, 1978 on primitive actinopterygian denticles). Evidence of acrodin evolution within actinopterygian phylogeny, however, demonstrates that oropharyngeal denticle and tooth histology continues to be modified profoundly after the evolution of a tooth replacement programme. Moreover, this acrodin histological 'signature' also demonstrates how a tooth-like development pattern has 'leaked out' from the orobranchial cavity, and spread beyond the jaws in certain fossil and living taxa. Again, this counters the usually assumed evolutionary polarity of skin denticle growth modes

moving across the jaws to become teeth (c.f. Janvier 1981, as discussed in section 10.6). It seems that we must consider seriously that teeth evolved independently of skin denticles.

Finally, the huge phylogenetic gulf separating living agnathans and gnathostomes appears to obstruct any direct investigation of dental lamina evolution. Once again, the diversity of actinopterygian dentitions may provide valuable insights. Primitive, non-teleostean taxa have densely denticulate oropharyngeal denticle plates on the branchial arches (and elsewhere, such as the clavicle and cleithrum: Patterson, 1977), but no teeth resembling those of the functional mandibular margin. In teleosts, however, the gill bars bear a range of conditions, from edentulous surfaces through to the apparent presence of teeth developing from dental laminae (section 10.5). This spread of tooth-like conditions may represent repeated derivations of tooth replacement systems; even if it represents only the spreading domains of pre-existing systems, it probably includes ranges of comparative, intermediate denticle-generating conditions which are unobtainable from any other vertebrate taxon.

Summary

Established theories of tooth and jaw evolution are compared with new developmental and morphological data within a phylogenetic context. Teeth result from a series of developmental specializations affecting the morphology, histology, patterning and replacement of denticles in the oropharynx. Previously, tooth and dermal skeletal diversity has been explained in terms of regulatory changes affecting the odontode, a fundamental dermal skeletal unit. It is increasingly clear that these changes represent the altered developmental regulation of contributory tissues, especially pre-and post-migratory neural crest cells and resultant epithelial-mesenchymal interactions. Fossil evidence indicates that many of these changes occurred at loci early in vertebrate phylogeny, preceding the origin of jaws and challenging accepted assumptions about dermal denticles and armour preceding initial stages of dental evolution.

Keratinous toothlets in living agnathans probably represent an independent evolution of oral feeding structures. Phosphatic conodont 'teeth' may represent early and independent manifestations of primitive stages in odontogenic developmental evolution. Whorl-like patterns of teeth in living chondrichthyans may represent a primitive morphological feature of all gnathostome dentitions, but these tooth families do not necessarily indicate a shared primitive (chondrichthyan) mode of tooth replacement. Models of tooth generation are discussed in the light of new developmental research. These concern tooth patterning, morphogenesis, and relation to the visceral skeleton. Combinatorial series of *Hox* and related gene expression boundaries probably have a fundamental role in patterning of the pharyngeal arches. These findings, combined with research concerning the segregation of neural crest cell fates, are combined to produce a revised, modular, tooth and skeletal developmental model.

Acknowledgements

We express our thanks to Georgie Koentges for stimulating discussions of work in press, and to Philippe Janvier, Anthony Graham, and Peter Thorogood for reading and commenting on a draft of this chapter. M.I.C. was funded by BBSRC Fellowship B/94/AF/1945; M.M.S. research was funded by NERC grants (GR3/8543, 10272) in collaboration with Paul Smith and Ivan Sansom.

References

Akimenko, M.-A., Johnson, S. L., Westerfield, M. and Ekker, M. (1995). Differential induction of four *msx* genes during fin development and regeneration in zebrafish. *Development*, **121**, 347–357.

Aldridge, R. J., Briggs, D. E. G., Smith, M. P., Clarkson, E. N. K. and Clark, N. D. L. (1993). The anatomy of conodonts. *Philosophical Transactions of the Royal Society of London Series B*, **338**, 405–421.

Bee, J. and Thorogood, P. V. (1980). The role of tissue interactions in the skeletogenic differentiation of avian neural crest cells. *Developmental Biology*, **78**, 47–62.

Bolker, J. A. and Raff, R. A. (1996). Developmental genetics and traditional homology. *BioEssays*, **18**, 489–494.

van der Brugghen, W. and Janvier, P. (1993). Denticles in thelodonts. *Nature*, **364**, 107.

Cappetta, H. (1987). *Chondrichthyes II. Handbook of Paleoichthyology*, volume 3B, ed. H.-P. Schultze. Stuttgart: Gustav Fischer.

Carroll, R. L. (1988). *Vertebrate Palaeontology and Evolution*. New York: W. H. Freeman and Company.

Clack, J. A. (1994). *Acanthostega gunnari*, a Devonian tetrapod from Greenland; the snout, palate and ventral parts of the braincase, with a discussion of their significance. *Meddelelser om Groenland, Geoscience*, **31**, 3–24.

Cloutier, R. and Ahlberg, P. E. (1995). Sarcopterygian interrelationships: how far are we from a phylogenetic censensus? *Geobios*, **19**, 241–248.

Coates, M. I. (1996). The Devonian tetrapod *Acanthostega gunnari* Jarvik: postcranial anatomy, basal tetrapod interrelationships, and patterns of skeletal evolution. *Royal Society of Edinburgh, Transactions (Earth Sciences)*, **87**, 363–421.

Couly, G. F., Coltey, P. M. and Le Douarin, N. M. (1993). The triple origin of the skull in higher vertebrates: a study in quail-chick chimeras. *Development*, **117**, 409–429.

Denison, R. H. (1978). *Placodermi. Handbook of Paleoichthyology*, ed. H.-P. Schultze. Stuttgart: Gustav Fischer.

Denison, R. H. (1979). *Acanthodii. Handbook of Paleoichthyology*, ed. H.-P. Schultze. Stuttgart: Gustav Fischer.

Dunkle, D. A. and Schaeffer, B. (1973). *Tegeolepis clarki* (Newberry), a Palaeonisciform from the Upper Devonian Ohio Shale. *Palaeontographica*, **143**(A), 151–158.

Dzik, J. (1991). Evolution of oral apparatuses in the conodont chordates. *Palaeontologica Polonica*, **36**, 265–323.

Forey, P. and Janvier, P. (1993). Agnathans and the origin of jawed vertebrates. *Nature*, **361**, 129–134.

Francillon-Vieillot, H., Trebeol, L., Meunier, F. J. and Slembrouck, J. (1994). Histological study of odontogenesis in the pharyngeal jaws of *Trachinotus teraia* Cuvier et Valenciennes 1932 (Osteichthyes, Teleostei, Carangidae). *Journal of Morphology*, **220**, 11–24.

Gans, C. and Northcutt, R. G. (1983). Neural crest and the origin of vertebrates: a new head. *Science*, **220**, 268–274.

Gardiner, B. G. (1984). The relationships of the palaeoniscid fishes, a review based on new specimens of *Mimia* and *Moythomasia* from the Upper Devonian of Western Australia. *Bulletin of the British Museum of Natural History (Geology)*, **37**, 173–428.

Gardiner, B. G. (1990). Placoderm fishes: diversity through time. In *Major Evolutionary Radiations*, eds. P. D. Taylor and G. P. Larwood, pp. 305–319. Oxford: Clarendon Press.

Gardiner, B. G. and Schaeffer, B. (1989). Interrelationships of lower actinopterygian fishes. *Zoological Journal of the Linnean Society*, **97**, 135–187.

Goodrich, E. S. (1907). On the scales of fishes, living and extinct, and their importance in classification. *Proceedings of the Zoological Society of London*, **1907**, 751–774.

Goujet, D. (1982). Les affinités des Placodermes, une revue des hypothèses actuelles. *Geobios*, (mémoire spécial), **6**, 27–38.

Graveson, A. C., Smith M. M. and Hall, B. K. (1997). Neural crest potential for development in a urodele amphibian; developmental and evolutionary significance. *Developmental Biology*, **188**, 34–42.

Hall, B. K. (1975). Evolutionary consequences of skeletal development. *American Zoologist*, **15**, 329–350.

Hall, B. K. (1995). Homology and embryonic development. *Evolutionary Biology*, **28**, 1–37.

Hertwig, O. (1879). Uber das Hautskelett der Fische. *Morphologische Jahrbuch*, **5**, 1–21.

Hickman, A. and Smith, M. M. (1994). Tooth development in the zebrafish *Brachidanio rerio*. *Journal of Morphology*, **220**, 396.

Holland, P. W. H. (1991). Cloning and evolutionary analysis of *msh*-like homeobox genes from mouse, zebrafish and ascidian. *Gene*, **98**, 253–257.

Holland, P. W. H. and Garcia-Fernandez J. (1996). *Hox* genes and chordate evolution. *Developmental Biology*, **173**, 382–395.

Hook, R. W. (1983). *Colosteus scutellus* (Newberry), a primitive temnospondyl amphibian from the Middle Pennsylvanian of Linton, Ohio. *American Museum Novitates*, **2770**, 1–41.

Hunt, P., Whiting, J., Nonchev, S., Sham, M. -H., Marshall, H., Graham, A., Cook, M., Allemann, R., Rigby, P. W. J., Gulisamo, M., Faiella, A., Boncinelli, E. N. and Krumlauf, R.

(1991). The branchial Hox Code and its implication for gene regulation, patterning of the nervous system and haed evolution. *Development*, Suppl., **2**, 63–77.

Huysseune, A. (1995). Phenotypic plasticity in the lower pharyngeal jaw dentition of *Astatoreochromis alluaudi* (Teleostei, Cichlidae). *Archives of Oral Biology*, **40**, 1005–1014.

Imai, H., Osumi-Yamashita, N., Ninomya, Y. and Eto, K. (1996). Contribution of early-emigrating midbrain crest cells to the dental mesenchyme of mandibular molar teeth in rat embryos. *Developmental Biology*, **176**, 151–165.

Imai, H., Osumi, N. and Eto, K. (1998). Contribution of foregut endoderm to tooth initiation of mandibular incisor in rat embryos. *European Journal of Oral Sciences*, **106**, 106:19–23.

Janvier, P. (1981). The phylogeny of the craniata, with particular reference to the significance of fossil 'agnathans'. *Journal of Vertebrate Paleontology*, **1**, 121–159.

Janvier, P. (1986). Les nouvelles conceptions de la phylogenie et de la classification des 'Agnathes' et des sarcopterygiens. *Océanis*, **12**, 123–138.

Janvier, P. (1993). Patterns of diversity in the skull of jawless fishes. In *The Skull*, volume 2, eds. J. Hanken and B. K. Hall, pp. 131–188. Chicago: University of Chicago Press.

Janvier, P. (1996). The dawn of the vertebrates: characters versus common ascent in the rise if current vertebrate phylogenies. *Palaeontology*, **39**, 259–288.

Jarvik, E. (1980). *Basic Structure and Evolution of Vertebrates*. London: Academic Press.

Jessen, H. (1980). Lower Devonian Porolepiformes from the Canadian Arctic with special reference to *Powichthys thorsteinssoni* Jessen. *Palaeontographica, Abteilung A.*, **167**, 180–214.

Jollie, M. (1984). Development of the head and pectoral skeleton of *Polypterus* with a note on scales (Pisces: Actinopterygii). *Journal of. Zoology, London*, **204**, 469–507.

Kerr, T. (1960). Development and stucture of some actinopterygian and urodele teeth. *Proceedings of the Zoological Society of London*, **133**, 401–422.

Koentjes, G. and Lumsden, A. S. (1996). Rhombencephalic neural crest segmentation is preserved throughout craniofacial ontogeny. *Development*, **122**, 3229–3442.

Krejsa, R. J., Bringas, P. Jr. and Slavkin, H. C. (1990a). The cyclostome model : an interpretation of conodont element structure and function based on cyclostome tooth morphology, function, and life history. *Courier Forschungsinstitut Senckenberg*, **118**, 473–492.

Krejsa, R. J., Bringas, P. Jr. and H. C. Slavkin (1990b). A neontological intepretation of conodont elements based on agnathan cyclostome tooth structure, function, and development. *Lethaia*, **23**, 359–378.

Langille, R. M. and Hall, B. K. (1988). Role of the neural crest in development of the trabeculae and branchial arches in embryonic sea lamprey, *Petromyzon marinus* (L.). *Development*, **102**, 301–310.

Lauder, G. V. (1985). Aquatic feeding in lower vertebrates. In *Functional Vertebrate Morphology*, eds. M Hildebrand, D. M. Bramble, K. F. Liem and D. B. Wake, pp. 210–229. Cambridge, MA: Harvard Univiversity Press.

Lauder, G. V. and Liem, K. F. (1983). The evolution and interrelationships of the actinopterygian fishes. *Bulletin of the Museum of Comparative Zoology*, **150**, 95–197.

Long, J. A. (1986). New ischnacanthid acanthodians from the Early Devonian of Australia, with comments on acanthodian relationships. *Zoological Journal of the Linnean Society*, **87**, 321–339.

Long, J. A. (1988). New palaeoniscid fishes from the Late Devonian and Early Carboniferous of Victoria. *Memoir of the Association of Australian Palaeontologists*, **7**, 1–63.

Lumsden, A. G. S. (1979). Pattern formation in the molar dentition of the mouse. *Journal de Biologie Buccale*, **7**, 77–103.

Lumsden, A. G. S. (1987). The neural crest contribution to tooth development in the mammalian embryo. In *Developmental and Evolutionary Aspects of the Neural Crest*, ed. P. F. A. Maderson, pp. 261–300. New York: Wiley.

Lumsden, A., Sprawson, N. and Graham, A. (1991). Segmental origin and migration of neural crest cells in in the hindbrain region of the chick embryo. *Development*, **113**, 1281–1291.

Mackenzie, A., Ferguson, M. W. J. and Sharpe, P. T. (1992). Expression patterns of the homeobox gene *Hox-8*, in the mouse embro suggest a role specifying tooth initiation and shape. *Development*, **115**, 403–420.

Maisey, J. G. (1986). Heads and tails: a chordate phylogeny. *Cladistics*, **2**, 201–256.

Maisey, J. G. (1989). Visceral skeleton and musculature of a Late Devonian shark. *Journal of Vertebrate Paleontology*, **9**, 174–190.

Mallatt, J. (1996). Ventilation and the origin of jawed vertebrates: a new mouth. *Zoological Journal of the Linnean Society*, **117**, 329–404.

Matsuo, I., Kurantani, S., Kimura, C., Takeda, N. and Aizawa, S. (1995). Mouse *Otx2* functions in the formation and patterning of avian cranial skeletal, connective and muscle tissues. *Genes and Development*, **9**, 2646–2658.

Mitsiadis, T. A., Lardelli, M., Lendahl, U. and Thesleff, I. (1995). Expression of *Notch 1, 2* and *3* is regulated by epithelial-mesenchymal interactions and retinoic acid in the developing mouse tooth and associated with determination of ameloblast cell fate. *Journal of Cell Biology*, **130**, 407–418.

Moss, M. L. (1968a). Bone, dentine and enamel and the evolution of vertebrates. In *Biology of the Mouth*, ed. P. Person, pp. 37–65. Washington, D. C.: American Association for the Advancement of Science.

Moss, M. L. (1968b). Comparative anatomy of vertebrate dermal bone and teeth. I. The epidermal co-participation hypothesis. *Acta Anatomica*, **71**, 178–208.

Moy-Thomas, J. A. and Miles, R. S. (1971). *Palaeozoic Fishes*, second edition. London: Chapman and Hall.

Nelson, G. J. (1970). Pharyngeal denticles (placoid scales of sharks), with notes on the dermal skeleton of vertebrates. *American Museum Novitates*, **2415**, 1–26.

Newth, D. R. (1956). On the neural crest of the lamprey embryo. *Journal of Embryology and Experimental Morphology*, **4**, 358–375.

Norell, M. A. (1992). Taxic origin and temporal diversity: the effect of phylogeny. In *Extinction and Phylogeny*, eds. M. Novacek and Q. Wheeler, pp. 89–118. New York: Columbia University Press.

Ørvig, T. (1967). Phylogeny of tooth tissues: evolution of some calcified tissues in early vertebrates. In *Structural and Chemical Organization of Teeth*, volume 1, ed. A. E. W. Miles, pp. 45–110. London: Academic Press.

Ørvig, T. (1977). A survey of odontodes (dermal teeth) from developmental, structural, functional, and phyletic points of view. In *Problems in Vertebrate Evolution*, eds. S. M. Andrews,

R. S. Miles and A. D. Walker, Linnaean Society Symposium 4, pp. 53–75. London: Academic Press.

Ørvig, T. (1978). Microstructure and growth of the dermal skeleton in fossil actinopterygian fishes: *Birgeria and Scanilepis*, *Zoologica Scripta*, **7**, 33–56.

Osborn, J. W. (1978). Morphogenic gradients: fields versus clones. In *Development, Function and Evolution of Teeth*, eds. P. M. Butler and K. A. Joysey, pp. 171–202. London: Academic Press.

Pander, C. H. (1856). Monographie der fossilen Fische des silurischen Systems der russisch-baltischen Gouverments, pp. 1–91. *Akademie der Wissenschaften, St. Petersburg*.

Parker, H. W. and Boeseman, M. (1954). The basking shark *Cetorhinus* in winter. *Proceedings of the Zoological Society of London*, **124**, 185–194.

Patterson, C. (1977). Cartilage bones, dermal bones and membrane bones, or the exoskeleton versus the endoskeleton. In *Problems in Vertebrate Evolution*, eds. S. M. Andrews, R. S. Miles and A. D. Walker, Linnean Society Symposium 4, pp. 77–121. London: Academic Press.

Patterson, C. (1992). Interpretation of the toothplates of chimaeroid fishes. *Zoological Journal of the Linnean Society*, **106**, 33–61.

Pearson, D. M. and Westoll, T. S. (1979). The Devonian actinopterygian *Cheirolepis* Agassiz. *Transactions of the Royal Society of Edinburgh*, **70**, 337–399.

Purnell, M. A. (1993). Feeding mechanisms in conodonts and the function of the earliest vertebrate hard tissues. *Geology*, **21**, 375–377.

Purnell, M. A. and von Bitter, P. H. (1992). Blade-shaped conodont elements functioned as cutting teeth. *Nature*, **359**, 629–631.

Qui, M., Bulfone, A., Martinez, S., Meneses, J. J., Shimamura, K., Pederson, R. and Rubenstein, J. L. R. (1995). Null mutation of *Dlx-2* results in abnormal morphogenesis of proximal first and second branchial arch derivatives and abnormal differentiation in the forebrain. *Genes and Development*, **9**, 2523–2538.

Reif, W-E. (1982). Evolution of dermal skeleton and dentition in vertebrates: the odontode-regulation theory. *Evolutionary Biology*, **15**, 287–368.

Repetski, J. E. (1978). A fish from the Upper Cambrian of North America. *Science*, **200**, 529–531.

Richter, M. and Smith, M. M. (1995). A microstructural study of the ganoine tissue in selected lower vertebrates. *Zoological Journal of the Linnean Society*, **114**, 173–212.

Rijli, F. M., Mark, M., Lakkaraju, S., Dierich, A., Dolle, P. and Chambon, P. (1993). A homeotic transformation is generated in the rostral branchial region of the head by disruption of *Hoxa-2*, which acts as a selector gene. *Cell*, **75**, 1333–1349.

Romer, A. S. (1956). *Osteology of the Reptiles*. Chicago: University of Chicago Press.

Sansom, I. J. (1992). The Palaeobiology of Panderodontidae and Selected other Conodonts. PhD thesis, University of Durham (unpublished).

Sansom, I. J., Smith, M. P., Armstrong, H. A. and Smith, M. M. (1992). Presence of the earliest vertebrate hard tissues in conodonts. *Science*, **256**, 1308–1311.

Sansom, I. J., Smith, P. M. and Smith, M. M. (1994). Dentine in conodonts. *Nature*, **368**, 591.

Satokata, I. and Mass, R. (1994). *Msx1* deficient mice exhibit cleft palate and abnormalities of craniofacial and tooth development. *Nature Genetics*, **6**, 348–355.

Schaeffer, B. (1975). Comments on the origin and basic radiation of the gnathostome fishes with particular reference to the feeding mechanism. *Colloquium Internationale, C.N.R.S.*, **218**, 101–109.

Schaeffer, B. (1977). The dermal skeleton in fishes. In *Problems in Vertebrate Evolution*, eds. S. M. Andrews, R. S. Miles and A. D. Walker, Linnean Society Symposium 4, pp. 25–52. London: Academic Press.

Schilling, T. and Kimmel, C. (1994). Segment and cell type lineage restrictions during pharyngeal arch development in the zebrafish embryo. *Development*, **120**, 483–494.

Schilling, T. F., Walker, C. and Kimmel, C. B. (1996). The *chinless* mutation and neural crest cell interactions in zebrafish jaw development. *Development*, **122**, 1417–1426.

Schultze, H-P. (1973). Crossopterygier mit heterozerker Schwanzflosse aus dem Oberdevon Kanadas, nebst einer Beschreibung von Onychodontida -Resten aus dem Mitteldevon Spaniens und aus dem Karbon der USA. *Palaeontographica Abteilung A*, **143**, 188–208.

Sellman, S. (1946). Some experiments on the determination of the larval tooth in *Ambystoma mexicanum*. *Odontologist Tidskrift* **54**, 1–28.

Sharpe, P. T. (1995). Homeobox genes and orofacial development. *Connective Tissue Research*, **32**, 17–25.

Sire, J-Y. and Huysseune, A. (1996). Structure and development of the odontodes in an armoured catfish, *Corydoras aeneus* (Siluriformes, Callichthyidae). *Acta Zoologica (Stockholm)*, **77**, 51–72.

Slavkin, H. C. and Diekwisch, T. (1996). Evolution in tooth biology: of morphology and molecules. *The Anatomical Record*, **245**, 131–150.

Slavkin, H. C., Krejsa, R. J., Fincham, A., Bringas, P. Jr., Santos, V., Sassano, Y., Snead, M. L. and Zeichner-David, M. (1991). Evolution of enamel proteins: a paradigm for mechanisms of biomineralization. In *Mechanisms and Phylogeny of Mineralization in Biological Systems*, eds. S. Suga and H. Nakahara, pp. 383–389. Berlin: Springer.

Smith, M. M. (1984). Petrodentine in extant and fossil dipnoan dentitions: microstructure, histogenesis and growth. *Proceedings of the Linnean Society N.S.W*, **107**, 367–407.

Smith, M. M. (1988). The dentition of Paleozoic lungfishes: a consideration of the significance of teeth, denticles and tooth plates for dipnoan phylogeny. In *Teeth Revisited: Proceedings of the VIIth International Symposium on Dental Morphology*, Paris 1986, eds. D. E. Russell, J.-P. Santoro and D. Sigogneau-Russell, D. *Mémoires du Muséum National d'Histoire Naturelle, Paris (Série C)*, **53**, 179–194.

Smith, M. M. (1992). Microstructure and evolution of enamel amongst osteichthyan fishes and early tetrapods. In *Structure, Function and Evolution of Teeth*, eds. P. Smith and E. Tchernov, pp. 1–19. London: Freund Publishing.

Smith, M. M. (1995). Heterochrony in the evolution of enamel in vertebrates. In *Evolutionary Change and Heterochrony*, ed. K. McNamara, pp 125–150. Chichester: Wiley.

Smith, M. M. and Coates, M. C. (1998). Evolutionary origins of the vertebrate dentition: phylogenetic patterns and developmental evolution. *European Journal of Oral Science*, **105**, 1–19.

Smith, M. M. and Hall, B. K. (1990). Developmental and evolutionary origins of vertebrate skeletogenic and odontogenic tissues. *Biological Reviews*, **65**, 277–374.

Smith, M. M. and Hall, B.K. (1993). A developmental model for evolution of the vertebrate exoskeleton and teeth: the role of cranial and trunk neural crest. *Evolutionary Biology*, **27**, 387–447.

Smith, M. M., Hickman, A., Amanze, D., Lumsden, A. and Thorogood, P. (1994). Trunk neural crest origin of caudal fin mesenchyme in the zebrafish *Brachydanio rerio*. *Proceedings of the Royal Society of London, Series B*, **256**, 137–145.

Smith, M. M., Sansom, I. J. and Smith, M. P. (1996). Teeth before armour: the earliest vertebrate mineralised tissues. *Modern Geology*, **20**, 1–17.

Smith, M. P., Sansom, I. J. and Repetski, J. (1996). Histology of the earliest fish. *Nature*, **380**, 702–704.

Takahashi, Y., Bontoux, M. and Le Douarin, N. M. (1991). Epithe-liomesenchymal interactions are critical for *Quox 7* expression and membrane bone differentiation in the neural crest-derived mandibular mesenchyme *EMBO Journal*, **10**, 2387–2393.

Thorogood, P. (1994). Craniofacial development and the neural crest. In *Early Fetal Growth and Development*, ed. R. H. T. Ward, S. K. Smith and D. Donnai, pp. 111–120. Chichester: Wiley.

Vastardis, H., Karimbux, N., Guthua, S.W., Seidman, J.G. and Seidman, C.E. (1996). A human *MSX1* homeodomain missense mutation causes selective tooth agenesis. *Nature Genetics*, **13**, 417–421.

Veitch, E., Begbie, J., Schilling, T. F., Smith, M. M. and Graham, A. (1999). Pharyngeal arch patterning in the absence of neural crest. *Current Biology*, **9**, 1481–1484.

Watson, D. M. S. (1937). The acanthodian fishes. *Philosophical Transactions of the Royal Society of London, Series B.*, **228**, 49–146.

Williams, M. E. (1990). Feeding behaviour in Cleveland Shale fishes. In *Evolutionary Paleobiology of Behaviour and Coevolution*, ed. A. J. Boucot, pp. 273–290. Amsterdam: Elsevier.

Williamson, W. C. (1849). On the microscopic structure of the scales and dermal teeth some ganoid and placoid fish. *Philosophical Transactions of the Royal Society of London*, **139**, 435–475.

Young, G. C., Karajute-Talimaa, V. N. and Smith, M. M. (1996). A new Late Cambrian vertebrate-like fossil from Australia. *Nature*, **383**, 310–312.

Zangerl, R. and Case, G. R. (1976). *Cobelodus aculeatus* (Cope), an anacanthus shark from Pennsylvanian black shales of North America. *Palaeontographica*, **A154**, 107–157.

Zangerl, R. (1981). *Chondrichthyes I. Handbook of Paleoichthyology*, ed. H.-P. Schultze. Stuttgart: Gustav Fischer.

11 Development and evolution of dentition patterns and their genetic basis

Z. Zhao, K. M. Weiss and D. W. Stock

11.1. Introduction

A fundamental question in developmental biology is how patterns are established, and in evolutionary biology how patterns are changed. The pattern of vertebrate dentitions (tooth morphology, number, and location) is of obvious importance to the survival of individual organisms and a great diversity of patterns exists among vertebrate taxa (Peyer, 1968). Furthermore, the abundance of teeth and jaws in the fossil record provides a tremendous amount of information about the evolutionary changes that have occurred in these patterns. However, despite the accumulation of histological and molecular information on the development of individual teeth (Thesleff *et al.*, 1996; Ruch, 1987), little is known about the control of the pattern of the dentition as a whole during embryonic development. Such information is crucial for understanding the mechanisms of evolutionary change in the vertebrate dentition. In this chapter, we review aspects of the evolution of dentition patterns, models that have been proposed for the developmental mechanisms responsible for these patterns and candidate genes for controlling them. Although the dentitions of mammals constitute only a small portion of the diversity observed in vertebrates, we will concentrate on this group because of the much greater availability of developmental and genetic data. For the purposes of this chapter, 'dentition pattern' refers to the number, location, and arrangement of differently shaped teeth, while 'dental pattern' is used to describe the crown structure of individual teeth, e.g. location, shape and size of cusps.

11.1.1. Axial definitions for the adult dentition and embryonic jaws

Various axial definitions for the adult dentition and embryonic mandible have been used in the literature (Table 11.1). However, these definitions can be inconsistent or confusing. In this chapter, the adult dentition is described as shown in Figure 11.1. The definitions for the embryonic jaws in this chapter are shown in Figure 11.2.

11.2. Homodonty and heterodonty

The dentition of mammals has a segmented organization, consisting of repeated units (teeth) arrayed along curvilinear axes (each jaw quadrant). Similar axially segmented organization characterizes other structures such as the vertebrae, ribs, limbs, and hindbrain (Bateson, 1894; Weiss, 1990, 1993). Structures consisting of a repetition of similar basic units were of great interest to morphologists even before the acceptance of the theory of evolution. For example, Richard Owen proposed an 'archetype' of the vertebrate skeleton in which serially homologous vertebrae along the length of the body axis were very similar to each other in structure (Panchen, 1994). Although such an archetype did not represent a common ancestor to Owen, Charles Darwin (1859) did interpret it as such. To Darwin, a common pattern of the evolution of segmented systems consisted of an origin as a series of morphologically similar subunits (often of large and variable number) followed by differentiation in morphology among elements and a greater standardization of their number and location. For example the archetypal vertebral column with similar vertebrae along its length would have evolved into the mammalian

Table 11.1. *Axial definitions for the adult dentition and embryonic jaws*

Adult dentition		Embryonic mandible	
This chapter	The literature	This chapter	The literature
Mesial	Anterior	Mesial	Anterior Distal
Distal	Posterior	Distal	Posterior Proximal Lateral*
Lingual	Internal	Lingual	Oral
Labial	Buccal External Vestibular	Labial	Buccal Aboral
		Oral	Dorsal[a] Upper[a]
		Aboral	Ventral[a] Lower[a]

Note: [a] Applied to the lower jaw

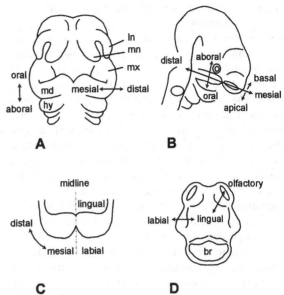

Figure 11.2A–D. **Diagrammatic illustration of axial definitions in embryonic jaws. A. Frontal view of the head. B. Side view of the head. Note that 'basal' and 'apical' are for the medial nasal process. C. Oral view of the mandibular arch. D. Oral view the upper part of the had (mandible removed). br, brain; hy, hyoid arch (the second branchial arch); ln, lateral nasal process; md, mandibular arch; mn, medial nasal process; mx, maxillary process.**

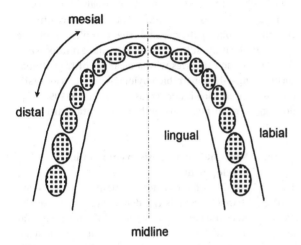

Figure 11.1. **Diagrammatic illustration of axial definitions in adult dentition.**

condition, with separate cervical, thoracic, lumbar, sacral and coccygeal regions.

The dentition appears to have undergone a similar pattern of evolution to the vertebral column (Weiss, 1990, 1993; Butler, 1995; Stock *et al.*, 1997). The early vertebrate dentition consists of numerous, similar, simple conical teeth. This condition is still found in many fish and non-mammalian tetrapods and is referred to as homodonty. The teeth of homodont species are constantly replaced during life, a condition known as polyphyodonty. Mammals in contrast have a regionally differentiated dental arch, with incisor, canine, premolar, and molar regions. (The homodont dentition of aquatic mammals, such as dolphins, is considered to be a

secondary condition derived from an ancestral heterodont form; Peyer, 1968.). This differentiation is typically repeated in each half of both the upper and lower jaws, along a linear axis from mesial to distal. In addition to the heterodonty of mammals, their teeth are typically replaced only once, a condition known as diphyodonty. Heterodonty was already present in some synapsid 'mammal-like' reptiles, while diphyodonty probably appeared early in mammal evolution (Peyer, 1968; Osborn, 1973; Kemp, 1982; Brown, 1983; Crompton and Luo, 1993).

Because of the differentiation among elements within the mammalian dentition, it shares with other segmented organ systems the phenomenon of homeotic variation both within and between species (Bateson, 1894; Butler, 1939, 1995). Homeotic changes involve the transformation of one member of a series into another, usually nearby, member. These same authors have also noted that inter- and intra-specific variations of teeth and other systems often affect more than one member of the series. The similarity of the dentition and other segmented organ systems in organization and variation raises the question of whether they share common genetic control of pattern formation, either in terms of the basic mechanisms or of the actual genes employed.

11.2.1. Acquisition of multi-cusped cheek teeth in mammals

Perhaps the most distinctive features of mammalian heterodonty are the crowns of the cheek teeth (molars and premolars), which are folded into diverse and complex shapes (Butler, 1939; Patterson, 1956; Jernvall, 1995). The basic morphological entity of the crown pattern is the cusp, which is recognizable as an essentially conical projection from the crown surface separated from adjacent projections by valleys (Butler, 1956). Cusps differ from other minor topographic features of the crown such as crenullations in that the former contain both dentine and enamel whereas the latter are exclusively the result of differences in enamel thickness. During development, cusps grow from the tips down and are independent centres of enamel and dentine formation within the same tooth (Butler, 1956; Snead *et al.*, 1988). Additional features of the crown surface include ridges (joined cusps) or crests (expanded cusps), and cingula, which are ridges passing around the base of the crown (often incompletely) and separated from the crown interior by a valley (Butler, 1956).

Multi-cusped mammalian cheek teeth have almost certainly evolved from single-cusped reptilian precursors. Extensive palaeontological studies since the latter part of the 19th nineteenth century have attempted to identify the homologue of the original single cusp in multi-cusped molars and to trace the evolution of specific cusps during the diversification of mammals (Osborn, 1888, 1907; Gregory, 1934; Butler, 1939, 1990, 1995; Patterson, 1956; Simpson, 1961; Vandebroek, 1961, 1967; Clemens, 1971; Hershkovitz, 1971). These studies have led to the widely accepted 'differentiation theory' in which additional cusps have arisen by budding or outgrowth from the crown of a tooth, rather than by clustering of multiple, originally distinct, conical teeth, as proposed by the rival 'concrescence theory' (see Osborn, 1907 and Peyer, 1968 for references). The former theory has been supported by data from embryological studies (Ooë, 1975, Schour, 1962).

There is general agreement that therian (placental and marsupial) mammals arose from an ancestor possessing a molar tooth pattern referred to as 'tribosphenic' (Osborn, 1888, 1907; Gregory, 1934; Patterson, 1956; Vandebroek, 1961, 1967; Butler, 1990, 1992; Jernvall, 1995). Tribosphenic upper molars possess a triangular arrangement of three main cusps, while the lower molars in a tribosphenic dentition are composed of an anterior 'trigonid' of three cusps and a posterior 'talonid' consisting of a basin surrounded by three additional cusps. Tribosphenic molars are still found in some modern groups and first appear in the fossil record in the early Cretaceous (Clemens, 1971; Butler, 1990, 1992). Such teeth represent the addition of a crushing function to a simpler form of tooth with only a shearing function. Shearing occurs with the occlusion of the trigonid in the space between upper molars, while the newly evolved protocone cusp of the upper molars occludes in the talonid basin to provide a crushing function (Butler, 1972).

Among the mammal-like reptiles, pelycosaurs of the Carboniferous and Permian and primitive therapsids of the Permian possessed single-cusped post-canine teeth (Romer, 1961; Kemp, 1982). Later therapsids of the Triassic possessed multi-cusped cheek teeth of various patterns, many of which paralleled molar evolution in mammals (Romer, 1961; Bown and Kraus, 1979). There have been numerous proposals for the derivation of tribosphenic molar teeth from Jurassic and earlier mammal molar patterns as well as from single cusped teeth (Osborn, 1888, 1907; Gregory, 1934; Butler, 1939, 1990, 1995; Patterson, 1956; Vandebroek, 1961, 1967; Clemens, 1971; Hershkovitz, 1971). The diversity of therapsid molar patterns and the complexity of the molars of the earliest fossil mammals, make it impossible to buttress such theories with a clear set of transitional forms along a demonstrable phylogenetic series of fossils (Simpson, 1961; Bown and Kraus, 1979; Rowe, 1993), but do suggest great flexibility within the dental patterning system.

11.2.2. Molar morphogenesis in embryonic development

The earliest sign of tooth development is a thickened epithelial band, known as the dental lamina, along the free margin of the jaw in the vicinity of the future dental arches, followed by the condensation of mesenchyme underneath it (Cohn, 1957; Schour, 1962; Ferguson, 1990; Schour, 1962; Cohn, 1957). The dental lamina invaginates into the mesenchyme at the sites of future teeth and gradually develops into bud-, cap-, and bell-shaped enamel organs (Figure 11. 3). During the bell stage the crown pattern of the molars is completed (See Introduction of this (volume; Cohn, 1957; Gaunt, 1964; Ruch, 1984, 1987). This morphodifferentiation is controlled by epithelial-mesenchymal interactions (Thesleff *et al.*, 1996; Butler, 1995; Ten Cate, 1995; Thesleff *et al.*, 1996). At the cap stage of molar morphogenesis, the inner enamel epithelium (IEE) of the enamel organ folds inward beside a condensed, rod-like, epithelial cell cluster know as the primary enamel knot (EK), which runs along the mesial–distal axis of the jaw. The folding of the IEE gives rise to two epithelial grooves, known as the enamel grooves (Schour, 1962), on the lingual and labial sides of the tooth germ. The enamel grooves, along with the associated mesenchyme, are the primary structures serving for

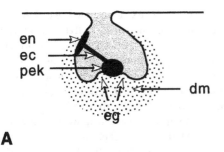

A

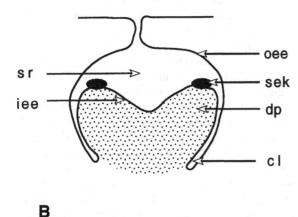

B

Figure 11.3. **Diagrammatic representation of the structure of molar germs at the cap (A) and bell (B) stages. cl, cervical loop; dm, dental mesenchyme; dp, dental papilla; eg, enamel groove; en, enamel navel; es, enamel cord; iee, inner enamel epithelium; oee, outer enamel epithelium; pek, primary enamel knot; sek, secondary enamel knot; sr, stellate reticulum.**

Table 11.2 *Regulatory factors expressed in the enamel knot and other embryonic organizers*

	EK	AER	ZPA	NOT	FP
Shh	+		+	+	+
Bmp2	+	+	+		
Bmp4	+	+	+		
Bmp7	+	+	+		
Fgf4	+	+			
Fgf9	+	+			
Msx2	+	+			

Note: AER, apical ectodermal ridge; Bmp, bone morphogenetic protein; EK, enamel knot; Fgf, fibroblast growth factor; FP, floor plate; NOT, notochord; Shh, sonic hedgehog; ZPA, zone of polarizing activity.

formation of the lingual and labial cusps (Figure 11.3 3A; Jernvall *et al.*, 1994; Butler, 1956; Ferguson, 1990; Jernvall *et al.*, 1994). Later during the bell stage, the primary EK disappears and secondary EKs appear on the future tip of each cusp. Unlike the longitudinal rod-like primary EK, the secondary EKs are cell-clusters that remain on the tip of the cusps until the crown pattern is nearly completed (Figure 11.3B; Jernvall *et al.*, 1994).

11.2.3. The EK as an organizing centre for the development of individual molars

The EK, which has been observed in many mammalian species, is a cluster of non-dividing cells derived from the dental epithelium. A potential role for this structure in crown pattern formation has long been long inferred (reviewed by Butler, 1956; Berkovitz, 1967; MacKenzie *et al.*, 1992), and efforts have been made to investigate it (Kirino *et al.*, 1973). The EK has been shown to express some signalling factors and the transcription factor *Msx2* (Table 11.2; MacKenzie *et al.*, 1992; Jernvall *et al.*, 1994,

1998; Vaahtokari *et al.*, 1996*a*; Åberg *et al.*, 1997; Thesleff and Jernvall, 1997; Jernvall *et al.*, 1998; Vaahtokari *et al.*, 1996*a*; Åberg *et al.*, 1997; Kettunen and Thesleff, 1998; MacKenzie *et al.*, 1992). In the expression of these genes, the EK resembles the apical ectodermal ridge (AER) and the zone of polarizing activity (ZPA) in the developing limb bud, the notochord, and the floor plate of the spinal cord (Table 11.2; Ingham, 1995; Hogan, 1996*a*,*b*; Martin, 1998; Colvin *et al.*, 1999), implicating the EK as an organizing centre in tooth growth and morphogenesis (Jernvall *et al.*, 1994; Vaahtokari *et al.*, 1996*a*; Thesleff and Jernvall, 1997). However, there is as yet no concrete experimental or functional evidence to demonstrate this, such as by means of specific disruption of EK formation, or ectopic grafting or heterogeneous recombination of EKs with other tissues. Another interesting feature of the primary EK is that it contains programmed cell death (apoptosis; Vaahtokari *et al.*, 1996*b*). The significance of this is not understood, although it has been suggested that apoptosis serves to remove the primary EK from the enamel organ at the late cap stage, causing cessation of growth-stimulating signals (Vaahtokari *et al.*, 1996*b*; Thesleff and Jernvall, 1997).

Of the molecules localized to the EK, FGF4, a mitogen, has been shown to stimulate division of both dental epithelial and mesenchymal cells *in vitro*. It has been suggested that FGF4 is produced by the EK paracrinally to promote the adjacent IEE and dental mesenchymal cells to divide, and eventually to cause folding of the IEE (Jernvall *et al.*, 1994; Thesleff and Jernvall, 1997; Vaahtokari *et al.*, 1996*a*; Thesleff and Jernvall, 1997). *Msx2*, the only transcription factor found in the EK so far (MacKenzie *et al.*, 1992), has been shown to be regulated by BMP4 in tooth development (Vainio *et al.*, 1993; Chen *et al.*, 1996). BMP4 and BMP2 were shown to induce a subset of known molecular markers for the EK (*Msx2*, *p21*) in cultured dental epithelia (Jernvall *et al.*, 1998). These authors suggested that mesenchymal BMP4 serves to induce the EK and that this may involve the induction of

Lef1, based on the requirement of epithelial expression of the latter gene for tooth development to progress beyond the bud stage and the ability of BMP4 to induce *Lef1* expression (Kratochwil *et al.*, 1996). Subsequently, *Bmp4* is expressed in the EK and may be responsible for apoptosis in this structure (Jernvall *et al.*, 1998).

Later formed 'secondary EKs' may play a similar role to the primary EK in the formation of subsequent cusps within the same tooth (Jernvall *et al.*, 1994), although it is not known how the number or arrangement of EKs, the critical features for control of crown pattern, are determined.

11.2.4.   Other enamel organ structures in molar development

Additional morphological structures in the cap stage enamel organ of molars are the enamel cord (also known as the enamel septum or enamel stratum) and enamel navel (Butler, 1956; Ferguson, 1990; Berkovitz, 1967). The enamel cord, a condensed cell cluster, connects the EK and the enamel navel (Figure 11. 3A). The enamel navel, a patch of cells in the outer enamel epithelium (OEE) at the labial side of the tooth germ, expresses the *Msx2*, *Dlx2*, and *Dlx3* genes (MacKenzie *et al.*, 1992; Zhao *et al.*, unpublished personal observations). Except for the EK, the roles of these transient structures in tooth development are not clear. The prevalent presumption is that the enamel navel and enamel cord in connection with the EK specify the position of the first buccal cusp, which serves as a reference point for the later-developing cusps (Butler, 1956; Ruch *et al.*, 1982; MacKenzie *et al.*, 1992).

11.3.   The mechanisms for epithelial folding

Whatever the ultimate mechanism for crown shape determination, we know that molars develop as independent morphological units, and that the mechanism is thus self-contained within the tooth bud once the latter has begun to develop (e.g., Butler, 1995). The initial folding of the IEE at the cap stage is coincident with the appearance of the primary EK. Proliferation of the IEE bilateral to the EK gives rise to the cap-shaped enamel organ with the enamel grooves and the cervical loops (Figure 11 .4A,B). At the same time, the dental mesenchyme is also in a high mitotic state, suggesting that the mesenchyme may also play a role in epithelial folding (Jernvall *et al.*, 1994; Vaahtokari *et al.*, 1996a). The proliferation and condensation of the dental mesenchyme have also been suggested to keep the EK relatively stationary within the tooth germ (Jernvall *et al.*, 1994). It has been proposed that growth factors, such as

FGF4, produced by the EK act to promote cell division in the IEE and the mesenchyme (Jernvall *et al.*, 1994; Thesleff and Jernvall, 1997). Later at the bell stage the primary EK disappears and the adjacent IEE becomes mitotically active. With the appearance of the secondary EKs, differential proliferation in the IEE further deepens the valley between the two enamel grooves and consequently elicits cusps along with the downgrowth of the cervical loops (Figure 11.4B; Jernvall *et al.*, 1994; Butler, 1956). An increased mitotic index of the IEE during the period of the epithelial folding was also demonstrated by Ruch and his co-workers (see Ruch, 1987) who suggested that the activity was related to cusp formation.

When the epithelial folding is in progress and the enamel organ increases its size, cells within the enamel organ differentiate to produce and secrete hydrophilic glycosaminoglycans into the extracellular spaces. These compounds cause the accumulation of fluid in the intercellular spaces and therefore an increase in pressure in the interior of the enamel organ (Ferguson, 1990; Butler, 1956; Ferguson, 1990). Since the cells are connected with each other through tight desmosomal contacts, they become star-shaped as a result of the accumulating fluid around them, and therefore are referred to as the stellate reticulum (SR). It is believed that the pressure from the SR is applied to both IEE and OEE to maintain the shape and volume of the enamel organ and to prevent the IEE folding from distortion (Figure 11. 4C; Butler, 1956).

The differential rates of growth and the uneven distribution of growth and signalling factors (e.g., in EKs compared to with surrounding cells) described above suggest that processes analogous to those termed 'self-organizing', as will be discussed later (Bateson, 1894; Webster and Goodwin, 1996; Weiss *et al.*, 1998a), may be at work in the development of crown morphology. Such systems form complex patterns, such as that of cusps and ridges on molar crowns, based on the kinetics of the interacting factors rather than any strict mapping between combinations of genes expressed and specific sites in the tooth germ. Osborn (1993) has simulated the development of crown shape with such an approach, using a simple computer model of morphogenic processes due to dynamic interactions among mechanical forces.

Consistent with the idea that crown morphogenesis is a quantitative interaction among regulatory factors, rather than trait-by-trait combinatorial mechanism of gene expression, is the appearance of transitory diastemal tooth rudiments in the maxillae of mice and voles (e.g., Peterkova *et al.*, 1993, 1998,; Witter *et al.*, 1996; Keranen *et al.*, 1999; Witter *et al.*, 1996). These structures show gene expression patterns indicating their dental nature, but regress through apoptosis before passing the

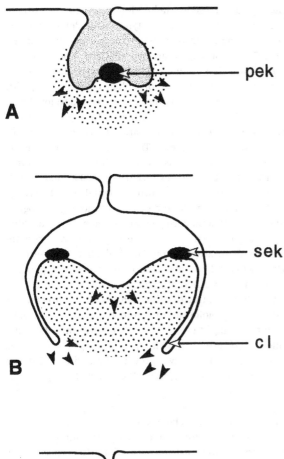

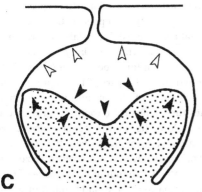

Figure 11.4. **Roles of cell proliferation (A, B) and pressure (C) in molar morphogenesis. A. Cap stage. B,C. Bell stage. cl, cervical loop; pek, primary enamel knot; sek, secondary enamel knot. Arrowheads in A and B indicate directions of tissue growth. Open and solid arrow heads in C indicate pressure on the enamel epithelia.**

rudiments resulting as a by product of evolutionary alterations in a quantitative pattern-generating mechanism responsible for the murine dental formula (e.g. Weiss *et al.*, 1998a). In either case, this interesting work might lead to a better understanding of processes by which teeth initiate or remain tooth-free, in the dental lamina.

Models of quantitative interaction processes have been able to describe the location and occurrence of primary tooth germs observed experimentally in the alligator (Kulesa *et al.*, 1996). Whatever the mechanisms are that cause the development of crown morphology, no existing theory adequately explains how the number and arrangement of cusps is produced in the first place.

11.4.   Tooth shape variation within and among mammalian dentitions

The tooth classes (incisor, canine, premolar, and molar) have characteristic shapes in many mammalian dentitions (Peyer, 1968). Such dentitions typically contain chisel-shaped incisors, conical canines, and multi-cusped cheek teeth. The crown patterns of molars tend to be more complex than those of premolars. Within the incisor, premolar, and molar tooth classes (mammals are generally assumed to have only a single canine) there are graded patterns of shape, size, and cusp development such that teeth most closely resemble their neighbours and become progressively more dissimilar with increasing distance along the jaw axis (Butler, 1939).

Much of the diversity of dentition patterns among mammals involves differences in the number, size, shape, and arrangement of cusps of the cheek teeth (Jernvall, 1995). New features in evolution often appear in multiple members of a tooth class (Butler, 1939, 1995). The slope of gradients in the form of teeth within a given class may also change during evolution (Butler, 1939). In contrast to correlated changes among members of a tooth class, different tooth classes exhibit a greater degree of evolutionary independence from each other. Variations from the more typical shape of a tooth class include the tusks of elephants (incisors), and walruses (canines), and the reduced, incisiform canines of cows (Peyer, 1968). Homeotic changes across tooth classes, such as the molarization of premolars in horses, have also occurred in evolution (Butler, 1978). It is generally agreed that most of the changes in tooth number in mammals have involved a reduction from the ancestral condition (Peyer, 1968). The teeth that are lost in evolution tend to be those at the margins of tooth classes. Whole tooth classes may also be lost, such as the canines

bud or earlier stages. One interpretation is that these are remnants specifically homologous to ancestral structures (canines and premolars) that existed in rodent ancestors (Peterkova *et al.*, 1996, 1998; Ruch *et al.*, 1998). Alternatively, homology between specific rudiments and specific teeth in the ancestral dentition may not exist, with the

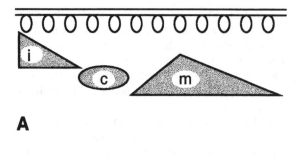

A

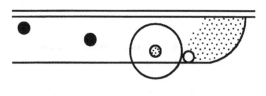

B

Figure 11.5A,B. **Diagrammatic illustration of the models for dentition pattern control. A. Field model. The ovals represent tooth primoria. i, incisor field; c, canine field; m, molar field. B. Clone model The stippled region represents the leading edge of the growing tissue. The black dots represent mature teeth. The stippled dot represents a growing tooth. The small open circle represents a cell clone that has developed to the critical size to form a tooth. The large circle represents the zone of inhibition. (Modified from Butler, 1939 and Lumsden, 1979.)**

and the premolars of rodents, and all of the upper incisors and canines of sheep and cows. An exception to the general pattern of tooth loss is the acquisition in dolphins of a large number of essentially homodont teeth.

11.5.   Theories for heterodont dentition patterning

Several hypotheses, largely based on comparative anatomy and palaeontology of adult dentitions, have been proposed for the mechanisms responsible for the patterning of heterodont dentitions (for reviews see Butler, 1978, 1995; Osborn, 1978; Lumsden, 1979; Ruch, 1987; Butler, 1978, 1995; Stock *et al.*, 1997; Weiss *et al.*, 1998a).

11.5.1.   Gradient models
Butler (1939) adapted a version of the morphogenetic gradient theory (see Wolpert, 1969 for example) to explain heterodont patterning. Similar models were also proposed later by Patterson (1956) and van Valen (1970).

Such models consider the tooth primordia to be initially identical, with final tooth type being determined by morphogenetic substance(s) in the mesenchyme of the jaw. These morphogens are hypothesized to be present in a concentration gradient along the mesial/distal axis of the jaw, affecting tooth morphology by interacting with the tooth primordia. In other words, teeth develop different morphologies because of their differing locations in this external field, at which particular concentrations of morphogen are present. The morphogenetic field is divided into subregions to specify the incisor, canine, and molar morphologies in Butler's (1939) model, with each subregion having its own gradient (Figure 11. 5A).

One of the important early evolutionary biologists and founders of modern genetics, William Bateson (1894, 1913) argued that quantitative processes were responsible for qualitatively segmented traits like the dentition. He discussed both mechanical analogies and possible chemical processes that might be involved (in the terms available to him at the time), in his effort to persuade evolutionary biologists to consider how discretely variable traits might evolve. A mathematical model of chemical processes, which provided the basis of morphogenic models was provided by Turing (1952). For a perspective on this topic, see Weiss *et al.* (1998a).

11.5.2.   Clone model
In contrast to the above model, Osborn (1973, 1978) adapted the progress zone theory (Summerbell *et al.*, 1973) to dentition pattern formation. He proposed that the three different tooth types represent three separate clones of cells that arise from differentially programmed stem progenitors. After a certain number of divisions the cell clone is large enough for a tooth primordium to be formed. The growth rate of a part of the clone specifies the initial growth rates of the primordium, which in turn determines the shape of the tooth. As growth progresses, the clone loses its 'morphogenetic vigour', resulting in shape changes in successively formed teeth. During the formation of the first primordium, a zone of inhibition is generated surrounding it. The zone of inhibition delays the formation of a second primordium, and eventually defines spaces between individual teeth. After a certain number of cell cycles, the cell clone loses its potency to form a primordium, and thus tooth number is constrained (Figure 11. 5B).

One of the challenges for the gradient model is that the morphogenetic material(s) have not been identified. Although some genes encoding developmental regulatory factors are expressed in the developing jaws in a graded fashion, none of them has been confirmed unambiguously to function as a morphogen (to be discussed in the following sections).

Retinoids, among the candidates for this type of role in embryonic development, regulate gene expression through binding to receptors (Petkovich, 1992). Kronmiller and Beeman (1994) detected a concentration difference of all-*trans* retinoic acid between incisor and molar regions and an opposite gradient of retinol in the same mandible. The receptors and binding proteins for retinoids are also present in the developing jaws (Ruberte *et al.*, 1991, 1992; Dollé *et al.*, 1990; Ruberte *et al.*, 1991, 1992). Experiments involving administration of retinoids to pregnant mice (Kalter and Warkany, 1961; Knudsen, 1965, 1967; Kalter and Warkany, 1961) or *in vitro* application of retinoids to cultured embryonic mandibles (Kronmiller *et al.*, 1992; Zhao *et al.*, unpublishedpersonal observations) have produced agenesis of molars, fusion of tooth germs, and supernumerary incisors in the offspring or the explants. In one case, incisors were found in the general molar region of retinoic acid-treated mouse mandibular arches in vitro (Kronmiller *et al.*, 1995). Experimental inactivation of retinoic acid and retinoid X nuclear receptors cause has no effects on dentition patterns (Krezel *et al.*, 1996; Mendelsohn *et al.*, 1994; Lohnes *et al.*, 1994; Kastner *et al.*, 1994; Lohnes *et al.*, 1994; Mendelsohn *et al.*, 1994; Krezel *et al.*, 1996).

Experimental evidence supporting the clone model was reported by Lumsden (1979), who obtained complete molar dentitions after grafting fragments of mouse mandibles from various stages to the anterior chamber of the eye of adult mice. Such isolated tooth rudiments are unlikely to have been under the influence of morphogenetic fields postulated to exist in the jaw. In the development of the dentition of the alligator (Westergaard and Ferguson, 1986, 1987) and lizard (Osborn, 1971) different tooth families initiate in accordance with jaw growth and the distance between previously formed teeth, supporting the zone of inhibition postulated by the clone model. However, according to the clone model, the sequential appearance of molars from mesial to distal and premolars from distal to mesial in mammals (Osborn, 1978) should be accompanied by growth of the jaw in both directions. The mesial to distal growth of the jaw is not indicated by patterns either of the mitotic index or of the expression of the *Msx1* gene that is often associated with growing tissues (MacKenzie *et al.*, 1992; Johnston and Bronsky, 1995).

The gradient model emphasizes that dentition pattern is specified by extrinsic factors in a quantitative manner. Such gradient mechanisms have been demonstrated in the development of other systems (Lawrence and Struhl, 1996; Nüsslein-Volhard, 1996). In contrast, the clone model stresses that tooth type is intrinsically determined within each clone of progenitor cells in a qualitative fashion. However, it is conceivable that different clones, if they truly exist, could be distinguished by different levels of gene expression or different rates of cell division.

11.5.3. Neural crest cells versus epithelium in dentition patterning

Tooth formation is controlled by continuous interaction between the epithelium of the developing jaws, and mesenchyme that includes cells that have migrated to the branchial arches from the neural crest (NC). The NC is a unique or even a defining characteristic of vertebrates (Northcutt and Gans, 1983), and is a major contributor to the exoskeleton, of which teeth are a part (Smith and Hall, 1993). In the mouse, migration of cranial NC cells occurs between E8.0 (4–6 somites) and E9.0 (14–16 somites; Nichols, 1981, 1986; Serbedzija, *et al.*, 1992), while teeth are initiated at about E11 (lower first molar) and E12 (incisors; this varies slightly among different strains; Ruch, 1987).

It has been demonstrated in amphibians (reviewed by Slavkin, 1974) and mammals (Lumsden, 1988) that neural crest cells play a critical role in tooth formation. NC cells migrate to the first arch from different rostral-caudal regions of the neural fold, at different developmental stages from the same region, and through numerous migration routes. Different populations of cells arrive at different locations within the first branchial arch (Nichols, 1981, 1986; Serbedzija, *et al.*, 1992; Trainor and Tam, 1995; Serbedzija, *et al.*, 1992; Osumi-Yamashita *et al.*, 1994; Trainor and Tam, 1995; Imai *et al.*, 1996). Do these aspects of NC cell migration bring signals to the developing jaws for the specification of tooth type and location? Recently, Köntges and Lumsden (1996) have demonstrated that the NC cells from the caudal midbrain populate the mesial elements of the chicken jaw, while the cells from the rostral hindbrain (rhombomere 1 and 2) contribute to the distal structures.

Tissue recombination experiments demonstrated that both epithelium and mesenchyme possess the capacity to control tooth morphology (Avery, 1954; Dryburgh, 1967; Miller, 1969, 1971; Avery, 1954; Dryburgh, 1967; Kollar and Baird, 1969, 1970a,b; Schmitt *et al.*, 1999). This may depend on the developmental stages of the embryo. The odontogenetic inductive capacity is in the epithelium at early stages (E9–E11) and later switches to the mesenchyme at the time when dental epithelium invaginates into the mesenchyme at E12.5 (Mina and Kollar, 1987). Lumsden (1988) suggested that the NC cells are not pre-programmed before migration, and the dental inductive potency is pre-patterned in the epithelium. He also showed that the trunk NC cells can form teeth when combined with jaw epithelium (Lunsden, 1988). A recent report of a transformation of tooth type induced by alteration of an epithelial signalling system (BMPs) lends support to Lumsden's (1988) proposal (Tucker *et al.*, 1998*b*; see below).

11.6.   Control of dentition patterns at the molecular level

11.6.1.   Combinatorial expression of homeobox genes

Homeobox genes encode transcription factors containing a DNA-binding homeodomain. These genes have been shown to play an important role in many aspects of embryogenesis including tooth development (Weiss *et al.*, 1994; 1995; Slavkin and Diekwisch, 1996; Stock *et al.*, 1997). There is good evidence that segmental identity along the longitudinal axis of vertebrates is at least in part controlled by combinatorial expression of the Hox family of homeobox-containing genes (Burke *et al.*, 1995). There are four clusters of Hox genes (about 39 genes) in mammals; within each cluster, the individual genes are expressed sequentially along the anterior-posterior body axis to control position along the vertebral column (Burke *et al.*, 1995). Combinatorial patterns of Hox gene expression have also been proposed to specify a regulatory-genetic 'code' that determines segment identity in the rhombomeres of the hindbrain, the branchial arches, and the limb bud (Hunt and Krumlauf, 1991). The most anterior boundary for Hox gene expression is rhombomere 2 in the central nervous system and the second branchial arch in the facial region (Hunt and Krumlauf, 1991). Thus, expression of Hox genes is found in neither the first branchial arch, nor the nasal processes that give rise to the jaws, nor in the forebrain, midbrain, and rostral hindbrain from which neural crest cells migrate to the developing jaws and teeth (Nichols, 1981, 1986; Trainor and Tam, 1995; Serbedzija, *et al.*, 1992; Osumi-Yamashita *et al.*, 1994; Trainor and Tam, 1995; Imai *et al.*, 1996). Interestingly, however, nested expression of other homeobox gene families, such as Dlx, Otx, and Emx, may be involved in regional specification of the forebrain and midbrain (reviewed by Rubenstein and Puelles, 1994). Some of the genes are also expressed in the developing jaws and teeth (to be discussed later).

It is believed that dentition patterns are specified before the dental primordia are morphologically discernible at E11.5 in the mouse (Lumsden, 1988; Lumsden and Buchanan, 1986). Homeobox genes expressed in the developing jaws include members of the Dlx, Msx, Prx, Six, Prox, Barx, and Tlx gene families, as well as *goosecoid* (*gsc*; MacKenzie *et al.*, 1991, 1992; Gaunt *et al.*, 1993; Raju *et al.*, 1993; Oliver *et al.*, 1993, 1995; Robinson and Mahon, 1994;

Simeone *et al.*, 1994; Leussink *et al.*, 1995; Tissier-Seta *et al.*, 1995; Qiu *et al.*, 1997; Weiss *et al.*, 1998*b*; Qiu *et al.*, 1997; Simeone *et al.*, 1994; Gaunt *et al.*, 1993; Leussink *et al.*, 1995; Raju *et al.*, 1993; Tissier-Seta *et al.*, 1995; Oliver *et al.*, 1993, 1995).

Three Msx genes have been isolated to date in mammals. *Msx1* and *Msx2* are reciprocally expressed in various growing structures including teeth (MacKenzie *et al.*, 1991, 1992), while *Msx3* expression is limited to the central nervous system (Shimeld *et al.*, 1996; Wang *et al.*, 1996). The expression of those former genes in the facial structures (the nasal processes and branchial arches) resembles that in the limb bud, and suggests that they are involved in control of tissue growth (Mina *et al.*, 1995; MacKenzie *et al.*, 1991, 1992; Maas and Bei, 1997).

Six mammalian Dlx genes, the homologues of the *Drosophila Distal-less* (*Dll*) gene, have been identified. These genes are organized in closely linked pairs, with each pair located on the same chromosome as one of the four Hox clusters (Ruddle *et al.*, 1994; Simeone *et al.*, 1994; Stock *et al.*, 1996*a*; McGuinness *et al.*, 1996; Nakamura *et al.*, 1996; Stock *et al.*, 1996*a*). The intriguing similarity in the arrangement of Dlx and Hox genes raises the question of whether the former genes control dentition patterns in a fashion analogous to that of Hox genes in other segmentally arranged structures. Expression of the Dlx genes shows overlapping but distinct patterns in the developing mandibular arch and maxillary process, suggesting a possible role in patterning the first branchial arch and dentition (Qiu *et al.*, 1997; Zhao *et al.*, unpublished observations) and dentition (Qiu *et al.*, 1997). In contrast to the Hox genes, where *trans*-paralogues (genes in similar positions on different chromosomes) often share more similar axial levels of expression than neighbouring *cis*-paralogues (genes on the same chromosome), Dlx *cis*-paralogues appear to share the greatest similarities in expression in the developing jaws (Zhao *et al.*, unpublishedpersonal observations). The correlation between Dlx gene expression and dentition patterning will be discussed later. All six Dlx genes are also subsequently expressed in the developing teeth (incisors and molars, in the case of mice), with dynamic patterns that vary among the members of the gene family (Weiss *et al.*, 1994, 1995, 1998*a*; Robinson and Mahon, 1994; Weiss *et al.*, 1994, 1995, 1998*a*; Zhao *et al.*, unpublishedpersonal observations).

Combinatorial involvement of homeobox genes in dentition patterning was hypothesized, based on their expression patterns (Weiss *et al.*, 1994, 1995, 1998*a*). Sharpe (1995) and his colleagues (Thomas *et al.*, 1998; Thomas and Sharpe, 1998; Thomas *et al.*, 1998) have suggested an 'odontogenetic homeobox code' model in which incisor, canine, and molar fields are defined by spatial combinatorial expression of a subset of the homeobox genes expressed in the developing jaws. In their latest version of the model (Thomas et al., 1998), incisors are controlled by *Msx1* and *Msx2*, while upper and lower molars are determined by *Dlx1/Dlx2* and *Dlx5/Dlx6*, respectively. *Barx1* is also involved in the molar code. The *gsc* gene was included in an earlier model (Sharpe, 1995).

One approach to investigate dentition patterning mechanisms is to compare gene expression and function between the upper and lower dentition. Dentition patterns on the upper and lower jaws are similar in terms of tooth type and location in most mammals including the mouse (Peyer, 1968), in spite of extensive divergence of dentition patterns among species and during evolution, suggesting an underlying similarity in controlling mechanisms between the upper and lower jaws. However, there are differences in the embryonic development of the lower jaw and upper jaws. The lower jaw develops from the mandibular part of the first branchial arch, while the upper jaw forms from two components, the medial nasal process and the maxillary process, which give rise to the premaxilla and maxilla respectively (Gaunt, 1964; Johnston and Bronsky, 1995; Gaunt, 1964). Morphogenesis of the jaws and formation of their cartilagous and bony elements also require differential expression of regulatory genes (Kuratani et al.et al., 1997). It is possible that the genetic systems that control dentition patterns may be independent of or overlapping with the systems that control jaw development, and an hypothesis that upper and lower dentition patterns are controlled by independent genetic systems has been proposed (Qiu et al.et al., 1997; Thomas et al., 1998; Thomas and Sharpe, 1998; Thomas et al., 1998).

The upper incisors reside in the premaxilla whereas the other teeth (upper canines, premolars, and molars) are in the maxilla (Peyer, 1968; Gaunt, 1964). Lumsden and Buchanan (1986) obtained incisors but no molars from *in vitro* cultured explants of murine medial nasal processes isolated before they fused with maxillary processes (E10). In the mouse mandible between E10.5– and E11.5, *Msx1* and *Msx2* are expressed in the mesial area including the presumptive incisor forming region, but not in the molar region, while in the upper jaw they are expressed in both incisor (medial nasal process) and molar forming (maxillary process) loci (MacKenzie et al., 1992). When *Msx1* was knocked out in the mouse genome by homologous recombination, both molar (of each jaw) and incisor (not clear whether in one or both jaws) tooth germs were arrested at the bud stage (Satokata and Maas, 1994; Maas and Bei, 1997).

Since the teeth are affected by the lack of *Msx1* at a stage before tooth morphology is recognizable, it remains unclear whether tooth shape is altered. Nevertheless, the positions of the tooth buds seem unchanged. Application of exogenous BMP4 to the molar buds of the *Msx1*- deficient mice *in vitro* can rescue molar morphogenesis from the bud to cap stage (Chen et al., 1996), suggesting that signals for molar teeth are present in the tissues even without *Msx1*. In the *Msx2* knockout, tooth shape is not affected, but the formation of enamel (Maas and Bei, 1997). *gsc* is expressed in the mandible, but apparently not in the upper jaw (either the tooth - bearing side of the medial nasal process or the maxillary process; Gaunt et al., 1993; Sharpe, 1995). The null mutation of *gsc* shows no dental phenotypes (Rivera-Pérez et al., 1995; Yamada et al., 1995).

Four Dlx genes, *Dlx3, Dlx5, Dlx6*, and *Dlx7*, are expressed in the lower jaw but not in the upper jaw (Robinson and Mahon, 1994; Simeone et al., 1994; Qiu et al., 1997; Weiss et al., 1998b; Zhao et al., unpublishedpersonal observations). Among them *Dlx3* and *Dlx7* are not expressed in the tooth forming region in the mandibular arch (Weiss et al., 1998b; Zhao et al., unpublishedpersonal observations). The relationship between the expression patterns of these genes and dentition pattern differs between the upper and lower jaws. *Dlx1* and *Dlx2* are expressed in both jaws. Their expression is negative in the mesial region of the mandible and the medial nasal process, the areas where incisors develop, according to Lumsden and Buchanan (1986). This pattern is coincident with primitive molar regions in this particular species (mouse). However, *Dlx2* has an almost identical expression pattern with that of *Dlx1* in the upper jaw (the maxillary process), but in the mandible its expression domain is broader than that of *Dlx1*, encompassing virtually the entire mandible (Weiss et al., 1998b; Zhao et al., unpublishedpersonal observations), including the diastemal region where canines and premolars develop in species with more complete dentition (Peyer, 1968). Neubüser et al. (1997) suggested that two discrete expression domains of the transcription factor *Pax9* in each mandibular half appearing prior to the manifestation of tooth germs correspond to presumptive molar and incisor regions. If the mesial expression domains of *Pax9* do in fact mark presumptive incisor regions, a result consistent with an earlier delineation of these regions by Lumsden (1982), then the *Dlx2* expression domain also includes the presumptive incisors.

If so, the Dlx code for the lower incisors and diastema would differ from that of the upper one. The experimental deletion of either *Dlx1* or *Dlx2* has no effect on tooth development or morphology, but the simultaneous loss of both genes causes a loss of upper molars that do not develop beyond the initiation stage, with other teeth remaining unaffected (Qiu et al., 1995; 1997; Thomas et *al.*, 1997). These results led to a hypothesis that *Dlx1* and *Dlx2* are involved in upper molar programming, while the lower molars are controlled by *Dlx5* and *Dlx6* (Qiu *et al.*, 1997; Thomas *et al.*, 1997, 1998; Thomas and Sharpe, 1998). An alternative explanation, as Peters and Balling (1999) pointed out, could be that *Dlx1* and *Dlx2* control both maxillary and mandibular molars, while *Dlx5* and *Dlx6* are functionally redundant with *Dlx1* and *Dlx2* in the lower jaw. Deletion of *Dlx5* does not affect tooth morphology, but enamel mineralization (Depew *et al.*, 1999; Acampora *et al.*, 1999). Compound knockout of the Dlx genes would answer the question.

Barx1, a homeobox gene, is expressed in the distal regions of upper and lower jaws, mainly associated with the molar regions (Tissier-Seta *et al.*, 1995; Thomas and Sharpe, 1998). Among tooth germs, it is only expressed in the developing molars (Tissier-Seta *et al.*, 1995). Recently, *Barx1* expression has been shown to be inhibited by BMP4 (Tucker *et al.*, 1998b). This result, along with the mesial restriction of *Bmp4* expression in the mandibular arch, suggests a mechanism for the restriction of *Barx1* expression to molar-forming regions. Furthermore, inhibition of BMP signalling in the mandibular arch led to apparent transformations of incisors toward a molar morphology, with a corresponding mesial expansion of the *Barx1* expression domain (Tucker *et al.*, 1998b). Additional work is required to determine whether *Barx1* plays a direct role in tooth type specification.In summary, although a number of homeobox genes exhibit restriction of expression to the presumptive region of particular tooth types, differences between the upper and lower jaws exist in several cases. Whether these genes are actually involved in tooth type specification remains to be demonstrated. Other families of genes encoding transcription factors.

Another large family of transcription factors is the HMG (High Mobility Group)-domain proteins. Two HMG-box genes, *Lef1* and *Tcf1*, are expressed in the first branchial arch and tooth germs (Oosterwegel *et al.*, 1993). In mice lacking *Lef1*, the development of molars and incisors is arrested at the bud stage. The null mutants of *Lef1* also lack hair and whiskers. *Lef1* is expressed early in the epidermis in spots where whiskers will develop. Overexpression of *Lef1* causes ectopic teeth to grow in the lip furrow; similar changes have been produced by tissue recombination experiments (van Genderan *et al.*, 1994). Further analyses of the *Lef1* −/− mice revealed that *Lef1* is an essential epithelial factor for early tooth development, but is not involved in specification of tooth type (Kratochwil *et al.*, 1996).

Another group of HMG-box genes is the Sox family, related to the sex-determining region Y gene (*Sry*; Laudet *et al.*, 1993). We have detected the expression of 14 of the approximately 20 described Sox genes from mouse tooth germs using reverse transcription-coupled polymerase chain reaction (RT-PCR; Stock *et al.*, 1996b). *In situ* hybridization studies have confirmed that *Sox4*, *Sox6*, and *Sox9* are indeed expressed in developing tooth germs (Zhao *et al.*, unpublished personal observations). *Sox6* and *Sox9* have been shown to be expressed in the developing jaws (Wright *et al.*, 1995). The roles of Sox genes in dentition patterning and tooth development are not known.

11.6.2.   Extracellular signalling factors

In addition to the transcription factors discussed above, a number of gene families encoding intercellular signalling molecules are likely to be involved in the control of dentition patterns. One such group is the TGF-β superfamily of ligands that bind to membrane receptors with serine/threonine kinase domains (Kingsley, 1994; Massagué, 1996). This superfamily can be divided into a number of smaller families or subfamilies based on sequence similarities. One such subfamily is the Bone Morphogenetic Proteins (BMPs), of which there are greater than 10 members (Hogan, 1996a,b; Wozney, 1998). A variety of BMPs have been shown to be expressed during first branchial arch and tooth development (Åberg *et al.*, 1997). Individual genes exhibit expression patterns that are overlapping yet distinct and which together span all phases of tooth development from initiation through the deposition of hard tissues. Some genes are restricted to a single tissue layer (*Bmp3*, *Bmp5*, *Bmp6*), while others (*Bmp2*, *Bmp4*, *Bmp7*) switch between epithelial and mesenchymal layers (Åberg *et al.*, 1997). Bmp knockout mutations have not shed much light on the role of the genes in tooth development, either because of early death (*Bmp2*, Zhang and Bradley, 1996; *Bmp4*, Winnier *et al.*, 1995) or the lack of effects on tooth development, possibly due to functional redundancy among family members (*Bmp5*, Kingsley *et al.*, 1992; *Bmp6*, Solloway *et al.*, 1998; *Bmp7*, Dudley and Robertson, 1997). Nevertheless, experiments involving the application of purified proteins in organ culture and the analysis of mice deficient for other genes have led to placement of several BMPs in proposed signalling pathways likely to be involved in the control of dentition patterns.

Early expression of *Bmp4* (mesial epithelium and mesenchyme of the mandibular process) and *Bmp2* (more distal mandibular mesenchyme) has been suggested to play a role in positioning the sites of tooth formation by antagonizing the induction of *Pax9* (a proposed early marker of odontogenic mesenchyme) by FGF8 (Neubüser *et al.*, 1997; see below). Such a role in inhibiting tooth induction appears to contradict other reports that BMP4 is capable of inducing a subset of the responses of odontogenic mesenchyme to odontogenic epithelium in culture (including expression of *Msx1*, *Msx2*, *Lef1*, and *Bmp4* itself; Vainio *et al.*, 1993; Chen *et al.*, 1996; Kratochwil *et al.*, 1996; Tucker *et al.*, 1998a). A possible resolution of this paradox is that an early role in inhibiting tooth induction is followed by a later role as an inductive signal from the epithelium to the mesenchyme (Neubüser *et al.*, 1997). *Bmp4* and *Msx1* are likely to act in a positive feedback loop which serves to elevate the expression of both in the dental mesenchyme (Chen *et al.*, 1996; Tucker *et al.*, 1998a).

The three mammalian TGFβs (TGFβ1, TGFβ2, TGFβ3) form a subfamily within the TGFβ superfamily and all three are expressed in tooth germs (Chai *et al.*, 1994; D'Souza and Litz ,1995). However, no dental defects were observed in mice lacking TGFβ1 (D'Souza and Litz, 1995) or TGFβ3 (Proetzel *et al.*, 1995). Cell and organ culture experiments have suggested that these genes play a role in the regulation of cell proliferation and differentiation in tooth germs (Bègue-Kirn *et al.*, 1992; 1994; Chai *et al.*, 1994).

Activins and inhibins are dimeric molecules whose subunits are encoded by three members of the TGFβ superfamily. Two of these genes, activin/inhibin-βA and activin/inhibin-βB form a related family, while a third, inhibin-α, is only distantly related (Kingsley, 1994). Inactivation of activin-βB does not result in a tooth phenotype, nor does it exacerbate the phenotype of activin-βA knockout mice (Schrewe *et al.*, 1994; Matzuk *et al.*, 1995a; Heikinheimo *et al.*, 1998). In contrast, in mice homozygous for an inactivation of activin-βA, incisors and mandibular molars do not progress beyond the bud stage, while the maxilllary molars are normal (Matzuk *et al.*, 1995a; Ferguson *et al.*, 1998). Interestingly, the teeth affected are the reciprocal of those affected by the *Dlx1/Dlx2* double knockout (Qiu *et al.*, 1997; Thomas *et al.*, 1997), suggesting developmental differences between the maxillary molars and other teeth (Ferguson *et al.*, 1998). The significance of such differences for the control of dentition pattern in development and evolution remains to be determined. Tissue recombination experiments suggest that activin (expressed in the dental mesenchyme beginning at E11.5) plays a role in mesenchyme-mesenchyme interaction, as well as mesenchyme to epithelium signalling (Ferguson *et al.*, 1998). Mice lacking type IIA or type IIB activin receptors have a similar phenotype to activin-βA deficient mice, while mice lacking the activin binding protein follistatin have a weakly penetrant phenotype of arrested lower incisor development (Matzuk *et al*, 1995b,c; Ferguson *et al.*, 1998).

Fibroblast growth factors (FGFs) comprise a family of intercellular signalling molecules with close to 20 members (Yamaguchi and Rossant, 1995; Ohbayashi *et al.*, 1998). Their effects are mediated by binding to a family of at least four high -affinity receptors with tyrosine kinase domains (Yamaguchi *et al.*, 1995). At least five FGFs have been shown to be expressed in tooth germs and/or the first branchial arch (Kettunen and Thesleff, 1998), as are three of the four mammalian FGF receptors (Kettunen *et al.*, 1998). The potential role of FGF4 as a mitogenic signal from the EK was described above. FGF4 is also capable of preventing apoptosis in dental tissues in the presence of BMP4 in the culture medium, suggesting a role in preventing premature apoptosis in the EK (Vaahtokari *et al.*, 1996b; Jernvall *et al.*, 1998). Interestingly, no FGF receptors are expressed in the primary EK, which may be the basis for its lack of proliferation and its eventual apoptosis despite its production of FGFs that promote and inhibit these processes, respectively, in other tissues (Kettunen *et al.*, 1998). FGF8 is likely to provide a signal for tooth initiation, based on its expression in the epithelium of tooth-forming regions prior to the morphological appearance of tooth germs and its ability to induce the expression of *Pax9*, a proposed early marker of odontogenic mesenchyme (Neubüser *et al.*, 1997). With respect to the pattern of the dentition, the antagonistic effects of BMP2 and BMP4 on FGF8 induction of *Pax9* expression suggest that the location of teeth may be determined, at least in part, by the overlap of FGF and BMP expression domains (Neubüser *et al.*, 1997). In other words, teeth may form in regions where FGF8 is present and BMPs are not. The expression and activities of FGF9 overlap those of both FGF4 and FGF8, suggesting that it may play roles in both tooth initiation and signalling from the EK (Kettunen and Thesleff, 1998).

The vertebrate genes *Sonic hedgehog* (*Shh*), *Desert hedgehog*, and *Indian hedgehog* are members of a recently identified family of signalling molecules (Ingham, 1995). The only one of these genes expressed in tooth-forming regions is *Shh*, which is expressed in the epithelium of the first branchial arch and later becomes restricted to the dental lamina and tooth germs (both incisors and molars; Bitgood and McMahon, 1995; Vaahtokari *et al.*, 1996a). Mice homozygous for an inactivation of *Shh* exhibited severe craniofacial defects, precluding an analysis of a possible role of this gene in patterning the dentition (Chiang *et al.*, 1996). The Shh signalling system regulates the expression of target genes through Gli zinc-finger-containing transcription factors (Ruiz i Altaba, 1997). Mice with a null mutation of *Gli2* exhibit fusion of maxillary incisors (Hardcastle *et al.*, 1998), a condition similar to that found in humans heterozygous for a mutation of *SHH* (see below). Interactions among Hedgehogs, BMPs and FGFs play important roles in pattern formation in various structures (Ingham, 1995). The proposed roles of BMPs and FGFs in determining tooth location described above suggest that additional interactions among signalling factors are likely to be involved in patterning the dentition.

11.6.3.   Spontaneous mutations with dental phenotypes

Various dental variations and abnormalities in tooth number, shape, and position have been described in many mammalian species and humans (Miles and Grigson, 1990; Burzynski and Escobar, 1983; Miles and Grigson, 1990). Variations in tooth number and crown shape appear to occur more frequently than tooth transposition (Miles and Grigson, 1990; Burzynski and Escobar, 1983; Miles and Grigson, 1990; Kantaputra and Gorlin, 1992). In most cases the data were obtained from analysis of animal skulls so that the linkage with either physical or genetic background is not clear. In some cases, disease, irregular eruption, retention of deciduous teeth, and various environmental factors may be the causes. Nevertheless, some dental pattern anomalies show clear genetic association in humans (McKusick, 1994; OMIM, 1997) and mice (Grüneberg, 1965). In humans, there are about 100 reported tooth number anomalies or syndromic conditions that appear to be familial. These are heterogeneous, usually only involve some of the teeth of a given type, vary considerably among affected members of the same family, and are typically syndromic (associated with numerous other, non-dental anomalies).

The well-described mouse mutations with dental phenotyes are *Tabby* (*Tb*), *Crinkled* (*Cr*), and *Crooked-tail* (*Cd*) (Grüneberg, 1965; Sofaer, 1969b). All of these share similar phenotypes of smaller teeth (incisors and molars), reduced, fused, or missing cusps, and missing third molars. Missing incisors in both upper and lower jaws frequently occur in *Tabby* homozygotes (*Tb/Tb*) and hemizygotes (*Tb/-*). In *Cd/Cd*, lower incisors are smaller, but the upper ones appear normal. In *Tb/Tb* and *Cd/Cd* mice, developing tooth germs of incisors and third molars were observed in the embryos (Sofaer, 1969a; Grewal, 1961; Sofaer, 1969a), suggesting that the mutations affect later stages of tooth development.

Enamel appears normal in these mutant mice, and the phenotypes of reduced and missing teeth have been interpreted as a failure of growth at critical stages, probably due to deficient growth of the dental mesenchyme (Grewal, 1961; Miller, 1978; Sofaer, 1969a; Miller, 1978;). However, it is interesting, with respect to dentition patterning, that supernumerary lower molars mesial to the first molars are found in *Tb* mice, including heterozygotes (Sofaer, 1969b). Sofaer (1969a) interpreted this phenotype as a progressive suppression of the first molar, which allows the dental lamina to extend mesially and gives rise to an additional tooth. Recently, the *Tabby*

mutation has been shown to result from an alteration in a novel, potentially transmembrane, protein known as ectodysplasin A (Srivastava *et al.*, 1997).

The molecular genetic bases of a number of other dental anomalies have also been determined. One type of hypodontia with missing second premolars and third molars of the permanent dentition has been associated with a missense mutation of the human *MSX1* gene (Vastardis *et al.*, 1996). The affected teeth in these individuals are the ones that develop last in the tooth class, raising the possibility that a process common to the development of all teeth is being affected, with the last teeth to develop being closest to the threshold of tooth loss (Thesleff, 1996). The loss of the last members of dentition to develop is also frequently the case in wild animals and humans with or without a known genetic cause (Sofaer, 1975; Gaunt, 1964; Sofaer, 1975; Burzynski and Escobar, 1983). Mice homozygous for *Small eye* (a semi-dominant mutation of *Pax6*) frequently exhibit a supernumerary incisor on one or both sides of the upper jaw, distal to the normal incisors (Kaufman *et al.*, 1995). The dental phenotype as well as other craniofacial defects varies in different genetic background (Quinn *et al.*, 1997), suggesting an involvement of multiple genetic factors. Mutations in the human *SHH* gene result in holoprosencephaly, a syndrome characterised by defects in midline structures of the forebrain and face. In mild cases of holoprosencephaly, including at least one caused by a missense mutation in *SHH*, three rather than four premaxillary incisors may be present, with the central one located on the midline (Roessler *et al.*, 1996). Rieger syndrome (RIEG) is an autosomal-dominant human disorder with some craniofacial defects. The dental abnormalities include missing maxillary distal incisors and microdontia of the mandibular mesial incisors (Semina *et al.*, 1997). The human *RIEG* gene and murine *Otlx2* (Mucchielli *et al.*, 1997) encode homeodomain proteins, and latter is shown to be expressed in tooth development.

Although the genetically-characterised dental anomalies described above suggest a role for these genes in some aspects of dentition patterning, no such genes have been shown to play a direct role in the specification of tooth type. Nevertheless, further production and characteriszation of such mutations should prove to be a powerful approach for elucidating the genetic basis of multiple aspects of dentition patterning.

11.7.   Conclusion and perspectives

The shape of individual teeth within the mammalian dentition is intimately related to their position. Neither the gradient field nor clone models that have been put forward to explain this and other observations are satisfactory explanations of all aspects of patterning of the dentition. Although it has been proposed that different regions of the dentition are specified by the expression of different combinations of homeobox genes (Sharpe, 1995; Weiss, 1990, 1993; Weiss et al., 1994, 1995, 1998a,b; Sharpe, 1995; Thomas et al., 1997;Thomas and Sharpe, 1998; Thomas et al., 1997), they have not been definitively demonstrated to play such a role. It is possible that patterning the dentition and the jaw is by independent or partially related molecular systems. Convincing molecular candidates for morphogens, which would probably have to act at very early stages of the development of the dentition, are also lacking.

The development of individual teeth is controlled by epithelial-mesenchymal interactions, and so it is also likely that this is the case for the patterning of the dentition as a whole. The primary signals for dentition formation are believed to be exerted from the epithelium, while the mesenchyme plays an important role in dental morphogenesis. The location of teeth, as well as tooth type, may be in part controlled by interactions between growth factors in the oral epithelium

Substantial progress in the study of individual tooth pattern formation has been made by the identification of the EK as a potential organizing centre, and its associated signalling factors, SHH, BMP2, BMP4, BMP7, FGF4, and FGF9. If the EKs regulate crown patterns by controlling the rates of proliferation of surrounding cells, and their appearance is regulated by the dynamics of proliferation in the enamel organ, the specific cusps are unlikely to be coded for by specific genes, but, instead, by the quantitative and/or temporal dynamics of the interaction of these various signalling factors. If this is the case, there may not be any strict homology among these 'obviously' homologous structures (see Webster and Goodwin, 1996; Weiss et al., 1998a; Webster and Goodwin, 1996). In other words, specific cusps may not be programmed by specific genes but instead the *entire* cusp pattern may arise as a consequence of the dynamic process itself. This would mean that while the process might be homologous among species, there may not be any strict homology among, or Darwinian historical continuity between, the specific cusps that have long been said to be homologous. This is an important point for evolutionary biologists, both systematists and those wishing to infer causal mechanisms from comparative and evolutionary data.

We still do not know what aspects of the dentition are programmed by specific genetic 'codes' and what parts are the result of the parameters (rather than the components) of quantitative processes. However, in this regard, we know of no genes whose evolutionary arrival

corresponds to the origin of a new dental trait. Regulation and protein dynamics are liable to be responsible for the evolution of dental patterning, rather than gene duplication or genome evolution events, even though the evolutionary results include discrete, qualitative changes in that pattern. This general point about the possible role of quantitative processes rather than specific, classically Darwinian programming, was raised by Bateson a century ago, in terms appropriate to that time period, in his objection to the strong gradualistic views of evolution held by Darwinian biologists (Bateson, 1894; Webster and Goodwin, 1996).

While we have not yet solved the problem of the genetic control of dentition patterning in development or evolution, significant advances have been made in the last few years in identifying genes that encode regulatory factors (signalling factors, receptors, and transcription factors) in the developing jaws and teeth as well as in the development of models and research strategies for pursuing the problem. At least we now have many candidate genes, and some direct functional evidence to work with. The rate of elucidation of genetic mechanisms for the control of pattern formation in other systems suggests that the next few years will witness similar advances in our understanding of the patterning of the dentition.

Summary

The evolution of the mammalian dentition, with its regionally differentiated tooth types (incisors, canines, premolars and molars) and defined tooth number, from precursors with more uniformly shaped teeth of variable number resembles that of other segmented systems, such as the vertebral column. This raises the possibility of similarities in the control and evolutionary modification of pattern in these systems. All of the teeth in the mammalian dentition are believed to have evolved from single-cusped teeth, with multi-cusped premolars and molars arising by outgrowth or budding from the occlusal surface of the primitive tooth crown.

Tooth morphogenesis in embryonic development is largely due to folding of the inner enamel epithelium of the enamel organ. An epithelial structure known as the enamel knot may play an important role in controlling this process. The expression of a number of signalling factors, such as *Shh*, *Bmps*, and *Fgf4*, and a transcription factor, *Msx2*, makes it a potential signalling centre in a developing tooth germ, similar to other known embryonic organizers.

Tooth type and location are determined in the developing jaws prior to morphological manifestation of tooth primordia. During the same time, the jaw also undergoes basic pattern formation with axial and regional differentiation. Dentition patterning is believed to be controlled by epithelial-mesenchymal interaction. The primary signals are believed to be provided initially by the epithelium, while the role played by the mesenchyme remains to be elucidated. Candidates for epithelial signalling factors include SHH, BMP2, BMP4, and FGF8. Different populations of neural crest cells migrate into different regions of the jaw, and a number of transcription factors, including the Msx, Dlx, and HMG box-containing gene families which are mainly expressed in the mesenchyme show overlapping but distinct expression patterns in the jaw. However, it is not clear yet whether patterning of both the jaws and dentition is achieved by the same molecular system, or by independent or partially related systems.

A number of spontaneous and induced mutations in animals and humans affect dental patterns but none has been demonstrated to play a direct role in specification of tooth type. Although none of the hypothetical models for patterning the dentition (e.g. field, clone, and odontogenetic homeobox code) has been established by convincing experimental data, continuing genetic and molecular studies should provide more information on the mechanisms of dentition pattern formation in the near future.

Acknowledgements

Work reported in this paper was supported in part by grants from the US National Institute for Dental Research (DE 10871), and the US National Science Foundation (SBR 9804907), and by research support from Penn State University.

References

Åberg, T., Wozney, J. and Thesleff, I. (1997). Expression patterns of bone morphogenetic proteins (*Bmps*) in the developing mouse tooth suggest roles in morphogenesis and cell differentiation. *Developmental Dynamics*, 210, 383–396.

Acampora, D., Merlo, G. R., Paleari, L., Zerega, B., Postiglione, M. P., Mantero, S., Bober, E., Barbieri, O., Simeone, A. and Levi, G. (1999). Craniofacial, vestibular and bone defects in mice lacking the Distal-less-related gene *Dlx5*. *Development* 126, 3795–3809.

Avery, J. K. (1954). Primary inductions of tooth formation. *Journal of Dental Research*. 33, 702 (abstract).

Bateson, W. (1894). *Materials for the Study of Variation, Treated with Special Regard to Discontinuity in the Origin of Species*. London: Macmillan.

Bateson, W. (1913). *Problems of Genetics*. New Haven: Yale University Press.

Bègue-Kirn, C., Smith, A. J., Ruch, J. V., Wozney, J. M., Purchio, A., Martmann, D. and Lesot, H. (1992). Effects of dentin proteins, transforming growth factor β1 (TGFβ1) and bone morphogenetic protein 2 (BMP2) on the differentiation of odontoblast in vitro. *International Journal of Developmental Biology*, 36, 491–503.

Bègue-Kirn, C., Smith, A. J., Loriot, M., Kupferle, C., Ruch, J. V. and Lesot, H. (1994). Comparative analysis of *TGFβs*, *BMPs*, *IGF1*, *msxs*, *fibronectin*, *osteonectin* and *bone sialoprotein* gene expression during normal and *in vitro*-induced odontoblast differentiation. *International Journal of Developmental Biology*, 38:405–420.

Berkovitz, B. K. B. (1967). An account of the enamel cord in *Setonix brachyurus* (Marsupialia) and on the presence of an enamel knot in *Trichosurus vulpecula*. *Archives of Oral Biology* 12, 49–59.

Bitgood, M. J. and McMahon, A. P. (1995). *Hedgehog* and *bmp* genes are coexpressed at many diverse sites of cell-cell interaction in the mouse embryo. *Developmental Biology*, 172, 126–138.

Bown, T. M. and Kraus, M. J. (1979). Origin of the tribosphenic molar and metatherian and eutherian dental formulae. In *Mesozoic Mammals: the First Two-Thirds of Mammalian History*, (eds. J. A. Lillegraven, Z. Kielan-Jaorowska and W. A. Clemens, pp. 172–191. Berkeley: University of California Press.

Brown, K. S. (1983). Evolution and development of the dentition. *Birth Defects*, 19, 29–66.

Burke, A., Nelson, C., Morgan, B. and Tabin, C. (1995). *Hox* genes and the evolution of vertebrate axial morphology. *Development* 121: 333–346.

Burzynski, N. J. and Escobar, V. H. (1983). Classification and genetics of numeric anomalies of dentition. *Birth Defects* 19, 95–106.

Butler, P. M. (1939). Studies of the mammalian dentition. Differentiation of the postcanine dentition. *Proceedings of the Zoological Society of London Series B*, 109, 1–36.

Butler, P. M. (1956). The ontogeny of molar pattern. *Biological Reviews*, 31, 30–71.

Butler, P. M. (1972). Some functional aspects of molar evolution. *Evolution*, 26, 474–483.

Butler, P. M. (1978). The ontogeny of mammalian heterodonty. *Journal de Biologie Buccale*, 6, 217–227.

Butler, P. M. (1990). Early trends in the evolution of tribosphenic molars. *Biology Reviews*, 65, 529–552.

Butler, P.M. (1992). Tribosphenic molars in the Cretaceous. In: *Structure, Function and Evolution of Teeth*, eds. P. Smith and E. Tchernov, pp. 125–138. London: Freund Publishing .

Butler, P. M. (1995). Ontogenetic aspects of dental evolution. *International Journal of Developmental Biology*, 39, 25–34.

Chai, Y., Mah, M., Crohin, C., Groff, S., Bringas, P. Jr, Le, T., Santos, V. and Slavkin, H. C. (1994). Specific transforming growth factor- subtypes regulate embryonic mouse Meckel's cartilage and tooth development. *Developmental Biology*, 162, 85–103.

Chen, Y., Bei, M., Woo, I., Satokata, I. and Maas, R. (1996). *Msx1* controls inductive signalling in mammalian tooth morphogenesis. *Development*, 122, 3035–3044.

Chiang, C., Litingtung, Y., Lee, E., Young, K. E., Corden, J. L.,

Westphal, H. and Beachy, P. A. (1996). Cyclopia and defective axial patterning in mice lacking *Sonic hedgehog* gene function. *Nature*, 383, 407–413.

Clemens, W. A. (1971). Mesozoic evolution of mammals with tribosphenic dentitions. In *Dental Morphology and Evolution*, ed. A. A. Dahlberg, pp. 181–192. Chicago: University of Chicago Press.

Cohn, S. A. (1957). Development of the molar teeth in the albino mouse. *American Journal of Anatomy*, 101, 295–320.

Colvin, J. S., Feldman, B., Nadeau, J. H., Goldfarb, M. and Ornitz, D. M. (1999). Genomic organization and embryonic expression of the mouse fibroblast growth factor 9 gene. *Developmental Dynamics*, 216, 72–88.

Crompton, A. W. and Luo, Z. (1993). Relationships of the Liassic mammals *Sinoconodon*, *Morganucodon oehleri*, and *Dinnetherium*. In *Mammal Phylogeny*, vol. 1, eds. F. S. Szalay, M. J. Novacek and M. C. McKenna, pp. 30–44. Berlin: Springer.

Darwin, C. (1859). *The Origin of Species*. London: Murray.

Depew, M., J., Liu, J. K., Long, J. E., Presley, R., Meneses, J. J., Pedersen, R. A. and Rubenstein, J. L. R. (1999). *Dlx5* regulates regional development of the branchial arches and sensory capsules. *Development*, 126, 3831–3846.

Deuschle, F. M., Geiger, J. F. and Warkany, J. (1959). Analysis of an anomalous oculodentofacial pattern in newborn rats by maternal hypervitaminosis A. *Journal of Dental Research*, 18, 149–155.

Dollé, P., Ruberte, E., Leroy, P., Morriss-Kay, G. and Chambon, P. (1990). Retinoic acid receptors and cellular retinoid binding proteins. I. A systematic study of their differential pattern of transcription during mouse organogenesis. *Development*, 110, 1133–1151.

Dryburgh, L. C. (1967). Epigenetics of early tooth development in the mouse. *Journal of Dental Research*, 46, 1264 (abstract).

D'Souza, R. N. and Litz, M. (1995). Analysis of tooth development in mice bearing a TGF-1 null mutation. *Connective Tissue Research*, 32, 41–46.

Dudley, A. T. and Robertson, E. J. (1997). Overlapping expression domains of bone morphogenetic protein family members potentially account for limited tissue defects in BMP7 deficient embryos. *Developmental Dynamics*, 208, 349–362.

Ferguson, C. A., Tucker, A. S., Christensen, L., Lau, A. L., Matzuk, M. M. and Sharpe, P. T. (1998). Activin is an essential early mesenchymal signal in tooth development that is required for patterning of the murine dentition. *Genes and Development*, 12, 2636–2649.

Ferguson, M. (1990). The dentition through life. In *The Dentition and Dental Care*, ed. R. J. Elderton, pp. 1–29. Oxford: Heinemann.

Gaunt, S. J., Blum, M. and De Robertis, E. M. (1993). Expression of the mouse *goosecoid* gene during mid-embryogenesis may mark mesenchymal cell lineages in the developing head, limbs and body wall. *Development*, 117, 769–778.

Gaunt, W. A. (1964). The development of the teeth and jaw of the albino mouse. *Acta Anatomica*, 57, 115–151.

Gregory, W. K. (1934). A half-century of trituberculy: the Cope-Osborn theory of dental evolution, with a revised summary of molar evolution from fish to man. *Proceedings of the American Philosophical Society*, 73, 169–317.

Grewal, M. S. (1961). The development of an inherited tooth defect in the mouse. *Journal of Embryology and Experimental Morphology*, 10, 202–211.

Grüneberg, H. (1965). Genes and genotypes affecting the teeth of the mouse. *Journal of Embryology and Experimental Morphology*, 14, 137–159.

Hardcastle, Z., Mo, R., Hui, C-C. and Sharpe, P. T. (1998). The Shh signalling pathway in tooth development: defects in *Gli2* and *Gli3* mutants. *Development*, 125, 2803–2811.

Heikinheimo, K., Bègue-Kirn, C., Ritvos, O., Tuuri, T. and Ruch, J. V. (1998). Activin and bone morphogenetic protein (BMP) signalling during tooth development. *European Journal of Oral Science*, 106 (Supplement 1), 167–173.

Hershkovitz, P. (1971). Basic crown patterns and cusp homologies of mammalian teeth. In *Dental Morphology and Evolution*, ed. A. A. Dahlberg, pp. 95–150. Chicago: University of Chicago Press.

Hogan, B. L. M. (1996a). Bone morphogenetic proteins in development. *Current Opinion in Genetics and Development*, 6, 432–438.

Hogan, B. L. M. (1996b). Bone morphogenetic proteins: multifunctional regulators of vertebrate development. *Genes and Development*, 10, 1580–1594.

Hunt, P. and Krumlauf, R. (1991). Deciphering the Hox code: clues to patterning branchial regions of the head. *Cell*, 66, 1075–1078.

Imai, H., Osumi-Yamashita, N., Ninomiya, Y. and Eto, K. (1996). Contribution of early-emigrating midbrain crest cells to the dental mesenchyme of mandibular molar teeth in rat embryos. *Developmental Biology*, 176, 151–165.

Ingham, P. W. (1995). Signalling by hedgehog family proteins in *Drosophila* and vertebrate development. *Current Opinion in Genetics and Development*, 5, 492–498.

Jernvall, J. (1995). Mammalian molar cusp patterns: developmental mechanisms of diversity. *Acta Zoologica Fennica*, 198, 1–61.

Jernvall, J., Kettunen, P., Karavanova, I., Martin, L. B. and Thesleff, I. (1994). Evidence for the role of the enamel knot as a control center in mammalian tooth cusp formation: non-dividing cells express growth factor *Fgf-4* gene. *International Journal of Developmental Biology*, 38, 463–469.

Jernvall, J., Åberg, T., Kettunen, P., Keränen, S. and Thesleff, I. (1998). The life history of an embryonic signalling center: BMP-4 induces *p21* and is associated with apoptosis in the mouse tooth enamel knot. *Development*, 125, 161–169.

Johnston, M. C. and Bronsky, P. T. (1995). Prenatal craniofacial development: new insights on normal and abnormal mechanisms. *Critical Reviews in Oral Biololy and Medicine*, 6, 25–79.

Kalter, H. and Warkany, J. (1961). Experimental production of congenital malformations in strains of inbred mice by maternal treatment with hypervitaminosis A. *American Journal of Pathology*, 38, 1–21.

Kantaputra, P. N. and Gorlin, R. J. (1992). Double dens invaginatus of molarized maxillary central incisors, premolarization of maxillary lateral incisors, multituberculism of the mandibular incisors, canines and first premolars, and sensorineural hearing loss. *Clinical Dysmorphology*, 1, 128–136.

Kastner, P., Grondona, J. M., Mark, M., Gansmuller, A., LeMeur, M., Décimo, D., Vonesch, J. L., Dollé, P. and Chambon, P. (1994). Genetic analysis of RXRα developmental function: convergence of RXR and RAR signalling pathways in heart and eye morphogenesis. *Cell*, 78, 987–1003.

Kaufman, M. H. (1992). *The Atlas of Mouse Development*. London: Academic Press.

Kaufman, M. H., Chang, H-H. and Shaw, J. P. (1995). Craniofacial abnormalities in homozygous *Small eye (Sey/Sey)* embryos and newborn mice. *Journal of Anatomy*, **186**, 607–617.

Kemp, T. S. (1982). *Mammal-like Reptiles and the Origin of Mammals.* London: Academic Press.

Keranen, S., Kettunen, P., Aberg, T., Thesleff, I., Jernvall, J. (1999). Gene expression patterns associated with suppression of odontogenesis in mouse and vole diastema regions. *Development, Genes and Evolution*, **209**, 495–506.

Kettunen, P. and Thesleff, I. (1998). Expression and function of FGFs-4, -8, and -9 suggest functional redundancy and repetitive use as epithelial signals during tooth morphogenesis. *Developmental Dynamics*, **211**, 256–268.

Kettunen, P., Karavanova, I. and Thesleff, I. (1998). Responsiveness of developing dental tissues to fibroblast growth factors: expression of splicing alternatives of FGFR1, -2, -3, and of FGFR4; and stimulation of cell proliferation by FGF-2, -4, -8, and -9. *Developmental Genetics*, **22**, 374–385.

Kingsley, D. M. (1994). The TGF-β superfamily: new members, new receptors, and new genetic tests of function in different organisms. *Genes and Development*, **8**, 133–146.

Kingsley, D. M., Bland, A. E., Grubber, J. M., Marker, P. C., Russell, L. B., Copeland, N. G. and Jenkins, N. A. (1992). The mouse *short ear* skeletal morphogenesis locus is associated with defects in a bone morphogenetic member of the TGF superfamily. *Cell* **71**, 399–410.

Kirino, T., Nozue, T. and Inoue, M. (1973). Deficiency of enamel knot in experimental morphology. *Okajimas Folia Anatomica Japonica*, **50**, 117–132.

Knudsen, P. A. (1965). Fusion of upper incisors at bud stage or cap stage in mouse embryos with exencephaly induced by hypervitaminosis A. *Acta Odontologica Scandinavica*, **23**, 549–565.

Knudsen, P. A. (1967). Congenital malformations of lower incisors and molars in exencephalic mouse embryos, induced by hypervitaminosis A. *Acta Odontologica Scandinavica*, **25**, 669–691.

Kollar, E. J. and Baird, G. R. (1969). The influence of the dental papilla on the development of tooth shape in embryonic mouse tooth germs. *Journal of Embryology and Experimental Morphology*, **21**, 131–148.

Kollar, E. J. and Baird, G. R. (1970a). Tissue interactions in embryonic mouse tooth germs. I. Reorganization of the dental epithelium during tooth-germ reconstruction. *Journal of Embryology and Experimental Morphology*, **24**, 159–171.

Kollar, E. J. and Baird, G. R. (1970b). Tissue interactions in embryonic mouse tooth germs. II. The inductive role of the dental papilla. *Journal of Embryology and Experimental Morphology*, **24**, 173–186.

Köntges, G. and Lumsden, A. (1996). Rhombencephalic neural crest segmentation is preserved throughout craniofacial ontogeny. *Development*, **122**, 3229–3242.

Kratochwil, K., Dull, M., Fariñas, I., Galceran, J. and Grosschedl, R. (1996). *Lef1* expression is activated by BMP-4 and regulates inductive tissue interactions in tooth and hair development. *Genes and Development*, **10**, 1382–1394.

Krezel, W., Dupe, V., Mark, M., Dierich, A., Kastner, P. and Chambon, P. (1996). RXRγ mull mice are apparently normal and compound RXRα$^{+/-}$/RXRβ$^{-/-}$/RXRγ$^{-/-}$ mutant mice are viable. *Proceedings of the National Academy of Sciences*, **93**, 9010–9014.

Kronmiller, J. E. and Beeman, C. S. (1994). Spatial distribution of endogenous retinoids in the murine embryonic mandible. *Archives of Oral Biology*, **39**, 1071–1078.

Kronmiller, J. E., Upholt, W. B. and Kollar, E. (1992). Alteration of murine odontogenic patterning and prolongation of expression of epidermal growth factor mRNA by retinol in vitro. *Archives of Oral Biology*, **37**, 129–138.

Kronmiller, J. E., Nguyen, T. and Berndt, W. (1995). Instruction by retinoic acid of incisor morphology in the mouse embryonic mandible. *Archives of Oral Biology*, **40**, 589–595.

Kulesa, P., Cruywagen, G., Lubkin, S., Maini, P., Sneyd, J., Ferguson, M. and Murray, J. (1996). On a model mechanism for the spatial patterning of teeth primordia in the alligator. *Journal of Theoretical Biology*, **180**, 287–296.

Kuratani, S., Matsuo, I. and Aizawa, S. (1997). Developmental patterning and evolution of the mammalian viscerocranium: genetic insights into comparative morphology. *Developmental Dynamics*, **209**, 139–155.

Laudet, V., Stehelin, D. and Clevers, H. (1993). Ancestry and diversity of the HMG box superfamily. *Nucleic Acids Research*, **21**, 2493–2501.

Lawrence, P. A. and Struhl, G. (1996). Morphogens, compartments, and pattern: lessons from *Drosophila*? *Cell*, **85**, 951–961.

Leussink, B., Brouwer, A., El Khattabi, M., Poelmann, R. E., Gittenberger-de Groot, A. C. and Meijlink, F. (1995). Expression patterns of the paired-related homeobox genes *MHox/Prx1* and *s8/Prx2* suggest roles in development of the heart and the forebrain. *Mechanisms of Development*, **52**, 51–64.

Lohnes, D., Mark, M., Mendelsohn, C., Dollé, P. Dierich, A., Gorry, P., Gansmuller, A. and Chambon, P. (1994). Function of the retinoic acid receptors (RARs) during development. (I) Craniofacial and skeletal abnormalities in RAR double mutants. *Development*, **120**, 2723–2748.

Lumsden, A. G. S. (1979). Pattern formation in the molar dentition of the mouse. *Journal de Biologie Buccale*, **7**, 77–103.

Lumsden, A. G. S. (1982). The developing innervation of the lower jaw and its relation to the formation of tooth germs in mouse embryos. In *Teeth* ed. B. Kurten, pp. 32–43. New York: Columbia University Press.

Lumsden, A. G. S. (1988). Spatial organization of the epithelium and the role of neural crest cells in the initiation of the mammalian tooth germ. *Development*, **103S**, 155–169.

Lumsden, A. G. S. and Buchanan, J. A. G. (1986). An experimental study of timing and topography of early tooth development in the mouse embryo with an analysis of the role of innervation. *Archives of Oral Biology*, **31**, 301–311.

Maas, R. and Bei, M. (1997). The genetic control of early tooth development. *Critical Reviews in Oral Biology and Medicine*, **8**, 4–39.

MacKenzie, A., Leeming, G. L., Jowett, A. K., Ferguson, M. W. J. and Sharpe, P. T. (1991). The homeobox gene Hox 7.1 has specific regional and temporal expression patterns during early murine craniofacial embryogenesis, especially tooth development *in vivo* and *in vitro*. *Development*, **111**, 269–285.

MacKenzie, A., Ferguson, M. W. J. and Sharpe, P. T. (1992). Expression patterns of the homeobox gene, *Hox-8*, in the mouse embryo suggest a role in specifying tooth initiation and shape. *Development*, **115**, 403–420.

Martin, G. R. (1998). The role of FGFs in the early development of vertebrate limbs. *Genes and Development*, **12**, 1571–1586.

Massagué, J. (1996). TGFβ signalling: receptors, transducers, and Mad proteins. *Cell*, **85**, 947–950.

Matzuk, M. M., Kumar, T. R., Vassalli, A., Bickenbatch, J. R., Roop, D. R., Jaenisch, R. and Bradley, A. (1995a). Functional analysis of activins during mammalian development. *Nature*, **374**, 354–356.

Matzuk, M. M., Kumar, T. R. and Bradley, A. (1995b). Different phenotypes for mice deficient in either activin or activin receptor type II. *Nature*, **374**, 356–360.

Matzuk, M. M., Lu, N., Vogel, H., Selheyer, K., Roop, D. R. and Bradley, A. (1995c). Multiple defects and prenatal death in mice deficient in follistatin. *Nature*, **374**, 360–363.

McGuinness, T., Porteus, M. H., Smiga, S., Bulfone, A., Kingsley, C., Qiu, M., Liu, J. K., Long, J. E., Xu, D. and Rubenstein, J. L. R. (1996). Sequence, organization, and transcription of the *Dlx-1* and *Dlx-2* locus. *Genomics*, **35**, 473–485.

McKusick, V. A. (1994). *Mendelian Inheritance in Man*, eleventh edition. Baltimore: Johns Hopkins Unversity Press.

Mendelsohn, C., Lohnes, Décimo, D., Lufkin, T., LeMeur, M., Chambon, P. and Mark, M. (1994). Function of the retinoic acid receptors (RARs) during development. (II) Multiple abnormalities at various stages of organogenesis in RAR double mutants. *Development*, **120**, 2749–2771.

Miles, A. E. W. and Grigson, C. (1990). *Colyer's Variations and Diseases of the Teeth of Animals*. Cambridge: Cambridge University Press.

Miller, W. A. (1969). Inductive changes in early tooth on the chick chorioallantois. *Journal of Dental Research*, **48**, 719–725.

Miller, W. A. (1971). Early dental development in mice. In *Dental Morphology and Evolution*, ed. A. A. Dahlberg, pp. 31–44. Chicago: University of Chicago Press.

Miller, W. A. (1978). The dentitions of Tabby and Crinkled mice (an upset in mesodermal:ectodermal interaction). In *Development, Function and Evolution of Teeth*, eds. P. M. Butler and K. A. Joysey, pp. 99–109. London: Academic Press.

Mina, M. and Kollar, E. J. (1987). The induction of odontogenesis in non-dental mesenchyme combined with early murine mandibular arch epithelium. *Archives of Oral Biology*, **32**, 123–127.

Mina, M., Gluhak, J., Upholt, W. B., Kollar, E. J. and Rogers, B. (1995). Experimental analysis of Msx-1 and Msx-2 gene expression during chick mandibular morphogenesis. *Developmental Dynamics*, **202**, 195–214.

Mucchielli, M., Mitsiadis, T. A., Raffo, S., Brunet, J., Proust. J. and Goridis, C. (1997). Mouse Otlx2/RIEG expression in the odontogenic epithelium precedes tooth initiation and requires mesenchymal-derived signals for its maintenance. *Developmental Biology*, **189**, 275–284.

Nakamura, S., Stock, D. W., Wydner, K. L., Bollekens, J. A., Takeshita, K., Nagai, B. M., Chiba, S., Kitamura, T., Freeland, T. M., Zhao, Z., Minowada, J., Lawrence, J. B., Weiss, K. M. and Ruddle, F. H. (1996). Genomic analysis of a new mammalian Distal-less gene: *Dlx7*. *Genomics*, **38**, 314–324.

Neubüser, A., Peters, H., Balling, R. and Martin, G. R. (1997). Antagonistic interactions between FGF and BMP signalling pathways: a mechanism for positioning the sites of tooth formation. *Cell*, **90**, 247–255.

Nichols, D. H. (1981). Neural crest formation in the head of the mouse embryo as observed using a new histological technique. *Journal of Embryology and Experimental Morphology*, **64**, 105–120.

Nichols, D. H. (1986). Formation and distribution of neural crest mesenchyme to the first pharyngeal arch region of the mouse embryo. *American Journal of Anatomy*, **176**, 221–231.

Northcutt, R. G. and Gans, C. (1983). The genesis of neural crest and epidermal placodes: a reinterpretation of vertebrate origins. *Quarterly Review of Biology*, **58**, 1–28.

Nüsslein-Volhard, C. (1996). Gradients that organize embryo development. *Scientific American*, **275**, 54–61.

Ohbayashi, N., Hoshikawa, M., Kimura, S., Yamasaki, M., Fukui, S. and Itoh, N. (1998). Structure and expression of the mRNA encoding a novel fibroblast growth factor, FGF-18. *Journal of Biological Chemistry*, **273**, 18161–18164.

Oliver, G., Sosa-Pineda, B., Geisendorf, S., Spana, E. P., Doe, C. Q. and Gruss, P. (1993). Prox 1, a prospero-related homeobox gene expressed during mouse development. *Mechanisms of Development*, **44**, 3–16.

Oliver, G., Gehr, R., Henkins, N. A., Copeland, N. G., Cheyette, B. N. R., Hartenstein, V., Zipursky, S. L. and Gruss, P. (1995). Homeobox genes and connective tissue patterning. *Development*, **121**, 693–705.

OMIM (Online Mendelian Inheritance in Man) (1997). Search of World Wide Web site as of 31 March 1997.

Ooë, T. (1975). La lame dentaire latérale et la niche de l'émail. *Acta Anatomica*, **92**, 259–271.

Oosterwegel, M., van de Wetering, M., Timmerman, J., Kruisbeek, A., Destree, O., Meijlink, F. and Clevers, H. (1993). Differential expression of the HMG box factors TCF-1 and LEF1 during murine embryogenesis. *Development*, **118**, 439–448.

Osborn, H. F. (1888). The evolution of mammalian molars to and from the tritubercular type. *American Naturalist*, **22**, 1067–1079.

Osborn, H. F. (1907). *Evolution of Mammalian Molar Teeth to and from The Triangular Type*, ed. W. K. Gregory. New York: Macmillan.

Osborn, J. W. (1971). The ontogeny of tooth succession in *Lacerta vivipara* Jacquin (1787). *Proceedings of the Royal Society of London*, **179**, 261–289.

Osborn, J. W. (1973). The evolution of dentitions. *American Scientist*, **61**, 548–559.

Osborn, J. W. (1978). Morphogenetic gradients: fields versus clones. In *Development Function and Evolution of Teeth*, eds. P. M. Butler and K. A. Joysey, pp. 171–201. New York: Academic Press.

Osborn, J. W. (1993). A model simulating tooth morphogenesis without morphogens. *Journal of Theoretical Biology*, **165**, 429–455.

Osumi-Yamashita, N., Ninomiya, Y., Doi, H. and Eto, K. (1994). The contribution of both forebrain and midbrain crest cells to the mesenchyme in the frontonasal mass of mouse embryos. *Developmental Biology*, **164**, 409–419.

Panchen, A. L. (1994). Richard Owen and the concept of homology. In *Homology: the Hierarchical Basis of Comparative Biology*, ed. B. K. Hall, pp. 21–62. New York: Academic Press.

Patterson, B. (1956). Early Cretaceous mammals and the evolution of mammalian molar teeth. *Fieldiana (Geology)* **13**, 1–105.

Peterkova, R., Peterka, M., Ruch, J. V. (1993). Morphometric analysis of potential maxillary diastemal dental anlagen in three strains of mice. *Journal of Craniofacial Genetics and Developmental Biology*, **13**, 213–222.

Peterkova, R., Peterka, M., Vonesch, J. L., Tureckova, J., Viriot, L.,

Ruch, J. V., Lesot, H. (1998). Correlation between apoptosis distribution and ABMP-2 and BMP-4 expression in vestigial tooth primordia in mice. *European Journal of Oral Sciences*, **106**, 667–670.

Peters, H. and Balling, R. (1999). Teeth: where and how to make them. *Trends in Genetics*, **15**, 59–65.

Petkovich, M. (1992). Regulation of gene expression by vitamin A: the role of nuclear retinoic acid receptors. *Annual Review of Nutrition*, **12**, 443–471.

Peyer, B. (1968). *Comparative Odontology*. Chicago: University of Chicago Press.

Price, J., Bowden, D., Wright, J., Pettenati, M. and Hart, T. (1998). Identification of a mutation in *DLX3* associated with trichodento-osseous (TDO) syndrome. *Human Molecolar Genetics*, **7**, 563–569.

Proetzel, G., Pawlowski, S. A., Wiles, M. V., Yin, M., Boivin, G. P., Howles, P. N., Ding, J., Ferguson, M. W. and Doetschman, T. (1995). Transforming growth factor-β3 is required for secondary palate fusion. *Nature Genetics*, **11**, 409–414.

Qiu, M., Bulfone, A., Martinez, S., Meneses, J. J., Shimamura, K., Pedersen, R. A. and Rubenstein, J. L. R. (1995). Null mutation of *Dlx-2* results in abnormal morphogenesis of proximal first and second branchial arch derivatives and abnormal differentiation in the forebrain. *Genes and Development*, **9**, 2523–2538.

Qiu, M., Bulfone, A., Ghattas, I., Meneses, J. J., Christensen, L., Sharpe, P. T., Presley, R., Pedersen, R. A. and Rubenstein, J. L. R. (1997). Role of the *Dlx* homeobox genes in proximodistal patterning of the branchial arches: Mutations of *Dlx-1* and *Dlx-2* alter morphogenesis of proximal skeletal and soft tissue structures derived from the first and second arches. *Developmental Biology*, **185**, 165–184.

Quinn, J. C., West, J. D. and Kaufman, M. H. (1997). Genetic background effects on dental and other craniofacial abnormalities in homozygous small eye (*Pax6^{sey}/Pax6^{sey}*) mice. *Anatomy and Embryology*, **196**, 311–321.

Raju, K., Tang, S., Dubé, I. D., Kamel-Reid, S., Bryce, D. M. and Breitman, M. L. (1993). Characterization and developmental expression of *Tlx-1*, the murine homologue of *HOX11*. *Mechanisms of Development*, **44**, 51–64.

Rivera-Pérez, J. A., Mallo, M., Gendron-Maguire, M., Gridley, T. and Behringer, R. R. (1995). *goosecoid* is not an essential component of the mouse gastrula organizer but is required for craniofacial and rib development. *Development*, **121**, 3005–3012.

Robinson, G. W. and Mahon, K. A. (1994). Differential and overlapping expression of *Dlx-2* and *Dlx-3* suggest distinct roles for Distal-less homeobox genes in craniofacial development. *Mechanisms of Development*, **48**, 199–215.

Roessler, E., Belloni, E., Gaudenz, K., Jay, P., Berta, P., Scherer, S. W. Tsui, L-C. and Muenke, M. (1996). Mutations in the human *Sonic Hedgehog* gene cause holoprosencephaly. *Nature Genetics*, **14**, 357–360.

Romer, A. S. (1961). Synapsid evolution and dentition. In *International Colloquium on the Evolution of Mammals*. VI. *Acad. Wetensch. Lett. Sch. Kunsten Belgie, Brussels, part I*, pp. 9–56. Brussels: Paleis der Academien.

Rowe, T. (1993). Phylogenetic systematics and the early history of mammals. In *Mammal Phylogeny*, vol. 1, eds. F. S. Szalay, M. J. Novacek and M. C. McKenna, pp. 129–145. Berlin: Springer.

Rubenstein, J. L. R. and Puelles, L. (1994). Homeobox gene expression during development of the vertebrate brain. *Current Topics in Developmental Biology*, **29**, 1–63.

Ruberte, E., Dolle, P., Chambon, P. and Morriss-Kay, G. (1991). Retinoic acid receptors and cellular retinoid binding proteins. II. Their differential pattern of transcription during early morphogenesis in mouse embryos. *Development*, **111**, 45–60.

Ruberte, E., Freiderich, V., Morriss-Kay, G. and Chambon, P. (1992). Differential distribution patterns of CRABP I and CRABP II transcripts during mouse embryogenesis. *Development*, **115**, 973–987.

Ruch, J. V. (1984). Tooth morphogenesis and differentiation. In *Dentine and Dentinogenesis*, ed. A. Linde, pp. 47–79. Boca Raton: CRC Press.

Ruch, J. V. (1987). Determinisms of odontogenesis. *Cell Biology Review*, RBC **14**, 1–81.

Ruch, J. V. (1995). Tooth crown morphogenesis and cytodifferentiations: candid questions and critical comments. *Connective Tissue Research*, **32**, 1–8.

Ruch, J. V., Lesot, H., Karcher-Djuricic, V., Meyer, J. M. and Olive, M. (1982). Facts and hypotheses concerning the control of odontoblast differentiation. *Differentiation*, **21**, 7–12.

Ruddle, F. H., Bentley, K. L., Murtha, M. T. and Risch, N. (1994). Gene loss and gain in the evolution of the vertebrates. *Development*, **120**, (Suppl), 155–161.

Ruiz i Altaba, A. (1997). Catching a Gli-mpse of Hedgehog. *Cell*, **90**, 193–196.

Satokata, I. and Maas, R. (1994). *Msx1* deficient mice exhibit cleft palate and abnormalities of craniofacial and tooth development. *Nature Genetics*, **6**, 348–356.

Schmitt, R., Lesot, H., Vonesch, J. L. and Ruch, J. V. (1999). Mouse odontogenesis *in vitro*: the cap-stage mesenchyme controls individual molar crown morphogenesis. *International Journal of Developmental Biology*, **43**, 255–260.

Schour, I. (1962). *Oral Histology and Embryology*, eighth edition. Philadelphia: Lea & Febiger.

Schrewe, H., Gendron-Maguire, M., Harbison, M. L. and Gridley, T. (1994). Mice homozygous for a null mutation of activin βB are viable and fertile. *Mechanisms of Development*, **47**, 43–51.

Semina, E. V., Reiter, R., Leysens, N. J., Alward, W. L. M., Small, K. W., Datson, N. A., Seigel-Bartelt, J., Bierke-Nelson, D., Bitoun, P., Zabel, B. U., Carey, J. C. and Murray, J. C. (1997). Cloning and characterization of a novel *bicoid*-related homeobox transcription factor gene, RIEG, involved in Rieger syndrome. *Nature Genetics*, **14**, 392–399.

Serbedzija, G. N., Bronner-Fraser, M. and Fraser, S. E. (1992). Vital dye analysis of cranial neural crest cell migration in the mouse embryo. *Development*, **116**, 297-307.

Sharpe, P. T. (1995). Homeobox genes and orofacial development. *Connective Tissue Research*, **32**, 17–25.

Shimeld, S. M., McKay, I. J. and Sharpe, P. T. (1996). The murine homeobox gene *Msx-3* shows highly restricted expression in the developing neural tube. *Mechanisms of Development*, **55**, 201–210.

Simeone, A., Acampora, D., Pannese, M., D'Esposito, M., Stornaiuolo, A., Gulisano, M., Mallamaci, A., Kastury, K., Druck, T., Huebner, K. and Boncinelli, E. (1994). Cloning and characterization of two members of the vertebrate *Dlx* gene family. *Proceedings of the National Academy of Sciences*, **91**, 2250–2254.

Simpson, G. G. (1961). Evolution of mesozoic mammals. In *International Colloquium on the Evolution of Mammals*. VI. *Acad. Wetensch. Lett. Sch. Kunsten Belgie, Brussels, part I*, pp. 57–95. Brussels: Paleis der Academien.

Slavkin, H. C. (1974). Embryonic tooth formation. A tool for developmental biology. In *Oral Sciences Reviews*, eds. A. H. Melcher and G. A. Zarb, vol. 4. Copenhagen: Munksgaard.

Slavkin, H. and Diekwisch, T. (1996). Evolution in tooth developmental biology: of morphology and molecules. *Anatomical Record*, **245**, 131–150.

Smith, M. M. and Hall, B. K. (1993). A developmental model for evolution of the vertebrate exoskeleton and teeth: the role of cranial and trunk neural crest. *Evolutionary Biology*, **27**, 387–448.

Snead, M., Luo, W., Lau, E. C. and Slavkin, H. C. (1988). Spatial- and temporal-restricted pattern for amelogenin gene expression during mouse molar tooth organogenesis. *Development*, **104**, 77–85.

Sofaer, J. A. (1969a). Aspects of the tabby-crinkled-downless syndrome. I. The development of tabby teeth. *Journal of Embryology and Experimental Morphology*, **22**, 181–205.

Sofaer, J. A. (1969b). Aspects of the tabby-crinkled-downless syndrome. II. Observations on the reaction to changes of genetic background. *Journal of Embryology and Experimental Morphology*, **22**, 207–227.

Sofaer, J. A. (1975). Genetic variation and tooth development. *British Medical Bulletin*, **31**, 107–110.

Solloway, M. J., Dudley, A. T., Bikoff, E. K., Lyons, K. M., Hogan, B. L. M. and Robertson, E. J. (1998). Mice lacking *Bmp6* function. *Developmental Genetics*, **22**, 321–339.

Srivastava, A. K., Pispa, J., Hartung, A. J., Du, Z., Ezer, S., Jenks, T., Shimada, T., Pekkanen, M., Mikkola, M. L., Ko, M. S. H., Thesleff, I., Kere, J. and Schlessinger, D. (1997). The Tabby phenotype is caused by mutation in a mouse homologue of the *EDA* gene that reveals novel mouse and human exons and encodes a protein (ectodysplasin-A) with collagenous domains. *Proceedings of the National Academy of Sciences, USA*, **94**, 13069–13074.

Stock, D. W., Ellies, D. L., Zhao, Z., Ekker, M., Ruddle, F. H. and Weiss, K. M. (1996a). The evolution of the vertebrate *Dlx* gene family. *Proceedings of the National Academy of Sciences*, **93**, 10858–10863.

Stock, D. W., Buchanan, A. V., Zhao, Z. and Weiss, K. M. (1996b). Numerous members of the *Sox* family of HMG box-containing genes are expressed in mouse teeth. *Genomics*, **37**, 234–237.

Stock, D. W., Weiss, K. M. and Zhao, Z. (1997). Patterning of the mammalian dentition in development and evolution. *Bioessays* **19**, 481–490.

Summerbell, D., Lewis, J. H. and Wolpert, L. (1973). Positional information in chick limb morphogenesis. *Nature*, **244**, 492–496.

Ten Cate, A. R. (1995). The experimental investigation of odontogenesis. *International Journal of Developmental Biology*, **39**, 5–11.

Thesleff, I. (1996). Two genes for missing teeth. *Nature Genetics*, **13**, 379–380.

Thesleff, I and Jernvall, J. (1998). The enamel knot: a putative signalling detection center regulating tooth development. *Cold Spring Harbor Symposia on Quantitative Biology*, **LXII**, 257–267.

Thesleff, I., Vaahtokari, A., Vainio, S. and Jowett, A. (1996). Molecular mechanisms of cell and tissue interactions during early tooth development. *Anatomical Record*, **245**, 151–161.

Thomas, B. L. and Sharpe, P. T. (1998). Patterning of the murine dentition by homeobox genes. *European Journal of Oral Sciences*, **106** (Supplement 1), 48–54.

Thomas, B. L., Tucker, A. S., Qiu, M., Ferguson, C. A., Hardcastle, Z., Rubenstein, J. L. R. and Sharpe, P. T. (1997). Role of *Dlx-1* and *Dlx-2* genes in patterning of the murine dentition. *Development*, **124**, 4811–4818.

Thomas, B. L., Tucker, A. S., Ferguson, C., Qiu, M., Rubenstein, J. L. R. and Sharpe, P. T. (1998). Molecular control of odontogenic patterning: positional dependent initiation and morphogenesis. *European Journal of Oral Sciences*, **106** (Suppl. 1), 44–47.

Tissier-Seta, J.-P., Mucchielli, M.-L., Mark, M., Mattei, M.-G., Goridis, C. and Brunet, J.-F. (1995). *Barx1*, a new mouse homeodomain transcription factor expressed in cranio-facial ectomesenchyme and the stomach. *Mechanisms of Development*, **51**, 3–15.

Trainor, P. and Tam, P. (1995). Cranial paraxial mesoderm and neural crest cells of the mouse embryo: co-distribution in the craniofacial mesenchyme but distinct segregation in branchial arches. *Development*, **121**, 2569–2582.

Tucker, A. S., Al Khamis, A. and Sharpe, P. T. (1998). Interactions between *Bmp-4* and *Msx-1* act to restrict gene expression to odontogenic mesenchyme. *Developmental Dynamics*, **212**, 533–539.

Turing, A. (1952). The chemical basis of morphogenesis. *Philosophical Transactions of the Royal Society of London, Series B*, **237**, 37–72.

Vaahtokari, A., Åberg, T., Jernvall, J., Keränen, S. and Thesleff, I. (1996a). The enamel knot as a signalling center in the developing mouse tooth. *Mechanisms of Development*, **54**, 39–43.

Vaahtokari, A., Åberg, T. and Thesleff, I. (1996b). Apoptosis in the developing tooth: association with an embryonic signalling center and suppression by EGF and FGF-4. *Development* **122**, 121–129.

Vainio, S., Karavanova, I., Jowett, A. and Thesleff, I. (1993). Identification of BMP-4 as a signal mediating secondary induction between epithelial and mesenchymal tissues during early tooth development. *Cell*, **75**, 45–58.

Vandebroek, G. (1961). The comparative antomy of the teeth of lower and non-specialized mammals. In: *International Colloquium on the Evolution of Mammals*. VI. *Acad. Wetensch. Lett. Sch. Kunsten Belgie, Brussels, part I*, pp. 215–320. Brussels: Paleis der Academien.

Vandebroek, G. (1967). Origin of the cusps and crests of the tribosphenic molar. *Journal of Dental Research*, **46**, (Supplement), 796–804.

van Genderen, C., Okamura, R., Farinas, I., Quo, R-G., Parslow, T., Bruhn, L. and Grosschedl, R. (1994). Development of several organs that require inductive epithelial-mesenchymal interactions is impaired in *LEF1*-deficient mice. *Genes and Development*, **8**, 2691–2703.

van Valen, L. (1970). An analysis of developmental fields. *Developmental Biology*, **23**, 456–477.

Vastardis, H., Karimbux, N., Guthua, S. W., Seidman, J. G. and Seidman, C. E. (1996). A human *MSX1* homeobox missense mutation causes seletive tooth agenesis. *Nature Genetics*, **13**, 417–421.

Wang, W., Chen, X., Xu, H. and Lufkin, T. (1996). *Msx3*, a novel murine homologue of the *Drosophila msh* homeobox gene restricted to the dorsal embryonic central nervous system. *Mechanisms of Development*, **58**, 203–215.

Webster, G. and Goodwin, B. (1996). *Form and Transformation*. Cambridge: Cambridge University Press.

Weiss, K. M. (1990). Duplication with variation: metameric logic in evolution from genes to morphology. *Yearbook of Physical Anthropology*, **33**, 1–23.

Weiss, K. M. (1993). A tooth, a toe, and a vertebra: the genetic dimensions of complex morphological traits. *Evolutionary Anthropology*, **2**, 121–134.

Weiss, K. M., Bollekens, J. Ruddle, F. H. and Takashita, K. (1994). Distal-less and other homeobox genes in the development of the dentition. *Journal of Experimental Zoology*, **270**, 273–284.

Weiss, K. M., Ruddle, F. H. and Bolleken, J. A. (1995). *Dlx* and other homeobox genes in the morphological development and evolution of the dentition. *Connective Tissue Research*, **31**, 1–6.

Weiss, K. M., Stock, D. W. and Zhao, Z. (1998a). Dynamic interactions and the evolutionary genetics of dental patterning. *Critical Reviews in Oral Biology and Medicine*, **9**, 369–398.

Weiss, K., Stock, D., Zhao, Z., Buchanan, A., Ruddle, F. and Shashikant, C. (1998b). Perspectives on genetic aspects of dental patterning. *European Journal of Oral Sciences*, **106** (Supplement 1), 55–63.

Westergaard, B. and Ferguson, M. W. J. (1986). Development of the dentition in *Alligator mississippiensis*. Early embryonic development in the lower jaw. *Journal of Zoology*, **210**, 575–597.

Westergaard, B. and Ferguson, M. W. J. (1987). Development of the dentition in *Alligator mississippiensis*. Later embryonic development in the lower jaw of embryos, hatchlings and young juveniles. *Journal of Zoology*, **212**, 191–222.

Winnier, G., Blessing, M., Labosky, P. A. and Hogan, B. L. M. (1995). Bone morphogenetic protein-4 is required for mesoderm formation and patterning in the mouse. *Genes and Development*, **9**, 2105–2116.

Witter, K., Misek, I., Peterka, M. and Peterkova, R. (1996). Stages of odontogenesis in the field vole (*Microtus agrestis*, Rodentia) – a pilot study. *Acta Veterinaria Brno*, **65**, 285–296.

Wolpert, L. (1969). Positional information and the spatial pattern of cellular differentiation. *Journal of Theoretical Biology*, **25**, 1–47.

Wozney, J. M. (1998). The bone morphogenetic protein family: multifunctional cellular regulators in the embryo and the adult. *European Journal of Oral Sciences*, **106** (Supplement 1), 160–166.

Wright, E., Hargrave, M. R., Christiansen, J., Cooper, L., Kun, J., Evans, T., Gangadharan, U., Greenfield, A. and Koopman, P. (1995). The *Sry*-related gene *Sox9* is expressed during chondrogenesis in mouse embryos. *Nature Genetics*, **9**, 15–20.

Yamada, G., Mansouri, A., Torres, M., Stuart, E. T., Blum, M., Schultz, M., De Robertis, E. M. and Gruss, P. (1995). Targeted mutation of the murine *goosecoid* gene results in craniofacial defects and neonatal death. *Development*, **121**, 2917–2922.

Yamaguchi, T. P. and Rossant, J. (1995). Fibroblast growth factors in mammalian development. *Current Opinion in Genetics and Development*, **5**, 485–491.

Zhang, H. and Bradley, A. (1996). Mice deficient for BMP2 are nonviable and have defects in amnion/chorion and cardiac development. *Development*, **122**, 2977–2986.

12 Evolution of tooth attachment in lower vertebrates to tetrapods

P. Gaengler

12.1. Introduction

During the evolution of teeth, different types of periodontal attachment have been developed. It is likely that teeth have either evolved from conodont elements (Figure 12.1) or from denticles of the odontode type (Figure 12.2) in the earliest agnathic chordates (Krejsa and Slavkin, 1987; Smith and Hall, 1990) before the dermal armour and tooth-like placoid scales appeared during the evolution of fishes (Gross, 1966). A new developmental model has been proposed on the basis of separation of odontogenic from osteogenic and chondrogenic neural crest potential (Smith and Hall, 1993). It is explained as a modular development, where each interactive morphogenetic unit may be commited to form either cartilage, bone, odontodes, or teeth (see Chapter 10).

The mode of tooth attachment in lower vertebrates constitutes a rather small spectrum which ranges from the loose fibrous type seen in the elasmobranchs to solid ankylosis between the tooth and the underlying jaw (Peyer, 1968). However, other highly complex types of attachment structures were developed mainly in bony fishes at a relatively late stage in the evolution of teeth. The mammalian periodontium could be regarded as the most complex of all types of tooth attachment, particularly in regard to structure and function (Shellis, 1982; Osborn, 1984; Berkovitz et al., 1992, 1995). However, even in the thecodont mammals there are major differences in periodontal reactivity between rodent, carnivorous, herbivorous and omnivorous dentitions because of apparent differences in the fine structure of the periodontium, in the mode of tooth replacement and in the eruption and growth rate of teeth. Therefore, the phylogenetic development of more and more complex tooth attachment structures relates to the variations in:

Form and function of teeth (homodont dentition, specialized homodont dentition, heterodont dentition)

Tooth replacement pattern (polyphyodonty, monophyodonty, diphyodonty, semidiphyodonty)

Eruption patterns, growth rate and mode of abrasion and attrition of teeth (continuously growing teeth, slow but continuous eruption, no continuous eruption).

In this sense, the traditional classification of tooth attachment into three types (acrodont, pleurodont, thecodont) does not fully reflect the biological variations encountered. A great variety of direct union of the tooth to bone (ankylosis) and of fibrous tooth attachments is detectable in the classes of agnatha, chondrichthyes, osteichthyes, amphibia, reptilia and mammalia.

The evolutionary origin of the attachment mechanism in odontodes is well described. Ørvig (1967) suggested the term odontode and Reif (1982) developed this odontode concept including dermal denticles and teeth. This concept was followed by Smith and Hall (1993) including the 'bone of attachment' within the fundamental odontogenic unit. In contrast to this concept the homology of conodont elements with vertebrate denticles, and perhaps teeth, is still disputed, and little is known about attachment mechanisms (Slavkin et al., 1991; see Chapter 10).

Osborn (1984) introduced a terminology of different types of tooth attachment beginning with an acrodont ankylosis, two types of protothecodont ankylosis, pleurodont ankylosis and ending with a gomphosis. In the thecodont ankylosis the roots of teeth are embedded in the jaw, whereas the thecodont gomphosis characterizes

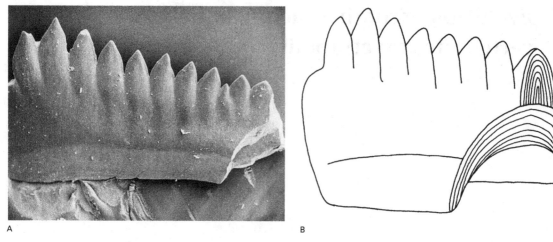

Figure 12.1A,B. Conodonts: earliest craniate teeth. A. Compound conodont element from Lower Carbonian (Serpuchov, Russia, × 250). B. Schematic structure of a compound conodont element.

the socketed teeth of alligators and crocodiles. Osborn (1984) also distinguished hard tissues taking part in the tooth attachment of non-mammalian vertebrates. Open-ended cylinders of mineralized tissue formed by cells differentiated from the base of the dental papilla were called attachment dentine. Protocement is developing on the outer surface of the root base and the bone of attachment is formed on the adjacent walls of bone surrounding the tooth base. On the base of the concept of differentiation clades (clones), where groups of identical cells have the same probability of growing and differentiating, Osborn (1984, 1993) suggested predictable homologies between the attachment tissues in reptiles and mammals. Attachment dentine disappears, protocement could be homologous with cement and the bone of attachment with alveolar bone.

A phylogenetic classification that takes into account the variations of the type of tooth attachment, type of cementum, formation of alveolar bone and form of epithelial attachment reflects the evolution of the periodontium in its broadest sense in the phylogeny of the dentition (Gaengler and Metzler, 1988, 1989, 1990, 1992).

12.2.   Variations of tooth attachment

12.2.1.   Protoacrodonty (odontodes or odontode generations)

The dermal bone of some Paleozoic species (Silurian agnathans; see Chapter 5) is composed of the first dentine-like structures (Müller, 1985). The odontodes are attached by acellular or cellular bone as part of the dermal armour incorporating sensory organs and protecting the underlying soft tissue (Figure 12.2). The prin-

cipal features include a superficial layer of odontodes, a vascular bone layer and a basal laminar bone layer. The protoacrodont attachment represents the typical pattern of all agnathans.

The odontode developmental unit gave rise to both dermal denticles and teeth (Reif, 1982). Smith and Hall (1993) included the bone of attachment, which is also part of each placoid denticle in thelodonts and sharks, because as these fish do not have dermal bone it is the basis for attachment of the soft tissues. Odontodes and, perhaps, the conodont elements support the developmental model that in agnathan fish some oropharyngeal denticles could have functioned as teeth before jaws had evolved (see Chapter 10).

12.2.2.   Acrodonty

True acrodont attachment represents the direct bony or fibrous attachment of teeth to the crest of the jaws by means of collagenous fibres or mineralized connective tissue. Indeed, both types of attachment may be found in the same jaw. The connective tissue matrix may be produced by either osteoblasts or odontoblasts. This is seen in the acanthodians (Denison, 1979) and other osteichthyan fish recorded first in the Early Silurian. Acanthodians and placoderms are well-documented gnathostomes with thin perichondral bone layers in the endocranium and the visceral skeleton (Denison, 1978; see Janvier, 1996. p. 274 for a recent account of the evolution and polarity of transformation of these early vertebrate skeletal tissues). According to Smith and Coates (1998; Chapter 10) it is likely that the origins of jaws and teeth are relatively separate events. Therefore, the acrodonty could represent the first coupling of the jaw cartilages and bones with teeth or tooth-like dermal bones.

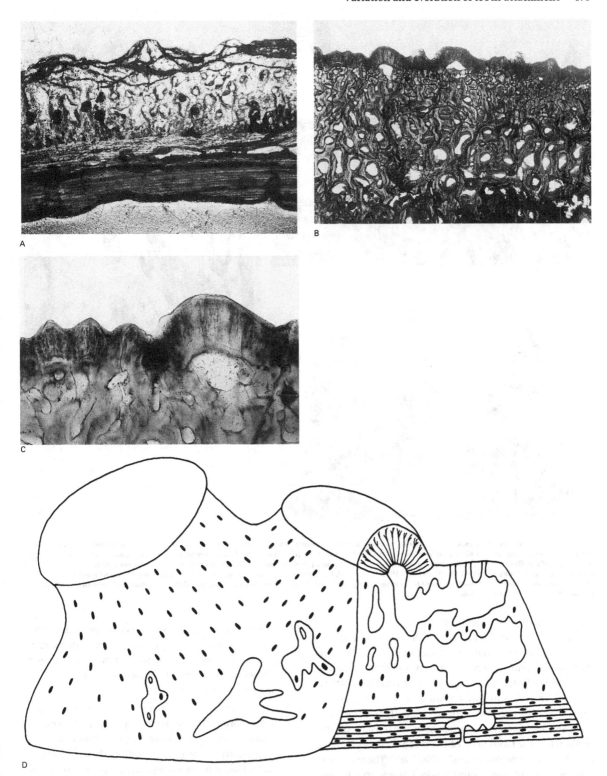

Figure 12.2A–D. **Protoacrodont. A. Ground, longitudinal section through a dermal bone of** *Holoptychius* **(Upper Devonian, Fram 3, River Pascha, Kuprava, Latvia) with a superficial bone layer, an intermediary vascular bone layer and a basal lamellar bone layer (× 75). B. Ground, longitudinal section through an odontode element of** *Glyptolepis* **(Upper Devonian, Fram 2, Stolbovo, district St Petersburg, Russia) with a superficial layer of pallial dentine, an intermediary vascular bone layer and a basal lamellar bone layer (× 75). C. Higher magnification of odontodes of Glyptolepis demonstrating the pallial dentine (× 300). D. Schematic structure of the protoacrodont.**

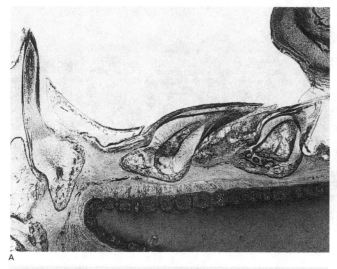

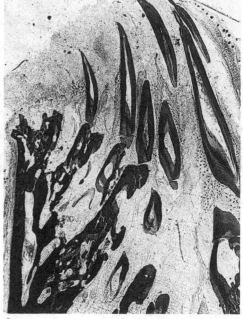

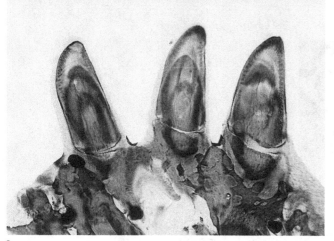

Figure 12.3A–C. **Acro-protothecodont. A. Ground, transverse section through the jaw of a shark (*Carcharhinus menisorrha*) demonstrating the highly specialized continuous fibrous attachment of the teeth to a connective tissue membrane overlying and moving around the jaw cartilage (H and E, × 16). B. Ground, longitudinal section through the dentition of *Pomacanthus paru* (Osteichthyes), the teeth are attached with concentric fibres to a cylindrical base (H and E, × 150). C. Ground, longitudinal section through the teeth of *Caranx* (Osteichthyes) demonstrating concentric fibre attachment to an intervening cylindrical base of mineralized tissue which is attached by fibres to a shallow groove in the jaw (H and E, × 300).**

12.2.3.   Acro-protothecodonty

The acro-protothecodont is the attachment type with the greatest biological variation (Figures 12.3, 12.4). The principal attachment of teeth to the jaws is more complex compared with the primitive acrodont condition. There is a protothecodontal fibre apparatus attaching the teeth to the crest of the jaw, with or without, intermediate pedicles. The pedicles develop as an extension of the teeth, while the basal bone of the tooth serves as bone of attachment to the jaw bone (Shellis, 1982). There is great variation in the acro-protothecodont tooth attachment which is characteristic for most species in the whole class of fishes (Zangerl, 1978; Schneider, 1987). It is proposed to subdivide this type into six groups (Figure 12.5):

1. Dentine of the tooth is attached on one side to the jaw bone with fibres and the trabecular dentine in the pulpal tissue is attached to bone with collagen fibres which insert into non-mineralized predentine matrix.

2. The tooth is attached on one side to the jaw bone with fibres and is supported by an outer layer of partially mineralized dentine. This creates a hinged attachment. This modification of fibrous attachment is present in a number of carnivorous fishes and allows a great range of movement.

3. The tooth has a bony or osteodentine base which is attached to the jaw (bone or cartilage) by fibres

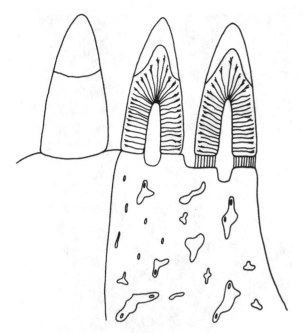

Figure 12.4. **Schematic structure of the fibrous and bony acrodont attachment.**

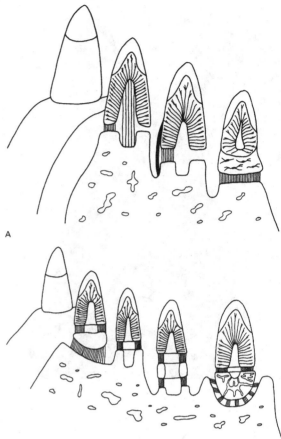

A

B

Figure 12.5A,B. **Schematic structures of the acro-protothecodontal attachment demonstrating the different groups. A. Group 1–3 (see text). B. Group 4–6 (see text).**

(e.g. sharks). These elasmobranch fishes exhibit a highly specialized continuous fibrous attachment of the teeth to a rather thick membrane of connective tissue (perichondrium) thought to move around the jaw cartilage.

4. The tooth is attached with concentric fibres to a cylindrical base of dentine (pedicel) which represents an extension of the tooth. There is a sharp boundary between the jaw bone and the pedicel, but they are fused together.

5. The tooth is connected to the bone with two narrow rings of fibrous tissue either side of a cylindrical base of mineralized tissue, aligned with the tooth (pedicel of dentine). There are concentric fibres between the tooth and this base, and between this base and the jaw.

6. The tooth is connected by concentric fibres to a solid cellular or acellular bone base. There is a primitive thecodontal attachment of this structure to a shallow groove in the jaw.

12.2.4. Pleurodonty

This refers to a bony and/or fibrous attachment of the tooth to the inner margin of the bone of the jaw (Bystrow, 1938, 1939). The teeth are homodont, haplodont and polyphyodont. They have a conical shape as in most fishes (Figure 12.6). Recent amphibian teeth, where

present, are small and their attachment to underlying bone is moving from the crest (acrodont) to the inner margin (pleurodont).

12.2.5. Thecodonty

Reptiles (archaeosaurs of the Permian period; recent, e.g. crocodiles) are the first group whose teeth lie in true sockets. This is a thecodont gomphosis in which (unlike mammals) successive teeth occupy the same socket.

All placental mammalian teeth are socketed. The attachment of teeth to the jaw is accomplished by means of a fibrous ligament (Westergaard, 1983). The varieties of acro-protothecodont pedicles and bone-of-attachment structures were converted into the unique thecodontal principal construction consisting of root cementum, periodontal ligament and alveolar bone (Osborn, 1984). However, little is known about the stages in the evolutionary process and the homologies of the cementum and periodontal ligament are unknown. Osborn (1984)

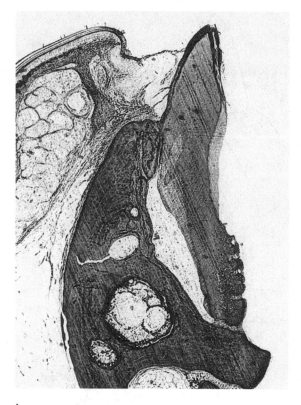

A

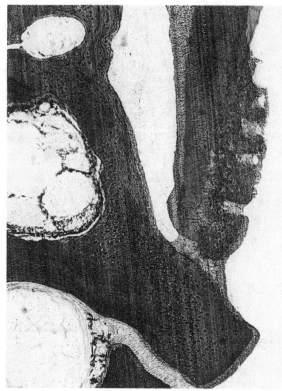

B

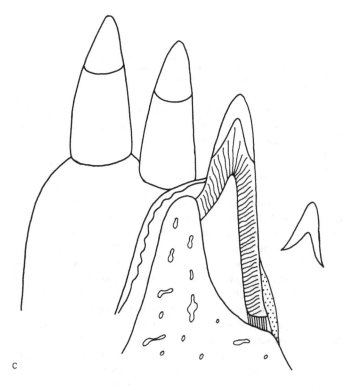

C

Figure 12.6A. **Ground, longitudinal section through a tooth in a lizard (*Lacerta trilineata*, Reptilia), note the bony attachment at the crest (H and E, ×75). B. Higher magnification of the tooth base surrounded with mineralized cementoid tissue and attached to bone with a primitive fibre apparatus (H and E, ×300). C. Schematic structure of the pleurodontal attachment to the inner margin of the jaw bone.**

predicted homologies between the attachment tissues in reptiles and mammals, where 'protocement' could be homologous with cementum and the bone of attachment with the alveolar bone formed by the follicle. Smith and Hall (1990) proposed that the Sharpey's fibre-containing basal tissue of thelodont scales was a form of cementum. The evolution of the thecodont attachment originated in the Permian period about 270 million years ago in a group of mammal-like reptiles (Hopson, 1964) and described by Osborn (1984) as a so-called 'mixed ankylosis /gomphosis'.

The different varieties of cement are decisive for the principal structure of tooth attachment, because the thecodont cement tissues are formed in different ways and consist of acellular-afibrillar, acellular-fibrillar or cellular-fibrillar structures (Schroeder and Listgarten, 1977). According to the presence or absence of cells and to the source of collagen fibres, the classification of root cementum distinguishes between acellular afibrillar cementum, acellular extrinsic fibre cementum, cellular mixed stratified cementum, and cellular intrinsic fibre cementum (Schroeder, 1986).

The range of variation in the thecodont attachment seems to be rather small compared with that in non-thecodont attachments. However, the morphogenesis of the crown with six patterns of heterodont variations in rodent, carnivorous, herbivorous and omnivorous dentitions is strongly associated with the specific structure and function of the periodontal attachment mechanisms (Gaengler and Metzler, 1992). This classification of six morpho-functional groups is based on the invagination hypothesis concerning different invagination patterns of the dental epithelial-mesenchymal tissue complex during the bell stage in the odontogenesis of rodent, carnivorous, herbivorous and omnivorous dentitions (Gaengler, 1986).

The structural basis of differentiation of tooth attachment and the variations range from moderate to extreme invaginations of the dental epithelial-mesenchymal tissue complex, through partial, or complete reduction of enamel, or **rotation**. Therefore, six groups of the periodontium can be distinguished (Figure 12.7):

1. Single-rooted teeth with no, or only moderate, invagination of dental epithelium during the bell stage of development, and a combination of acellular afibrillar cementum, acellular extrinsic fibre cementum, cellular mixed stratified cementum and cellular intrinsic fibre cementum (incisors and most canines of all omnivorous, carnivorous and herbivorous dentitions).

2. Multirooted teeth with moderate invagination of dental epithelium, and a combination of mainly acellular extrinsic fibre cementum and cellular, mixed stratified cementum (premolars and molars of all omnivorous and carnivorous dentitions).

3. Multirooted teeth with deep invaginations of dental epithelium and the development of coronal cementum, and with considerable formation of cellular, mixed stratified root cementum as the main mechanism of continuous eruption (non-continuously growing premolars and molars of all herbivorous dentitions).

4. Continuously growing teeth with a periodontal ligament only at the cementum-covered aspect of the tooth (incisors of all rodents, lagomorphs and some other mammals, some molars, e.g. guinea pig and rabbit).

5. Teeth with secondary loss of enamel, and with a circumferential periodontal ligament and a thin layer of acellular extrinsic fibre cementum (all dolphins, whales and armadillos).

6. Multirooted teeth with partial loss of enamel (only at the cusps), considerable formation of cellular, mixed stratified root cementum (apically) as part of continuous eruption (some molars of rodent dentitions, e.g. rats).

12.3.   Variations of epithelial integrity

In the various types of tooth attachment, the junctional epithelium of gingiva attaches to different structures. Epithelial integrity means that there is a direct union of epithelial cells to the ectodermally derived structure of enamel. In teeth lacking enamel the junctional epithelium attaches to cementum and/or dentine. From a biological point of view this is a unique 'gap' in the epithelial integrity (Gaengler, 1986). The following patterns can be distinguished.

12.3.1.   Acrodontal, acro-protothecodontal and pleurodontal attachment

1. Junctional epithelium attaches to pallial dentine structures such as enameloid, which is hypermineralized with respect to the underlying dentine. This is characteristic for most fish and amphibian teeth.

2. Epithelial integrity is evident in reptilian and in some adult amphibian teeth because of the epithelial attachment to ectodermally derived enamel.

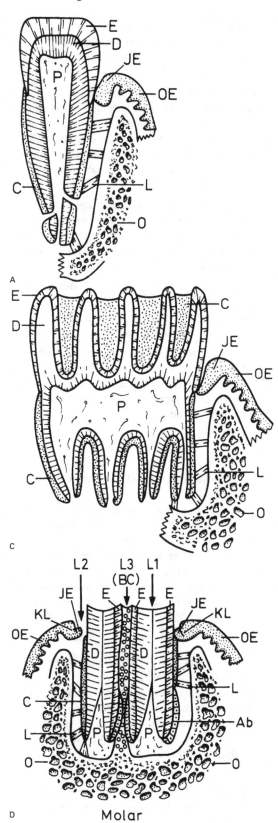

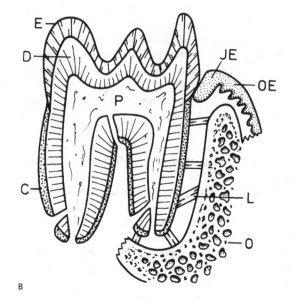

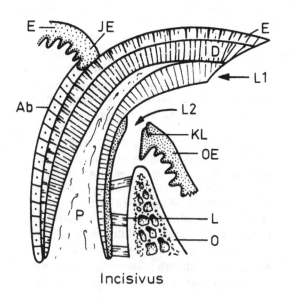

Molar

Incisivus

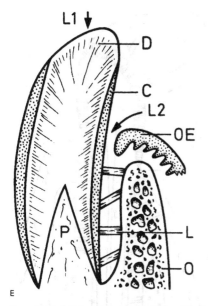

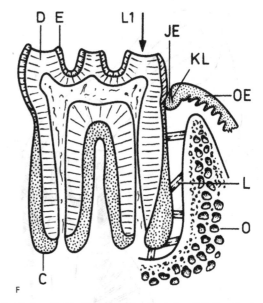

Figure 12.7A–F. **Schematic representation of the invagination hypothesis of heterodonty with six patterns of variation. A. Pattern 1: no invagination. B. Pattern 2: slight invagination. C. Pattern 3: multiple invagination, with the formation of crown cementum. D. Pattern 4: partial reduction of enamel, with no invagination (incisors) or with multiple invagination and rotation (molars). E. Pattern 5: total reduction of enamel with no invagination representing secondary homodonty. F. Pattern 6: partial reduction of enamel with moderate invagination. E, enamel; D, dentine; P, pulp; C, cementum; JE, junctional epithelium; OE, oral epithelium; KL, keratine lamella; L, periodontal ligament; O, alveolar bone; Ab, ameloblasts; L1–L3, 'gaps' in the epithelial integrity. (From Gaengler, 1995).**

12.3.2.   Thecodontal attachment

1. Junctional epithelium attaches to prismatic enamel creating a long-lasting epithelial integrity of high regenerative potential (Figures 12.8, 12.9).

2. Epithelial attachment to enamel plus to coronal cementum (overlap of enamel by cementum).

3. Epithelial attachment to enamel in slow but continuously erupting teeth where the oral gingival epithelium is forming a downgrowing keratin lamella separating attaching cells of the junctional epithelium (e.g. in rat molars).

4. Loss of epithelial integrity, attachment of junctional epithelium to cementum or dentine on one aspect in all continuously growing teeth (Figure 12.10) and in teeth with enamel reduction (tusks).

12.4.   **Evolutionary considerations**

The evolution of the mammalian periodontium including the development of a root is separated by a long period of geological time from the condition in basal vertebrate dermal armour, with a primitive direct union to bone, or some of the more complex acro-protothecodont structures seen in fish teeth. Cartilaginous and bony fishes demonstrate clearly that teeth need specialized attachment structures. The higher specialization in teeth of marsupials and placental mammals is associated with a thecodont gomphosis type of tooth attachment. Continuous growth of mammalian teeth, the slow but continuous eruption of teeth in most herbivorous dentitions, and carnivorous, or omnivorous, dietary behaviour are features that can be correlated to the main structural elements of tooth support: cementum, periodontal ligament and alveolar bone. It is well documented that the turnover of periodontal structures, the patterns of degeneration and regeneration, the features of periodontal diseases and, last but not least, the results of experimental procedures differ greatly from species to species (Page and Schroeder, 1982; Gaengler et al., 1988; Hoffmann et al., 1988; Metzler, 1990). Nevertheless there are unique features characterizing more or less all variations of tooth attachment in health and disease.

The *protoacrodont* is composed of the first dentine-like structures of odontodes or odontode generations. It may be distinguished between skin denticles, oral denticles, and teeth of the jaws and pharyngeal arches. Even the bone of attachment could be attributed to the odontode structure (Smith and Hall, 1993).

The true *acrodont* represents the ankylosis or fibrous attachment of teeth to the crest of the jaw.

Very early during the evolution of tooth attachment the *acro-protothecodont* condition exhibits different attach-

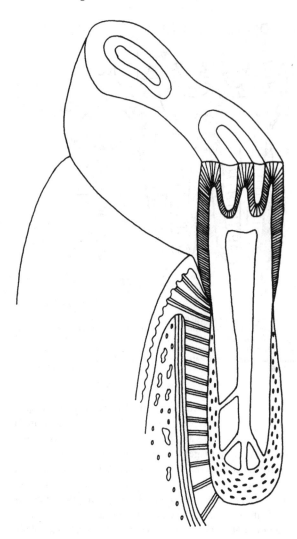

Figure 12.8. **Schematic structure of the thecodontal attachment in herbivorous dentitions.**

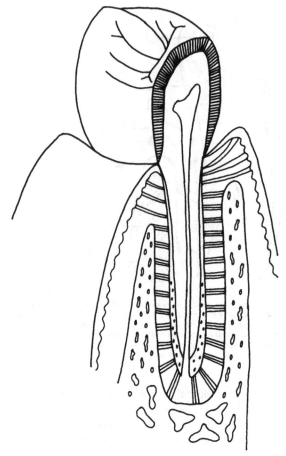

Figure 12.9. **Schematic structure of the thecodontal attachment in omnivorous (and carnivorous) dentitions.**

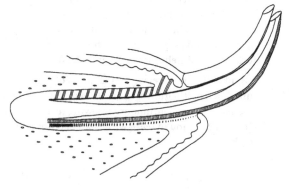

Figure 12.10. **Schematic structure of the thecodontal attachment in rodent dentitions.**

ment types with the greatest biological variation involving mainly five tissues: dentine, bone of attachment, fibre apparatus, jaw bone and junctional epithelium. The mode of dentine attachment in chondrichthyans and osteichthyans constitutes a spectrum which ranges from the osteodentine base seen in the elasmobranchs, to a flexible non-mineralized predentine matrix in bony fishes and amphibians. The bone of attachment represents the biomineralization of the tooth supporting connective tissue, and osteoblasts take part in the matrix secretion. Apart from the bone of attachment the pedicel develops as an extension of the dentine tooth. Great variations exist in the form and in the composition of pedicels and the bone of attachment ranging from a completely separated cylindrical base to a solid cellular or acellular bony base. This leads to some doubts as to

which structures represent those homologous with the thecodont cementum, periodontal ligament, and lining alveolar bone. Although Osborn (1984) proposed a hypothetical sequence showing the evolution of the thecodont attachment from an acrodont ankylosis, where an intervening mineralized tissue 'protocement' could be homologous with cement and the bone of attachment with the alveolar bone lining the socket.

The fibre apparatus in the acro-protothecodont is well developed and in some cases highly specialized, as demonstrated for example in the hinged attachment in a number of carnivorous fishes. The junctional epithelium attaches mainly to pallial dentine structures such as enameloid. Therefore, the epithelial attachment creates a unique 'gap' in the epithelial integrity. In a number of bony fishes the bone of attachment is located in a shallow or deeper groove resembling a primitive *protothecodont attachment* by the fibre apparatus to the jaw bone.

In *pleurodonty* isolated, in-between, mineralized attachment structures disappear and ankylosis and fibre attachment are combined. However, there is an intermediary hard tissue which occurs on the *outside* of dentine. It was described as cementoid around the basal part of the tooth (Gaengler and Metzler, 1992). This is in contrast to the development of the acrodont and pleurodont ankylosis which occurs as a continuation of hard tissue formation on the *inside* of the pulp until the gap between the dentine and the jaw bone is filled with cellular bone (Howes, 1979). The junctional epithelium in reptilian and adult amphibian teeth attaches to ectodermally derived aprismatic enamel. Therefore, this is the first time in the evolution of tooth attachment that the epithelial integrity is evident. Recently it has been discussed that enameloid has an epithelial contribution as afibrillar acellular cementum may have (Smith and Hall, 1993). Nevertheless, the very specific attachment mechanism of junctional epithelium to ectodermally derived enamel during the tooth eruption is a relatively late event in the evolution of tooth attachment (Schroeder and Listgarten, 1977) and the ability of junctional epithelium to attach to non-ectodermally derived cementum or dentine (or even to dental calculus) is a common feature in all thecodont dentitions (Gaengler, 1986; Gaengler et al., 1988; Lehm and Gaengler, 1990).

The evolutionary selection for six groups of the *thecodont* type of tooth attachment was an important event in the phylogeny of mammals. These six patterns correlate directly with the different forms (heterodonty) and functions of the specialized teeth according to the invagination hypothesis (Gaengler, 1986).

The continuous tooth replacement in non-mammalian vertebrates is seemingly more related to permanent body growth than to any other functional influences such as tooth wear (see Chapter 13). The change from polyphyodonty to diphyodonty and the further development of heterodonty may be attributed to the decrease in the growth rate of mammals (Osborn, 1993).

The functional and morphological consequences of diphyodonty and semidiphyodonty (as in humans with 32 permanent teeth, but only 20 replaced teeth) are, therefore, four main mechanisms of tooth eruption, of abrasion and attrition, corresponding to the mode of thecodont attachment:

1. Carnivorous dentitions with practically no wear and no continuous eruption

2. Herbivorous dentitions with extended wear over the lifetime and slow but continuous eruption

3. Omnivorous dentitions with less wear and very slow but continuous eruption

4. Rodent dentitions (and some non-rodent teeth) with continuous growth of some (or all) teeth.

Therefore, the thecodont gomphosis is the main modifying factor of different eruption mechanisms (for review, see Berkovitz, 1990). In very slow and slow but continuously erupting teeth, the growth into stable occlusal conditions is due to the development and the composition of cementum and to the remodelling of the alveolar bone. Even in the human dentition this slow but continuous eruption over the whole lifetime is well documented (Ainamo and Ainamo, 1984). In permanently growing teeth there is a need for reduction of enamel, either by rotation of the tooth and/or by secondary loss of enamel. The partial reduction of dental epithelium is associated with the reduction of the periodontal ligament which is attached only at the cementum-covered aspect of the tooth. The total reduction of dental epithelium and, therefore, the secondary loss of enamel has no consequences for the fully structured circumferential periodontal ligament.

The normal eruption rate and the normal permanent growth of teeth are determining factors of the migration rate of the periodontal ligament, of the development of cementum and of alveolar bone remodelling. These features are, without a doubt, decisive for the regeneration potential of the lifelong functioning diphyodont dentition (Gaengler and Merte, 1983; Hoffmann et al., 1988; Lehm and Gaengler, 1990).

Because of the 'phylogenetic memory' and of the genetic information remaining quiescent in the genome, perturbations of tissue interactions can alter the gene expression. This has been demonstrated by the ability of chick epithelium combined with mouse molar mesenchyme to participate in odontogenesis (Kollar and Fischer, 1980). It is likely that all pathological features developing during the natural history of periodontal diseases and also the regenerative patterns are perturbations of the intrinsic conserved pattern of tooth attachment or tooth replacement. Degenerative changes and the regenerative potential are different in various types of dentition. Even in thecodont mammals there are major differences in periodontal reactivity between rodent, carnivorous, herbivorous and omnivorous dentitions because of apparent differences in the fine structure of the periodontium and in the eruption and growth rate of teeth.

Pathological reactions of the mammalian tooth support like ankylosis, recession of gingiva, attachment of junctional epithelium to dentine and/or cementum, downgrowth of pocket epithelium, root resorption, reduction of periodontal ligament, remodelling of alveolar bone and hypercementosis may be reverting to older phylogenetic patterns of tooth attachment. This is why comparative periodontology may provide an important contribution in elucidating the natural history of periodontal diseases. From an evolutionary standpoint, periodontitis could be directly related to the effects of reduced attrition and abrasion of teeth (Ainamo, 1972; Ainamo and Ainamo, 1984). The less the wear, and the less the functional eruption, the more periodontal inflammatory destruction has a chance to be maintained. The slow but continuous eruption seen throughout life as response to attrition (and occurring both occlusally and interproximally) could be a major factor in preventing periodontal breakdown and dental caries (Gaengler, 1991). There is no doubt that periodontal diseases (and dental caries) are infectious in nature, caused by plaque bacteria and associated with other local influences and, mainly, with host responses. There are many factors contributing to the disease progression or stagnation, the natural level of tooth function due to the different structures of the thecodont attachment being one of the important conditions.

Ancient human populations show high occlusal and approximal wear rates (Levers and Darling, 1983), and the prevalence of periodontal diseases is more related to age, wear and local immunological factors and has no correlation with carbohydrate-rich dietary influences (Gaengler, 1995).

Therefore, from a comparative odontological point of view, it is evident and absolutely normal that omnivorous teeth show abrasion and attrition even at a young age, whereby formation of secondary dentine and slow but continuous eruption are sufficient responses. In this sense the pulp chamber represents a kind of depository for formation of secondary dentine when it is needed (in contrast to carnivorous teeth where it is not needed). Ongoing periodontal regeneration of slowly erupting omnivorous teeth could counteract the disease progression of periodontitis. Therefore, the dynamics of regeneration and degeneration of the periodontium in health and disease may be related to the 'phylogenetic memory', to the feeding type, and to normal or abnormal function including the potential of reverting to older phylogenetic patterns of tooth attachment.

Summary

During evolution, different types of tooth attachment have been developed. On the basis of the comparative histology of periodontal tissues, protoacrodontal, acrodontal, acro-protothecodontal, pleurodontal and thecodontal structures can be distinguished that depend upon the area of attachment (crestal, marginal or socked type) and the mode of attachment (ankylosis, fibrous or combined type). The greatest variations in tooth attachment have originated in acro-protothecodont bony fishes and in pleurodont reptiles, whereas the selection for a true socketed type was an important event in the evolution of mammals. The detailed structures of cementum, of the fibre apparatus and of the junctional epithelium vary from type to type and within one type. These principal structures are decisive for reaction patterns of degeneration and regeneration. Due to 'phylogenetic memory', changes of the mammalian periodontium in health and disease could be interpreted as copies of phylogenetically older patterns.

References

Ainamo, J. (1972). Relationship between occlusal wear of the teeth and periodontal health. *Scandinavian Journal of Dental Research*, 80, 505–509.

Ainamo, A. and Ainamo, J. (1984). The dentition is intended to last a lifetime. *International Dental Journal*, 2, 87–92.

Berkovitz, B. K. B. (1990). The structure of the periodontal ligament: an update. *European Journal of Orthodontics*, 12, 51–76.

Berkovitz, B. K. B., Holland, G. R. and Moxham, B. J. (1992). *A Colour Atlas and Textbook of Oral Anatomy*, second edition, St Louis: Mosby Year Book.

Berkovitz, B. K. B., Moxham, B. J. and Newman, H. N. (1995). *The Periodontal Ligament in Health and Disease*, second edition, London: Mosby-Wolfe.

Bystrow, A. P. (1938). Zahnstrukturen der Labyrinthodonten. *Acta Zoologica*, 19, 387–425.

Bystrow, A. P. (1939). Zahnstrukturen der Crossopterygier. *Acta Zoologica* 20, 387–425.

Denison, R. (1978). *Handbook of Paleoichthyology*, Vol. 2, *Placodermi*. Stuttgart: Gustav Fischer.

Denison, R. (1979). *Handbook of Paleoichthyology*, Vol. 5, *Acanthodii*. Stuttgart: Gustav Fischer.

Gaengler, P. (1986). Klinische und experimentelle Aspekte der vergleichenden Odontologie und Periodontologie. *Nova Acta Leopoldina Neue Folge* 58, 262, 525–537.

Gaengler, P. (1991). Mikromorphologische Grundlagen der

Degeneration und Regeneration des Periodonts - Schlußfolgerungen für die Kieferorthopädie. In *Kieferorthopädischer Gewebeumbau*, ed. W. Harzer, pp 41–48. Berlin: Quintessenz.

Gaengler, P. (1995). *Lehrbuch der konservierenden Zahnheilkunde.* Berlin: Ullstein Mosby.

Gaengler, P. and Metzler, E. (1988/1989/1990). Die periodontale und endodontale Differenzierung der Odontogenese: Der phylogenetische Weg zum Thekodont, Teil I–III. *Freiberger Forschungshefte.* C 419, 122–129, C 427, 85–94, C 436, 70–92.

Gaengler, P. and Metzler, E. (1992). The periodontal differentiation in the phylogeny of teeth: an overview. *Journal of Periodontal Research*, 27, 214–225.

Gaengler, P., Hoyer, I., Kosa, W., Metzler, E. and Schuder, S. (1988). Degeneration und Regeneration des Endodonts und Periodonts als Kopie phylogenetischer Zahnentwicklungsmuster. *Zahn- Mund- und Kieferheilkunde; mit Zentralblatt*, 76, 4–13.

Gaengler, P. and Merte, K. (1983). Effects of force application on periodontal blood circulation: a vital microscopic study in rats. *Journal of Periodontal Research*, 18, 86–92.

Gross, W. (1966). Kleine Schuppenkunde. *Neues Jahrbuch für Geologie und Paläontologie Abhandlungen*, 125, 29–49.

Hoffmann, T., Gaengler, P., Genco, R. and Schreiber, D. (1988). Zur vergleichenden Morphogenese der resorptiven knöchernen Hyperregeneration des Periodonts und Endodonts. *Zahn- Mund- und Kieferheilkunde; mit Zentralblatt*, 76, 704–710.

Hopson, J. A. (1964). Tooth replacement in cynodont, dicynodont and therocephalian reptiles. *Proceedings of the Zoological Society of London*, 142, 625–654.

Howes, R. I. (1979). Root morphogenesis in ectopically transplanted pleurodont teeth of the iguana. *Acta Anatomica*, 103, 400–408.

Janvier, P. (1996). *Early Vertebrates.* Oxford: Clarendon Press.

Kollar, E. J. and Fisher, C. (1980). Tooth induction in chick epithelium: expression of quiescent genes for enamel synthesis. *Science*, 207, 993–995.

Krejsa, R. J. and Slavkin, H. C. (1987). Earliest craniate teeth identified. *Journal of Dental Research*, 66, 144.

Lehm, A. and Gaengler, P. (1990). Mikromorphologische Reaktionsmuster des menschlichen Periodonts. *Zahn- Mund- und Kieferheilkunde; mit Zentralblatt*, 78, 273–280.

Levers, B. G. H. and Darling, A. I. (1983). Continuous eruption of some adult human teeth of ancient populations. *Archives of Oral Biology*, 28, 401–408.

Metzler, E. (1990). *Die periodontale Differenzierung in der Odontogenese: Der phylogenetische Weg zum Thekodont.* Dissertation, Erfurt Medical Academy.

Müller, A. H. (1985). *Lehrbuch der Paläozoologie. III. Vertebraten.* Jena: Gustav Fischer.

Ørvig, T. (1967). Phylogeny of tooth tissues: evolution of some

calcified tissues in early vertebrates. In *Structural and Chemical Organization of Teeth*, vol. 1, ed. A. E. W. Miles, pp. 45–110. London: Academic Press.

Osborn, J. W. (1984). From reptile to mammal: evolutionary considerations of the dentition with emphasis on tooth attachment. *Symposia of the Zoological Society of London*, 52, 549–574.

Osborn J. W. (1993). A model simulating tooth morphogenesis without morphogens. *Journal of Theoretical Biology*, 65, 429–445.

Page, R. C. and Schroeder, H. E. (1982). *Periodontitis in Man and other Animals: a Comparative Review.* Basel: Karger.

Peyer, B. (1968). *Comparative Odontology.* Chicago: University of Chicago Press.

Reif, W. E. (1982). Evolution of dental skeleton and dentition in vertebrates: the odontode-regulation theory. *Evolutionary Biology*, 15, 287–368.

Schneider, J. (1987). Grundlagen der Morphogenie, Taxonomie und Biostratigraphie isolierter Xenacanthodier-Zähne (Elassmobranchii). *Freiberger Forschungshefte*, C 419, 71–80.

Schroeder, H. E. (1986). *Handbook of Microscopic Anatomy*, vol. V/5, *The Periodontium.* Berlin: Springer.

Schroeder, H. E. and Listgarten, M. A. (1977). *Fine Structure of the Developing Epithelial Attachment of Human Teeth.* Basel: Karger.

Shellis, R. P. (1982). Comparative anatomy of tooth attachment. In *The Periodontal Ligament in Health and Disease*. eds. B. K. B. Berkovitz, B. J. Moxham and H. N. Newman, pp. 3–24. Oxford: Pergamon Press.

Slavkin, H. C., Krejsa, R. J., Fincham, A., Bringas, P. Jr., Santos, V., Sassano, Y., Snead, M. L. and Zeichner-David, M. (1991). Evolution of enamel proteins: a paradigm for mechanisms of biomineralization. In *Mechanisms and Phylogeny of Mineralization in Biological Systems*, eds. S. Suga and H. Nakahara, pp. 383–389. Berlin: Springer.

Smith, M. M. and Coates, M. I. (1998). Evolutionary origins of the vertebrate dentition: phylogenetic patterns and developmental evolution. *European Journal of Oral Sciences*, 106, 482–500.

Smith, M. M. and Hall, B. K. (1990). Development and evolutionary origins of vertebrate skeletogenic and odontogenic tissues. *Biological Reviews*, 65, 277–373.

Smith, M. M. and Hall, B. K. (1993). A developmental model for evolution of the vertebrate exoskeleton and teeth: the role of cranial and trunk neural crest. *Evolutionary Biology*, 27, 387–447.

Westergaard, B. (1983). A new detailed model for mammalian dentitional evolution. *Zeitschrift für zoologische Systematik und Evolutionsforschung*, 21, 68–78.

Zangerl, R. (1978). *Handbook of Paleoichthyology*, vol. 3A, Chondrichthyes I. Stuttgart: Gustav Fischer.

13 Tooth replacement patterns in non-mammalian vertebrates

B. K. Berkovitz

13.1. Introduction

The complexities involved in the development of a mammalian tooth are slowly being unravelled. Involved in the epithelial-mesenchymal interactions are factors such as homeobox genes, growth factors and extracellular matrix molecules (Chapters 1–3). When one then attempts to project control mechanisms relating to the development of individual teeth to the ontogeny of the dentition as a whole, the topic indeed appears daunting. Not only has a tooth to develop, but it must also erupt into its correct position within the jaw. For most non-mammalian vertebrates (NMV), the teeth also are continually being replaced (polyphyodontism), involving resorption of the predecessors. Although teeth can be seen to undergo wear, tooth replacement is not a direct response to wear as, in animals such as the shark, replacement occurs too rapidly for wear to occur to any significant degree. As NMV generally grow throughout life, replacement is considered to be more related to this growth process as successional teeth erupt into progressively larger jaws. This is associated with progressively larger replacement teeth being formed (which may also exhibit a different morphology) and/or an increase in tooth number. Temporary gaps within the dentitions are usually apparent (Figure 13.1) and their extent will reflect the time interval between loss of a tooth and its replacement by a successor.

Apart from very rare examples, such as the piranha fish, the teeth of NMV do not occlude or shear past each other, and can only function in grasping the food, which is generally swallowed whole and head first. Assuming teeth play a role in the feeding mechanism, strategies will have evolved to ensure that the dentitions are suitably functional. In this context, the studies of Edmund have been particularly important in stimulating research into tooth replacement mechanisms. In order to avoid the possibility of a number of teeth in adjacent positions being shed simultaneously and resulting in regions of the jaw being temporarily edentulous (thereby possibly hindering the food-gathering process), Edmund showed that tooth replacement was not haphazard, but represented a carefully controlled mechanism. If one considers a model of the lower jaw of a green lizard (*Lacerta viridis*), the arrangement of the teeth appears random, especially when comparing adjacent teeth (Figure 13.1A). However, if alternate teeth are compared, a recognisable pattern emerges. Thus, starting for instance at tooth position 16 (arrow), this tooth has recently erupted and, as yet, no successor is evident. Progressing anteriorly through tooth positions 14, 12, 10, 8, 6, the replacing teeth are at younger stages of development. Thus, it can be assumed that these teeth will successively erupt in an alternating sequence from the back to the front of the jaw. A similar pattern exists for the teeth in the odd-numbered tooth positions. Figure 13.1B shows the same model as in Figure 13.1A but waves of tooth replacement passing through alternate tooth positions have been superimposed. Edmund (1960, 1962, 1969) demonstrated in a comprehensive survey of living (and extinct) reptiles that, in virtually all forms, waves of replacement were seen to pass continually along the jaws, usually from back to front, and affecting teeth in *alternate tooth positions*. A wave affecting the teeth in even-numbered positions would be followed by a wave passing through the odd-numbered series. If a wave in one series starts when a wave in the other series has run about half its course, then this ensures that when exfoliation is occurring in one series, the teeth of the adjacent series are functional and have a large proportion of their life left. In this way,

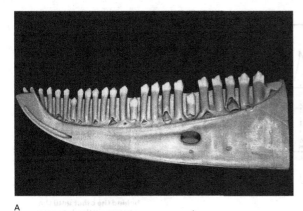 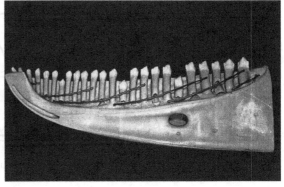

A   B

Figure 13.1a. **Model of complete dentition of lower jaw of a green lizard (*Lacerta viridis*). Tooth 16 (from the front) is arrowed. (Courtesy Dr J. S. Cooper.) B. Same model as A with the waves of tooth replacement passing through alternate positions being indicated by oblique lines. (Courtesy Dr J. S. Cooper.)**

two adjacent teeth are rarely, if ever, absent from the dentition at the same time, thus ensuring the retention of what may be considered a functional dentition. The presence of a well-controlled tooth replacement mechanism must also reflect tight control of the order of tooth development and eruption across the whole dentition.

Edmund (1960, 1962, 1969) put forward his 'Zahnreihen' theory in an attempt to account for the replacement patterns seen. He postulated that throughout life, impulses pass along the jaws from front-to-back at regular intervals, stimulating the development of teeth at each successive tooth locus. The time interval between successive stimuli, measured in terms of the number of tooth positions apart, could determine the number of teeth in each tooth Zahnreihe and the direction of the subsequent replacement wave. Even though teeth are initiated in a front to back sequence, if the Zahnreihen spacing is greater than 2 (and less than 3), then after a few generations replacement waves will be established which are seen to pass through alternate tooth positions and which run from back to front (Figure 13.2). This was found to be by far the most common pattern in reptiles. With a Zahnreihen spacing of 2.0, the developing successors will all be at about the same stage of development and almost perfect alternation of waves of replacement would be seen. With a Zahnreihen spacing of less than 2 (and greater than 1), alternate waves of replacement would pass from front to back of the jaws. For purposes of mathematical modelling, Zahreihen spacing can be considered in terms of fractions of whole numbers.

The purpose of this article is to review various data related to patterns of tooth replacement in NMV and, where possible, relate it to tooth development. In this way, it is hoped to stimulate further consideration of the wider aspects of developmental control of whole denti-

tions. As will be evident, there are no more than a handful of longitudinal studies investigating tooth replacement patterns in the living animal and attempting to relate it to the initial sequence of tooth development.

For a simplified account of the dental anatomy of some of the NMV described in this review, the reader is referred to Berkovitz *et al.*, (1978).

13.2.   **Tooth replacement in sharks**

The teeth of sharks differ from those of other NMV in that there is a continuous fibrous attachment, all the teeth being attached to a common sheet of connective tissue. Hence, teeth in sharks are replaced either as more or less whole rows or in groups of individual teeth, depending on whether the outward movement is blocked by other teeth which overlap them (Strasburg, 1963; Moss, 1967). Sharks possess a number of rows of successional teeth (each at a slightly younger stage of development) below each functional tooth, suggesting that tooth replacement is rapid. This rapidity has been demonstrated in longitudinal studies where teeth have been marked by clipping or with dyes. Tooth rows have been reported to be replaced about every 10–14 days in young lemon sharks, *Negaprion brevirostris* (Moss, 1967; Boyne, 1970; Reif *et al.*, 1978) and immature dogfish, *Mustelis canis* (Ifft and Zinn, 1948), while Reif *et al.* (1978) give a replacement time of 28 days for the nurse shark, *Ginglymostoma cirratum*. When young lemon sharks were deprived of food for a period of about 3 weeks, the overall pattern of replacement was not affected but the period taken to shed teeth was slightly increased (Moss, 1967)

The studies referred to above were all short term,

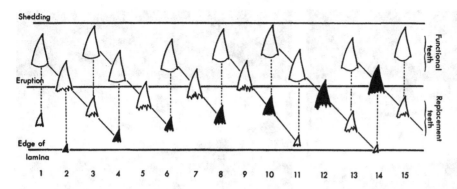

Figure 13.2. **Schematic diagram of a reptilian jaw showing oblique rows of teeth (*Zahnreihen*) on the dental lamina. The first tooth of the dentition is formed at position 1 and has moved a little way across the lamina by the time the second tooth forms at position 2. These two teeth both continue to move across the lamina and the third forms at position 3, and so on. In this way, an oblique row of teeth at successive stages of development is formed on the lamina, and further similar rows (*Zahnreihen*), marked by continuous lines, follow one behind the other until the whole dentition is built up. (Courtesy Dr J. S. Cooper, 1962 and *British Journal of Herpetology* 3:214–217.)**

being carried out for no more than a few weeks. Luer *et al.* (1990) investigated replacement times in three juvenile nurse sharks over 3 years, particular attention being paid to water temperature. They found the most rapid rates of tooth replacement occurred in the summer (water temperature between 27 and 29 °C) when a row was shed every 9–21 days, while the slowest rates of 51–70 days were seen in winter when the water temperatures were 19–22 °C. Perhaps somewhat surprisingly, rates of replacement did not slow down as the animals aged. This finding differs from that of Wass (1973) who observed that the time taken to replace a tooth row in young sandbar sharks, *Carcharhinus milberti* was 18 days, whereas that for older animals slowed to 38 days.

By determining the average increment in tooth width with successive tooth replacements in the pelagic shark, *Isistius brasiliensis*, Strasburg (1963) related this to total body length, giving rise to a Strasburg plot (Figure 13.3). It was calculated that, between body lengths of 140–501 mm, there were 15 replacements. With each successive replacement, tooth width in young lemon sharks increased by 0.14 mm (Moss, 1967), allowing a rough estimate of 125 replacements for the lemon shark during growth to a maximum size of 250 cm. In young dogfish, *Mustela canis*, the average increase in tooth width was 0.03 mm from which it was calculated that six tooth row replacements occurred during a body length growth of 10 cm. (Moss, 1972). In a detailed study of morphogenesis, pattern formation and function of the dentition of the Pacific bull-head shark, *Heterodotus*, Reif (1976) described the changing morphology of the teeth with growth and calculated conservatively that in the course of a 10–year life span, this shark produces between 2000 and 4000 teeth. He also noted that teeth of the first generation in this shark were rudimentary and non-functional (Reif, 1976).

13.3.   Tooth replacement in bony fish

Unlike sharks, bony fish (as well as amphibians and reptiles) have teeth with individual attachments to the jaws. Thus, teeth are replaced individually, rather than as whole tooth rows. Tooth replacement studies have been limited to fish where the teeth are arranged in discrete rows (for consideration of the condition where teeth are arranged in batteries, see Chapters 12, 16). Miller and Radnor (1973) analysed skeletal material in the bowfin, *Amia calva*. Each tooth-bearing bone had its own tooth series with often a distinct difference in tooth morphology and replacement pattern, but symmetry between sides was rare. Although some examples were seen of alternate waves of tooth replacement, many instances were seen of apparently random patterns of tooth replacement and these increased with age. In several bones, the maxillae in particular, series with every third or fourth tooth being at a similar stage of development were observed.

Wakita *et al.* (1977) studied tooth replacement in the premaxillae and dentaries of *Prionurus microlepidotus* Lacepede from serial sections. No symmetry could be recognised in any specimen, either between right and left sides or upper and lower jaw quadrants, leading these authors to conclude that tooth replacement proceeded independently in each quadrant. Perhaps their most interesting observation related to the manner of replacement at each tooth locus. Each functional tooth was accompanied by two developing replacing teeth at different stages of development. Whereas one would expect the two replacing teeth to represent members of the same series, this was not the case. One replacing tooth was located lingually and the other labially. At each tooth locus, replacement alternated so that a functional tooth that originally developed lingually was replaced by

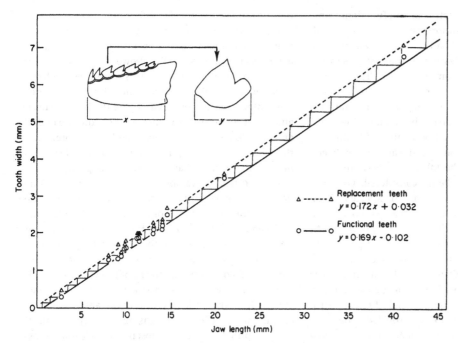

Figure 13.3. **Strasburg plot of replacement tooth widths and functional tooth widths against jaw lengths, measurements being taken from radiographs. Jaw length (x) was measured between the symphysis and the articulation facet; tooth width (y) was measured as the greatest anteroposterior dimension of the tooth.** (From Berkovitz and Shellis, 1978 and courtesy of *Journal of Zoology*, London.)

Replacement teeth
$y = 0.172x + 0.032$

Functional teeth
$y = 0.169x - 0.102$

a tooth which had developed labially and vice versa. During the resorptive process, therefore, a replacing tooth developing lingually would move labially to resorb the functioning tooth (which itself had developed labially), while a tooth developing labially would need to move lingually to effect resorption of the functional predecessor (which itself had developed lingually). Thus, the randomness of replacement occurring in each quadrant was contrasted with the regular alternation of replacement at each locus, suggesting that some control was effected through inhibition mechanisms of one developing tooth germ on its neighbour. It was also observed that when a functional tooth was assessed to have been lost earlier than expected (i.e. through trauma), tooth replacement at the affected locus proceeded faster than that in the two adjacent loci.

Berkovitz and Moore (1974, 1975) undertook what seems to be the only detailed longitudinal study of tooth replacement in teleosts, studying the rainbow trout, *Salmo gairdneri*. The dentitions of 11 fish were recorded with wax impressions taken from anaesthetised fish twice weekly for periods of up to 9 months. From the raw data relating to the presence or absence of teeth, it is possible to draw lines linking the eruption times of teeth in adjacent, and alternate tooth positions and any other series of regularly spaced points (e.g. every third tooth) to represent waves of replacement. Which of these lines has true biological significance awaits clarification (see also Osborn, 1970; DeMar, 1972, 1973), but, in a series of

papers Osborn (1972, 1973, 1974b, 1977) has argued convincingly that it is unlikely to be the Zahnreihen. However, alternate replacement waves and Zahnreihen spacing are terms commonly used by researchers and are useful in allowing for general comparisons. Such replacement waves also more easily illustrate the functional state of the dentition as a whole at a given moment. Replacement waves could be identified in all trout, passing through alternate tooth positions (Berkovitz and Moore, 1974, 1975). In the dentary, these waves passed steeply from back to front (i.e. having a Zahnreihen spacing of more than 2.0 tooth positions and less than about 2.7; Edmund, 1960), whereas, in the upper marginal toothrow, the waves were flatter (i.e. having a Zahnreihen spacing of approximately 2.0 tooth positions). With such patterns of tooth replacement, it was observed that virtually no instances occurred over the prolonged experimental period where two adjacent teeth were absent simultaneously in a dentition. If this could be an indication of functionality of a dentition, then the trout dentition is always maintained in a highly functional state (see DeMar, 1973 and Osborn, 1975 for further consideration of what constitutes a 'functional' dentition).

As an example of the life cycles of teeth, generation times in trout (defined as the period between the first appearance of a tooth in the mouth and the appearance of its successor) for all tooth sites (including those on the tongue) averaged about 8 weeks for the smaller fish and 13 weeks for the larger fish (Berkovitz and Moore 1974,

1975). The mean ratio of functional: non-functional time (i.e. the time each tooth position was occupied by a tooth compared with the time when the position was unoccupied prior to replacement) varied for the first generation of teeth in the dentary from 1.1 to 4.0. These ratios directly reflected the mean percentage of teeth to be expected on the jaws of a fish at any one time and thus showed considerable variation (i.e. between just over 50% and just under 80% functional occupancy of tooth positions respectively). Such biological variation between animals clearly needs to be taken into account when mathematical models are being considered. Over the experimental period, there was an unexpected shortening of generation times, mainly as the result of a shortening of the non-functional time between loss of a tooth and the eruption of the successor. It was not possible to determine whether this was caused by the experimental regime. Close inspection of tooth replacement records commonly revealed small variations in successive replacement waves. Combining longitudinal studies with a histological one, Berkovitz (1977a) provided a detailed time sequence for the life cycle of teeth in the rainbow trout.

Assuming that life cycles of teeth generated within a dentition do not differ too radically from each other, it may be surmised that the order in which teeth initially appear will have a bearing in establishing the subsequent pattern of tooth replacement. Different interpretations have been forwarded to account for the establishment of the alternating pattern of tooth replacement reported for NMV. As stated earlier, one theory suggests that the teeth are generated in succession from the front to the back of the jaws, the replacement pattern reflecting the rate of tooth initiation and the rate of migration of the teeth upwards into a functional position (Edmund 1960, 1962, 1969, Figure 13.2). Another view implies that replacement patterns are incidental to the necessary order of eruption imposed by a field of inhibition surrounding each developing tooth. This would imply that teeth would develop (at least in part) in an alternate manner and, if the zone of inhibition was even, allow the teeth to be evenly spaced (e.g. Lawson, et al., 1971; Osborn, 1971, 1973).

Berkovitz studied tooth development in the rainbow trout in order to see its association with the subsequent development of the alternating pattern of tooth replacement (Berkovitz, 1977b). The initial sequence of development for the dentary was that the first 10 teeth developed in an alternate sequence, the teeth occupying the odd-numbered tooth positions developed first, being rapidly followed by a wave passing through the even-numbered positions. The sequence of development was 3, 5, 1, 7, 4, 9, 2, 6, 10, 8, 11, 12. For the upper marginal tooth row of the rainbow trout (Berkovitz, 1978) the first

tooth to appear in the maxillary bone was in the first tooth position and there was an almost perfect wave of tooth development passing back through the odd-numbered tooth positions closely followed by a wave affecting the even-numbered positions. This closely resembled the situation seen in the dentary (although the first tooth in the dentary was in tooth locus 3) and the initial eruption and replacement pattern followed the sequence of development. However, for the eight teeth in the premaxilla, the sequence of development was entirely different. The first tooth to arise was seen in tooth position 4, the remaining teeth developing in sequence in both directions. As the maxillary and premaxillary teeth both show similar alternating waves of tooth replacement, similar replacement waves can clearly be derived from a very different sequence of tooth initiation. Whether the different sequences between maxillary and premaxillary teeth reflected different patterns of growth in the underlying bones awaits clarification. Other findings associated with tooth development were the absence of a true dental lamina (the teeth arising directly within the surface epithelium) and the absence of an initial series of rudimentary, non-functional teeth (particularly seen during tooth development in reptiles; see section 13.5).

Piranhas (*Serrasalmus*: Characidae) are unusual among fishes in that they use their teeth to cut ingestible pieces of flesh from large prey. The marginal blade-like teeth (six on the premaxilla and seven on the dentary) are closely interlocked, forming a continuous serrated blade in each jaw quadrant (Shellis and Berkovitz, 1976), and teeth cannot be shed individually. From an examination of skeletal material it was concluded that, not only were teeth within each quadrant shed simultaneously and that upper and lower quadrants from each half of the jaw were in synchrony with each other, but that right and left halves of the dentition were out of phase (Roberts, 1967; Berkovitz, 1975). This would appear to ensure that the apparent disadvantage of temporarily being completely edentulous in one half of the jaws would be compensated for by having a functional dentition in the other half.

Berkovitz and Shellis (1978) studied tooth replacement in a longitudinal study of eight piranhas maintained for periods of up to 13 months. The main observations were that the teeth within any jaw quadrant were shed and replaced almost simultaneously and that the time between exfoliation of one tooth row and its replacement by the succeeding row was very rapid (about 5 days). As the teeth remained functional for about 100 days, the short period when teeth are absent is not a serious impediment to feeding. Unlike the previous observations on skeletal material, the pattern of tooth replacement was highly variable. All four quadrants may be in phase (so that for a few days no teeth were present), all four quadrants may be out of phase with each other, diagonally opposing quadrants (i.e. upper left and lower right) may be in synchrony. With increasing age, the rate of tooth replacement decreased and, conversely, the functional lifespans of successive generations of teeth increased. The results of a Strasburg plot from cross-sectional data with tooth width against jaw length indicated that teeth had been replaced about 27 times in the oldest skeletal material with a jaw length of 45 mm (Figure 13.3). In the largest dentary examined (45 mm), no replacement teeth were present and the functional teeth showed heavy wear. This suggests that in old specimens replacement teeth may fail to develop.

Not only are the marginal teeth in each jaw quadrant of the piranha replaced simultaneously, as might be expected they also appear to be initiated simultaneously, there being a slight anterior–posterior gradient (Berkovitz, 1975; Berkovitz and Shellis, 1978). This extremely rare sequence of development in this rather unusual dentition (possibly related to functional similarity with sharks) is almost alone in being consistent with Edmund's Zahnreihen theory. Simultaneous development also occurs for the smaller and less compressed teeth on the ectopterygoid bones of the piranha (Shellis and Berkovitz, 1976).

Berkovitz (1980) undertook a longitudinal study to investigate the stability of the patterns of tooth replacement seen in piranhas. Two fish used in the initial study of Berkovitz and Shellis (1978) were kept for 3 more years and tooth replacement patterns then observed during the following 10-month period. Although the teeth in each quadrant were still replaced simultaneously, the timing of replacement had changed so that the initial patterns of 3 years earlier had changed. In addition to these age changes, the teeth in the older specimens functioned for about 125 days compared with about 100 days 3 years earlier, a change to be expected from the Strasburg plot (Figure 13.3).

Teeth also occur in limited numbers in the pharyngeal arches of fish. Their replacement patterns have been studied in cyprinid and coitid fishes and have been shown to occur in waves passing through alternate tooth positions (Evans and Deubler, 1955; Nakajima, 1979, 1984, 1987).

Only one experimental study has been undertaken to determine factors controlling tooth initiation in adult fish. Tuisku and Hildebrand (1994) studied the effect of unilateral section of part of the mandibular nerve in the cichlid fish, *Tilapia mariae*. This procedure resulted in the complete absence of teeth in the dentary on the operated side, while continuous tooth replacement occurred normally on the unoperated side, implying that nerves may have a role in maintaining tooth initiation..

13.4.   Tooth replacement in amphibians

Gillette (1955) analysed skeletal material of the frog, *Rana pipiens* and observed that teeth in alternate tooth positions were at similar stages of development. Furthermore, functional teeth generally alternated with empty loci whose teeth were in the process of replacement. Thus, replacement waves were flat, with a Zahnreihen spacing of 2.0. He deduced that total life cycle time for teeth was about 90 days. In the frog, *Hemiphractus proboscideus*, Shaw (1989) reported that functional teeth in general were also only present in alternate loci, while between them either no functional tooth was present or a tooth undergoing resorption, and a successional tooth was beginning to erupt. By examining alizarin red whole mounts, Lawson (1966) found a similar pattern of tooth replacement in the frog, *Rana temporaria*, but the percentage of spaces was lower (26% as opposed to 50% in true alternation). Lawson explained this partial obscuring of the true alternating pattern as being a consequence of the rapid speed at which the final stages of resorption and loss of the functional teeth occur and may result if only specimens with the majority of teeth in the terminal stages in the tooth life cycle are examined.

Goin and Hester (1961), studying tooth replacement in the skulls of the green tree frog, *Hyla cinerea*, also observed alternating waves of replacement, but these were more irregular than those seen by Gillette (1955), some waves sloping backwards, others forwards and others remaining rather flat. This was evidenced by groups of functional teeth in even-numbered tooth loci alternating with groups in the odd-numbered positions, the number of groups varying between 4 and 13. Similar variation was seen in three other types of frog.

Examination of skeletal material of the red-backed salamander, *Plethodon cinereus* (Lawson et al., 1971) again revealed waves of replacement passing through alternate tooth positions, although there was variation in the pattern, some specimens showing near perfect alternation, some having waves passing from back to front, while in others the waves passed from front to back. Early tooth development and replacement have been studied histologically in *Xenopus laevis* up to the time of metamorphosis, when about 20 teeth are present in a single row on the premaxilla and maxilla (compared with about 60 in old adults). Shaw (1979, personal communication) observed that the teeth developed rapidly in an alternating sequence commencing at the back of the jaw and passing forwards. In the vast majority of cases (55 out of 60), the even-numbered tooth positions developed and erupted first, commencing at position 20. The first set of teeth took 26 days to develop, erupt and become ankylosed and were functional for 7 days. The main phase of resorption was very rapid, taking only about a day. Replacement was assessed to occur about every 16 days . The odd-numbered series was about 8 days out of phase with the even-numbered series. Shaw (1985) also determined the time sequence of replacement cycles in a longitudinal study of adult *Xenopus*. He reported a complete development cycle took between 60 and 70 days, with the tooth being in a functional position for about 25 days and a gap being present between replacements for about 10–15 days. (The remaining period of about 25–30 days in the life cycle is occupied by the pre-eruptive developmental stages).

Miller and Rowe (1973) described observations on tooth replacement patterns in the adult mudpuppy, *Necturus maculosus*, which indicated important seasonal variations in the process. They found that teeth were not replaced during the winter period of hibernation and also for much of the spring and summer (or at least it was very slow at these latter times of the year). This was supported by studies where animals kept in the dark and at cold temperatures to simulate hibernation conditions ceased to replace their teeth. Tooth replacement was most active in late summer and autumn. Replacement patterns were very irregular and classical waves passing through alternate tooth positions were rarely seen. In a number of animals, replacement waves affected every third tooth.

An exception to the general rule of alternate replacement in amphibia relates to the findings of Bolt and DeMar (1983) for the extinct microsaurs *Euryodus* and *Cardiocephalus*, where there was evidence of simultaneous replacement of teeth.

Wake (1976, 1980) compared the fetal and adult dentitions in viviparous caecilians (aquatic, limbless and blind). Some of the first-formed teeth were rudimentary and underwent resorption without becoming functional. In addition to differences in tooth morphology related to differences in the feeding mechanism, the numerous tooth rows of the fetus (nine rows in *Typhlonectes*) were replaced by a more or less single tooth row in the adult. Clearly, complex and as yet unknown control mechanisms must be involved in this transition which Wake believed could not be explained by existing theories. Histological examination also revealed evidence of alternation in the pattern of tooth replacement.

If a segment of the mandible and its associated teeth are excised in adult urodeles, regeneration will occur. Furthermore, this process can occur more than once and is more rapid following re-regeneration (Graver, 1978). New teeth were added in alternate sequence at the anterior end of the tooth row in the regenerate and re-regenerate jaw (Graver, 1978). Regenerated teeth were of normal adult size and could be replaced without the presence of surrounding bone, which grew in later and

was often incomplete (Howes, 1978). If a digit is also transplanted to the regeneration site in the maxilla of frogs, this resulted in an earlier bony union but did not affect the rate of regeneration of teeth, which still became ankylosed to a separately induced bony base rather than to the transplanted bone of the digit (Howes and Eakers, 1984).

13.5.   Tooth replacement in reptiles

The most comprehensive accounts of the patterns of tooth replacement in reptiles as determined by analysis of skeletal material are those of Edmund (1960, 1962, 1969). Surveying a large cross-section of different species (including fossils), he demonstrated that the vast majority showed replacement waves passing through alternate tooth positions in a back-to-front direction (although there was little synchrony between the halves of the jaws). In some groups of snakes (elapids), replacement waves passed from front to back, while in the viperid-crotalids, there was a tendency for simple alternation. In a large series of crocodile skulls, Edmund (1962) established that, although waves of replacement passed through alternate tooth positions in adult animals from back to front, in the young the direction was reversed. His explanation for this change was that teeth at the back of the jaws remained in function for longer periods compared with anterior teeth. He noted that, as age advances, the replacement waves were more irregular. A rare exception to the rule of alternate tooth replacement was present in the fossil reptile *Trilophosaurus*, where teeth were said to be replaced in a direct sequence from front to back (Edmund, 1960).

In almost every modern reptile (Edmund, 1960, 1969), each functional tooth was accompanied by at least one or two replacement teeth, and in lizards and snakes three or even four developing replacements were not uncommon (with even more for the fangs of snakes). The teeth of some reptiles change in morphology with continuous replacement (Cooper, 1963).

Cooper (1963) confirmed the presence of alternating waves of replacement passing from back to front within a group of lizards, *Lacerta*. Tooth numbers increased from 63 at hatching to about 83 after 3 years and males had on average four more teeth than females. This sexual dimorphism (together with the larger head) might be related to the fact that the male lizard grips the body of the female within its jaws during copulation.

Evans (1985) demonstrated alternate tooth replacement in the Lower Jurassic lepidosaur, *Gephyrosaurus bridensis*, but deduced that replacement was slower in the middle region of the jaws and very slowly posteriorly. Miller and Radnor (1970) examined the skulls of 15 young *Caiman sclerops* to determine the degree of variability in tooth replacement patterns. Whereas all skulls showed evidence of wave replacement affecting alternate tooth positions, considerable variability existed. Bilateral symmetry was never encountered in the dentary, but was present in three maxillae. Apparent crossing over of replacement waves was observed, which the authors interpreted in terms of precocious or retarded development of one or two individual teeth.

The dentition of the African lizard, *Agama agama*, is unusual in that tooth replacement only occurs in the nine anterior pleurodont teeth (five upper and four lower), while the remaining posterior acrodont teeth are not replaced although they increase in number by addition to the back of the tooth row. In this lizard, Cooper *et al.* (1970) found the teeth were replaced in a successive sequence from front to back. Replacement was limited to three (possibly four) generations of teeth. Unlike many reptiles, the replacement teeth were very much larger than their predecessors and there was little symmetry of replacement. Longitudinal studies of tooth replacement in living reptiles have been undertaken by Edmund (1962, alligator), Cooper (1963, lizard), Cooper (1966, slow-worm), Miller and Rowe (1973, mudpuppy), and Kline and Cullum (1984, 1985, iguana). Edmund (1960) undertook one of the first longitudinal studies of tooth replacement, radiographing two young specimens of *Alligator mississippiensis* at monthly intervals for up to 3 years. The results showed the presence of replacement waves passing through alternate tooth positions and in a back-to-front direction. The waves were irregular and this feature increased with age. The average functional life for an anterior tooth was about 9 months, while that for a posterior tooth was about 16 months. The average life cycle time from initial development to exfoliation for a tooth in a young alligator was estimated to be about 2 years.

Tooth replacement in the lizard family, *Lacerta*, was investigated by Cooper (1963) utilising a wax impression technique. He found that the functional life of teeth in *Lacerta lepida* increased with age from 2–3 weeks at hatching to about 38 weeks in the mature animal. As this animal can live for up to 20 years, Cooper calculated (conservatively) that there may be as many as 80 replacements during life. He reported the existence of the classic pattern of waves of replacement passing through alternate tooth positions and that bilateral symmetry was a conspicuous feature of replacement patterns. However, portions of waves showing a flat gradient or even local reversal of wave motion were frequently evident anteriorly. Such biological variation needs to be taken into

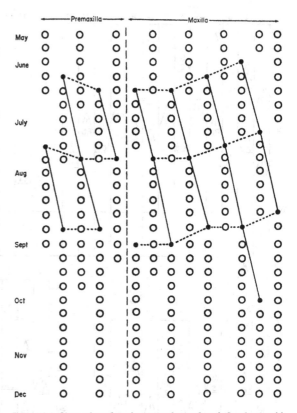

Figure 13.4. Chart of tooth replacement from a female 3 to 4-year-old slow-worm (*Anguis fragilis*). Open circles represent ankylosed functional teeth, black dots represent newly erupted teeth and unoccupied tooth positions are left blank. Each horizontal row is a record showing the circumstances at every tooth position, i.e. functional tooth present, tooth newly erupted, or vacant space. The records were taken at weekly intervals and arranged consecutively from the top to the bottom of the page. The suture between maxilla and premaxilla is indicated by a broken vertical line. *Zahnreihen* are marked by continuous lines and the replacement sequences by dotted lines. (Courtesy Dr J. S. Cooper, 1966 and *Journal of Zoology*, London.)

account when undertaking mathematical modelling of replacement patterns (DeMar, 1972, 1973; Osborn, 1972, 1974b). The form of the replacement wave in *Lacerta* showed species variations. In the more primitive species, *Lacerta lepida*, the acceleration of waves anteriorly was not so pronounced as in the more advanced species *Lacerta muralis nigriventris*, and the difference in length of functional life between anterior and posterior teeth became more marked in the advanced species (Cooper, 1963).

Cooper (1966) studied tooth replacement in the slow-worm, *Anguis fragilis*, using a wax impression technique and also direct visual observation through a microscope at weekly intervals. Three adult females were studied following hibernation from May to December. Tooth numbers are constant in the slow-worm with, from the second year of life, nine on the premaxilla, nine on the maxilla and 10 on the dentary. Replacement waves were

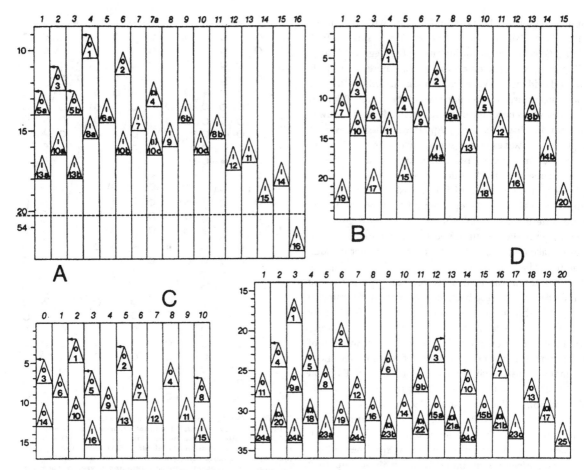

Figure 13.5. **Diagrams of embryonic tooth initiations (tips of triangles) establishing the lower jaw tooth positions present at hatching, indicated by italic numbers (abscissa). Arrows signify that teeth are placed in the direction of the arrow compared with the next tooth generation, making the tooth family rather arbitrary. Small numbers in triangles represent tooth initiation sequence. Capital numbers: 0, resorptive teeth; I, first set of functional teeth. Brackets indicate variability, as some teeth are either resorptive, transitory/normally functioning, or not formed. A. Tuatara (*Sphenodon punctatus*). Ordinate, rough estimate of weeks after egg laying. Position 7a is not present at hatching if tooth 4 is resorptive and tooth 10c is not formed. B. Sand lizard (*Lacerta agilis*). Ordinate, days after egg laying. C. Slow-worm (*Anguis fragilis*). Ordinate, days after the latest stage without visual odontogenic tissue mobilisation. Position 0 is not present at hatching. D. Alligator (*Alligator mississippiensis*). Ordinate, days after egg laying. Total tooth initiation pattern up to hatching in Westergaard and Ferguson (1987). (From Westergaard, 1988 and Courtesy of *Societas pro Fauna et Flora Fennica. Memoranda.*)**

close to a pattern of simple alternate replacement (Figure 13.4), representing a Zahnreihen spacing of approximately 2.0, although sometimes sloping slightly from back to front, other times sloping from front to back. Bilateral symmetry was seen in tooth replacement patterns as well as some symmetry between patterns in the upper and lower jaws. Such symmetry and synchrony may be important in maintaining a dentition that is functional. It seemed most likely that replacement ceased during hibernation. The functional life of all teeth appeared similar, irrespective of their size and position in the jaws, and there was little change related to age of the specimens.

As the pattern of tooth replacement may be established as a result of the order in which the teeth are initiated, it is of interest to note some data of Westergaard (1986, 1988) relating to the early sequence of development for the dentary teeth in the slow-worm (Figure 13.5C). This shows that an initial series of non-functional rudimentary teeth (which undergo resorption) develop in a seemingly complex sequence which does not show alternation. However, the first functional teeth develop rapidly in the odd-numbered positions in an alternating sequence from back to front.

Tooth replacement has been studied in young green iguanas (*Iguana iguana*) over periods of up to two and a half years using wax impressions taken weekly (Kline and Cullum, 1984). Waves of tooth replacement were seen to pass through alternate tooth positions from back to front, with a Zahnreihen spacing calculated to be 2.4.

Unlike previous observations in alligators and lizards (Edmund, 1962; Cooper, 1963), Kline and Cullum (1984) did not find any differences in the life span of anterior compared with posterior teeth. During the course of the study, mean tooth width in the dentary increased from 0.85 mm to 1.31 mm, while that in the maxilla increased from 0.88 mm to 1.54 mm.

Having demonstrated the fundamental nature of alternate tooth replacement in reptiles, Edmund (1960) interpreted some observations of tooth development in the crocodile by Woederman (1919, 1921), incorrectly it would seem (see Osborn, 1971), to formulate his Zahnreihen theory that teeth were initiated at the front of the jaw and passed through successive tooth positions to the back (Figure 13.2). As a knowledge of this sequence is fundamental to an understanding of the topic, studies on the sequence of tooth development in reptiles are of considerable importance and have been published for lizards (Osborn, 1971; Westergaard, 1988), the alligator (Westergaard and Fergusson, 1986, 1987, 1990) and the tuatara, Sphenodon (Westergaard, 1986). All these studies, together with those previously described for fish and amphibia, show Edmund's theory to account for the establishment of tooth replacement patterns to be incorrect in that teeth never develop in a continuous sequence from front to back (apart from the piranha fish: Berkovitz and Shellis, 1978). Neither is there any evidence supporting the view of a 'signal' passing along this route (see also Osborn, 1971,1972, 1973; Westergaard and Ferguson, 1987). Among the common features observed were the presence initially of numerous rudimentary teeth, some only containing dentine, which underwent resorption or which functioned for only a very short period of time. The earliest of these teeth arose directly from the oral epithelium, before the appearance of the dental lamina.

Osborn (1971) found that the first tooth bud to be initiated in the lower jaw (the dental determinant) of the common lizard, Lacerta vivipara, was situated towards the back, at the site of tooth position 11. Behind the dental determinant, teeth were budded off in succession as space became available. In front, however, interstitial growth between teeth allowed for the later development of an intervening row. Thus, an initial row of teeth formed in the odd-numbered positions (i.e. 9, 7, 5, 3), closely followed by those in the even-numbered positions 10, 8, 6, 4, 2. Because of the fundamental nature of this alternation, Osborn proposed that the teeth were evenly separated from each other as a result of a regular zone of inhibition set up around each developing tooth germ which temporarily prevented a neighbouring tooth from arising. The sequential (as opposed to alternate) initiation of tooth buds behind the dental determinant reflected the presence of 'tip' growth (rather than interstitial growth). Whereas Osborn considered zones of inhibition to be regular, Weishampel (1991) has produced mathematical models with more variable zones of inhibition. Because his study was based on somewhat incomplete material, Osborn (1993) later acknowledged that a minor change in the initial sequence of development was necessitated following constructive comments from Westergaard and Fergusson (1987).

In the most detailed study of tooth development probably in any animal, Westergaard and Ferguson (1986, 1987, 1990) demonstrated that in the alligator, Alligator mississippiensis, although there were later areas where there was clear evidence of alternation, the overall sequence did not show perfect regularity, especially where the first-formed rudimentary teeth (which number approximately 19) were concerned. Thus, the first three tooth germs in the lower jaw arose in tooth loci 3, 6 and 12, respectively. The authors highlighted important regional differences in interstitial growth which, together with the concepts of inhibition and fields of form determination, had resulted in Westergaard (1980, 1983, 1986) proposing a tooth position theory of tooth development. Up to five generations of teeth may develop prior to hatching. Although there was no close correlation, the initiation pattern in the jaws was seen to be an approximate indicator of the eruption pattern, which showed waves passing through more or less alternate tooth positions. The first four waves were almost horizontal, the following ones more typically sloping from back to front (Westergaard and Ferguson, 1987, 1990).

Unlike the situation described by Osborn for Lacerta vivipara, the irregular nature of tooth initiation of rudimentary teeth and the anterior situation of the dental determinant has also been shown to occur in the sand lizard, Lacerta agilis and the slow-worm, Anguis fragilis: evidence of more regular alternation, but from front to back, was apparent in the tuatara, Sphenodon punctatus (Figure 13.4; Westergaard, 1986, 1988). These studies emphasised that interstitial growth was variable and may create space for one or more intervening tooth positions to develop and it was only with the subsequent formation of the dental lamina that replacing teeth developed at the same positions as their predecessors.

In a series of papers, Osborn (1970, 1971,1973, 1974a, 1977, 1978, 1979, 1984, 1993) and Osborn and Crompton (1973) have attempted to relate patterns of tooth replacement and tooth development to the evolution of the mammalian dentition from a reptilian ancestor. Among the main changes to be accounted for are the loss of alternate waves of tooth replacement, the change from polyphyodontonty to diphyodonty and the development of heterodonty. An important concept that was introduced was that of the clone (or clade), a group of identical cells all having the same probability of growing and dividing in particular directions. Thus, in the dentary of a reptile, the whole dentition arose from a single clone of cells, the dental determinant (perhaps two clones for the upper jaw, one in the premaxilla and one in the maxilla). The shape potential of tooth primordia was envisaged as being determined by cell ancestry. The sequences in which teeth were initiated had to match the observed gradients in tooth shape. Gradients in tooth shape may rise to a peak of complexity and then fall within a clone, but can never fall and then rise. If the latter occurred, then this was indicative of the presence of an additional clone. The change from a reptilian to a mammalian dentition was thought to coincide with an increase in the number of clones to three, each allowing for heterodonty of shape: an incisor clone, a canine clone and a molar clone. Fossil evidence seemed to indicate that the evolution of two, and then three clones in the upper jaw occurred ahead of that in the lower. The homologies of the initial clones were unclear.

In mammals, deciduous teeth from the incisor clone develop in sequence from front to back, deciduous molar teeth from the molar clone develop in sequence from back to front, while permanent molars develop in sequence from front to back (Osborn, 1970). The shape potential at the anterior margin of the molar clone must be different from that at the posterior margin, as is also that of the older dental lamina from the deciduous molars which gives rise to the successional premolars. The presence of bony crypts separating adjacent deciduous teeth was thought to suppress the process of inhibition and therefore alternate tooth replacement.

Osborn (1993) has reinterpreted the data concerning the order of tooth initiation in alligators by Westergaard and Ferguson (1986, 1987, 1990) who, as previously described, demonstrated the irregular nature of tooth initiation from a single dental determinant, especially with regard to the initial rudimentary teeth. Osborn (1993) proposed that the teeth do not conform to a monotonic gradient in shape, and that the alligator dentition probably contains three clones (see also Kieser et al., 1993). If this were accepted, it would help explain the apparent initial irregular sequence of tooth development. A similar interpretation of three clones was suggested for the dentition of the fossil pelycosaur, Dimetrodon (Osborn, 1973). Aspects of the evolution of the mammalian dentition are also discussed by Ziegler (1971) and Westergaard (1980, 1983, 1986).

Most recently, Sneyd et al. (1993) have proposed a mathematical model for the initiation and spatial positioning in relation to the dental determinant in the alligator. The model makes use of biological data derived from previous developmental studies (Westergaard and Ferguson, 1986, 1987). The model also incorporates mechano-chemical and reaction-diffusion mechanisms and demonstrates that simpler and more intuitive approaches to pattern formation can be misleading in the study of pattern formation in more complex situations, such as the one studied.

Although attempts have been made to interpret fossil dentitions at the reptilian–mammalian boundary, because of the incomplete nature of the material and the understandable lack of suitably juvenile specimens, conclusions derived from such studies as to the number and nature of tooth replacement have been open to different interpretations (e.g. Kermack, 1956; Hopson, 1971, 1980; Parrington, 1971; Osborn and Crompton, 1973; Osborn, 1974a; Brink, 1977; Gow, 1985).

When the biological mechanisms involved in initiation and pattern formation of individual teeth are eventually unravelled (see Chapters 1, 2), it will then be necessary to focus on control mechanisms for the dentition as a whole. As an illustration of the possible complexities, one could consider the remarkable dentition of certain adult gobiid fish, Sicyopterus japonicus and Sicydium plumieri (Mochizuki and Fukui, 1983; Kakizizawa et al., 1986: Mochizuki et al., 1991). Beneath each functional tooth in the upper jaw, there may be a graduated series of up to about 45 successional teeth, arranged in a complex spiral. Tooth replacement occurs about every 9 days in an alternating pattern to maintain a complete functional series of teeth. Furthermore, the functional teeth are not shed but resorbed and subsequently degraded (Figure 13.6).

13.6.   Concluding remarks

A review of the literature on tooth replacement has revealed widespread evidence of tooth replacement occurring in waves passing through alternate tooth positions. However, examples exist where such patterns cannot be recognised. With replacement affecting alternate tooth positions, the pattern is variable and successive waves may show local variations such that, with time, the pattern changes. Only a handful of longitudinal studies have been undertaken to determine the nature and time course of tooth replacement, with little or no experimental work existing to determine factors controlling replacement patterns, such as when a tooth is shed and when its successor erupts. Accidental early loss (and conversely prolonged retention) of a tooth appears likely to produce local changes in the form of a replacement developing tooth germ sets up in its own vicinity, which temporarily prevents an adjacent tooth from developing in synchrony. The molecular biology of any such inhibitory mechanism awaits explanation. The dentitions of non-mammalian vertebrates, with their powers of regeneration, would appear to provide suitable experimental models for understanding control mechanisms in biology. The present lack of data makes any explanation concerning the evolutionary change from a reptilian to a mammalian dentition speculative.

Summary

Tooth replacement is considered in non-mammalian vertebrates. There is continuous replacement in the vast majority of dentate animals and the process appears to be related to continuous body growth (and unrelated to tooth wear). With each replacement, there is a slight increase in tooth size. In elasmobranchs, the teeth are shed in rows and, in young sharks, this may occur approximately every 3 weeks. In the remaining non-mammalian vertebrates, the most commonly recognisable pattern of tooth replacement occurs in waves that pass through alternate tooth positions: the waves more commonly pass from back to front. An obvious advantage of this pattern of tooth replacement is that no extensive region of the tooth-containing jaw is ever devoid of teeth. Indeed, it is extremely rare to find even two adjacent teeth missing at the same time. Rare variations from the alternate pattern of tooth replacement are encountered such that teeth may be shed more or less simultaneously (e.g. piranha fish), or no distinct alternating pattern may be discerned. The rate of tooth replacement slows down with age, presumably related to the increased size of the teeth. With respect to the number of times teeth can be replaced, it has been calculated for example that, during the life of the piranha (*Serrasalmus*), teeth may be replaced about 30 times, while in some lizards, there may be about 80 replacements at each tooth position. The few longitudinal studies on tooth replacement that have been undertaken confirm the patterns deduced by studying osteological material. These have additionally provided information concerning the temporal sequences associated with the life cycle of the teeth and also indicate that slight variations may exist with successive tooth replacements such that, with time, the patterns of tooth replacement can change. Few studies exist which relate the patterns of tooth replacement to those which order the initiation of teeth, but such studies reflect the obvious relationship between the two.

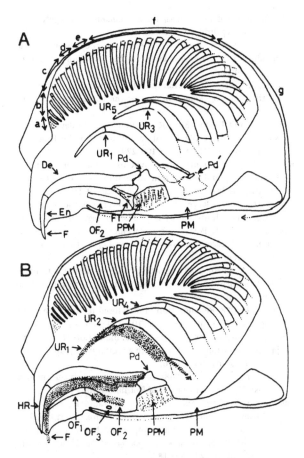

Figure 13.6A,B. **Diagram showing successions of upper jaw of the gobiid fish *Sicydium plumieri*. A. A succession which includes a functional tooth (F). B. A succession which includes a half-erupted replacement tooth (HR). Dotted parts in B show positions of a functional tooth, an unerupted replacement tooth (UR$_1$) which moves to the position of a half-erupted replacement tooth in the next replacement, and an old functional tooth (OF$_2$) under resorption respectively. De, dentine; En, enameloid (crown); FT, fibrous tissue; OF$_1$, an old functional tooth in front of HR; OF$_2$, an old functional tooth in front of F; OF$_3$, an old functional tooth in front of OF$_1$, Pd, pedicel; Pd': developing pedicel; PM, premaxillary; PPM, projection of premaxillary; UR$_1$, UR$_2$, UR$_3$, etc.; unerupted replacement teeth; a, b, c, d, e, f, g, different developmental stages of the upper jaw teeth. (From Mochizuki *et al.*, 1991 and Courtesy of *Natural History Research*.)**

wave, while removal of the nerve supply stops teeth being produced altogether in fish. Continuous tooth replacement in non-mammalian vertebrates is probably related to body growth (and unrelated to tooth wear) and allows for a mechanism whereby tooth size can be increased.

Only a few studies exist detailing the order in which teeth are initiated and these, not surprisingly, demonstrate variations which may be related to jaw growth. Implicated by some researchers in establishing the pattern of tooth initiation is a sphere of inhibition that a

References

Berkovitz, B. K. B. (1975). Observations on tooth replacement in piranhas (Characidae). *Archives of Oral Biology*, **20**, 53–56.

Berkovitz, B. K. B. (1977*a*). Chronology of tooth development in the rainbow trout (*Salmo gairdneri*). *Journal of Experimental Zoology*, **200**, 65–70.

Berkovitz, B. K. B. (1977*b*). The order of tooth development and eruption in the rainbow trout (*Salmo gairdneri*). *Journal of Experimental Zoology*, **201**, 221–226.

Berkovitz, B. K. B. (1978). Tooth ontogeny in the upper jaw and tongue of the rainbow trout (*Salmo gairdneri*). *Journal de Biologie Buccale*, **6**, 205–215.

Berkovitz, B. K. B. (1980). The effects of age on tooth replacement patterns in piranhas (Pisces: Characidae). *Archives of Oral Biology*, **25**, 833–835.

Berkovitz, B. K. B. and Moore, M. H. (1974). A longitudinal study of replacement patterns of teeth on the lower jaw and tongue in the rainbow trout *Salmo gairdneri*. *Archives of Oral Biology*, **19**, 1111–1119.

Berkovitz, B. K. B. and Moore, M. H. (1975). Tooth replacement in the upper jaw of the rainbow trout (*Salmo gairdneri*). *Journal of Experimental Zoology*, **193**, 221–234.

Berkovitz, B. K. B. and Shellis, R. P. (1978). A longitudinal study of tooth succession in piranhas (Pisces: Characidae), with an analysis of the replacement cycle. *Journal of Zoology, London*, **184**, 545–561.

Berkovitz, B. K. B., Holland, G. R. and Moxham, B. J. (1978). *A Colour Atlas and Textbook of Oral Anatomy*, pp. 195–201. London: Wolfe Medical.

Bolt, J. R. and DeMar, R. E. (1983). Simultaneous tooth replacement in *Euryodus* and *Cardiocephalus*. *Journal of Paleontology*, **57**, 911–923.

Boyne, P. J. (1970). Study of the chronologic development and eruption of teeth in elasmobranchs. *Journal of Dental Research*, **49**, 556–560.

Brink, A. S. (1977). A model of tooth replacement in the 'mammal-like reptile' *Diademodon*. *South African Journal of Science*, **73**, 138–143.

Cooper, J. S. (1963). *Dental Anatomy of the Genus* Lacerta. PhD thesis, University of Bristol.

Cooper, J. S. (1966). Tooth replacement in the slow-worm (*Anguis fragilis*). *Journal of Zoology, London*, **150**, 235–248.

Cooper, J. S., Poole, D. F. G. and Lawson, R. (1970). The dentition of agamid lizards with special reference to tooth replacement. *Journal of Zoology, London*, **162**, 85–98.

DeMar, R. (1972). Evolutionary implications of Zahnreihen. *Evolution*, **26**, 435–450.

DeMar, R. (1973). The functional implications of the geometrical organizations of dentitions. *Journal of Paleontology*, **47**, 452–461.

Edmund, A. G. (1960). Tooth replacement phenomena in the lower vertebrates. *Royal Ontario Museum. Life Sciences. Contributions*, **52**, 1–190.

Edmund, A. G. (1962). Sequence and rate of tooth replacement in the crocodile. *Royal Ontario Museum. Life Sciences. Contributions*, **56**, 1–42.

Edmund, A. G. (1969). Dentition. In *Biology of the Reptiles*, eds. C. Gans, A. d'A. Bellairs and T. S. Parsons, pp.117–200. New York: Academic Press.

Evans, H. E. and Deubler, E. E. Jr (1955). Pharyngeal tooth replace-ment in *Semotilus atromaculatus* and *Clinostomus elongatus*, two species of cyprinid fishes. *Copeia* 31–41.

Evans, S. E. (1985). Tooth replacement in the Lower Jurassic lepidosaur *Gephyrosaurus bridensis*. *Neues Jahrbuch für Geologie und Paläontologie*, **7**, 411–420.

Gillette, R. (1955). The dynamics of continuous succession of teeth in the frog (*Rana pipiens*). *American Journal of Anatomy*, **96**, 1–36.

Goin, C. J. and Hester, M. (1961). Studies on the development, succession and replacement in the teeth of the frog *Hyla cinerea*. *Journal of Morphology*, **109**, 279–287.

Gow, C. E. (1985). Dentitions of juvenile *Thrinaxodon* (Reptilia: Cynodontia) and the origin of mammalian diphyodonty. *Annals of the Geological Survey of South Africa*, **19**, 1–17.

Graver, H. T. (1978). Re-regeneration of lower jaws and the dental lamina in adult urodeles. *Journal of Morphology*, **157**, 269–280.

Hopson, J. A. (1971). Postcanine replacement in the gomphodont cynodont *Diademodon*. *Zoological Journal of the Linnean Society*, **50**, (Supplement 1), 1–21.

Hopson, J. A. (1980). Tooth functions and replacement in early mesozoic ornithischian dinosaurs: implications for aestivation. *Lethaia*, **13**, 93–105.

Howes, R. I. 1978. Regeneration of ankylosed teeth in the adult frog premaxilla. *Acta Anatomica*, **101**, 179–186.

Howes, R. I. and Eakers, E. C. (1984). Augmentation of tooth and jaw regeneration in the frog with a digital transplant. *Journal of Dental Research*, **63**, 670–674.

Ifft, J. D. and Zinn, D. J. (1948). Tooth succession in the smooth dogfish *Mustelus canis*. *Biological Bulletin*, **95**, 100–106.

Kakizizawa, Y., Kajiyama, N., Nagai, K., Kado, K., Fujita, M., Kashiwaka, Y., Imai, C., Hirama, A. and Yorioka, M. (1986). The histological structure of the upper and lower teeth in the gobiid fish *Sicyopterus japonicus*. *Journal of the Nihon University School of Dentistry*, **28**, 175–187.

Kermack, K. A. (1956). Tooth replacement in mammal-like reptiles of the suborders Gorgonopsia and Therocephalia. *Philosophical Transactions of the Royal Society of London, Series B*, **240**, 95–133.

Kieser, J. A., Klapsidis, C., Law, L. and Marion, M. (1993). Heterodonty and patterns of tooth replacement in *Crocodylus niloticus*. *Journal of Morphology*, **218**, 198–201.

Kline, L. W. and Cullum, D. R. (1984). A long-term study of the tooth replacement phenomenon in the young green *Iguana iguana*. *Journal of Herpetology*, **18**, 176–185.

Kline, L. W. and Cullum, D. R. (1985). Tooth replacement and growth in the young green *Iguana iguana*. *Journal of Morphology*, **186**, 265–269.

Lawson, R. (1966). Tooth replacement in the frog *Rana temporaria*. *Journal of Morphology*, **119**, 233–240.

Lawson, R., Wake, D. B. and Beck, N. T. (1971). Tooth replacement in the red-backed salamander, *Plethodon cinereus*. *Journal of Morphology*, **134**, 259–269.

Luer, C. L., Blum, P. C. and Gilbert, P. W. (1990). Rate of tooth replacement in the nurse shark *Ginglymosota cirratum*. *Copeia* 182–191.

Miller, W. A. and Radnor, C. J. P. (1970). Tooth replacement patterns in young *Caimen sclerops*. *Journal of Morphology*, **130**, 501–510.

Miller, W. A. and Radnor, C. J. P. (1973). Tooth replacement patterns in the Bowfin (*Amia calva*-Holostei). *Journal of Morphology*, **140**, 381–395.

Miller, W. A. and Rowe, D. J. (1973). Preliminary investigation of variations in tooth replacement in adult *Necturus maculosus*. *Journal of Morphology*, **140**, 63–76.

Mochizuki, K. and Fukui, S. (1983). Development and replacement of upper jaw teeth in gobiid fish *Sicyopterus japonicus*. *Japanese Journal of Ichthyology*, **30**, 27–36.

Mochizuki, K., Fukui, S. and Gultneh, S. (1991). Development and replacement of teeth on jaws and pharynx in a gobiid fish *Sicydium plumieri* from Puerto Rico, with comments on resorption of upper jaw teeth. *Natural History Research*, **1**, 41–52.

Moss, S. A. (1967). Tooth replacement in the lemon shark, *Negaprion brevirostris*. In *Sharks, Skates and Rays*, eds. P. W. Gilbert, R. F. Mathewson and D. P. Rall, pp. 319–329. Baltimore: Johns Hopkins University Press.

Moss, S. A. (1972). Tooth replacement and body growth rates in the smooth dogfish, *Mustelus canis* (Mitchill). *Copeia* 808–811.

Nakajima, T. (1979). The development and replacement pattern of the pharyngeal dentition in a Japanese cyprinid fish, *Gnathopogon coerulescens*. *Copeia* 22–28.

Nakajima, T. (1984). Larval vs adult pharyngeal dentition in some Japanese cyprinid fishes. *Journal of Dental Research*, **63**, 1140–1146.

Nakajima, T. (1987). Development of pharyngeal dentition in the coitid fishes, *Misgurnus anguillicaudatus* and *Cobitis biwae* with a consideration of evolution of cypriniform dentitions. *Copeia* 208–213.

Osborn, J. W. (1970). New approach to Zahnreihen. *Nature*, **225**, 343–346.

Osborn, J. W. (1971). The ontogeny of tooth succession in *Lacerta vivipara* Jacquin (1787). *Proceedings of the Royal Society of London, Series B*, **179**, 261–289.

Osborn, J. W. (1972). On the biological improbability of Zahnreihen as embryological units. *Evolution*, **26**, 601–607.

Osborn, J.W. (1973). The evolution of dentitions. *American Scientist*, **61**, 548–559.

Osborn, J. W. (1974a). On tooth succession in *Diademodon*. *Evolution*, **28**, 141–157.

Osborn, J. W. (1974b). On the control of tooth replacement in reptiles and its relationship to growth. *Journal of Theoretical Biology*, **46**, 509–527.

Osborn, J. W. (1975). Tooth replacement: efficiency, patterns and evolution. *Evolution*, **29**, 180–186.

Osborn, J. W. (1977). The interpretation of patterns in dentitions. *Journal of the Linnean Society*, **9**, 217–229.

Osborn, J. W. (1978). Morphogenetic gradients: fields versus clones. In *Development, Function and Evolution of Teeth*, eds. P. M. Butler and K. A.Joysey, pp. 171–201. London: Academic Press.

Osborn, J. W. (1979). A cladistic interpretation of morphogenesis. *Journal de Biologie Buccale*, **6**, 327–337.

Osborn, J. W. (1984). From reptile to mammal: evolutionary considerations of the dentition with emphasis on tooth attachment. *Symposium of the Zoological Society of London*, **52**, 449–474.

Osborn, J. W. (1993). A model simulating tooth morphogenesis without morphogens. *Journal of Theoretical Biology*, **65**, 429–445.

Osborn, J. W. and Crompton, A. W. (1973). The evolution of mammalian from reptilian dentitions. *Brevoria*, **399**, 2–18.

Parrington, F. R. (1971). On the upper triassic mammals *Philosophical Transactions of the Royal Society of London, Series B*, **282**, 177–204.

Reif, W.-E. (1976). Morphogenesis, pattern formation and function of the dentition of *Heterodontus* (Selachii). *Zoomorphologie*, **83**, 1–47.

Reif, W. E., McGill, D. and Motta, P. (1978). Tooth replacement rates of the sharks *Trikis semifasciata* and *Ginglymostoma cirratum*. *Zoologisches Jahrbüch Anat.*, **99**, 151–156.

Roberts, T. R. (1967). Tooth formation and replacement in characoid fishes. *Stanford Ichthyological Bulletin*, **4**, 231–247.

Shaw, J. P. (1979). The time scale of tooth development and replacement in *Xenopus laevis* (Daudin). *Journal of Anatomy*, **129**, 323–342.

Shaw, J. P. (1985). Tooth replacement in adult *Xenopus laevis* (Amphibia: Anura). *Journal of Zoology, London*, **207**, 171–179.

Shaw, J. P. (1989). Observations on the polyphyodont dentition of *Hemiphractus proboscideus* (Anura: Hylidae). *Journal of Zoology, London*, **217**, 499–510.

Shellis, R. P. and Berkovitz, B. K. B. (1976). Observations on the dental anatomy of piranhas (Characidae) with special reference to tooth structure. *Journal of Zoology, London*, **180**, 69–84.

Sneyd, J., Atri, A., Fergusson, M. J., Lewis, M. A., Seward, W. and Murray, J. D. (1993). A model for the spacial patterning of teeth primordia in the alligator: initiation of the dental determinant. *Journal of Theoretical Biology*, **165**, 633–658.

Strasburg, D. W. (1963). The diet and dentition of *Isistius brasiliensis*, with remarks on tooth replacement in other sharks. *Copeia* 33–40.

Tuisku, F. and Hildebrand, C. (1994). Evidence for a neural influence on tooth germ generation in a polyphyodont species. *Developmental Biology*, **165**, 1–9.

Wake, M. H. (1976). The development and replacement of teeth in viviparous caecilians. *Journal of Morphology*, **148**, 33–64.

Wake, M. H. (1980). Fetal tooth development and adult replacement in *Dermosis mexicanus* (Amphibia: Gymnophiona): fields versus clones. *Journal of Morphology*, **166**, 203–216.

Wakita, M., Itoh, K. and Kobayashi, S. (1977). Tooth replacement in the teleost fish *Prionurus microlepidotus* Lacepede. *Journal of Morphology*, **153**, 129–142.

Wass, R. C. (1973). Size, growth, and reproduction of the sandbar shark, *Carcharhinus milberti*, in Hawaii. *Pacific Science*, **27**, 305–318.

Weishampel, D. B. (1991). A theoretical morphological approach to tooth replacement in lower vertebrates. In *Constructional Morphology and Evolution*, eds. N. Schmidt-Kitter and K. Vogel, pp. 295–310. Berlin, Springer.

Westergaard, B. (1980). Evolution of the mammalian dentition. *Societé Geologique de France. Memoires, N. S.*, **139**, 191–200.

Westergaard, B. (1983). A new detailed model for mammalian dentitional evolution. *Zeitschrift für Zoologische Systematik und Evolutionsforschung* **21**, 68–78.

Westgaard, B. (1986). The pattern of embryonic tooth initiation in reptiles. In *Teeth Revisited*, Proceedings of the VII International Symposium on Dental Morphology, Paris, eds. D. E. Russel, J-P. Santori and D. Sigogneau-Russel, *Mémoires Museum National de Histoire Naturelle (Series C)*, **53**, 55–63.

Westergaard, B. (1988). Early dentition development in the lower jaws of *Anguis fragilis* and *Lacerta agilis*. *Societas pro Fauna et Flora Fennica. Memoranda*, **64**, 148–151.

Westergaard, B. and Ferguson, M. W. J. (1986). Development of the dentition in *Alligator mississippiensis*. Early embryonic development in the lower jaw. *Journal of Zoology, London*, **210**, 575–597.

Westergaard, B. and Ferguson, M. W. J. (1987). Development of the dentition in *Alligator mississippiensis*. Later development in the lower jaws of hatchlings and young juveniles. *Journal of Zoology, London*, **212**, 191–222.

Westergaard, B. and Ferguson, M. W. J. (1990). Development of the dentition in *Alligator mississippiensis*: upper jaw dental and craniofacial development in embryos, hatchlings, and young juveniles, with a comparison to lower jaw development. *American Journal of Anatomy*, **187**, 393–421.

Woerdeman, M. W. (1919). Beitrage zur Entwicklungsgeschichte von Zahnen und Gebiss der Reptiilien. I. Die Anlage und Entwicklung des embryonalen Gebisses als Ganzes und seine Beziehung zur Zahnleiste. *Archiv für mikroskopielle Anatomie*, **92**, 104–192.

Woerdeman, M. W. (1921). Beitrage zur Entwicklungsgesschichtee von Zahnen und Gebiss der Reptilien. IV. Uber die Anlage und Entwicklung der Zahne. *Archiv für mikroskopielle Anatomie*, **95**, 265–395

Ziegler, A. C. (1971). A theory of the evolution of therial dental formulas and replacement patterns. *Quarterly Review of Biology*, **46**, 226–249.

14 The evolution of tooth shape and tooth function in primates

P. M. Butler

14.1. Introduction

Evolution proceeds by selection of phenotypic variations (in this case, of functioning dentitions), and so indirectly of the genotypes that produce them. Evidence for the course of evolution is derived largely from the fossil record, and much of it is provided by teeth. To understand evolution we need to know not only what changes have in fact taken place but also why these changes were of advantage in the functional life of the animal and have thus been selected; in other words, what is the adaptive significance of evolutionary change? In this chapter I will review some of the changes of tooth shape that have taken place in the course of primate evolution, and which appear to be explicable as adaptations to changes of function. Dental adaptations may be considered at two levels. First, there are characters that integrate the teeth into a masticatory system, involving the jaws and their muscles. Then there are modifications of the system in adaptation to functional demands. Most often, these demands are put upon the system by the types of foods eaten, and by feeding behaviour, but for some teeth, especially the incisors and canines, these functions may be unconnected with feeding (for example, grooming and fighting).

14.1.1. Brief introduction to fossil primates

Primates were derived from primitive insectivorous eutherians in the Cretaceous, and I will use one of these, *Kennalestes* (Kielan-Jaworowska, 1968, 1981), as a basis of comparison to show the direction of evolutionary change. The early palaeontological history of primates has recently been reviewed by Rose (1995) and Rose *et al.* (1994). The oldest form that has been classified as a primate is *Purgatorius*, from the terminal Cretaceous and early Palaeocene of North America. It belongs to the Plesiadapiformes, a mainly Palaeocene group that was until recently included in the Primates, owing to dental resemblances. However, plesiadapiformes lack certain primate characters of the skull and limbs, and they are now regarded as a separate, but related group (Cartmill, 1972; Martin, 1986; Andrews, 1988; Beard, 1993). True primates appeared as immigrants in North America and Europe at the beginning of the Eocene, and the order probably originated in Asia or Africa. Two early groups are recognized: Adapidae and Omomyidae. The omomyids have been regarded as related to the tarsiers (Tarsiiformes), but direct links have yet to be established. The Anthropoidea (or Simiiformes) share some non-dental characters with *Tarsius*, but Gingerich (1976, 1981a) pointed to resemblances to adapids in the teeth. The diversity of primitive anthropoids in the Eocene and Oligocene of North Africa (Godinot, 1994; Simons, 1992) seems to indicate that the suborder originated on that continent. At the present time, the Anthropoidea are divided into the Platyrrhini (Ceboidea) in South America and the Catarrhini in Asia and Africa. Within the latter group, the Old World monkeys (Cercopithecoidea) and apes (Hominoidea) originated in Africa and spread to Eurasia in the Miocene.

14.1.2. Regional differentiation of the dentition

The heterodont dentition of mammals had already been established in the Jurassic (see Lillegraven *et al.*, 1979). It reflects a regional distribution of function: the anterior teeth, being farther from the articulation of the jaw, have a greater range and speed of movement, whereas the molars, situated nearer to the muscles, can exert greater occlusal pressure. The anterior teeth are therefore involved mainly in the prehension and preliminary

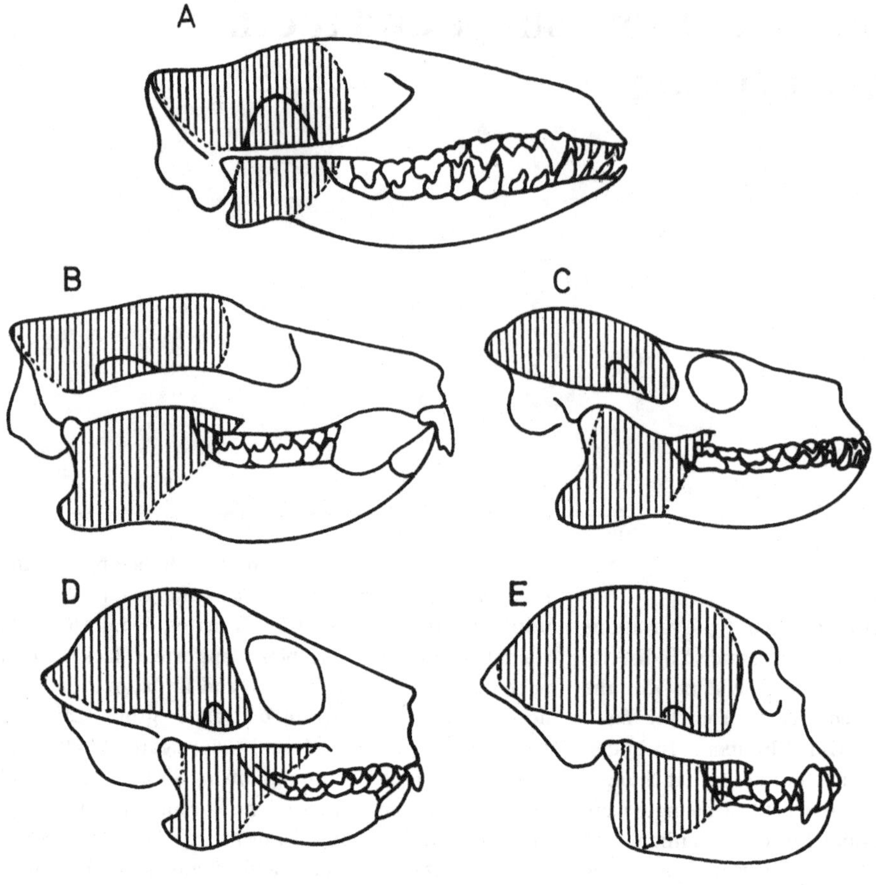

Figure 14.1. **Skull and mandible of (A)** *Kennalestes,* **(B)** *Plesiadapis* **(C)** *Notharctus* **(D)** *Necrolemur* **(E)** *Cebus,* **to illustrate shortening of the face in primates. Position of adductor jaw muscles shaded.**

treatment of food, preparatory to its mastication by the molars. Anterior teeth are best placed to function by using the kinetic energy of closure rather than sustained pressure (Olson, 1961). In primitive insectivorous mammals the force was concentrated in the canines, while the incisors were adapted for the precision movements of picking up small objects. The premolars, especially the more posterior ones, usually have elevated cusps used in puncture-crushing food taken in at the side of the mouth (Hiiemae and Kay, 1973).

Permanent teeth anterior to the molars have deciduous predecessors, adapted to the needs of the juvenile animal in which the jaw is shorter. The premolar–molar transition takes place farther forward in the deciduous dentition, so that the last deciduous premolar (usually called dP 4 for evolutionary reasons) is always molariform. The deciduous premolars and permanent molars form a continuous morphological series, and the break in the sequence of patterns between premolars and molars in the permanent dentition is due to replacement of the deciduous premolars by the permanent ones. Functionally, the molars belong with the premolars to

the permanent dentition, whereas developmentally they belong with the deciduous teeth to the primary dentition.

There is an evolutionary trend in primates to shorten the face and jaws (Figure 14.1). This is probably an adaptation to arboreal life, as it is associated with reduction of the olfactory organs and forward rotation of the orbits for stereoscopic vision. The prehensile hand, another arboreal adaptation, is used to convey food to the mouth. There is also a reduction in the number of teeth, perhaps to reduce crowding in the shorter jaw. No true primate has a third incisor, and the first premolar is retained only in some adapids and variably in the most primitive omomyids (*Teilhardina, Steinius*). In catarrhines the second premolar has also been lost.

Shortening the jaws brings the incisors nearer to the muscles and so enables them to exert a greater sustained force. The incisors have accordingly become specialized in various ways, in adaptation to methods of feeding. The gnawing incisors of the aye-aye (*Daubentonia*) are an extreme example.

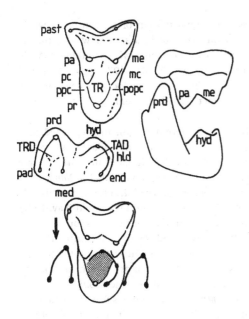

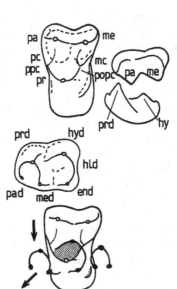

Figure 14.2. **Comparison of second upper molars of *Kennalestes* (left) and a primitive primate, *Teilhardina* (right). Anterior (mesial) is to the left. In both cases the teeth are shown in crown and buccal views, with an occlusal diagram with the teeth in centric occlusion; end, entoconid; hld, hypoconulid; hyd, hypoconid; mc, metaconule; me, metacone; med, metaconid; pa, paracone; pad, paraconid; past, parastyle; pc, paraconule; popc, postprotocrista; ppc, preprotocrista; pr, protocone; prd, protoconid; TAD, talonid; TR, trigon; TRD, trigonid. Arrows on occlusal diagrams show direction of movement of the lower tooth during phases 1 and 2.**

14.2.   The primitive primate molar pattern

Like other eutherians, primates have been derived from Cretaceous ancestors with tribosphenic molars (Bown and Kraus, 1979; Butler, 1990; Figure 14.2 left). In these teeth, the upper molar is triangular, with an elevated paracone and metacone and a lingual protocone, surrounding the trigon basin. The mesial and distal margins of the trigon basin are formed from crests of the protocone (pre- and post-protocrista), and minor cusps, paraconule and metaconule, develop on these. The margin of the tooth buccal to the paracone and metacone is occupied by the stylar shelf, which primitively bore cusps; these have disappeared in most eutherians, except for the parastyle at the mesiobuccal corner of the tooth. The lower molar consists of a mesial trigonid and a distal talonid. The trigonid is an elevated triangular prism, with three cusps, protoconid buccally, and paraconid and metaconid lingually. It occluded in the embrasure between two upper molars. The talonid, much lower than the trigonid, has three cusps surrounding a basin: hypoconid buccally, hypoconulid distally and entoconid lingually. In occlusion, the hypoconid passed between paracone and metacone of the upper molar, into the trigon basin, and the talonid basin received the protocone.

The occlusal relations of tribosphenic molars have been investigated by the study of wear facets on fossil teeth (Crompton, 1971; Butler, 1977; Crompton and Kielan-Jaworowska, 1978), and by cineradiogaphic study of mastication (Crompton and Hiiemae, 1970; Hiiemae and Kay, 1973; Kay and Hiiemae, 1974). Early eutherians were probably insectivorous. The food was pierced with sharp cusps and cut with crests that operated like the blades of scissors (Crompton, 1971; Every, 1974; Osborn and Lumsden, 1978), and crushed between protocone and talonid. The spread of flowering plants during the later Cretaceous opened the way to a diet which included plant material such as fruit, seeds and shoots (Szalay, 1972). Change in the physical nature of the food led to adaptive modifications of the molars (Lucas, 1979; Chapter 20, section 20.4). Primates are one of a number of groups in which this dietary shift took place.

What seems to be the most primitive type of primate molar occurs, with minor variations, both in plesiadapiformes, e.g. *Purgatorius*, *Palenochtha*, *Palaecthon*, *Pronothodectes* and in some early primates, e.g. the adapid *Donrussellia* and the omomyid *Teilhardina*. Comparison with *Kennalestes* (Figure 14.2) indicates a reduction of piercing and cutting in favour of crushing and grinding. The paracone and metacone on the upper molar and the trigonid on the lower molar were reduced in height, and there was less penetration between upper and lower cusps. Chewing was more transverse, and the trigonid no longer penetrated into the interdental embrasure. Opposing surfaces on the slopes of the cusps were less steep, so that food between the teeth was subjected to greater vertical force. The cusps moved in grooves in which food was reduced by shearing between parallel-moving surfaces ('rolling crush' of Osborn and Lumsden, 1978). The trigon basin was enlarged by broadening the lingual part of the tooth, and the talonid basin was widened.

During the power stroke of chewing (Hiiemae, 1978; Weijs, 1994) the lower molars travel lingually and upwards across the upper molars. In primitive tribosphenic molars this movement ends when the lingual surface of the hypoconid meets the buccal surface of the protocone, crushing the food. The lower teeth are then at their highest position, and are said to be in centric occlusion. They then separate for the opening stroke (Crompton and Kielan-Jaworowska, 1978). In mammals such as primates, in which the power stroke is less steeply inclined, the lingual movement is continued past the centric position, so that the hypoconid retains contact with the protocone during the early phase of the opening stroke (Figure 14.2). Two sets of wear facets can be distinguished, according to whether they are formed when the lower molar is buccal or lingual to the centric position (Mills, 1963, 1973, 1978; Butler, 1973). Kay and Hiiemae (1974) called the two parts of the power stroke phase 1 and phase 2. In primates the phase 2 movement is directed more anteriorly than the phase 1 movement, as the centre of rotation of the jaw shifts towards the opposite side of the mouth; the tip of the hypoconid travels mesiolingually, towards the notch between the protocone and the paraconule. The protoconid also travels mesiolingually, along the distal cingulum of the more anterior upper molar, and when in more advanced primates a hypocone develops from this cingulum (Hunter and Jernvall, 1995; Chapter 19), protoconid–hypocone occlusion parallels hypoconid–protocone occlusion.

14.3.   Evolution of the hypocone

A distolingual cusp, or hypocone, evolved in many orders of mammals, mainly in the later Eocene and Oligocene, and especially in those that were partly or entirely herbivorous (Hunter and Jernvall, 1995). Among primates a hypocone developed independently in omomyids (e.g. *Microchoerus*), adapids (e.g. *Adapis*) and anthropoids (absent in the Late Eocene *Oligopithecus*). It is lacking among living primates only in some lemuriforms (Lemuridae, Cheirogalidae), in *Tarsius*, and in marmosets; in the last case its absence is secondary (Kay, 1994).

Usually the hypocone appears as a small cusp at the lingual end of the distal cingulum (Figure 14.3B). It first contacts the paraconid and then (in phase 2 of occlusion) the protocone (Figure 14.3G). Eventually, the cingulum expands to form a talon which increasingly overlaps the trigonid. As the hypocone enlarges, the paraconid disappears, and the hypocone makes contact with the metaconid, passing between it and the entoconid of the more mesial tooth in transverse jaw movement. Its buccal surface is worn by the protoconid in phase 2. The hypo-

cone forms the lingual margin of a talon basin which occludes with the trigonid in a similar way to that in which the trigon basin occludes with the talonid (Mills, 1963; Kay and Hiiemae, 1974; Kay, 1977; Figure 14.3G). In primitive catarrhines the hypocone develops an additional contact with the enlarged hypoconulid (Butler, 1986).

A hypocone of different origin, distinguished as a pseudohypocone, occurs in the adapid subfamily Notharctinae (Figure 14.3E,F). It arises in *Cantius* from a distal crest of the protocone, known as the *Nannopithex* fold, which also occurs in several omomyids. This fold occludes with the paraconid and protoconid (Kay, 1977), like the distal cingulum of other primates, but it stands higher on the crown, and so occludes with a lower trigonid. When fully developed in *Notharctus* the pseudohypocone has the same functions as a true hypocone, the paraconid having merged with the metaconid.

The *Nannopithex* fold of primates has been homologized, probably incorrectly, with the 'postprotocingulum' of plesiadapiformes, a distolingual crest on the protocone that connects with the lingual end of the distal cingulum (Figure 14.3A). The prehypocone crest, which in many anthropoids joins the hypocone to the trigon, is a secondary development and not a relic of the *Nannopithex* fold.

14.4.   Molars

14.4.1.   Differences between the molars

In primitive eutherians such as *Kennalestes* M^2 is wider and proportionately shorter than M^1, and M^3 is much shorter, with the metacone considerably reduced. These differences survived in primitive primates, but there is a widespread tendency for the molars to become more equal in size. This was perhaps associated with elevation of the jaw joint, which would have the effect of equalizing occlusal pressure along the tooth row (Greaves, 1974; Weijs, 1994). On the other hand, reduction of posterior molars in ceboids, leading to loss of the third molars in marmosets (Ford, 1980), is associated with a forward shift of the zygomatic arch: the largest molar is situated below the zygoma root, where the masseter muscle can exert most force (Bluntschli, 1911).

Differences between the lower molars reflect their occlusal relations (Figure 14.3J). The trigonid of M_1 occludes with the distal half of P^4. In primitive primates the paraconid of M_1 is therefore larger and farther separated from the metaconid than on M_2. The talonid of M_3 is extended in length by a distally prominent hypoconulid; this occludes with the distal surface of M^3 in place of the

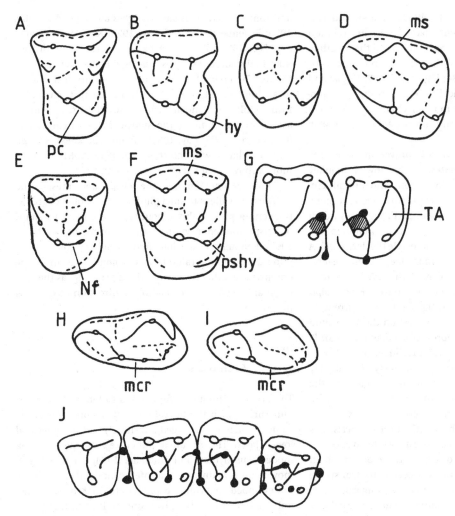

Figure 14.3A–F. **Left upper molars. A.** *Palaechthon* (plesiadapiform). **B.** *Galago.* **C.** *Aotus* (ceboid). **D.** *Alouatta* (folivorous ceboid). **E.** *Cantius* (adapid). **F.** *Notharctus* (adapid). pc, postprotocingulum; hy, hypocone; ms, mesostyle; Nf, *Nannopithex* fold; pshy, pseudohypocone. G. Occlusal diagram of a primate with large hypocone; lower molar (cusps black) in phase 2 (note that the protoconid bites against the talon (TA) in the same way that the hypoconid bites against the trigon). H. I. Lower molars of (H) *Adapis* and (I) *Lemur*, to show metacristid (mcr). J. Generalized occlusal diagram (centric occlusion) to show the relationship of lower molar cusps (black) to P[4] and the upper molars (note that the trigonid of M$_1$ occludes with the distal part of P[4], and the enlarged hypoconulid of M$_3$ occludes distally to the metacone of M[3]).

trigonid of the missing M$_4$. The hypoconulid on M$_3$ thus enlarges in correlation with the development of a metacone and hypocone on M[3]. When large, it can be comparable in size with the hypoconid and protoconid, giving the tooth a 'three-lobed' appearance, e.g. in most ceropithecoids (but not *Cercopithecus* and *Erythrocebus*; Swindler, 1983). There is less difference between M$_3$ and M$_2$ in hominoids, where the hypoconulid is enlarged on both teeth. However, when the hypoconulid is reduced on M$_2$ of humans it is often retained on M$_3$ (Dahlberg, 1945).

14.4.2. Dietary adaptations of molars
The relationship of molar form to diet was investigated by Kay (1975). He classified the diets of living primates according to whether they consisted largely (>45%) of insects (and other small animals), fruit (and soft plant material such as flowers and gums), and leaves (plant material with a high cellulose content). He measured

features of M$_2$ related to shearing, crushing and grinding. After elimination of the allometric effect of body size, he found that frugivores have smaller and lower-crowned molars, with shorter shearing crests, than insectivores or folivores. The last two categories are distinguishable by size: folivores are larger, with body weights exceeding 500 g. The lower energy content of leaves necessitates a more bulky food intake, and a long digestive process is required to break down cellulose cell walls (Szalay, 1972; Kay and Hylander, 1978; Janis and Fortelius, 1988). Nearly all primates eat some fruit, and according to the proportion eaten, frugivory intergrades into insectivory on the one hand and folivory on the other. Early fossil primates fall mostly in the insectivore range (Szalay, 1975; Kay and Covert, 1984; Williams and Covert, 1994); the main trend of evolution was towards frugivory, and folivory is a derived condition in which shearing adaptations have been secondarily increased.

Related species may differ in food preference, e.g.

lemurs (Kay et al., 1978) and ceboids (Anapol and Lee, 1994). A shearing quotient, based on the relation between the total length of crests on M_2 and the length of the tooth, is correlated with the proportion of insects or leaves in the diet (Kay, 1984; Kay et al., 1987; Ungar and Kay, 1995). The interspecific differences in shearing quotients are probably related to biomechanical differences in foods and how they are processed (Lucas, 1979; Chapter 20).

Shearing efficiency has been improved in more than one way in different primates. In the howler monkey (Alouatta) a mesostyle develops from the buccal cingulum, opposite the notch between paracone and metacone. The V-shaped hypoconid passes through this notch in phase 1, and it shears against crests that join the paracone and metacone to the mesostyle (Butler, 1996). A similar adaptation occurs in Indriidae, and in several fossil genera, e.g. Plesiadapis (Gingerich, 1976), Notharctus (Gregory, 1920) and the Pleistocene lemuroid Megaladapis (Thenius, 1953). Another shearing adaptation is strengthening of the metacristid, a distal crest on the metaconid which shears with the preprotocrista (Figure 14.3H,I). This occurs in Lemuridae and Indriidae, and also in Adapis (Maier 1980). A cusp, the metastylid (probably equivalent to cusp 7 in humans) may develop on the metacristid. In an experiment in which Galago crassicaudatus and three species of lemur were fed carrots, Sheine and Kay (1982) found that the length of the metacristid is significantly correlated with masticatory effectiveness.

The bilophodont teeth of cercopithecoids, in which buccal and lingual cusps are connected by transverse ridges (lophs), are unlike those of tapirs, kangaroos, etc., in that the ridges are not shearing blades. Shearing is performed by mesial and distal crests on the buccal and lingual cusps. The colobines eat mainly leaves, but the lophs function as wedges to break large, tough seeds (Lucas and Teaford, 1994). Cercopithecines, by contrast, eat more fruit, and many are ground feeders, eating grass and roots. Their lophs serve to crush and grind food in the intervening valleys (Kay, 1978). The lophs guide the chewing movement, in which phase 2, instead of being mesiolingual as in other primates, continues the transverse movement of phase 1; the hypoconid passes distally to the protocone, between protocone and hypocone (Mills, 1973, 1978; Maier, 1977a; Butler, 1986). Delson (1975) suggested that the bilophodont molars of cercopithecoids originally evolved as an adaptation to leaf eating. There was a parallel development of bilophodonty in the lemuroid family Indriidae, again associated with folivory (Maier, 1977b).

Adaptation for crushing and grinding is increased in primates that eat seeds and hard fruits (durophagy), e.g. the orang-utan, some cercopithecoids such as Cercocebus,

and some platyrrhines such as Cebus and Callicebus (Jolly, 1970; Rosenberger and Kinzey, 1976; Kay, 1981; Anapol and Lee, 1994). They have flattened crowns, with low, marginally situated cusps, and wide, shallow crushing basins. The enamel is thick, in compensation for heavy wear due to high occlusal loads. In contrast, enamel is thin on folivorous molars, where the exposure of dentine assists in maintaining the sharpness of cutting edges (Rensberger, 1973; Kay, 1981). Enamel thickness is inversely correlated with crest length (Kay, 1981). Human molars show durophagous characteristics.

Thick enamel rounds the cusps and crests, and reduces the valleys to sulci (Butler, 1986). Some degree of thickening presumably accounts for the bulbous shape of cusps (bunodonty) in early anthropoids (Godinot, 1994). Enamel crenulations and subsidiary folds in the talonid, trigon and talon basins of hominoids and some omomyids would presumably aid in grinding, as they are often arranged transversely to the direction of chewing movement.

14.5.   Premolars

The primitive premolar function is to tenderize food by puncturing with the buccal cusps (paracone and protoconid), which are arranged alternately in the upper and lower jaws. The so-called puncture-crushing chewing mode is used (Kay and Hiiemae, 1974), in which jaw movement is mainly vertical. Metacones and talonids are undeveloped in primitive premolars. In Kennalestes the only lingual cusp is the protocone on P^4, and this is also the case in the most primitive plesiadapiformes, e.g. Purgatorius and Palaechthon (Figure 14.4A). In early euprimates there is a protocone on P^3, developed from the lingual cingulum, and a metaconid on P_4, on a distolingual crest of the protoconid. When larger, the metaconid stands lingual to the protoconid, the two cusps forming a transverse ridge that bites between transverse paraconeprotocone ridges on P^4 and P^3, to form a crushing mechanism. The function has extended forward by one tooth in ceboids, where there is a protocone on P^2 and a metaconid on P_3 (Figure 14.4B).

Some shearing takes place between the mesiodistally running longitudinal crests of the paracones and protoconids, and that shearing function has been emphasized in some primates. In indriids, for example (Figure 14.4D), the premolars form a serrated cutting edge used to slice rind from fruit (Kay and Hylander, 1978); the sectorial premolars of the extinct Archaeolemur probably had a similar function (Szalay and Delson, 1979). The most highly specialized shearing premolars are those of the

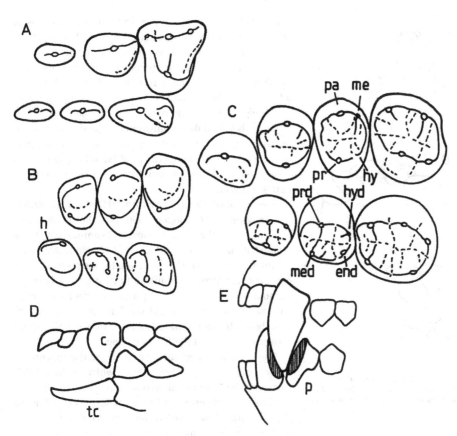

Figure 14.4A. **Upper and lower premolars of** *Palaechthon* **(plesiadapiform). B. The same,** *Aotus* **(ceboid); h, honing surface. C. Human upper canine. M 1/, and P/3–M/1, showing partial molarization of the posterior premolars. D. Anterior dentition of** *Indri* **(lemuroid); c, upper canine; tc, lower incisors forming tooth comb. E. Anterior dentition of** *Cebus* **(an anthropoid), showing honing of the upper canine by the anterior premolar (p) and the lower canine.**

plesiadapiform family Carpolestidae (Rose, 1975). By contrast, in durophagous forms like *Cebus* and *Pongo* the premolars are adapted for crushing, with blunt cusps, thick enamel, and relatively large protocones.

Molarization incorporates the last premolars into the region of molar function. It involves development on P^4 of a metacone, postprotocrista and hypocone (if present on the molar), and enlargement of the talonid and development of an entoconid on P_4. Degrees of molarization can be seen in the adapids *Notharctus* and *Adapis*, and in the plesiadapiformes *Plesiolestes* and *Microsyops*. Among living primates, it is most advanced in *Galago* and *Hapalemur* (Maier, 1980). Rudimentary molar cusps also occur on human premolars (Figure 14.4C).

The anterior lower premolar of anthropoids is adapted for honing the upper canine. P_2 of platyrrhines and P_3 of catarrhines are modified in the same way: the posterior crest of the canine is kept sharp by shearing against a forwardly extended ridge on the premolar, with thick enamel (Zingeser, 1969; Every, 1970; Ryan, 1979; Walker, 1984; Figure 14.4B,E). The metaconid is reduced or absent on the honing tooth. In humans, owing to the small canine, P_3 resembles P_4. Specialized shearing between P_2 and the upper canine developed also in strep-

sirhines. P^2 is higher than P^3 in lorisoids, and it may function in sharpening P_2. In *Phaner* P^2 looks like a second canine.

14.6.   Canines and incisors

In *Kennalestes* the incisors in each jaw are small, simple teeth, arranged in a longitudinal row anterior to the canine (Figure 14.5A). The canines are larger, but they resemble anterior premolars in shape, having two roots. Upper incisors and canine are vertical; those of the lower jaw are somewhat procumbent. Shortening of the face in primates resulted in bringing the incisors closer to, and eventually between the canines, and enabled the anterior teeth to exert greater force.

In plesiadapiformes the anterior incisors are enlarged to form a pair of forceps. I_1 is very procumbent, and I^1 has additional cusps that would improve the grip (Gingerich, 1976; Rose and Beard, 1993). These teeth may have been used for picking up insects. In the case of *Plesiadapis*, which was probably plant-eating, Gingerich (1976) suggested that the incisors were adapted for cutting stems.

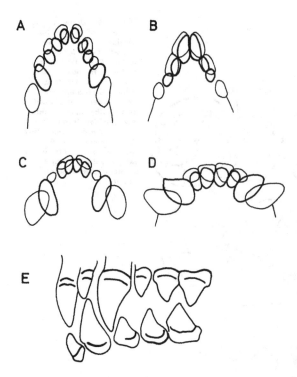

Figure 14.5A–D. **Anterior upper and lower incisors and canines superimposed; lower teeth with heavier outline. A. *Kennalestes*. B. *Ourayia* (omomyid). C. *Notharctus* (adapid). D. *Cebus* (anthropoid). E. Anterior dentition of *Tarsius*.**

Forceps-like incisors also developed in omomyids (Figure 14.5B), which were probably mainly insectivorous. With enlargement of the incisors, the canines and anterior premolars are reduced (Rose and Bown, 1984). *Tarsius* uses its anterior teeth to kill prey that it catches with its hands (Maier, 1984). Though I^1 is enlarged, I_1 remains small, and the canines are not reduced, but form with I^1 the main killing teeth (Figure 14.5E).

In adapids, and especially in anthropoids, the incisors are broadened and arranged more transversely to form a horizontal incisive edge (Figure 14.5C,D). At first, as in the adapid *Notharctus* (Rosenberger *et al.*, 1985) I^1 was widened to shear with both I_1 and I_2. The incisal shear was subsequently extended to the I^2–I_2 contact (Figure 14.5D); when the lower canine is small, as in man and many female anthropoids, it becomes involved with I^2 in the incisal function (Greenfield, 1993). Incisive shear was probably primarily a frugivorous adaptation, for such actions as biting pieces from fruit held in the hand, though it is also used in nipping tough leaves and stems. In living anthropoids incisor width is greater in frugivores than in folivores (Hylander, 1975; Kay and Hylander, 1978; Eaglen, 1984). Sharpness is maintained, especially in frugivores, by thinning of the lingual enamel, which is virtually absent in some cercopithe-

cines (Shellis and Hiiemae, 1986) and in some marmosets that use their incisors to gouge bark (Rosenberger, 1978). Incisors are used in a variety of feeding methods. Ungar (1994) classified their functions into nipping, incising, crushing, scraping and stripping. Where incisors are adapted for gripping and tearing, or for stripping leaves from twigs, the incisal edge is blunt and wears flat.

There is sexual dimorphism in canine size, relative to molar size, in most anthropoids and some adapids (Fleagle *et al.*, 1980; Gingerich, 1981b; Kelley, 1995). It is associated with the amount of competition between males (Kay *et al.*, 1988).

The 'tooth comb' of lemuriforms (Martin, 1972; Szalay and Seligsohn, 1977; Eaglen, 1980, 1986; Rose *et al.*, 1981) is formed from the lower incisors and canines, which are elongated and horizontal (Figure 14.4D). (In indriids the canine is absent, and the tooth comb consists of incisors only; Luckett, 1986.) In addition to its use for grooming, the tooth comb is employed by many lemurs for scraping bark and collecting gum and sap. The upper incisors are usually small and peg-like, and leaf-eaters use the upper lip against the lower teeth in plucking.

The most modified anterior teeth are the rodent-like incisors of the aye-aye (*Daubentonia*), which Luckett (1986) showed to be unreplaced second milk incisors. They are rootless teeth of permanent growth, deeply embedded in the jaw (the lower tooth reaches back beyond the molars), and with enamel confined to the labial surface. They are used for gnawing into wood to obtain insect larvae. The origin of this adaptation is unknown: possibly it evolved from a condition like that of *Phaner* and *Allocebus*, where enlarged, protuberant upper incisors are used in conjunction with the tooth comb to prise exudates from tree trunks (Martin, 1972).

Summary

All teeth are basically similar, formed in essentially the same way from enamel organ and papilla, with crown, root, enamel, dentine and so forth, but the diversity of their shapes shows that the morphogenetic processes by which they are produced are very modifiable. These processes are still imperfectly understood, but rapid progress is being made concerning the identification of the molecular gene products involved, and their distribution in the jaw and in the individual tooth germs (see Chapter 2). Teeth are serially repeated (meristic) structures, like vertebrae and the segments of insects, and their development probably has something in common with these (Weiss, 1993; Butler, 1995). The number of genes involved is undoubtedly large. Combinations of genes have been selected, so that the teeth have not evolved independently, but in correlation with their neighbours and occlusal opponents. They are parts of a functional system that necessarily has remained operational throughout evolutionary change.

I have tried to interpret in functional terms the main trends of primate dental evolution from a primitive eutherian such as *Kennalestes*. The changes seem to be the result of adaptation to arboreal life in which increasing amounts of plant material, especially fruit, were added to the primitive insectivorous diet. Arboreal adaptations included a shorter face, associated with more powerful anterior teeth, and a prehensile hand that could be used in feeding. Adaptations to frugivory resulted in lower molar cusps, expansion of crushing-grinding surfaces (including the hypocone), and transverse arrangement of the incisors. Many primates advanced further, to leaf eating, with increased body size and greater emphasis on cutting crests, or to hard-object feeding, with thickened enamel and flat-crowned teeth. These changes took place more than once, in different groups of primates, and to different extents. Parallelism and convergence are widespread in dental evolution, adding to the difficulties of phylogenetic analysis. Similar animals, exposed to similar selective pressures, are likely to evolve in the same direction.

References

Anapol, F. and Lee, S. (1994). Morphological adaptations to diet in platyrrhine primates. *American Journal of Physical Anthropology*, 94, 239–261.

Andrews, P. (1988). A phylogenetic analysis of the Primates. In *The Phylogeny and Classification of the Tetrapods*, ed. M. J. Benton, pp. 143–175. Oxford: Clarendon Press.

Beard, K. C. (1993). Phylogenetic systematics of the Primatomorpha, with special reference to Dermoptera. In *Mammal Phylogeny, Placentals*, eds. F. S. Szalay, M. J. Novacek and M. C. McKenna, pp. 129–150. Berlin: Springer.

Bluntschli, H. (1911). Das Platyrrhinengebiss und die Bolksche Hypothese von der Stammesgeschichte des Primatengebiss. *Verhandlungen Anatomische Gesellschaft*, 25, 120–136.

Bown, T. M. and Kraus, M. J. (1979). Origin of the tribosphenic molar and metatherian and eutherian dental formulae. In *Mesozoic Mammals: the First Two-Thirds of Mammalian History*, eds. J. A. Lillegraven, Z. Kielan-Jaworowska and W. A. Clemens, pp. 172–181. Berkeley: University of California Press.

Butler, P. M. (1973). Molar wear facets of Early Tertiary North American primates. *Symposium of the Fourth International Congress of Primatology*, 3, 1–27.

Butler, P. M. (1977). Evolutionary radiation of the cheek teeth of Cretaceous placentals. *Acta Palaeontologica Polonica*, 22, 241–271.

Butler, P. M. (1986). Problems of dental evolution in the higher primates. In *Major Topics in Primate and Human Evolution*, eds. B. Wood, L. Martin and P. Andrews, pp. 89–106. Cambridge: Cambridge University Press.

Butler, P. M. (1990). Early trends in the evolution of tribosphenic molars. *Biological Reviews*, 65, 529–552.

Butler, P. M. (1995). Ontogenetic aspects of dental evolution. *International Journal of Developmental Biology*, 39, 25–34.

Butler, P. M. (1996). Dilambdodont molars: a functional interpretation of their evolution. *Palaeovertebrata*, 25, 205–213.

Cartmill, M. (1972). Arboreal adaptation and the origin of the order Primates. In *The Functional and Evolutionary Biology of the Primates*, ed. R. Tuttle, pp. 97–122. New York: Aldine-Atherton.

Crompton, A. W. (1971). The origin of the tribosphenic molar. In *Early Mammals*, eds. D. M. and K. A. Kermack, *Zoological Journal of the Linnaen Society*, 50, (Supplement 1): 65–87.

Crompton, A. and Hiiemae, K. (1970). Molar occlusion and mandibular movements during occlusion in the American opossum *Didelphis marsupialis* (L.). *Zoological Journal of the Linnean Society*, 49, 21–47.

Crompton, A. W. and Kielan-Jaworowska, Z. (1978). Molar structure and occlusion in Cretaceous therian mammals. In *Development, Function and Evolution of Teeth*, eds. P. M. Butler and K. A. Joysey, pp. 249–287. London: Academic Press.

Dahlberg, A. A. (1945). The changing dentition of man. *Journal of the American Dental Association*, 321, 676–690.

Delson, E. (1975). Evolutionary history of the Cercopithecidae. In *Approaches to Primate Biology*, ed. F. S. Szalay. *Contributions to Primatology*, 5, 167–217.

Eaglen, R. H. (1980). Toothcomb homology and toothcomb function in extant strepsirhines. *International Journal of Primatology*, 1, 275–286.

Eaglen, R. H. (1984). Incisor size and diet revisited: the view from a platyrrhine perspective. *American Journal of Physical Anthropology*, 64, 263–275.

Eaglen, R. H. (1986). Morphometrics of the anterior dentition in strepsirhine primates. *American Journal of Physical Anthropology*, 71, 185–201.

Every, R. G. (1970). Sharpness of teeth in man and other primates. *Postilla*, 143, 1–30.

Every, R. G. (1974). Thegosis in prosimians. In *Prosimian Biology*, eds. R. D. Martin, G. A. Doyle and A. C. Walker, pp. 579–619. London: Duckworth.

Fleagle, J. G., Kay, R. F. and Simons, E. L. (1980). Sexual dimorphism in early anthropoids. *Nature*, 287, 328–330.

Ford, S. M. (1980). Callitrichids as phyletic dwarfs, and the place of Callitrichidae in Platyrrhini. *Primates*, 2, 31–43.

Gingerich, P. D. (1976). Cranial anatomy and evolution of early Tertiary Plesiadapidae (Mammalia, Primates). *University of Michigan Papers in Paleontology*, 15, 1–141.

Gingerich, P. D. (1981a). Early Cenozoic Omomyidae and the evolutionary history of tarsiiform primates. *Journal of Human Evolution*, 10, 345–374.

Gingerich, P. D. (1981b). Cranial morphology and adaptations in Eocene Adapidae. 1. Sexual dimorphism in *Adapis magnus* and *Adapis parisiensis*. *American Journal of Physical Anthropology*, 56, 217–234.

Godinot, M. (1994). Early North African primates and their significance for the origin of Simiiformes (= Anthropoidea). In

Anthropoid Origins, eds. J. G. Fleagle and R. F. Kay, pp. 235–295. New York: Plenum Press.

Greaves, W. S. (1974). Functional implications of mammalian jaw joint position. *Forma et Functio*, **7**, 363–376.

Greenfield, L. O. (1993). A tooth at the border of two morphogenetic fields. *Human Evolution*, **3**, 187–204.

Gregory, W. K. (1920). On the structure and relations of *Notharctus*, an American Eocene primate. *Memoirs of the American Museum of Natural History*, N.S. **3**, 51–243.

Hiiemae, K. M. (1978). Mammalian mastication: a review of the activity of the jaw muscles and the movements they produce in chewing. In *Development, Function and Evolution of Teeth*, eds. P. M. Butler and K. A. Joysey, pp. 359–398. London: Academic Press.

Hiiemae, K. M. and Kay, R. F. (1973). Evolutionary trends in the dynamics of primate mastication. *Symposium of the Fourth International Congress of Primatology*, **3**, 24–68.

Hunter, J. P. and Jernvall, J. (1995). The hypocone as a key innovation in mammalian evolution. *Proceedings of the National Academy of Sciences, USA*, **92**, 10718–10722.

Hylander, W. L. (1975). Incisor size and diet in anthropoids with special reference to Cercopithecidae. *Science*, **89**, 1095–1098.

Janis, C. M. and Fortelius, M. (1988). On the means whereby mammals achieve increased functional durability of their dentitions, with special reference to limiting factors. *Biological Reviews*, **63**, 197–230.

Jolly, C. (1970). The seed-eaters: a new model of hominid differentiation based on a baboon analogy. *Man*, **5**, 5–26.

Kay, R. F. (1975). The functional adaptations of primate molar teeth. *American Journal of Physical Anthropology*, **43**, 195–216.

Kay, R. F. (1977). The evolution of molar occlusion in the Cercopithecidae and early catarrhines. *American Journal of Physical Anthropology*, **46**, 327–352.

Kay, R. F. (1978). Molar structure and diet in extant Cercopithecidae. In *Development, Function and Evolution of Teeth*, eds. P. M. Butler and K. A. Joysey, pp. 309–339. London: Academic Press.

Kay, R. F. (1981). The nut-crackers: a new theory of the adaptations of the Ramapithecinae. *American Journal of Physical Anthropology*, **55**, 141–151.

Kay, R. F. (1984). On the use of anatomical features to infer foraging behavior in extinct primates. In *Adaptations for Foraging in Non-Human Primates*, eds. P. C. Rodman and J. G. H. Cant, pp. 21–53. New York: Columbia University Press.

Kay, R. F. (1994). 'Giant' tamarin from the Miocene of Colombia. *American Journal of Physical Anthropology*, **95**, 333–353.

Kay, R. F. and Covert, H. H. (1984). Anatomy and behavior of extinct primates. In *Food Acquisition and Processing in Primates*, eds. D. J. Chivers, B. A. Wood and A. Bilsborough, pp. 467–508. New York: Plenum Press.

Kay, R. F. and Hiiemae, K. M. (1974). Jaw movement and tooth use in Recent and fossil primates. *American Journal of Physical Anthropology*, **40**, 227–256.

Kay, R. F. and Hylander, W. L. (1978). The dental structure of mammalian folivores with special reference to Primates and Phalangeroidea (Marsupialia). In *Biology of Arboreal Folivores*, ed. G. G. Montgomery, pp. 173–191. Washington: Smithsonian Institution.

Kay, R. F., Sussman, R. W. and Tattersall, I. (1978). Dietary and dental variations in the genus *Lemur*, with comments concerning dietary–dental correlations among Malagasy primates. *American Journal of Physical Anthropology*, **49**, 119–128.

Kay, R. F., Madden, R. H., Plavcan, J. M., Cifelli, R. L. and Deaz, J. G. (1987). *Stirtonia victoriae*, a new species of Miocene Columbian primate. *Journal of Human Evolution*, **16**, 173–196.

Kay, R. F., Plavcan, J. M., Wright, P. C. and Glander, K. E. (1988). Sexual selection and canine dimorphism in New World monkeys. *American Journal of Physical Anthropology*, **77**, 385–397.

Kelley, J. (1995). Sexual dimorphism in canine shape among great apes. *American Journal of Physical Anthropology*, **96**, 365–389.

Kielan-Jaworowska, Z. (1968). Preliminary data on the Upper Cretaceous Eutherian mammals from Bayn Dzak, Gobi Desert. *Palaeontologia Polonica* **19**, 171–191.

Kielan-Jaworowska, Z. (1981). Skull structure in *Kennalestes* and *Asioryctes*. *Palaeontologia Polonica* **42**, 25–78.

Lillegraven, J. A., Kielan-Jaworowska, Z. and Clemens, W. A. (1979). *Mesozoic Mammals: the First Two-Thirds of Mammalian History*. Berkeley: University of California Press.

Lucas, P. W. (1979). The dental–dietary adaptations of mammals. *Neues Jahrbuch für Geologie und Paläontologie*, **8**, 486–512.

Lucas, P. W. and Teaford, M. F. (1994). Functional morphology of colobine teeth. In *Colobine Monkeys, their Ecology, Behaviour and Evolution*, ed. A. G. Davies and J. F. Oates, pp. 173–203. Cambridge: Cambridge University Press.

Luckett, W. P. (1986). Developmental evidence for tooth homologies in strepsirhine primates (Abstract). *Primate Report*, **14**, 122–123.

Maier, W. (1977a). Die Evolution der bilophodonten Molaren der Cercopithecoidea. *Zeitschrift für Morphologie und Anthropologie*, **68**, 26–56.

Maier, W. (1977b). Die bilophodonten Molaren der Indriidae (Primates): ein evolutionsmorphologische Erklären. *Natur und Museum*, **108**, 288–300.

Maier, W. (1980). Konstruktionsmorphologische Untersuchungen am Gebiss der rezenten Prosimiae (Primates). *Abhandlungen der Senckenbergischen Naturforschenden Gesellschaft*, **538**, 1–158.

Maier, W. (1984). Functional morphology of the dentition of the Tarsiidae. In *Biology of Tarsiers*, ed. C. Niemitz, pp. 45–58. Stuttgart: Gustav Fischer.

Martin, R. D. (1972). Adaptive radiation and behaviour of the Malagasy lemurs. *Philosophical Transactions of the Royal Society of London*, **264**, 295–352.

Martin, R. D. (1986). Primates: a definition. In *Major Topics in Primate and Human Evolution*, ed. B. Wood, L. Martin and P. Andrews, pp. 1–31. Cambridge: Cambridge University Press.

Mills, J. R. E. (1963). Occlusion and malocclusion of the teeth of primates. In *Dental Anthropology*, ed. R. Brothwell, pp. 29–51. Oxford: Pergamon Press.

Mills, J. R. E. (1973). Evolution of mastication in primates. *Symposium of the Fourth International Congress of Primatology*, **3**, 65–81.

Mills, J. R. E. (1978). The relationship between tooth patterns and jaw movements in the Hominoidea. In *Development, Function, and Evolution of Teeth*, ed. P. M. Butler and K. A. Joysey, pp. 341–354. London: Academic Press.

Novacek, M. J. (1986). The primitive eutherian dental formula. *Journal of Vertebrate Paleontology*, **6**, 191–196.

Olson, E. C. (1961). Jaw mechanisms: rhipidistians, amphibians, reptiles. *American Zoologist*, **1**, 205–215.

Osborn, J. W. and Lumsden, A. G. S. (1978). An alternative to 'thegosis' and a re-examination of the ways in which mammalian molars work. *Neues Jahrbuch für Geologie und Paläontologie*, **156**, 371–392.

Rensberger, J. M. (1973). An occlusion model for mastication and dental wear in herbivorous mammals. *Journal of Paleontology*, **47**, 515–528.

Rose, K. D. (1975). The Carpolestidae, Early Tertiary primates from North America. *Bulletin of the Museum of Comparative Zoology*, **147**, 1–74.

Rose, K. D. (1995). The earliest primates. *Evolutionary Anthropology*, **3**, 159–173.

Rose, K. D. and Beard, K. C. (1993). Exceptional new dentitions of the diminutive plesiadapiforms *Tinimomys* and *Niptomomys*, with comments on the upper incisors of Plesiadapiformes. *Annals of the Carnegie Museum*, **62**, 351–361.

Rose, K. D. and Bown, T. M. (1984). Gradual phyletic evolution at the generic level in early Eocene omomyid primates. *Nature*, **309**, 250–252.

Rose, K. D., Walker, A. C. and Jacobs, L. L. (1981). Function of the mandibular toothcomb in living and extinct mammals. *Nature*, **289**, 583–585.

Rose, K. D., Godinot, M. and Bown, T. M. (1994). The early radiation of Euprimates and the initial diversification of Omomyidae. In *Anthropoid Origins*, ed. J. G. Fleagle and R. F. Kay, pp. 1–28. New York: Plenum Press.

Rosenberger, A. L. (1978). Loss of incisor enamel in marmosets. *Journal of Mammalogy*, **59**, 207–208.

Rosenberger, A. L. and Kinzey, W. G. (1976). Functional patterns of molar occlusion in platyrrhine primates. *American Journal of Physical Anthropology*, **45**, 281–298.

Rosenberger, A. L., Strasser, E. and Delson, E. (1985). Anterior dentition of *Notharctus* and the adapid- anthropoid hypothesis. *Folia Primatologica*, **44**, 15–39.

Ryan, A. S. (1979). Tooth sharpening in primates. *Current Anthropology*, **20**, 121–122.

Sheine, W. S. and Kay, R. F. (1982). A model for comparison of masticatory effectiveness in primates. *Journal of Morphology*, **172**, 139–149.

Shellis, R. P. and Hiiemae, K. M. (1986). Distribution of enamel on the incisors of Old World monkeys. *American Journal of Physical Anthropology*, **71**, 103–113.

Simons, E. L. (1992). Diversity in the early Tertiary anthropoidean radiation in Africa. *Proceedings of the National Academy of Sciences, USA*, **89**, 10743–10747.

Swindler, D. R. (1983). Variation and homology of the primate hypoconulid. *Folia Primatologica*, **41**, 112–123.

Szalay, F. S. (1972). Paleobiology of the earliest primates. In *The Functional and Evolutionary Biology of Primates*, ed. R. Tuttle, pp. 3–34. New York: Aldine-Atherton.

Szalay, F. S. (1975). Phylogeny, adaptations and dispersal of the tarsiiform primates. In *Phylogeny of the Primates*, ed. W. P. Luckett and F. S. Szalay, pp. 357–404. New York: Plenum Press.

Szalay, F. S. and Delson, E. (1979). *Evolutionary History of the Primates*. New York: Academic Press.

Szalay, F. S. and Seligsohn, D. (1977). Why did the strepsirhine toothcomb evolve? *Folia Primatologica*, **27**, 75–82.

Thenius, E. (1953). Zur Gebiss-Analyse von *Megaladapis edwardsi* (Lemur, Mammal.). *Zoologischer Anzeiger*, **150**, 251–260.

Ungar, P. S. (1994). Pattern of ingestive behavior and anterior tooth use differences in sympatric anthropoid primates. *American Journal of Physical Anthropology*, **95**, 197–219.

Ungar, P. S. and Kay, R. F. (1995). The dietary adaptations of European Miocene catarrhines. *Proceedings of the National Academy of Sciences, USA*, **92**, 5479–5481.

Walker, A. (1984). Mechanisms of honing in the male baboon canine. *American Journal of Physical Anthropology*, **65**, 197–220.

Weijs, W. A. (1994). Evolutionary approach of masticatory molar patterns in mammals. *Advances in Comparative and Environmental Physiology*, **18**, 281–320.

Weiss, K. M. (1993). A tooth, a toe and a vertebra: the genetic dimensions of complex morphological traits. *Evolutionary Anthropology*, **2**, 121–137.

Williams, B. A. and Covert, H. H. (1994). New Early Eocene anaptomorphine primate (Omomyidae) from the Washakie Basin, Wyoming, with comments on the phylogeny and paleobiology. *American Journal of Physical Anthropology*, **93**, 323–340.

Zingeser, M. R. (1969). Cercopithecid canine tooth honing mechanisms. *American Journal of Physical Anthropology*, **31**, 205–214.

15 'Schultz's Rule' and the evolution of tooth emergence and replacement patterns in primates and ungulates

B. Holly Smith

15.1. Introduction

Paleontologists often reconstruct the sequence of eruption of teeth of fossil mammals (e.g., Stehlin, 1912; Gregory, 1920; Kellogg, 1936; Lamberton, 1938; Tattersall and Schwartz, 1974; Wallace, 1977; Kay and Simons, 1983; Gingerich, 1984; Lucas and Schoch, 1990; Smith, 1994; Martin, 1997). Evolutionary studies, however, have contradictory traditions about the meaning of such sequences. One tradition is that sequence of tooth eruption is a good phylogenetic character, capable of showing genetic relatedness among species (e.g., Tattersall and Schwartz, 1974; Schwartz, 1974; Byrd, 1981), and thus presumably non-adaptive or conservatively adaptive. A second school of thought is that sequence of eruption is an adaptive characteristic: either reflecting dental morphology (e.g., Slaughter et al., 1974), or facial architecture (Simpson et al., 1990), or life history (Schultz, 1935, 1956). Indeed Schultz saw the eruption sequence of teeth as highly adapted to rate of post-natal growth. The obvious question is, does sequence of tooth eruption inform us of species taxonomic affiliation, dental function, facial form, or life history? One approach to the question is to re-examine Schultz's hypothesis on the adaptive nature of tooth eruption sequence. Here, Schultz's ideas are tested against newly gathered data on primates and ungulates, with a few additional data on small insectivorous mammals.

15.2. The dentition in a life-history context

Reptiles tend to grow throughout life, changing prey, and prey size slowly over time (Dodson, 1975). To keep pace with growth and changing diet, reptiles erupt multiple waves of simple, ever-larger teeth (see Chapter 13). Mammals, however, share a system of tooth emergence and replacement that accommodates rapid growth to a fixed body size (see Pond, 1977; MacDonald, 1984). Fixed body size probably allowed mammals to evolve a lasting adult dentition with heterodont teeth that occlude in a complex manner (see Zeigler, 1971; Osborn and Crompton, 1973). The small mammalian neonate, however, still must eat through a considerable size transition. To do this, eutherians (placental mammals) form small 'primary' or 'deciduous' teeth in utero, teeth which erupt around the time of birth in precocial mammals or after birth in altricial mammals (Smith et al., 1994). Most eutherians learn to eat with these teeth while still supplemented with mother's milk. Larger, sturdier 'secondary' or 'permanent' teeth appear later, including molars that fill in the back of the jaws and replacements that erupt underneath deciduous teeth. Metatherians (marsupial mammals) probably once shared this strategy, but subsequently reduced tooth replacement to a single tooth in each quadrant (Luckett, 1993; Cifelli et al., 1996; Martin, 1997), coping with early growth stages by intense lactation (Pond, 1977). For marsupial or placental, however, development and appearance of teeth must be coordinated with growth. Mammals musts have teeth to be weaned and the permanent teeth that erupt must be able to acquire, prepare, and chew adult food for a lifetime.

The order of emergence of mammalian teeth is highly patterned and a tiny fraction of theoretically possible variants appear in life. In eutherians, the entire deciduous set is typically erupted, sometimes with the exception of the first premolar (see below), before permanent teeth appear. First molars are very commonly the first

	ADDITIONS →			REPLACEMENTS			
TUPAIA	M1	M2	M3	I1	I2	C	P P
AOTES	M1	M2	M3	I1	I2	P	P C
LEMUR, PYGATHRIX	M1	M2	I1	I2	M3	[C	P P]
MICROCEBUS	M1	M2	I1	I2	C	M3	P P
SAIMIRI	M1	M2	I1	I2	P	P	M3 C
TARSIUS	M1	M2	I1	I2	P	P	C M3
PITHECIA, CACAJAO	M1	I1	M2	I2	M3	P	P C
COLOBUS	M1	I1	I2	M2	P	P	[C M3]
MAJORITY OF MONKEYS AND ALL APES	M1	I1	I2	M2	P	P	[C M3]
HOMO (CHIEFLY WHITES)	[I1	M1]	I2	[P	C	P]	M2 M3

Figure 15.1. **Diagrammatic representation of the sequence of eruption of the permanent teeth in primates, redrawn from Schultz (1956, Figure 20). Replacement teeth are shown in light outline font to emphasize molar progression. Parentheses indicate that the particular teeth enclosed vary in position. Arrows follow progress of molars to later positions in sequence across the primate *scala naturae*.**

permanent tooth to emerge. Molars emerge from front-to-back and no mammal species, where tooth identification is clear, is known to alter this order. Other teeth also tend to a front to back appearance, although there are exceptions (see Ziegler, 1971; Osborn and Crompton, 1973). Eruption order is similar in the maxilla and mandible, but not identical, often with the mandibular tooth emerging slightly ahead of its maxillary isomere. Beyond these similarities, however, there is real variation to explain. In extant mammals, incisors tend to emerge together and premolars tend to emerge together, but this seems not to be the case in early mammals (see Martin, 1997). And although the permanent dentition most often begins with the first molar, it may end with any tooth, an incisor, canine, premolar, or molar, varying even among closely related species.

It is worth keeping in mind that the fundamental mammalian plan of tooth number, kind, and replacement evolved in some particular species which possessed some particular life history strategy and dento-facial morphology. Early eutherians were small, presumably rapidly growing, and had long dentaries with 11 tooth positions per quadrant (three incisors, one canine, four premolars, and three molars). Many lineages subsequently evolved larger body size, slower growth, shorter faces, and reduced tooth number. Mammals at extremes of body size or environment, notably the shrews, some bats, rodents, elephants, and sea mammals, greatly altered teeth and tooth replacement. Many lineages, however, including primates, ungulates, carnivores and larger insectivores continued to maintain fairly generalized eruption and replacement of teeth. Paradoxically, the breadth of mammals retaining a primitive system of

eruption and replacement testifies both to the flexibility and conservatism of the system.

15.2.1.   Schultz's hypotheses

Adolph Schultz (1935) and others (Krogman, 1930; Bennejeant, 1936) clarified the basic outline of tooth eruption in primates and left us with a much simplified arena for testing hypotheses. Schultz (1956, 1960) went furthest to find meaning in sequence of eruption, proposing a very limited pattern that varied in a regular way across primates.

To understand eruption sequence (Figure 15.1), Schultz contrasted molars (M1–M3), which are 'additions' to the deciduous dental arch, with 'replacement' teeth, the permanent incisors, canines and premolars that replace deciduous predecessors. When permanent teeth are so divided, primates showed a striking pattern across the *scala naturae* (Figure 15.1): at top, *Tupaia*, the tree shrew (then classified with primates) erupts all three molar teeth before replacing any deciduous teeth. Moving to slower-growing, longer-lived primates, molars scatter to later and later positions, but little else changes in the progression. Humans, particularly 'white' populations, are shown with a particularly extreme sequence in which M1 is no longer the first permanent tooth to erupt and the two other molars are very last to emerge.

Schultz himself was convinced of the underlying cause of the trend: 'There can be little doubt that these ontogenetic specializations represent necessary adaptation to the gradual prolongation of the period of postnatal growth' (Schultz, 1960:13). On one point, however, his illustration overpowered the text. The dashed-in molar progression strongly suggests that primates are characterized by late molar appearance. The accompanying text, however, proposes that long life places an extra load on deciduous teeth and that species adapt by replacing them relatively early. Thus, Schultz's adaptive model might dash in, not the backward drift of molars, but a forward march of replacement teeth.

What I will call *'Schultz's Rule' is the tendency for replacing teeth to come in relatively early in slow-growing, longer-lived species*. My first task is to try to replicate Schultz's observations on primates, and the second, to test his explanation against other mammals. If 'Schultz's Rule' is a powerful one, it should apply to more than just primates.

15.2.2.   An adaptive model

Figure 15.2 presents a simple model of tooth eruption sequence by contrasting rapidly and slowly growing mammals. In Figure 15.2A, the springbok *Antidorcas* erupts teeth in three sets: set 1 (deciduous), set 2 (molars), and set 3 (replacing teeth). In contrast, *Homo sapiens*

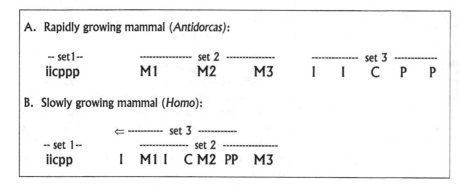

Figure 15.2. **Model of 'Schultz's Rule,' the adaptation of tooth eruption sequence to fast and slow growth.** Very rapidly growing species (case A) erupts the three sets of teeth sequentially: deciduous teeth, molars, replacement teeth. As growth and deveopment slows (case B), deciduous teeth wear out before growth is completed. Set 3 shifts to replace deciduous teeth relatively early and sets 2 and 3 become simultaneous. Little else changes.

(Figure 15.2B), replaces deciduous teeth before completing molar eruption. Although the *Homo* sequence appears to be quite different from that of *Antidorcas*, we can account for it by a simple shift of replacement teeth. Whereas *Antidorcas* erupts sets 2 and 3 sequentially, *Homo* erupts them simultaneously.

As Schultz suggested, the shift of replacement teeth may be related to growth rate and life span. *Antidorcas* is a small (34 kg) bovid that grows to adult size in about 2 years (Rautenbach, 1971; Nowak and Paradiso, 1983). In *Homo sapiens*, the slowest-growing mammal, however, some 20 years pass before the face can accommodate third molars. If humans followed the springbok pattern, deciduous teeth would have to function for some 20 years before replacement–far longer than their size and structure allow. One solution to the problem of slow growth would be to evolve larger deciduous teeth, but these teeth are formed in utero, presumably under some size constraint. A less costly solution is to shift the replacing set to a relatively early position.

The model in Figure 15.2 suggests that complexity of the human sequence (which switches back and forth between eruption of molars and replacement teeth) is superficial. The human sequence can be understood as a crossing over of sets of teeth adapting to different factors. In this view, the development of deciduous teeth and molars is primarily timed to match growth of the face, a relatively invariant adaptive problem. Development of replacement teeth, however, is on a sliding scale, adapting to demands on deciduous teeth.

Because early mammals were small and rapidly growing, Figure 15.2A or something close to it, is likely to be more 'primitive' in an evolutionary sense and Figure 15.2B is likely to be relatively 'advanced.' If tooth eruption is a functional system adapted to life history, we might expect to find a scattering of both 'primitive' and 'advanced' sequences through mammalian orders. Further, since the timing of events in life (e.g., age of maturation, life span) and the size of organisms (body weight, brain weight) are correlated (see Harvey and Clutton-Brock, 1985), 'primitive' sequences might characterize small mammals and 'advanced' sequences characterize larger ones. Thus, if the simple model presented is a good one, extant species with identical eruption sequences may be only distant relatives.

15.3.   Materials and methods

Sequences of eruption of mandibular permanent teeth are described in the Appendix for a series of extant mammals with fairly generalized dentitions.

15.3.1.   Definitions

Defining eruption sequence is not easy, since sequences of calcification, initial movement, emergence through alveolus or gingiva, and completed eruption are not necessarily in agreement (Garn and Lewis, 1963; Schwartz, 1974). In dentistry and many primate studies, the marker of tooth eruption preferred for clarity and replicability is *appearance of any part of the cusp or crown through the gingiva*, usually called 'tooth emergence.' The sequences reported here are 'emergence sequences' as much as possible. This criterion is easily met for primates, although not so easily met for other orders. For ungulates, several stages of eruption were often described in original studies (e.g., tooth just above bone, mid-eruption, fully erupted). Since alveolar emergence can long precede gingival emergence for molariform teeth (Garn and Lewis, 1963), 'just above bone' was not taken as erupted. It was most helpful when studies noted whether teeth were faceted and in wear (e.g., Wallace, 1977; Kay and Simons, 1983).

15.3.2.   Taxa

Emergence sequences of members of the orders Primates (12 species), Scandentia (1), and Soricomorpha (2) are contrasted with those of members of Artiodactyla (17), Perissodactyla (2), and Hyracoidea (1). The first group has adaptive similarity but not taxonomic coherence, since Primates and Scandentia (grouped in the grandorder Archonta) may be only distant relatives of soricomorph insectivores. Group one is referred to here in an informal sense as 'primates/insectivores.' Group two consists of herbivorous hoofed mammals, all members of the grandorder Ungulata. Within orders, species chosen have complete data for all permanent mandibular teeth and span, as much as possible, the largest available range of body size within wild taxa. Domestic mammals are excluded because artificial selection is presumed to have altered life history. Only the mandible is studied because ruminant artiodactyls lack maxillary incisors.

Complete data and sources appear in the Appendix. Higher level taxonomy follows Eisenberg (1981); genus and species names follow Wilson and Reeder (1993) for ungulates and Smith et al., (1994) for primates.

Within Primates, six of 11 families are known by complete data for gingival emergence sequence, but Callitrichidae are omitted here because members lack M3, an important tooth in patterns under investigation. For anthropoids, sequences are based on very substantial samples and fine-grained studies except for the leaf-eating monkey *Alouatta*, which is based on a single individual. Lemurids have been studied on a fine scale, but sample size is also small. Unfortunately, data for the smallest primates (e.g., *Microcebus*, *Tarsius*) are preferentially missing. Because comparable complete data could not be found for any tiny insectivorous primate, *Urotrichus* (the Japanese shrew mole, order Soricomorpha) and *Tupaia* (the tree shrew, order Scandentia) are substituted to give some representation to very quickly growing generalized mammals.

Within Artiodactyla, data were located for seven of eight families. No data were located for Tragulidae (chevrotains), the smallest living members of the order. The Camelidae is omitted here because extant members are domestic (camel, llama), have reduced tooth number (guanaco), or ever-growing incisors (vicuna). Within Perissodactyla, two of three families are represented, Equidae and Rhinocerotidae. The latter contains the slowest-growing ungulates besides the elephant, but is of limited diagnostic value because the family has eliminated all incisors, both deciduous and permanent. Data located for the Tapiridae (the tapir) were incomplete.

For ungulates, exceptional studies with excellent data include those of *Sus* (wild pig), *Tayassu* (peccary), and *Odocoileus* (white-tail deer), which are detailed, fine-scaled, and backed up by large numbers of subjects; studies of *Antilocapra* (pronghorn), *Bison*, and *Hemitragus* (tahr) are also very good. For other ungulates, data are of lower quality overall; data were sometimes sketchy, and some sequences had to be based on combinations of

observations of gingival emergence, radiographs of living individuals, and skeletonized individuals.

15.3.3. Methods

All data were taken from the literature, constructing sequences from original publications. The goal was to construct a gingival sequence of emergence, building a sequence of the order of teeth appearing through the gingiva and in wear. Sequences were compiled from (1) actual reports of sequences observed within individuals, and (2) mean ages of tooth emergence. The two methods (the most common sequence within individuals and sequences of means) are equivalent when samples are substantial and well spread out over juveniles (Smith and Garn, 1987). For three primate species, *Macaca nemestrina*, *Pan troglodytes*, and *Homo sapiens*, actual compilations of most likely sequences were available (Smith, 1994). Two sequences are given for *Homo sapiens*, from Australian Aborigine males and American white males, to represent the range of findings for human populations. When sexes differ (as in humans), male sequences were used.

15.3.4. Appendix

In the Appendix, special notation gives some extra information on variation, first premolars, and special dental adaptations: Whenever variants in sequence are known to be substantial, square brackets [ ] surround the tooth pair, following Schultz (1935). Parentheses ( ) mark cases in which the actual order has not been determined, i.e., the teeth are 'tied.'

Only five extant taxa studied here possess a mandibular first premolar: *Sus*, *Hippopotamus*, *Equus*, *Ceratotherium*, and *Procavia*. These mammals, along with most others that retain a first premolar, typically erupt only a single tooth in this position, and it is unclear to which series the tooth belongs, deciduous or permanent. Kindahl (1967), Zeigler (1971, 1972), and Luckett (1993), all of whom considered the question over a range of mammals, allocate it to the deciduous dentition. Luckett (1993) regards *Tapirus* as the only unequivocal case in which a first premolar (upper jaw only) has two tooth generations. Extinct archaeocetes are thought to have a replacing permanent P1 in the mandible (Kellogg, 1936; Uhen, 1996), although the evidence is not unequivocal. In the Appendix, emergence of p1 appears if timing is known (it tends to appear between the deciduous and permanent dentitions or with M_1), but the tooth is dropped from further analysis.

Lemuroidea modified mandibular incisors and canine into a procumbent 'tooth comb,' which erupts as a unit (Eaglen, 1985). Here the tooth comb is represented as a unit labeled 'tc' since it is unlikely individual teeth could erupt separately.

Also in the Appendix, two life history variables appear alongside emergence sequences: age of emergence of the mandibular first molar (from Smith et al., 1994) and life span (from Harvey and Clutton Brock, 1985 and Nowak and Paradiso, 1983). Age of M_1 emergence serves as an objective estimate of rate of postnatal growth: a low age of M_1 emergence indicates fast growth and a high age indicates slow growth. In primates, age at which M_1 erupts is highly associated with other measures of life history (Smith, 1989). In mammals studied here, emergence of M_1 ranges from about 1 month to over 6 years. Life span ranges from 11 to 100 years, although that of the little known Urotrichus may be much shorter.

For three genera with unknown age of M_1 emergence (the insectivore Urotrichus, the primate Alouatta, and the ungulate Okapia), post-natal growth rate was approximated by comparing other aspects of life history among species (Nowak and Paradiso, 1983). In one case (Tupaia), an unknown age of emergence of M_1 was estimated with some confidence. It is known that the entire permanent dentition erupts between about 1 and 3 months of age in Tupaia (Shigehara, 1980; Tsang and Collins, 1985; Hertenstein et al., 1987). Further, it is likely that M_1 appears after completion of the deciduous dentition (day 27) and before males begin puberty (day 40–50). Data on 24 primate species (Smith et al., 1994) were used to calculate an equation predicting age of M_1 emergence from age of completion of the deciduous dentition. The resulting equation predicts emergence of M_1 at day 31 in Tupaia, an estimate that fits well with other life history data for the genus.

15.3.5.   Analysis

A glance at the appendix will show that complete sequences of tooth emergence are hard to evaluate. The first 'analysis' of data is simply to diagram sequences in a manner that brings out patterns in the data. Diagrams (see Figures 15.4–15.7) focus on seven non-canine teeth shared by almost all taxa (I1–2, P3–4, M1–3). Canines are discarded because they have heterogeneous special functions, sometimes featured in male–male competition and sometimes serving only as an extra incisor. The I_3 and P_2, teeth that disappeared in the evolution of many mammalian taxa, are shown, but are collapsed with the tooth emerging nearest in time. Further, numbers are stripped from incisors because I_1 and I_2 always emerge in that order in present data. Numbers are stripped from premolars to simplify comparison and premolar order is discussed separately. This notation amounts to a special language of reporting tooth eruption sequence, but it is a language that makes patterns clearer.

The second analysis quantifies aspects of sequence and looks for associations with age of M_1 emergence, life

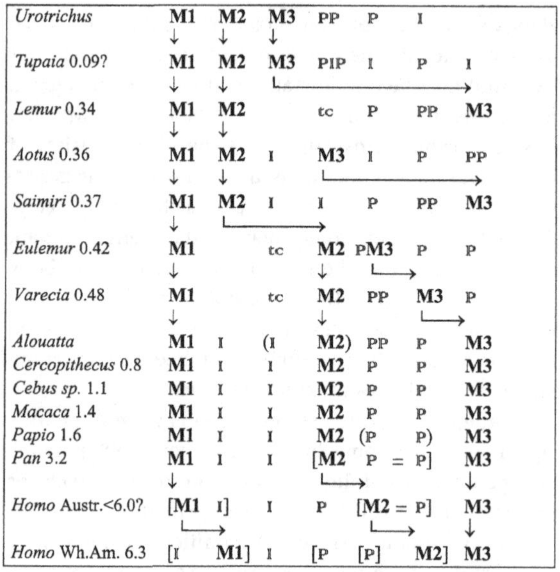

Urotrichus	M1↓	M2↓	M3↓	PP	P	I	
Tupaia 0.09?	M1↓	M2↓	M3 ↳	PIP	I	P	I →
Lemur 0.34	M1↓	M2↓		tc	P	PP	M3
Aotus 0.36	M1↓	M2↓	I	M3 ↳	I	P	PP →
Saimiri 0.37	M1↓	M2 ↳	I	I →	P	PP	M3
Eulemur 0.42	M1↓		tc	M2↓	PM3 ↳	P	P
Varecia 0.48	M1↓		tc	M2↓	PP	M3 ↳	P
Alouatta	M1	I	(I	M2)	PP	P	M3
Cercopithecus 0.8	M1	I	I	M2	P	P	M3
Cebus sp. 1.1	M1	I	I	M2	P	P	M3
Macaca 1.4	M1	I	I	M2	P	P	M3
Papio 1.6	M1	I	I	M2	(P	P)	M3
Pan 3.2	M1	I	I	[M2 ↳	P	= P]	M3↓
Homo Austr.<6.0?	[M1 ↳	I]	I	P	[M2 ↳	= P]	M3↓
Homo Wh.Am. 6.3	[I	M1]	I	[P	[P]	M2]	M3

Figure 15.3. **New (post-1960) data on sequences of gingival emergence of mandibular teeth in primates plus two small insectivorous mammals (members of Soricomorpha and Scandentia). When ranked by the age at which M1 emerges (at left), molars scatter to later positions, M3 greatly, M2 partly, and M1 in only the slowest growing mammal. The trend is imperfect at M3, as Schultz also found.**

span, brain weight, and body weight, all aspects of 'life history.'

Brain and body weights for primates (not shown here) are taken from Harvey and Clutton-Brock (1985) and for other mammals from Gingerich (n.d.), a compendium which will be described elsewhere. With the exception of the predicted M_1 emergence in Tupaia, all data analyses count time from birth. Repeating analyses with time counted post-conception did not change conclusions about emergence sequence, probably because most of the mammals studied here (excepting Tupaia and Urotrichus) are precocial (born well developed after a long gestation), and share many characteristics of the timing of events in development. Broad comparisons of altricial and precocial mammals, however, might need to count time from conception.

15.4.   **Results**

15.4.1.   **Can Schultz's findings be replicated?**

Patterns in eruption sequence emerged when Schultz (1956) ranked species by the scala naturae (see Figure 15.1). New data allow species to be ranked more objectively, here by the age at which a developmental marker is crossed. In Figure 15.3, species are listed by increasing

Urotrichus		**M1**				**M2**					**M3**	PP	P	I	
Tupaia 0.09?		**M1**				**M2**					**M3**	PIP	I	P	I
Lemur 0.34		**M1**				**M2**	tc	P	P		P**M3**				
Aotus 0.36		**M1**				**M2**				I	**M3**	I	P	PP	
Saimiri 0.37		**M1**				**M2**	I	I	P	PP	**M3**				
Eulemur 0.42		**M1**		tc		**M2**					P**M3**	P	P		
Varecia 0.48		**M1**		tc		**M2**				PP	**M3**	P			
Alouatta		**M1**	I	(I		**M2**)	PP	P			**M3**				
Cercopithecus 0.83		**M1**	I	I		**M2**	P	P			**M3**				
Cebus sp. 1.1		**M1**	I	I		**M2**	P	P			**M3**				
Macaca 1.4		**M1**	I	I		**M2**	P	P			**M3**				
Papio 1.6		**M1**	I	I		**M2**	P	P			**M3**				
Pan 3.2		**M1**	I	I		[**M2**	P	P]			**M3**				
Homo <6.0?		[**M1**	I]	I	P	[**M2**	P]				**M3**				
Homo 6.3	[I	**M1**]		I	[P	[P]	**M2**]				**M3**				

Figure 15.4. **Diagram of mandibular tooth emergence in primate/insectivores. Molars (boldface) are represented as stable and replacement teeth (outline font) as mobile. Extra conventions: all replacement teeth and teeth within a field (premolars, incisors) are kept together if possible and left shifts of replacement teeth are conservative; I3 and P2 are doubled with tooth nearest in time. When ordered by increasing age of M1 emergence (at left), replacement teeth sweep across molars to earlier positions.**

age of M_1 emergence, and emergence sequences are based on data published after Schultz's 1956 diagram.

For anthropoids, Schultz's sequences were primarily based on skeletal samples with jaw unspecified. Although new data are limited to mandibular gingival emergence, new and old data (Figures 15.1, 15.3) are in fair agreement. As Schultz showed, little variation has been described between catarrhine species (see Smith, 1994). Few new data have appeared since Schultz for colobine monkeys, but those available suggest colobines are more rapidly growing and have more primitive sequences than cercopithecines (Wolf, 1984). Large studies of living humans have substantiated Schultz on human sequence: the first incisor erupts before the first molar in many individuals, and may characterize some populations. In a study of 6000 American children, Smith and Garn (1987) found that $I_1 M_1$ characterizes just over half of males and about a third to half of females.

For prosimians and the tree shrew, Schultz's sequences were taken from Bennejeant (1936), and these data differ most from newer sources. More recently, Shigehara (1980) found premolars before incisors in *Tupaia*, a reversal of Schultz's diagram. *Urotrichus* also shows premolars before incisors (Usuki, 1967), as do some fast-growing ungulates, evidence that this is a real phenomenon. These two non-primate species, both small and rapidly growing, show eruption of all three molars before any tooth replacement. No living primate appears to maintain this primitive sequence as the dominant one (see also Schwartz, 1974), although it may appear as a variant. Lemurids all show some form of advancement, whether in early appearance of the incisor-canine tooth comb or M_3 (Eaglen, 1985).

In Figure 15.3, molars drift to late positions in step with life history evolution, as represented here by M_1 emergence. Starting at the top and going down the chart, all species retain M_1 as the first tooth until *Homo sapiens*. The second molar shows a perfect trend, drifting from position 2 to 4, 5, and 6 as M_1 emergence increases from 0.4 to 6.3 years. Exceptions from the trend appear only with M_3, which appears later than expected in *Lemur* and *Saimiri*. Rankings here are tentative, however, because *Lemur*, *Aotus*, and *Saimiri* erupt M_1 at very nearly the same time (0.34, 0.36, and 0.37 years) and a new study of any of these species might reorder them, improving the trend. But in any case, no rearrangement of taxa produces a perfect trend in M_3 appearance.

For a new perspective on Schultz's rule, Figure 15.4 diagrams sequence in a manner that emphasizes replacement teeth. In it, molars are kept in rigid alignment as if they are stable, attuned to facial growth; replacement teeth are shown as if mobile. Although more cumbersome, this diagram shows the sweep of the changes across taxa in a striking fashion. In Figure 15.4, replacement teeth appear to march forward across columns of molars. Incisors traverse the greatest distance, crossing from the very last teeth to the very first teeth. Premolars have a more modest range, from just after M_3 to just before M_2. With its resemblance to a scatter graph with a cloud of points, Figure 15.4 de-emphasizes imperfections and emphasizes an overall association between life history and position of replacement teeth in sequence. The answer to the question posed is clear: Schultz's primate pattern can be replicated, although lemurids show some specializations regarding the tooth comb.

15.4.2.   Do ungulates follow 'Schultz's Rule?'

Ungulates (Figure 15.5) span a very similar range in age of M_1 emergence and length of life as do the primates/ insectivores above (Figure 15.4). Discounting humans, M_1 emergence ranges from 0.09 to 3.2 years in the primate/ insectivore group and from 0.08 to 2.75 years in ungulates; life span is known to range from about 11 to 50+ years in each. Primates certainly contain many more slowly growing species, however. A glut of bovids and cervids emerge M_1 at about 0.2 year, whereas many anthropoid primates do so at one year or more. Thus, if

UNGULATES

Antidorcas 0.08	**M1**			**M2**					**M3**	I	II	P	P
Antilocapra 0.17	**M1**			**M2**					**M3**	I	P	P	IPI
Rangifer 0.17	**M1**		I	**M2**				II	**M3**	PP	P		
Odocoileus 0.18	**M1**		I	**M2**				II	**M3**	P	PP		
Muntiacus 0.2?	**M1**			**M2**				I	**M3**	I	PP	PI	
Hemitragus 0.21	**M1**			**M2**				I	[**M3**	I]	P	PPI	
Sylvicapra 0.23	**M1**			**M2**					**M3**	PP	P	I	II
Aepyceros 0.33	**M1**			**M2**				I	**M3**	I	PP	IP	
Okapia	**M1**			**M2**				I	(**M3**	I)	PI	PP	
Cervus 0.33?	**M1**			**M2**		I		I	**M3**	IPP	P		
Tayassu 0.40	**M1**I			**M2**	I	I	P	PP	**M3**				
Connochaetes 0.46	**M1**			**M2**				I	**M3**	I	P	PI	
Sus 0.47	**M1**			**M2**	I	P	PP	I	**M3**				
Taurotragus 0.53	**M1**			**M2**				I	**M3**	II	PP	P	
Bison 0.54	**M1**			**M2**				IP	**M3**	P	I	IP	
Procavia 0.54	**M1**	I	I	(**M2**	PP	P)			**M3**				
Giraffa 0.66	**M1**			**M2**				[I	**M3**]	[P	PP]	II	
Equus 0.88	**M1**			**M2**			I	PP	**M3**	I	PI		
Hippopotamus ~2	**M1**	I	I	**M2**	(PP)	P			**M3**				
Ceratotherium 2.75	**M1**P			[**M2**	P	P]			**M3**				

Figure 15.5. Diagram of mandibular tooth emergence in ungulates. Molars are represented as stable and replacement teeth as mobile. Conventions as in Figure 15.4. Replacement teeth in ungulates tend to shift to early positions with slower growth and development, but major exceptions appear in large bovids and the giraffe.

the two charts were combined, anthropoid primates would fit in near the bottom.

The most striking finding is that ungulate tooth succession is much like that of the primates/insectivores. Overall, replacement teeth range over similar positions in the two charts (Figures 15.4, 15.5). In each group, the first molar begins the sequence, but any tooth may end it. While incisors tend to emerge together and premolars tend to emerge together, incisors tend also to precede premolars except in very rapidly growing species. Like primates (Figure 15.4), ungulates (Figure 15.5) also replace teeth in a pattern relative to life history. Eruption of all three molars before any teeth are replaced is found in rapidly growing members *Antidorcas*, *Antilocapra*, and *Sylvicapra*. Moving down the chart towards large-bodied, slow-growing taxa, replacement teeth begin to cross over molars. Most taxa show some incisor emergence before M_3, and several even erupt premolars before M_3, including *Tayassu*, *Sus*, *Procavia*, *Hippopotamus*, and *Ceratotherium*, a finding also common in higher primates. The rhinoceros is especially interesting because its premolars emerge relatively early, as one would expect with slow growth, even though all incisors have been lost.

Despite similarities to primates/insectivores, ungulates follow Schultz's Rule more intermittently. Figure 15.5 (ungulates) is much less tidy than Figure 15.4 (primates/insectivores) and substantial exceptions appear. Two deer (*Odocoileus* and *Rangifer*) show very early emergence of I_1, more extreme even than the lemur tooth comb. The larger bovids *Connochaetes*, *Taurotragus*, and *Bison* delay M_1 until about one-half year, but fail to adapt emergence sequence significantly. A striking exception to the rule is found in the giraffe. Sources (sometimes tables and text within a source) disagree on the giraffe sequence;

it may match that of its relative the okapi (M_1 M_2 I_1 M_3 I_2 ...) or it may have an even more primitive sequence (M_1 M_2 M_3 I_1 I_2 ...). At 500 kg or more, the giraffe makes a spectacular exception to any idea that body size alone predicts eruption sequence. At 2 kg, the hyrax *Procavia* has a sequence far more 'advanced.' Overall, a tendency toward Schultz's rule can be seen in ungulates, but considerable unexplained variation remains.

On closer examination, Figure 15.5 gives the distinct impression that ungulates are heterogeneous in response of eruption sequence to life history. In particular, the 'generalized' and 'specialized' herbivores are difficult to force into the same box. The specialized group might be thought of as ruminant artiodactyls (antilocaprids, cervids, bovids, giraffids) and perissodactyls. The former evolved selenodont cheek teeth to consume a low quality diet and the latter evolved lophodont cheek teeth to consume an extremely low quality diet (see MacDonald, 1984). If these specialized herbivores are separated from more generalized ones, two relatively homogeneous groups emerge. Taxa with more generalized adaptations, including artiodactyls of suborder Suina (suids, tayassuids, hippos) and the order Hyracoidea, interleave perfectly with Soricomorpha, Scandentia, and Primates (Figure 15.6). While Suina and Hyracoidea show a fluid adaptation of sequence to life history, specialized ungulates (Figure 15.7) seem comparatively inert. Specialized ungulates do step up along with Schultz's Rule, but they take only one or two steps towards earlier tooth replacement, rather than the several expected from their life histories.

15.4.3. Quantitative measures

Ordering raw data as in Figures 15.3–15.7 is a first step, but, quantitative expressions of sequence patterns are

GENERALIZED MAMMALS

Urotrichus	M1			M2					M3	PP	P	I	
Tupaia 0.09	M1			M2					M3	PIP	I	P	I
Lemur 0.34	M1			M2		tc	P	P	PM3				
Aotus 0.36	M1			M2				I	M3	I	P	PP	
Saimiri 0.37	M1			M2	I	I	P	PP	M3				
Tayassu 0.40	M1ɪ			M2	I	I	P	PP	M3				
Eulemur 0.42	M1		tc	M2					PM3	P	P		
Sus 0.47	M1			M2	I	P	PP	I	M3				
Varecia 0.48	M1		tc	M2				PP	M3	P			
Procavia 0.54	M1	I	I	(M2	PP	P)			M3				
Alouatta	M1	I	(I	M2)	PP	P			M3				
Cercopithecus 0.83	M1	I	I	M2	P	P			M3				
Cebus sp. 1.1	M1	I	I	M2	P	P			M3				
Macaca 1.4	M1	I	I	M2	P	P			M3				
Papio 1.6	M1	I	I	M2	P	P			M3				
Hippopotamus ~2	M1	I	I	M2	(PP)	P			M3				
Pan 3.2	M1	I	I	[M2	P	P]			M3				
Homo <6.0?	[M1	I]	I	P	[M2	P]			M3				
Homo 6.3	[ɪ	M1]	I	[P	[P]	M2]			M3				

Figure 15.6. **Diagram of mandibular tooth emergence in generalized mammals (Soricomorpha, Scandentia, Primates, suine artiodactyls, Hyracoidea). Molars are represented as stable and replacement teeth as mobile. Conventions as in Figure 15.4. Generalized herbivores and primate/insectivores share a fluid adaptation of sequence to slow growth.**

SPECIALIZED UNGULATES

Antidorcas 0.08	M1			M2				M3	I	II	P	P
Antilocapra 0.17	M1			M2				M3	I	P	P	IPI
Rangifer 0.17	M1		I	M2			II	M3	PP	P		
Odocoileus 0.18	M1		I	M2			II	M3	P	PP		
Muntiacus 0.2?	M1			M2			I	M3	I	PP	PI	
Hemitragus 0.21	M1			M2			I	[M3	I]	P	PPI	
Sylvicapra 0.23	M1			M2				M3	PP	P	I	II
Aepyceros 0.33	M1			M2			I	M3	I	PP	IP	
Okapia	M1			M2			I	(M3	I)	PI	PP	
Cervus 0.33?	M1			M2		I	I	M3	IPP	P		
Connochaetes 0.46	M1			M2			I	M3	I	P	PI	
Taurotragus 0.53	M1			M2			I	M3	II	PP	P	
Bison 0.54	M1			M2			IP	M3	P	I	IP	
Giraffa 0.66	M1			M2			[I	M3]	[P	PP]	II	
Equus 0.88	M1			M2		I	PP	M3	I	PI		
Ceratotherium 2.75	M1P			[M2	P	P]		M3				

Figure 15.7. **Diagram of mandibular tooth emergence in specialized ungulates (ruminant artiodactyls, Perissodactyla). Molars are represented as stable and replacement teeth as mobile. Conventions as in Figure 15.4. Premolars adapt much as expected in Schultz's Rule, but slowly growing specialized herbivores resist shifting incisors forward in sequence. Except for *Equus*, all lack upper incisors or all incisors (*Ceratotherium*) and may take over incisor function with soft tissues.**

needed to allow statistical analysis. Several quantitative schemes might be tried, but since incisor position is crucial to Schultz's Rule, emergence of the first incisor was ranked relative to molars as 0.5 (I_1 emerges before M_1), 1.5 (I_1 emerges after M_1), 2.5 (after M_2), or 3.5 (after M_3). In this scheme the two human populations have advanced I_1 to ranks 0.5 and 1.5; ranks 2.5–3.5 characterize the other mammals. Advancement of P_3 relative to molars ('P_3 position') was quantified in the same manner, ranking P_3 as 1.5 in humans (P_3 emerging after M_1), and 2.5–3.5 (after M_2 or M_3) in other mammals. Given rank order, correlation can readily demonstrate trend and direction, although it is a weak statistical approach given the restricted number of assigned ranks.

For numerical analyses, lack of life history data for the little known *Urotrichus* restricts study to Archonta (primates plus the tree shrew) and Ungulata. Table 15.1 shows Spearman's correlation between rank of replacement tooth position (0.5–3.5) and rank of a series of life history variables (1–N). Schultz's Rule predicts *negative* correlations for Table 1 since replacement teeth are pre-dicted to drop rank as timing and size measures increase. As shown, 22 of 24 correlations computed over all taxa or within grandorders were negative and 14 were significantly so at $P<0.05$. Results for I_1 and P_3 tell a similar story: for all taxa combined, advance of replacing teeth is associated with a slowing of life history, as measured by increasing age of M_1 eruption and life span ($rho = -0.56$ to -0.75). Thus, as mammals grow up more slowly and live longer, their first incisors and third premolars appear earlier in sequence. Both incisors and premolars tend to shift, shown by a positive correlation of I_1 and P_3 position ($rho = 0.60$). When dissected into grandorders, however, the strength of the trend lies within archontans ($rho = -0.77$ to -0.87). Although in the proper direction, correlation is weaker for ungulates ($rho = -0.26$ to -0.63). Combining taxa, correlations of tooth position with size (body weight, brain weight) were uneven in direction ($rho = -0.16$ to 0.11), a sign of underlying heterogeneity.

To try to resolve heterogeneity in the data, Table 15.1 also groups species by adaptive categories 'generalist'

Table 15.1. *Spearman's rank order correlation between eruption sequence (appearance of I_1 or P_3 relative to mandibular molars[1])
and life history variables for extant taxa*

	Spearman's rho							
Taxon/group	Age M_1 emerges	(N)	Life span	(N)	Body weight	(N)	Brain weight	(N)
Appearance of I_1 relative to molars								
All taxa	−0.66*	(30)	−0.56*	(30)	0.03	(32)	−0.16	(29)
Archontans	−0.87*	(12)	−0.85*	(11)	−0.84*	(13)	−0.84*	(13)
Ungulates	−0.32	(18)	−0.26	(19)	−0.21	(19)	−0.08	(16)
Generalists	−0.87	(16)	−0.67	(15)	−0.49	(17)	−0.59	(17)
Specialists	−0.13	(14)	−0.46	(15)	−0.31	(15)	−0.23	(12)
Appearance of P_3 relative to molars								
All taxa	−0.75*	(31)	−0.60*	(31)	0.11	(33)	−0.20	(30)
Archontans	−0.77*	(12)	−0.77*	(11)	−0.80*	(13)	−0.83*	(13)
Ungulates	−0.63*	(19)	−0.40*	(20)	−0.09	(20)	−0.13	(17)
Generalists	−0.71	(16)	−0.61	(15)	−0.67	(17)	−0.76	(17)
Specialists	−0.59	(15)	−0.49	(16)	−0.41	(16)	−0.51	(13)

Note: [1]Appearance of I_1 coded as: 0.5, before M1; 1.5, after M1; 2.5, after M2; 3.5, after M3; appearance of P_3 coded as: 1.5, after M1; 2.5, after M2; 3.5, after M3.
*Significantly different from zero at $P<0.01$ in one-tailed test (negative associations predicted). No tests performed for 'generalists' and 'specialists' (ruminant artiodactyls, perissodactyls) because division was made *post hoc*, in an attempt to resolve heterogeneity in the data.

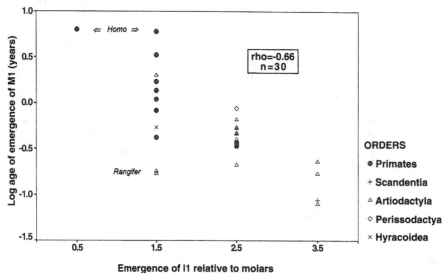

Figure 15.8. **Log (base 10) of the age of emergence of mandibular first molar for species different I1 positions:** 0.5, before M1; 1.5, after M1; 2.5, after M2, and 3.5, after M3. Two deer (*Rangifer* and *Odocoileus*) appear as outliers for sequence 1.5. Despite considerable overlap in timing, age at first molar emergence steps up as incisors shift to early positions.

and specialist' (the same division as in Figures 15.6 versus Figure 15.7). Tests were not performed because the division was made *post hoc*, but comparisons are still instructive. Dividing species into adaptive categories reduces heterogeneity, resulting in two groups with correlations of coherent direction and similar strength. Moreover, results suggest that any and all life history variables – M_1 emergence, life span, body weight, and brain weight – are associated with emergence sequence. Thus, within adaptive categories, species that have larger bodies, larger brains, grow up more slowly, and live longer, tend to shift replacement teeth forward in sequence of emergence. Premolar progression is very similar in the two adaptive categories, but incisor progression is not. 'Generalists' advance incisors strongly with life history

evolution (*rho* = −0.87), whereas specialists either eliminate incisors (the rhinoceros) or advance them little (*rho* = −0.13).

A stronger statistical approach to the influence of life history on tooth eruption is to sort taxa by sequence and compare life histories. In Figure 15.8 (supporting data in Table 15.2), the logarithm of age of M_1 emergence is graphed for each position of I_1 in the emergence sequence. Variance of age of M_1 emergence is particularly high for taxa with sequence 1.5 (M_1 I_1 M_2 M_3) where the distribution is expanded by a human population at the high end and by two deer (*Odocoileus* and *Rangifer*) at the low end. The 'primitive' sequence, M_1 M_2 M_3 I_1 (shortened to MMMI in Table 15.2) is associated with a median age of M_1 emergence of 0.13 years. With each shift forward in I_1,

Table 15.2. *Timing of life events (M₁ emergence and life span) in mammals with different emergence sequences*

Mandibular sequence code and string[a]	Age of emergence of M₁ (years)			Life span (years)		
	Median[b]	Range	N species[c]	Median[b]	Range	N species[c]
Incisor position						
0.5 I1 M M M	6.30	–	1	100	–	1
1.5 M I1 M M	1.10	0.17–6.00	11	31 (11)[d]	20–100	10
2.5 M M I1 M[e]	0.39	0.21–0.88	14	24	13–36	15
3.5 M M M I1	0.13	0.08–0.23	4	13	11–17	4
Premolar position						
1.5 M P3 M M	6.17	6.0–6.3	2	100	–	2
2.5 M M P3 M	0.88	0.34–3.33	13	29 (11)[d]	21–45	12
3.5 M M M P3	0.28	0.08–2.00	16	20	11–54	17

[a] Disregarding other teeth. Molars always erupt as M1 M2 M3.
[b] Kruskal–Wallis test finds medians significantly different at $P<0.01$.
[c] Each population of *Homo sapiens* entered separately.
[d] Hyrax life span (11 years) is an outlier, much shorter than expected given its protracted dental development.
[e] Without specialized ungulates, values are, respectively: 0.37, 0.34–0.47, 5, 25, 13–27, 5.

the pace of life events slows and M₁ is delayed, with median ages of emergence of 0.39, 1.10, and finally 6.3 years. Since variances are significantly different for sequence codes 1.5–3.5, non-parametric comparisons are in order. By the Kruskal–Wallis test, median age of M₁ emergence differs in the three sequence groups with N>1 ($P < 0.01$) as shown in Table 15.2. Also shown in Table 15.2, as I₁ shifts forward, life span increases from 13 to 24, 31, and 100 years. Repeating the exercise for P₃ position produced a very similar result, showing systematic increase in age of M₁ emergence and life span as P₃ advances relative to molars (Table 15.2).

Comparisons in Figure 15.8 beg one further question, whether the age of M₁ emergence associated with a particular sequence is the same regardless of taxon or adaptive group. To be brief, a series of two-way analyses of variance (ANOVAs) (with factors grandorder versus I₁ position, adaptive group versus I₁ position, grandorder versus P₃ position, adaptive group versus P₃ position) demonstrated no strong evidence for taxonomic or adaptive effects across the board – thus no real evidence of different 'set points' for M₁ timing. Until stronger evidence emerges, it appears that a given sequence is associated with a particular age of M₁ emergence whatever the grandorder. An exception can be made for specialized ungulates, where a significant interaction points to unusual values for one particular category. Specialized ungulates drift into moderately slow growth while maintaining the sequence MMIM, rather than the expected MIMM.

Given these analyses, it can be proposed that ungulates tend to follow Schultz's Rule in tooth replacement patterns, but that slowly growing specialized ungulates resist shifting incisor position.

15.4.4. Early replacement teeth or late molars?

Patterned changes in tooth eruption sequence observed here could be brought about by relatively early replacement teeth or relatively late molars *or both*; sequence data alone cannot determine which. The ultimate resolution of which teeth are late and which early must come from numerical data on ages of emergence of teeth. While some such data support early replacing teeth as a major contributor to sequence variation in primates (Smith, 1992), ungulate adaptations may be more varied. At least one genus, *Procavia*, the hyrax, shows several oddities in emergence timing. Its molars continue to erupt slowly over several years, completing eruption as late as 5 years of age (Steyn and Hanks, 1983), a value almost half its life span (Nowak and Paradiso, 1983). This is sufficiently out of proportion to other mammals to suggest that *Procavia* has initiated an elephant-like or kangaroo-like late progression of molars.

15.4.5. Does emergence sequence reflect facial architecture?

If Schultz's Rule holds good, life history should be a better predictor of eruption sequence than facial architecture or simply membership in a taxonomic group. In mammals collected here, primates/insectivores have a closed tooth row with either long or short faces, whereas ungulates tend to a more stereotyped long face, often with a large diastema between anterior and cheek teeth. Thus, if facial architecture is crucial, we might expect to see changes that correspond to face shape in the former and relatively fixed sequences of emergence within the latter. On the other hand, sequences that step up in pace with life history, no matter the taxonomic grouping, argue for a life history explanation.

A few comparisons cast doubt on the determinative

Table 15.3. *Primate/insectivores paired with their best match in mandibular emergence sequence among ungulates (showing seven tooth positions shared by all taxa, with I3 and P2 collapsed with tooth nearest in time)*

Taxon pairs, age M_1 emerges	Emergence sequence						
Urotrichus (early?)	M1	M2	M3	PP	P	I	
Sylvicapra 0.23	M1	M2	M3	PP	P	I	II
Tupaia ca. 0.09	M1	M2	M3	PIP	I	P	I
Antilocapra 0.17	M1	M2	M3	I	P	P	IPI
Aotus 0.36	M1	M2	I	M3	I	P	PP
Aepyceros 0.33	M1	M2	I	M3	I	PP	IP
Lemur 0.34	M1	M2	tc		P	P	PM3
Tayassu 0.40	M1I	M2	I	I	P	PP	M3
Eulemur 0.42	M1	tc	M2		PM3	P	P
Rangifer 0.17	M1	I	M2	II	M3	PP	P
Saimiri 0.37	M1	M2	I	I	P	PP	M3
Tayassu 0.40	M1I	M2	I	I	P	PP	M3
Cebus[a] 1.1	M1	I	I	M2	P	P	M3
Procavia 0.54	M1	I	I	M2	PP	P	M3
Pan 3.2	M1	I	I	M2	P	P	M3
Hippopotamus ~2	M1	I	I	M2	PP	P	M3
Homo 6.3 (elephants?)	I1	M1	I	P	P	M2	M3

No adequate match was found for *Alouatta* or *Homo*, but all other primates could be matched approximately choosing from 20 ungulates.
[a] Cercopithecine monkeys all match *Cebus* or *Pan*.
tc, tooth comb.

effect of gross facial form on emergence sequence. First, within primates there is little similarity among emergence sequences of *Saimiri*, *Cebus*, and *Homo*, although all have short faces. *Lemur* and *Papio*, the more dog-faced species, are equally mismatched in sequence (Figure 15.3). While it is true that mid-size ruminants seem relatively inert, ungulate emergence sequences step up through a considerable range despite comparatively stereotyped facial form (Figure 15.5).

In Table 15.3, primates/insectivores are paired with a best match in emergence sequence in ungulates. Close resemblance pairs the Japanese shrew mole with the duiker, the tree shrew with the pronghorn, the night monkey with the impala, the true lemur with the caribou, the squirrel monkey with the peccary, the cebus monkey with the hyrax, and the chimpanzee with the hippopotamus. It is difficult to imagine a hypothesis about facial morphology that could explain pairings such as these. Matched pairs, however, share aspects of life history, as shown by the correlation between their ages of M_1 emergence of $r = 0.65-0.98$ (depending on how matches are allocated in *Cebus*, *Pan*, *Procavia*, and *Hippopotamus*, which all share the same sequence). Humans have no match in ungulates, perhaps because elephants, the only ungulate to grow up as slowly as humans, have greatly modified tooth succession.

Although gross facial morphology has no simple and obvious explanatory power for emergence sequence, more precise hypotheses about dental function remain promising. The specialized herbivores, including ruminant artiodactyls and perissodactyls, for example, appear to slow growth without much advancement of incisors. These mammals make extensive use of the tongue and soft tissues in acquiring food (MacDonald, 1984) and incisors may be comparatively under used. Indeed, most of the Ruminantia have no upper incisors and the rhinoceros has no incisors whatsoever. Both dental function and dental morphology remain likely contributors to emergence sequences.

15.4.6. Can we predict life histories from sequences of tooth emergence?

If life history is a fundamental determinant of tooth eruption sequence, simple sequence strings preserved in teeth of fossil mammals might inform us whether a species lived and died on a fast time scale, like a small bovid, for example, or a slow one, like a monkey.

Table 15.4 contains test cases of eruption/emergence sequences for extinct mammals reconstructed from original illustrations or raw data in listed sources (although some uncertainty remains over the definitions of eruption). Life histories of extant taxa which share the same sequence characteristics are tabulated for comparison.

One extra caution is needed before proceeding. Life histories may have been distributed uniquely in long extinct mammal faunas. For example, long life and slow growth were probably rare or absent in very early mammals (Lillegraven *et al.*, 1987). If underlying life history distributions differ between the present and the past, it is safer to compare fossils with the total range known in extant mammals rather than apply specific means from the present to the past. Mean values would be highly dependent on the particular distribution of mammals living today and sampled in the present study.

In Table 15.4 each case is of interest because it is not clear how to choose a living model for life history. *Coryphodon* is a primitive ungulate-like mammal of large body size known from the Paleocene and Eocene of the northern hemisphere, only distantly related to extant mammals. Authors of an ontogenetic study compared *Coryphodon* to hippos based on body size and habitat (Lucas and Schoch, 1990). Judging from the hippo, we might expect that *Coryphodon* grew up very slowly, with M_1 emergence about 2 years of age and long life, perhaps 40-50 years. However, I_1 and P_3 positions like that of *Coryphodon* are found today in mammals emerging M_1 between 0.34-0.88 years, with life spans on the order of 21-35 years. If we limit comparison to generalized

Table 15.4. *Reconstructed sequences of tooth eruption/emergence for extinct species and ranges of life events in similar extant taxa.*[a]

Order Genus/species	Sequence/source[b]	Range in extant species matched for sequence code		
		Sequence code (I_1 position, P_3 position)[c]	Age of emergence of M_1 (years)[d]	Life span (years)
Pantodonta				
Coryphodon sp.	(M1 I3) p1 M2 P4 P3 (P2 I1) C (M3 I2)	2.5, 2.5	0.34–0.88	21–35
	Lucas and Schoch, 1990		0.34–0.47	21–27
Archaeoceti				
Dorudontinae	(M1 p1) M2 M3 P4 P3 P2 (P1? I3 C)	3.5?, 3.5	0.08–0.66	11–36
Zygorhiza kochii	I1 erupts sometime after M2 and probably after M3;		0.09–0.42	12–31
plus *Dorudon atrox*	I2 is after P4			
	Kellogg, 1936; Uhen, 1996			
Primates				
Adapis parisiensis	(M1 p1) M2 (I1 I2 M3) [P4 C] P3 P2	–, 3.5	0.08–0.66	11–36
	Stehlin, 1912; Gingerich, unpublished data		0.09–0.42	12–31
Notharctus tenebrosus	M1 M2 M3 (P4 P3 P2)	–, 3.5	0.08–0.66	11–36
	Gregory, 1920		0.09–0.42	12–31
Archaeolemur majori	M1 M2 I1 I2 M3 P4 P3 P2	2.5, 3.5	0.21–0.66	13–36
	Lamberton, 1938		0.36	13
Apidium phiomense	M1 M2 P2 P4 (P3 M3) C	–, 2.5?	0.34–3.33 (11)[e]	21–55
	Kay and Simons,			
Australopithecus africanus	M1 I1 I2 M2 P3 P4 C? M3	1.5, 2.5	0.54–3.33 (11)[e]	26–55
	Smith, 1994			

[a] Living archontans and ungulates with same I1 and P3 position in sequence.
[b] Sequence uncertain ( ); sequence varies, [ ]; teeth omitted if no basis for assigning order.
[c] As in Table 15.3.
[d] First line, all taxa; second line, generalized mammals only (removing specialized herbivores).
[e] *Procavia* has an unusually short lifespan (11 years); other species with comparable sequences have lifespans >20 years.

mammals, the estimate becomes more restricted (M_1 at 0.34–0.47 years and a 21–27 year life span), pointing to a life history considerably faster than a hippo, something more like a large deer or pig.

For several other cases in Table 15.4 more could be said if incisors could be placed relative to molars in sequence. *Zygorhiza* and *Dorudon* are both dorudontine archaeocetes (primitive whales) with body length something on the order of a mid-sized odontocete (toothed whale). Modern whales grow up fairly slowly, but are so transformed in adaptation from land mammal ancestors that they are questionable models for archaeocete life history. Since archaeocetes preserved primitive tooth replacement, however, their emergence sequence makes a place to start. As shown, the late appearance of premolars in archaeocetes tends to rule out the longest lived models. If comparison is limited to generalized mammals, life history of dorudontine archaeocetes is predicted to have run a fast to moderate pace (M_1 at 0.09–0.42 years, life span of 12–31 years).

The cases of two adapid primates *Adapis parisiensis* and *Notharctus tenebrosus* are instructive, although incisors have not been placed relative to M_3. Even in these Eocene primates, no hard data support a truly primitive emergence sequence, with all molars before replacement teeth. Based on incisor alveoli, one specimen of *Adapis parisiensis* observed by Gingerich (unpublished data) sug-

gests mandibular incisors were in place before M_3. If so, *Adapis parisiensis* had the sequence shown for *Archaeolemur*, a subfossil indriid from Madagascar (also in Table 15.4). The *Archaeolemur* sequence suggests a life history on pace with a large lemur or small anthropoid.

Apidium phiomense is an anthropoid known from the Oligocene of Egypt. Kay and Simons (1983) were able to demonstrate that P_4 and probably P_3 erupt in an advanced position, which in turn suggests a life history somewhat slowed, at least not on the fastest pace.

A last prediction can be made for *Australopithecus africanus*, the extinct early hominid. Its growth rate was originally compared to *Homo sapiens* (Mann, 1975); its dental development and emergence sequence, however, place it with great apes and monkeys (Smith, 1994). From Table 15.4, the *Australopithecus* emergence sequence suggests M_1 erupted from 0.54 to 3.33 years with life span ranging up to 55 years – in other words, matching a chimpanzee perhaps, but not a human.

Predicted life histories are just that until they are supported by other data. Fortunately, study of incremental lines in teeth promises to be able to establish the age of tooth eruption in many extinct species. One such study put M_1 emergence at about 3 years in *Australopithecus africanus* (Bromage and Dean, 1985). Until such detailed and time-consuming work is done on a large scale however, emergence sequences can provide some basis for choosing a living model for life history of an extinct mammal.

Other aspects of sequence than those studied here may prove productive in predicting life histories. For example, the tendency for some fossil mammals to erupt premolars before incisors is intriguing, but too few living mammals with this pattern have been sampled here to understand its meaning. Predictions should also improve as the data base for living mammals expands.

15.4.7.   Unexplained variation

Schultz's Rule has some power of explanation in present data, but it cannot explain all variance in emergence sequence. Such unexplained variance may give information about dental function or simply phylogenetic history. One unexplained feature is the characteristic 'backwards' emergence of premolars, the order $P_4 P_3$, in primates (Schwartz, 1974; Tattersall and Schwartz,1974), tree shrews (Shigehara, 1980), many insectivores, some carnivore families (Slaughter et al., 1974), and some other fossil mammals (Table 15.4). Most regard the distant ancestral order to be $P_3 P_4$, with the 'backwards' $P_4 P_3$ sequence the phylogenetic novelty (Slaughter et al., 1974, Tattersall and Schwartz, 1974). In primates, premolar order does revert to $P_3 P_4$ in apes and humans, a change which has been interpreted as a retardation of distal teeth in presence of slow growth (Smith, 1994). Looking at other mammals, however, no patterns emerge with respect to length of post-natal life. The ungulates erupt premolars in the 'front-to-back' order $P_3 P_4$ across the board (see Appendix), regardless of the pace of growth.

Whereas the $P_3 P_4$ sequence tends to characterize groups of related mammals, P_2 appearance is much more variable. Kay and Simons (1983) cautioned that P_2 is very variable in primates. Although it tends to appear together with other premolars, P_2 can jump position even in closely related species (see Appendix). In a study of insectivores and carnivores, Slaughter et al. (1974) simply concluded that the tooth had so little functional importance that it was more free to vary than other teeth.

Real exceptions to Schultz's Rule can be found in lemuriforms (lemurs and lorises) with tooth combs. Living lorisids (not included here because complete gingival sequences are unavailable) are said to replace some anterior teeth very early, even before emergence of M_1 (Schwartz, 1974). Clearly this cannot be because lorises grow up slowly; quite the opposite is true. Perhaps deciduous tooth combs are discarded early for functional reasons, as may also be true in Eulemur and Varecia. To generalize this finding, specialized dental functions should warn that Schultz's Rule might not apply.

Finally, patterns of emergence in other mammalian orders, notably the Carnivora and Chiroptera (the latter also included in Archonta), remain to be explained. Preliminary investigation finds very 'advanced' sequences in these orders. Such major differences between higher taxa may make eruption sequences useful in mammalian systematics. Even such major taxonomic differences, however, are likely to have adaptive explanations.

15.5.   Conclusions

To understand Schultz's Rule, one can imagine that replacement teeth (I1–P4) are on a sliding scale, adapting to demands on deciduous teeth. In slowly growing species, replacement teeth 'slide' to earlier positions to compensate for increased loads on deciduous teeth. The dampened response observed in specialized herbivores however, suggests that either their deciduous incisors manage to carry the load imposed by slow growth or that soft tissues have taken over some of the work of incisors.

Resemblance in tooth emergence sequence cross-cuts taxonomic group and facial form to a surprising degree. Sequences trend from (presumed) primitive (M M M P P I I) to (presumed) advanced (M I I M P P M) within two grandorders, Archonta (primates + tree shrew) and Ungulata. The large range of sequences in each grandorder is strong evidence that tooth eruption is a highly functional adaptive system. Thus, emergence sequence must be used with care in phylogenetic analysis.

Evidence suggests that sequence of eruption of teeth is a signature of life history, but this is not to say that phylogenetic history and dento-facial morphology leave no trace. Primates and the tree shrew share a characteristic 'backwards' emergence of the premolars P_4 and P_3 (Schwartz, 1974), a sequence of unknown significance for life history. It may be taxonomically useful, as may minor sequences involving teeth not studied here (deciduous teeth and permanent C, I_3, P_2). In addition, work in progress on Carnivora and Chiroptera finds more 'advanced' sequences than those described here, sequences that might reflect morphology and phylogeny. The hypothesis that face shape is correlated in a simple manner with emergence sequence, however, can be rejected. Despite considerable differences in facial architecture of archontans and ungulates, tooth eruption sequences are pages taken from the same book. Near matches in sequence characterize the tree shrew and the pronghorn, the peccary and the squirrel monkey, and the chimpanzee and the hippopotamus, pairs that share few characteristics of facial architecture. Similarity of the overall pace of life, however, can account for such matches.

Finally, emergence sequences must be seen as limited data. While they can show an orderly response to life history, they cannot discern how change is brought about – which teeth are early and which late. Allochronic analyses of actual ages teeth erupt could answer many more questions (Smith, 1992). Complete data of that type, however, are still relatively scarce, although basic sequences allow some predictions about life histories of extinct mammals. For such predictions, the most important datum appears to be position of incisors relative to molars, and secondly, position of premolars to molars. New approaches in histological study (Bromage and Dean, 1985; Dean, 1989; Swindler and Beynon, 1993) promise we will soon be able to test such hypotheses by reconstructing actual times of events in life history for extinct species. When such data are available, we will have a powerful tool for making inference about the evolution of life history in mammals.

Summary

Adolph Schultz's empiric search for a pattern in tooth emergence resulted in a general rule that is more widely applicable than he knew. 'Schultz's rule,' the tendency for replacing teeth to appear relatively early in slow-growing, long-lived species, applies to a broad range of mammals. The rule successfully orders new data from the orders Primates, Scandentia, and Soricomorpha, and the more generalized members of the grandorder Ungulata (Hyracoidea and suine Artiodactyla). The rule applies in a more limited way to the specialized Ungulata (Perissodactyla and ruminant Artiodactyla). As growth rate slows (M_1 emerges late) and life lengthens, generalized herbivores and omnivores fluidly adapt sequence of tooth emergence, whereas specialized herbivores adapt stiffly.

Acknowledgements

I thank Tristine L. Smart for major work on the mammal project, in both assembling and interpreting the data base on which this work depends. Philip Gingerich provided unpublished data on body and brain weight in extant mammals and tooth emergence in fossil primates. I thank John Mitani for his contributions to the mammal library research project. This research supported by National Science Foundation grants SBR-9408408 and SBR-925390 and a grant from the University of Michigan Vice-President for Research.

References

Attwell, C. A. M. (1980). Age determination of the blue wildebeest *Connochaetes taurinus* in Zululand. *South African Journal of Zoology*, **15**, 121–130.

Bennejeant, C. (1936). *Anomalies et Variations Dentaires chez les Primates*. Paris: Clermont-Ferrand.

Bromage, T. G. and Dean, M. C. (1985). Re-evaluation of the age at death of immature fossil hominids. *Nature*, **317**, 525–27.

Byrd, K. E. (1981). Sequences of dental ontogeny and callitrichid taxonomy. *Primates*, **22**, 103–118.

Caughley, G. (1965). Horn rings and tooth eruption as criteria of age in the Himalayan thar, *Hemitragus jemlahicus*. *New Zealand Journal of Science*, **8**, 333–351.

Chapman, D. I., Chapman, N. G. and Colles C. M. (1985). Tooth eruption in Reeves' muntjac (*Muntiacus reevesi*) and its use as a method of age estimates (Mammalia: Cervidae). *Journal of Zoology, London (A)*, **205**, 205–221.

Cifelle, R. L., Rowe, T. B., Luckett, W. P., Banta, J., Reyes, R. and Howes, R. I. (1996). Fossil evidence for the origin of the marsupial pattern of tooth replacement. *Nature*, **379**, 715–718.

Dean, M. C. (1989). The developing dentition and tooth structure in hominoids. *Folia primatologica*, **53**, 160–176.

Dodson, P. (1975). Functional and ecological significance of relative growth in *Alligator*. *Journal of Zoology, London*, **175**, 315–355.

Eaglen, R. H. (1985). Behavioral correlates of tooth eruption in Madagascar lemurs. *American Journal of Physical Anthropology*, **66**, 307–315.

Eisenberg, J. G. (1981). *The Mammalian Radiations*. Chicago: University of Chicago Press.

Garn, S. M. and Lewis, A. B. (1963). Phylogenetic and intra-specific variations in tooth sequence polymorphism. In *Dental Anthropology*, ed. D. R. Brothwell, pp. 53–73. Oxford: Pergamon.

Gingerich, P. D. (1984). Dentition of *Sivaladapis nagrii* (Adapidae) from the late Miocene of India. *International Journal of Primatology*, **5**, 63–79.

Gregory, W. K. (1920). The structure and relations of *Notharctus*, an American Eocene primate. *Memoirs of the American Museum of Natural History*, **3**, 49–243.

Hall-Martin, A. J. (1976). Dentition and age determination of the giraffe *Giraffa camelopardalis*. *Journal of Zoology, London*, **180**, 263–289.

Harvey, P. H. and Clutton-Brock, T. H. (1985). Life history variation in primates. *Evolution*, **39**, 559–81.

Hertenstein, B., Zimmermann, E. and Rahmann, H. (1987). Zur Reproduktion und ontogenetischen Entwicklung von Spitzhörnchen (*Tupaia belangeri*). *Zeitschrift fur Kölner Zoo*, **30**, 119–133.

Hillman-Smith, A. K. K., Owen-Smith, N., Anderson, J. L., Hall-Martin, A. J. and Selaladi, J. P. (1986). Age estimtion of the White rhinoceros (*Ceratotherium simum*). *Journal of Zoology, London (A)*, **210**, 355–379.

Hoover, R. L., Till, C. E. and Ogilvie, S. (1959). *The antelope of Colorado: a research and management study*. Boulder: State of Colorado Department of Game and Fish (Technical Bulletin 4).

Jaspers, R. and DeVree, F. (1978). Trends in the development of the skull of *Okapia johnstoni* (Sclater, 1901). *Acta Zoologica Pathologica Antverpiensia*, **71**, 107–129.

Jeffery, R. C. V. and Hanks, J. (1980). Age determination of eland

Taurotragus oryx (Pallas, 1766) in the Natal Highveld. *South African Journal of Zoology,* **16**, 113–122.

Kay, R. F. and Simons, E. L. (1983). Dental formulae and dental eruption patterns in Parapithecidae (Primates, Anthropoidea). *American Journal of Physical Anthropology,* **62**, 363–375.

Kellogg, R. (1936). *A review of the Archaeoceti.* Washington, DC: Carnegie Institution of Washington, Publication **482**. 1–366.

Kerr, M. A. and Roth, H. H. (1970). Studies on the agricultural utilization of semi-domesticated eland (*Taurotragus oryx*) in Rhodesia. *Rhodesian Journal of Agricultural Research,* **8**, 149–155.

Kindhal, M. (1967). Some comparative aspects of the reduction of the premolars in the Insectivora. *Journal of Dental Research,* **46**, 805–808.

Kirkpatrick, R. D. and Sowls, L. K. (1962). Age determination of the collared peccary by the tooth-replacement pattern. *Journal of Wildlife Management,* **26**, 214–217.

Krogman, W. M. (1930). Studies in growth changes in the skull and face of anthropoids. I. The eruption of the teeth in anthropoids and Old World apes. *American Journal of Anatomy,* **46**, 303–313.

Lamberton, C. (1938). Dentition de lait de quelques lémuriens subfossiles Malagaches. *Mammalia,* **2**, 57–80.

Laws, R. M. (1968). Dentition and ageing of the hippopotamus. *East African Wildlife Journal,* **6**, 19–52.

Lillegraven, J. A., Thompson, S. D., McNab, B. K. and Patton, J. L. (1987). The origin of eutherian mammals. *Biological Journal of the Linnaean Society,* **32**, 281–336.

Lucas, S. G. and Schoch, R. M. (1990). Ontogenetic studies of early Cenozoic *Coryphodon* (Mammalia, Pantodonta). *Journal of Paleontology,* **54**, 831–841.

Luckett, W. P. (1993). An ontogenetic assessment of dental homologies in therian mammals, in *Mammal Phylogeny*, Vol. 1, ed. F. S. Szalay, M. J. Novacek and M. C. McKenna, pp. 182–204. Berlin: Springer.

MacDonald, D. (1984). *The Encyclopedia of Mammals.* New York: Facts on File.

Mann, A. E. (1975). *Some Paleodemographic Aspects of the South African Australopithecines.* Philadelphia: University of Pennsylvania Press.

Martin, T. (1997). Tooth replacement in late Jurassic Dryolestidae (Eupantotheria, Mammalia). *Journal of Mammalian Evolution,* **4**, 1–18.

Matschke, G. H. (1967). Aging European wild hogs by dentition. *Journal of Wildlife Management,* **31**, 109–113.

Miller, F. L. (1972). Eruption and attrition of mandibular teeth in barren-ground caribou. *Journal of Wildlife Management,* **36**, 606–612.

Nowak, R. M. and Paradiso, J. L. (1983). *Walker's Mammals of the World.* Baltimore: Johns Hopkins University Press.

Osborn, J. W. and Crompton, A. W. (1973). The evolution of mammalian from reptilian dentitions. *Brev. Museum of Comparative Zoology,* **399**, 1–18.

Pond, C. M. (1977). The significance of lactation in the evolution of mammals. *Evolution,* **31**, 177–199.

Quimby, D. C. and Gaab, J. E. (1957). Mandibular dentition as an age indicator in Rocky Mountain elk. *Journal of Wildlife Management,* **21**, 435–451.

Rautenbach, I. L. (1971). Ageing criteria in the springbok, *Antidorcas marsupialis* (Zimmermann, 1780) (Artiodactyla: Bovidae). *Annals of the Transvaal Museum,* **27**, 83–133.

Roche, J. (1978). Denture et âge des camans de rochers (Genre *Procavia*). *Mammalia,* **42**, 97–103.

Roettcher, D. and Hofmann, R. R. (1970). The ageing of impala from a population in the Kenya Rift Valley. *East African Wildlife Journal,* **8**, 37–42.

Schultz, A. H. (1935). Eruption and decay of the permanent teeth in primates. *American Journal of Physical Anthropology,* **19**, 489–581.

Schultz, A. H. (1956). Postembryonic age changes. In *Primatologia,* vol. 1, ed. H. Hofer, A. H. Schultz and D. Starck, pp. 887–964. Basel: Karger.

Schultz, A. H. (1960). Age changes in primates and their modification in man. In *Human Growth*, vol. III, ed. J. M. Tanner, pp. 1–20. Oxford: Pergamon Press.

Schwartz, J. H. (1974). Dental development and eruption in the Prosimians and its bearing on their evolution. PhD dissertation, Columbia University.

Severinghaus, C. W. (1949). Tooth development and wear as criteria of age in white-tailed deer. *Journal of Wildlife Management,* **13**, 195–216.

Shigehara, N. (1980). Epiphyseal union, tooth eruption, and sexual maturation in the common tree shrew, with reference to its systematic problem. *Primates,* **21**, 1–19.

Simpson, S. W., Lovejoy, C. O. and Meindl, R. S. (1990). Hominoid dental maturation. *Journal of Human Evolution,* **19**, 285–297.

Singer, R. and Boné, E. L. (1960). Modern giraffes and the fossil giraffids of Africa. *Annals of the South African Museum,* **45**, 375–548.

Slaughter, B. H., Pine, R. H. and Pine, N. E. (1974). Eruption of cheek teeth in Insectivora and Carnivora. *Journal of Mammalogy,* **55**, 115–125.

Smith, B. H. (1989). Dental development as a measure of life history in primates. *Evolution,* **43**, 683–88.

Smith, B. H. (1992). Life history and the evolution of human maturation. *Evolutionary Anthropology,* **1**, 134–42.

Smith, B. H. (1994). Sequence of emergence of the permanent teeth in *Macaca, Pan, Homo,* and *Australopithecus*: Its evolutionary significance. *American Journal of Human Biology,* **6**, 61–76.

Smith, B. H. and Garn, S. M. (1987). Polymorphisms in eruption sequence of permanent teeth in American children. *American Journal of Physical Anthropology,* **74**, 289–303.

Smith, B. H., Crummett, T. L. and Brandt, K. L. (1994). Ages of eruption of primate teeth: a compendium for aging individuals or comparing life histories. *Yearbook of Physical Anthropology,* **37**, 177–231.

Smuts, G. L. (1974). Age determination in Burchell's zebra (*Equus burchelli antiquorum*) from the Kruger National Park. *Journal of the South African Wildlife Management Association,* **4**, 103–115.

Stehlin, H. G. (1912). Die Säugetiere des schweizerischen Eocaens. *Abhandlungen Schweizerische Pälaotologische Gesellschaft* (Zurich), **38**, 1165–1298.

Steyn, D. and Hanks, J. (1983). Age determination and growth in the hyrax *Procavia capensis* (Mammalia: Procaviidae). *Journal of Zoology, London,* **201**, 247–257.

Swindler, D. R. and Beynon, A. D. (1993). The development and microstructure of the dentition of *Theropithecus*. In *Theropithecus: the Rise and Fall of a Primate Genus*, ed. N. G. Jablonski, pp. 331–381. Cambridge: Cambridge University Press.

Tattersall, I. and Schwartz, J. H. (1974). Craniodental morphology and the systematics of the Malagasy lemurs (Primates, Prosimii). *Anthropological Papers of the American Museum of Natural History*, **52**, 139–192.

Tsang, W. N. and Collins, P. M. (1985). Techniques for hand-rearing tree-shrews (*Tupaia belangeri*) from birth. *Zoo Biology*, **4**, 23–31.

Uhen M. (1996). *Dorudon atrox* (Mammalia, Cetacea): form, function, and phylogenetic relationships of an Archaeocete from the Late Middle Eocene of Egypt. PhD dissertation, University of Michigan.

Usuki, H. (1967). Studies of the shrew mole (*Urotrichus talpoides*). III. Some problems about dentition with special reference to tooth replacment. *Journal of the Mammalogy Society of Japan*, **3**, 158–162.

Wallace, J. A. (1977). Gingival eruption sequences of permanent teeth in early hominids. *American Journal of Physical Anthropology*, **46**, 483–494.

Wegrzyn, M. and Serwatka, S. (1984). Teeth eruption in the European bison. *Acta Theriologica* **29**, 111–121.

Wilson, D. E. and Reeder, D. M. (1993). *Mammal Species of the World: A Taxonomic and Geographic Reference*, second edition. Washington, DC: Smithsonian Institution.

Wilson, V. J., Schmidt, J. L. and Hanks, J. (1984). Age determination and body growth of the Common duiker *Sylvicapra grimmia* (Mammalia). *Journal of Zoology, London*, **202**, 283–297.

Wolf, K. (1984). Reproductive competition among co-resident male silvered leaf-monkeys (*Presbytis cristata*). PhD dissertation, Yale University.

Zeigler, A. C. (1971). A theory of the evolution of therian dental formulas and replacement patterns. *Quarterly Review of Biology*, **46**, 226–249.

Zeigler, A. C. (1972). Processes of mammalian tooth development as illustrated by dental ontogeny in the mole *Scapanus latimanus* (Talpidae: Insectivora). *Archives of Oral Biology*, **17**, 61–76.

Appendix: complete mandibular emergence sequence (age of emergence of M₁ and life span (both in years) follow species name)

Note: parentheses ( ) surround teeth when actual sequence has not been resolved, i.e., teeth are 'tied;'

square parentheses [ ] surround teeth known to vary in sequence at >15%; = indicates variation is > 40%. Premolars omit numbers when there is no basis for assigning order. tc* tooth comb (I₁, I₂, C₁) emerges all at once.

I. GRANDORDER INSECTIVORA

ORDER SORICOMORPHA
TALPIDAE
Urotrichus talpoides   M1 M2 M3 P2 P3 P4 I1 C
Urotrichus pilirostris   M1 M2 M3 (P2 P3 P4) I1 C p1

II. GRANDORDER ARCHONTA

ORDER SCANDENTIA
TUPAIIDAE
Tupaia glis c. 0.09?, 12.4   M1 M2 M3 P2 I3 P4 (I1 C) P3 I2

ORDER PRIMATES
LEMURIDAE
Lemur catta 0.34, 27.1   M1 M2 tc* P4 P3 P2 M3
Eulemur fulvus 0.42, 30.8   M1 tc* M2 P2 M3 P4 P3
Varecia variegata 0.48, –   M1 tc* M2 P2 P4 M3 P3

CEBIDAE
Aotus trivirgatus 0.36, 12.6   M1 M2 I1 M3 I2 P4 P3 P2 C
Saimiri sciureus 0.37, 21   M1 M2 I1 I2 P4 P3 P2 M3 C
Alouatta caraya   M1 I1 (I2 M2) P P P4 M3
Cebus sp. 1.1, 40 (no data P2)   M1 I1 I2 M2 P4 P3 M3 C?
CERCOPITHECIDAE
Cercopithecus aethiops 0.8, 31   M1 I1 I2 M2 P4 P3 C M3
Macaca nemestrina (male) 1.4, 26.3 M1 I1 I2 M2 P4 P3 M3 C
Papio anubis (male) 1.6, –   M1 I1 I2 M2 C (P4 P3) M3
PONGIDAE
Pan troglodytes 3.2, 44.5   M1 I1 I2 [M2 P3=P4] M3 C
HOMINIDAE
Homo sapiens
Australian Aborig. <6.0?, 100 (male) [M1 I1] I2 C P3 [M2=P4] M3
White American 6.3, 100 (male)   [I1 M1] I2 [C P3 [P4] M2] M3

III. GRANDORDER UNGULATA

ORDER ARTIODACTYLA
SUBORDER RUMINANTIA
ANTILOCAPRIDAE
Antilocapra americana 0.17, 11.2   M1 M2 M3 I1 P3 P4 I2 P2 I3 C
BOVIDAE
Antidorcas marsupialis 0.08, 17.1   M1 M2 M3 I1 (I2 I3 C) P3 P4
Hemitragus jemlahicus 0.21, 21.8   M1 M2 I1 [M3 I2] P3 [P2 P4] I3
Sylvicapra grimmia 0.23, 14.3   M1 M2 M3 (P P P) I1 [I2 I3] C
Aepyceros melampus 0.33, 17.5   M1 M2 I1 M3 I2 (P2 P3) I3 C P4
Connochaetes taurinus 0.46, 21.4   M1 M2 I1 M3 I2 (P3 P4) I3 C
Taurotragus oryx 0.53, 23.5   M1 M2 I1 M3 I2 I3 C (P2 P3) P4
Bison bonasus 0.54, 20   M1 M2 I1 P2 (M3 P3) I2 I3 P4 C
CERVIDAE
Rangifer tarandus 0.17, 20.2   M1 I1 M2 I2 I3 M3 (P3 P2) P4
Odocoileus virginianus 0.18, 20   M1 I1 M2 (I2 I3 C) M3 P2 (P3 P4)
Muntiacus reevesi 0.2?, 17.6   M1 M2 I1 M3 I2 (P3 P2) P4 (I3 C)
Cervus elaphus 0.33, 26.7   M1 M2 I1 I2 M3 (I3 C) [P2 P3] P4
GIRAFFIDAE
Okapia johnstoni –, 33   M1 M2 I1 (M3 I2) (P3 I3) P4 P2 C
Giraffa camelopardalis 0.66, 36.2   M1 M2 [I1 M3] I2 [P4 P3 P2] I2 I3 C
SUBORDER SUINA
TAYASSUIDAE
Tayassu tajacu 0.40, 24.6   M1 C I3 M2 I1 I2 P P P M3
SUIDAE
Sus scrofa 0.47, 27   M1 p1 I3 C M2 I1 P3 P4 P2 I2 M3
HIPPOPOTAMIDAE
Hippopotamus amphibius ~2, 54.3   p1 M1 (I1 I2 C) M2 (P2 P3) P4 M3
ORDER PERISSODACTYLA
EQUIDAE
Equus burchelli 0.88, 35 (male)   p1 M1 M2 I1 (P2 P3) M3 I2 (C P4) I3
RHINOCEROTIDAE
Ceratotherium simum 2.75, 36   p1 M1 P2 (M2 P3 P4) M3
Order Hyracoidea
PROCAVIIDAE
Procavia capensis 0.54, 11   M1 (I1 I2) (M2 P P P) M3

Sources: I: Usuki (1967) II: Shigehara (1980); original primate literature cited in Smith & Garn (1987), Smith (1994) and Smith *et al.* (1994). III: Hoover *et al.* (1959); Rautenbach (1971); Caughley (1965); Wilson *et al.* (1984); Roettcher and Hofmann (1970); Attwell (1980); Jeffery and Hanks (1980) and Kerr and Roth (1970); Wegrzyn and Serwatka (1984); Miller (1972); Severinghaus (1949); Chapman *et al.* (1985); Quimby and Gaab (1957); Jaspers and DeVree (1978); Hall-Martin (1976) and Singer and Boné (1960); Kirkpatrick and Sowls (1962); Matschke (1967); Laws (1968); Smuts (1974); Hillman-Smith *et al.* (1986); Steyn and Hanks (1983) and Roche (1978).

Macrostructure and function

Part four

16 Developmental plasticity in the dentition of a heterodont polyphyodont fish species

A. Huysseune

16.1. Introduction

The dentition has long been, and still is, important to vertebrate biologists: to palaeontologists, because teeth often represent the sole fossil evidence available; to taxonomists, because of the estimated high taxonomic value of teeth; and to developmental biologists, because teeth provide an excellent experimental model to test developmental principles. Yet, relatively little attention has been paid in these various disciplines to intraspecific variation and the relative roles played by the genome and the environment to generate this variation. Can developmental processes leading to a species-specific, predictable outcome (i.e. a dental unit of specific size, shape, etc.) be modified in the course of an organism's ontogeny by changes in the external environment to produce a unit of different size, shape, etc., with respect to what is 'normal' for the species? In other words can the dentition exhibit environmentally induced phenotypic variation?

The subject of phenotypic plasticity has gained much interest in vertebrate biology in the past 15 years. Phenotypic plasticity, as opposed to genetic polymorphism, covers all types of environmentally induced phenotypic variation (Stearns, 1989) and is defined as the ability of an individual to respond to changes in the environment in terms of its anatomy, physiological state and/or behaviour (Smits, 1996). In this chapter I am primarily concerned with plasticity related to changes in prey type and/or the way it is processed. What changes can we expect in tooth number, position, size and shape in relation to a changing function?

Studies on dental variation have largely concentrated on mammals (cf. Miles and Grigson, 1990). Whilst intraspecific variation in number, size and shape of mammalian teeth in general and human teeth in particular, have

long been considered to be of genetic origin, or at least to involve genetic mechanisms, it is now increasingly clear that these variables are much more sensitive to environmental influences than was hitherto assumed (see, for example, Lauweryns et al., 1993). However, rather than changes driven by functional necessity, these influences concern developmental perturbations, of epigenetic origin (effect of maternal diet, pathologic conditions, irradiation, toxins, etc.) that act early in ontogeny and can generate variation in tooth number (hyperdontia, oligodontia; e.g. Melamed et al., 1994; Schalk-van der Weide et al., 1994) as well as size (Brook, 1984). Epigenetic effects are especially suspected in cases of asymmetrical anomalies (Sofaer, 1978). Environmental influences are perhaps most difficult to show for tooth shape since this is thought to have a strong genetic component, at least in mammals (Slavkin, 1988). The only variable that has long been recognized to be more sensitive to various environmental effects and the only variable where effects of altered (muscle) function have been amply documented is the position of teeth. Again, most studies concern mammals (see Miles and Grigson, 1990 for many examples). Altered masticatory function affects craniofacial growth and has secondary effects on the position of teeth, hence occlusion. In fact, nearly the entire field of human orthodontics has concentrated on the adaptation of the dentition to changes in growth of the craniofacial skeleton and the factors influencing rate and direction of that growth.

As a consequence, current ideas on variability of the dentition are largely based on work with diphyodont models (i.e. organisms that have two tooth generations: the deciduous and the permanent teeth) or even more specialized monophyodonts. Notable exceptions are the studies of Reif on shark dentitions, which are polyphyo-

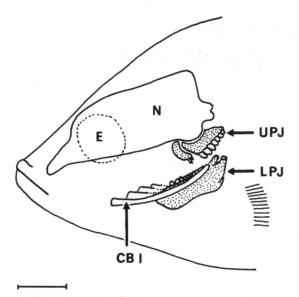

Figure 16.1. **Semi-schematic representation of the position of the paired upper pharyngeal jaws (UPJ; modified dorsal parts of branchial arches II to IV) and single lower pharyngeal jaw (LPJ; modified fused ceratobranchials of the fifth arch) within the head of an adult *Astatoreochromis alluaudi*. Also figured are the neurocranium (N), the position of the eye (E), ceratobranchial I (CBI) and the anterior portions of ceratobranchials II to IV. Scale bar represents 5 mm.**

dont (i.e. have many tooth generations; Reif, 1984). In this chapter, I will illustrate how dietary changes affect tooth variables through altered muscle activity in a polyphyodont species and discuss the developmental mechanisms that may be involved in generating the observed plasticity. Finally, I will stress the role of polyphyodonty versus diphyodonty in generating developmental plasticity.

16.2.   Phenotypic plasticity in the pharyngeal jaw apparatus in cichlid fishes

Cichlids form a speciose family of highly evolved freshwater fishes, with species numbers possibly exceeding one thousand, and with a main distribution in the tropics and subtropics of Central and Southern America and Africa. Several cichlid species have been found to be polymorphic or to exhibit phenotypic plasticity in morphological features (reviewed in Vandewalle *et al.*, 1994).

Part of the evolutionary success of cichlids has been ascribed to the possession of a second set of jaws, called pharyngeal jaws (paired upper and single lower pharyngeal jaw), which derive from modified second to fifth branchial arches (Vandewalle *et al.*, 1994; Figure 16.1).

Cichlids show an astonishing range of trophic groups (Witte and van Oijen, 1990). So-called pharyngeal crushers feed on molluscs by transporting the prey to the pharyngeal jaws, where it is crushed. The main muscles involved in the crushing action insert on the horns of the lower pharyngeal jaw (LPJ; Figure 16.2A,C). Kinematical analysis suggests a strong dominance of the crushing power stroke during which the LPJ is pulled dorsally and anteriorly to occlude against the posteriorly moving upper pharyngeal jaws (Liem and Kaufman, 1984). A biomechanical model for pharyngeal crushing in cichlids is still lacking. Galis's (1992) model for pharyngeal biting was developed for an insectivorous cichlid species and should only be extrapolated to species with a pharyngeal crushing function with great care.

Astatoreochromis alluaudi is a potentially mollusc-crushing cichlid for which phenotypic plasticity has been well established (Greenwood, 1965; Hoogerhoud, 1984, 1986a,b; Witte *et al.*, 1990; Smits, 1996). Two forms are found in the great African Lakes and tributary river systems: a form that feeds on soft food items (insects, plant debris, etc.) and has slender pharyngeal jaws with numerous fine and sharp teeth (Figure 16.2A,B), and a molluscivorous (durophagous) form that has stout (so-called hypertrophied) pharyngeal jaws with less teeth, of which many are large and molariform (Figure 16.2C,D). The horns of the LPJ in the durophagous specimens are much sturdier providing an enlarged insertion area for the more powerful crushing muscles (Smits, 1996). In Lake Victoria, the durophagous form starts to feed on snails at 40 mm standard length (SL) only; at smaller sizes, it still feeds on soft food items and very much resembles the soft food form (Hoogerhoud, 1986b). Progeny from the durophagous form, raised in tanks on soft food, exhibits the soft food phenotype (Greenwood, 1965; Hoogerhoud, 1986a,b; Smits, 1996).

The LPJ in both phenotypes has a triangular outline when viewed from above. Its dorsal, dentigerous (tooth-bearing) surface is almost completely covered with functional teeth showing a complex pattern of implantation. Teeth vary greatly in size and shape; the largest teeth are found in the central posteriormost part of the tooth-bearing region. Every functional tooth is ligamentously attached to a cylinder of attachment bone, itself anchored within the bone of the dorsal surface of the jaw (dentigerous plate) (Figure 16.2B,D, Figure 16.3A). Below the dentigerous plate is the medullary cavity which houses replacement teeth. In soft food specimens, these lie freely in a vast medullary cavity that contains relatively sparse bone (Figure 16.2B); in hard food specimens, trabeculae reinforcing the medullary cavity enclose bony crypts housing the replacement teeth (Figure 16.2D) (cf. Huysseune *et al.*, 1994).

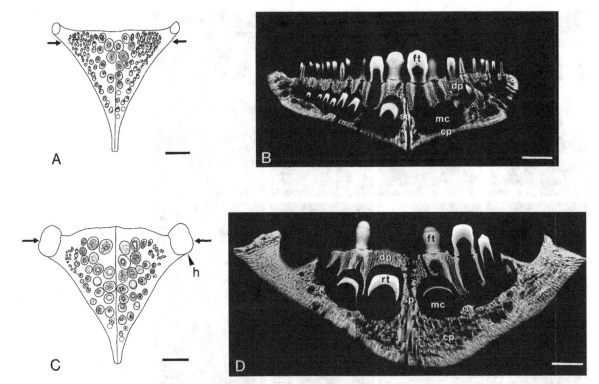

Figure 16.2. **Dorsal view of the LPJ of a pond-raised soft food specimen (A) and a wild-caught durophagous specimen (C) of *Astatoreochromis alluaudi* of similar size (85 mm S. L.). Arrows indicate the level of the transverse ground section from which microradiographs are presented in B (soft food specimen) and D (durophagous specimen), respectively. On cross-sections (B,D), each half of the LPJ can be considered as a triangle, with a dorsal dentigerous plate, a ventral cortical plate, and a medial sutural plate, the three of which surround a medullary cavity. Scale bars in A, C represent 2 mm; scale bars in B, D represent 1 mm; h, horn; ft, functional tooth; dp, dentigerous plate; sp, sutural plate; mc, medullary cavity of the jaw; cp, cortical plate; rt, replacement tooth.**

16.3.   Tooth development and replacement on the pharyngeal jaws

Cichlids, like most fish, and for that matter most non-mammalian vertebrates, have a polyphyodont dentition: numerous tooth generations replace each other throughout life. Because the replacement process is crucial in understanding how dietary changes impose their phenotypic effects on the dentition, I will briefly summarize how replacement teeth develop on the LPJ of *Astatoreochromis alluaudi*.

In both forms, the process is basically similar. Replacement teeth develop deep to the functional tooth within the medullary cavity of the jaw bone, at the end of a thin epithelial strand originating from the superficial epithelium and penetrating through the dentigerous plate via a gubernacular canal (Figure 16.3B; see below for a discussion on the developmental significance of this strand). The epithelial strand can develop either directly from the pharyngeal epithelium or from the reduced enamel epithelium of its predecessor. Once the replacement tooth starts to mineralize, resorption of the over-

lying part of the dentigerous plate is initiated. Successive stages of migration of replacement teeth coincide with ever more advanced stages of resorption of the attachment bone of the predecessor and part of the surrounding tissue (Figure 16.3C,D). Eventually, the functional tooth is lost and a wide canal is formed through which the replacement tooth migrates to the surface (Figure 16.3D). Deposition of new attachment bone starts while the new tooth still lies within the open crypt (Figure 16.3E). First the cylinder of more compact attachment bone is formed, the thickness of its wall diminishing towards the medullary cavity (Figure 16.3E). Subsequently cancellous bone is deposited along its outer circumference to anchor the new attachment bone into the dentigerous plate even before the replacement tooth reaches alignment with the surrounding functional teeth (Figure 16.3F). Lastly, cancellous bone is deposited at the base of the attachment bone cylinder to form a thin delimiting bony plate (Figure 16.3G). The position of the teeth in subsequent generations is determined both by the pathway followed by the gubernacular canal and the pathway of eruption, and it is not clear

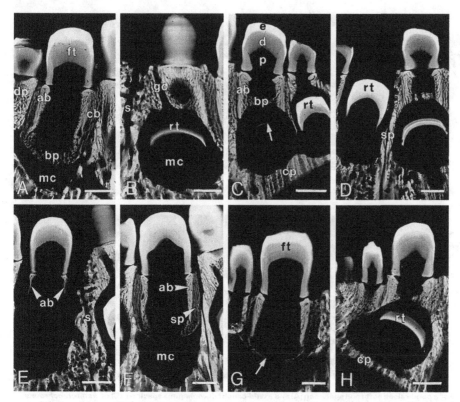

Figure 16.3A–H. **Details of microradiographs from transverse ground sections of the LPJ of *Astatoreochromis alluaudi*. Scale bars in A–F and H represent 0. 5 mm, in G 0. 25 mm. A. Well-fixed functional tooth in a hard food specimen of 101 mm SL sitting on a collar of attachment bone (ab) itself anchored within the dentigerous plate by means of cancellous bone (cb). The unmineralized ligament attaching the tooth to its attachment bone is not visible; bp, bony plate delimiting the cylinder of attachment bone. B. Hard food specimen of 85 mm SL Gubernacular canal (gc) through the posterior wall of the cylinder of attachment bone of a functional molariform tooth. Note the incipient mineralization of the replacement tooth in the medullary cavity below; s, suture between the halves of the LPJ. C. Same specimen as in B. Ongoing development and mineralization of adjacent replacement teeth and concomitant resorption of the cylinders of attachment bone of their predecessor. Note incipient mineralization of one of the two replacement teeth (arrow); e, enameloid; d, dentine; p, pulp cavity of the tooth. D. Same specimen as in B. The replacement tooth migrates to the surface of the LPJ through a wide canal. Note resorption of the dorsal part of the adjacent sutural plate. E. Hard food specimen of 85 mm SL A small amount of new attachment bone is deposited while the new tooth lies still deep within the open crypt. F. Hard food specimen of 101 mm SL Anchoring of young attachment bone into the surrounding dentigerous plate by means of cancellous bone. Where part of the suture has been eroded to allow passage of the replacement tooth, the dorsal part of the sutural plate is rebuilt by cancellous bone. G. Soft food specimen of 66 mm SL In a last step of the replacement cycle, a thin bony plate is formed (arrow) which delimits the cylinder of attachment bone from the medullary cavity of the LPJ. H. Hard food specimen of 101 mm SL; lateral region of the LPJ. Growth of this large-sized replacement tooth is seen to affect the attachment bone of its predecessor as well as of the adjoining functional tooth.**

whether the two necessarily coincide (personal observations). Sometimes, in each of the two phenotypes, growth of the replacement tooth is seen to provoke resorption of the cylinder of attachment bone of two adjacent functional teeth (Figure 16.3H).

16.4.   Changes in tooth number

One of the most striking features that distinguishes the LPJ dentition in the two phenotypes is the number of teeth. Counts of tooth number on the LPJ of more than 100 individuals from various sources are graphically represented in Figure 16.4. The specimens derive from

four sources: (1) wild-caught fish from Mwanza, Lake Victoria, i.e. a snail-eating population; (2) progeny from Lake Victoria fish, raised in aquaria for several generations, and subsequently released in ponds in Cameroon where they fed on soft food items (for details see Huysseune *et al.*, 1994); (3) descendants (F_8–F_{12}) from Lake Victoria fish, raised in tanks on a soft diet of minced liver and Tetramin; and (4) specimens of the same broods as (3) but raised on a diet of (relatively soft-shelled) snails. The data must be considered with some care. First, ranges of fish size are not comparable. Wild-caught specimens are limited to a range of standard lengths of about 60 to 125 mm because smaller (juvenile) fish are difficult to catch. The second data set is very limited. The two other data sets cover a much wider range of fish sizes, from 10 up to

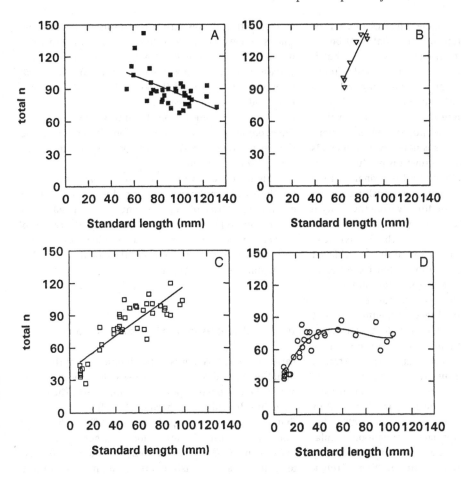

Figure 16.4 **Graph representing the total number of teeth on the lower pharyngeal jaw (total n) plotted against standard length. Full squares (A): wild-caught specimens of Lake Victoria (snail-eating); triangles (B): pond-raised specimens (feeding on soft prey); open squares (C): tank-reared animals fed a soft diet; open circles (D): tank-reared animals fed a hard diet. Linear regression lines added for A (*n = 37*; *r*²=0.301; *P*<0.01) B (*n = 8*; *r*²=0. 885; *P*<0.01) and C (*n = 42*; *r*²=0.733; *P*<0.01); polynomial regression line added for D (*n = 31*; *r*²=0.817; *P*<0.01).**

about 100 mm SL. For all data sets, I calculated linear regressions of tooth number on standard length, along with r^2 and P values. Logarithmic transformation of the data did not increase r^2 (percentage of variation explained by the regression), except for the fourth data set (fish raised experimentally on snails) where a polynomial regression (third degree) increased the r^2 value considerably.

What can be extrapolated from Figure 16.4? First, wild-caught specimens thriving on a diet of hard-shelled snails show a significant decrease in absolute tooth number as they get larger. The decrease is primarily due to the very high values found for the smaller fish. These values could be exceptional, but it is also possible that the Lake Victoria population shows a steep initial rise in tooth number, raising the value to a high level. A subsequent switch to a molluscivorous diet could then lead to a decrease in absolute number. In an earlier paper, dealing with a more limited data set, I found that tooth number on the LPJ in this population appears to be maintained as the fish get larger, or at least does not increase (Huysseune, 1995). Second, the pond specimens

(2) as well as the tank-reared specimens (3), both fed a soft diet, show an increase of tooth number with fish size. Third, tank-raised specimens fed with snails (4) initially show a rise in tooth number that closely matches that of the soft food group. However, at about 40 mm SL, the trajectory changes dramatically, tooth number appears to drop and the slope of the regression comes to resemble that of the regression for the wild-caught specimens, though at a lower level. So, as in the wild snail-eating population, a maintenance, or even decrease in absolute tooth number is observed. That tooth number is lower in the laboratory-reared specimens may be due to a steeper initial rise in the wild population since the latter normally switches to a molluscan diet at 40 mm SL as well (Hoogerhoud, 1986b).

In an attempt to explain these observations, it is necessary to introduce Reif's (1982) concept of a dental lamina. According to Reif all teeth of recent vertebrates form in a deep epithelial invagination, called the dental lamina. Related to this feature is his concept of tooth families, whereby one family represents a functional tooth and all its successors, linked by a shared dental lamina (that need not be a permanent feature but may be formed anew for every replacement tooth). Osborn's (1984) definition of tooth family is less strict in that it includes all the teeth developed at a particular tooth *position* (apparently implying no developmental relationship between them). The assignment of a tooth to a particular tooth family is necessary if we want to explain tooth number adjustment in molluscivorous specimens of *Astatoreochromis alluaudi*. This is not easy and depends on whether the epithelial downgrowth is produced from the reduced enamel epithelium of a functional tooth (a clear indication of a developmental relationship between this tooth and the new germ) or directly from the pharyngeal epithelium (in which case the germ should be considered the founder of a new tooth family). The identification of tooth families can only be achieved through detailed mapping based on extensive series of serial sections. If a true decrease in absolute tooth number does occur, there are two alternative solutions: non-replacement of functional teeth (i.e. elimination of supernumerary tooth families), either alone or in combination with – and outweighing – the addition of new tooth families. Earlier I noticed that in the posterior, central part of the dentigerous area – the region that is most essential in crushing – tooth number in the largest hard food specimen appears to be comparable to that in the smallest soft food specimen (Huysseune, 1995). This suggests that at least here normal (i.e. one-for-one) replacement continues to take place and the number of tooth families is maintained. I therefore suspect that, if replacement ceases, this will be in the more marginal areas of the dentigerous region. Reif (1984) showed that, in shark dentitions, supernumerary teeth develop during ontogeny, and assumed that they are subsequently eliminated through some process of counting the tooth family number. The author however admitted that nothing is known of a possible tooth counting mechanism. He simply stated that they are not replaced by a successor due to an inhibitory field mechanism, a somewhat unfortunate term he used, probably to refer to Osborn's (1978) clone theory. A fusion of neighbouring tooth families has never been observed in any recent shark (Reif, 1982). If, on the contrary, tooth number is not reduced during growth of durophagous specimens, the most simple solution appears to be suppression of formation of additional tooth families along the margins of the dentigerous area (cf. the large number of small teeth in marginal areas in soft food specimens, Figure 16.2A).

Inhibition of tooth germ formation, either within an existing tooth family, or to found a new family, must act on the level of initiation of the dental lamina, which, as we saw, is non-permanent in cichlids. Stated otherwise, some mechanisms must control the timing and position of the epithelial downgrowth. Various non-neural mechanisms have been suggested to be involved in tooth initiation (occurrence of growth factors, of retinoids, expression of homeobox genes, see Thesleff et al., 1995). Interestingly, Tuisku and Hildebrand (1994) showed that, in cichlids, tooth replacement in the lower (oral) jaw is inhibited after denervation, lending strong support to the hypothesis that local innervation can have an inductive (instructive or perhaps only permissive) role in (regenerative) odontogenesis.

Only once we have established the pattern of tooth families will we know whether this corresponds to assumed patterns of replacement as discussed above (one-for-one replacement in the more central region of the LPJ; new positions, if required, marginally). Only then will we be closer to an understanding of the control of tooth number in this system.

16.5.   Changes in tooth size and shape

Earlier on, measurements of tooth base area of the functional teeth, indicative for tooth size, have been carried out in the two phenotypes (Huysseune, 1995). The data – on eight soft food and 10 hard food adult animals – show that the size of the (approximately 30) teeth in the functionally most relevant region, is larger in hard food than in soft food specimens. Whereas the size of these teeth increases with fish length in the hard food specimens, there is evidence that, roughly speaking, only some of the largest teeth in soft food specimens grow substantially with fish size. So, besides maintaining (and possibly even decreasing) their total tooth number, molluscivorous specimens produce larger teeth.

Diet-related differences in pharyngeal tooth size reflecting not just differences in wear have already been reported in cichlids (Hoogerhoud, 1986b; Meyer, 1990a,b). A key issue is how larger teeth in the durophagous form can be accommodated on the jaw. The solution is provided at least partly, if not entirely, by volume increase of the LPJ proper (Huysseune et al., 1994). The dentigerous area is enlarged and potentially new dentigerous surface is added posteriorly (cf. the outcurved posterior region in hard food specimens). If, however, the increase in surface does not keep pace with that required for accommodating larger teeth and if tooth number does not decrease, then teeth must occupy a proportionally larger area of the dentigerous region (leaving less free space in between them). In my previous study I have been unable to demonstrate such an increased dental cover in hard food specimens, suggesting that larger teeth in the durophagous phenotype are accommodated solely through a proportional increase of the dentigerous area and maintenance or decrease of tooth number.

Tooth shape in both forms varies greatly according to position on the LPJ. Teeth in the hard food phenotype grade from distinct molariform (large, rounded, central cusp absent) to enlarged unicuspid teeth, with a small number of slender (non-enlarged) teeth present laterocaudally (Figure 16.2C). In the soft food phenotype, enlarged unicuspid teeth soon grade into slender teeth occupying a considerable part of the dentigerous area (Figure 16.2A). As long as no single formula is available to quantify tooth shape, and as long as tooth families cannot be readily identified, it will be very difficult to judge the extent of shape changes at individual positions. Among the 30 central posteriormost teeth, I found that the tooth base in soft food specimens appears to be more compressed but the difference with hard food specimens was not significant (Huysseune, 1995). Far more indicative for tooth shape is the shape of the tooth tip but this has not yet been mapped in quantitative terms. Barel *et al.* (1976) reported that, in conditions of extreme size increase, several *Haplochromis* species (i.e. other cichlids) can switch from bicuspid to unicuspid teeth. Interestingly, the reverse happens in urodeles, where the change at metamorphosis is from unicuspid to bicuspid (Smith and Miles, 1971); indeed many non-mammalian vertebrates show a gradual change in tooth morphology with successive replacements.

The mechanism that generates the various (pharyngeal) tooth shapes found in *A. alluaudi* and in cichlids in general (cf. Barel *et al.*, 1976, 1977; Witte and van Oijen, 1990; Casciotta and Arratia, 1993), remains obscure. In mammals, unravelling of the genetic basis for the determination of tooth shape (incisor, canine or molar) has made considerable progress over the last years (cf. the odontogenic homeobox code proposed by Sharpe, 1995; see Chapter 1) and links are being established between homeobox genes, growth factors and retinoids during successive stages of odontogenesis (Thesleff *et al.*, 1995; Kronmiller *et al.*, 1995). Shape of cichlid pharyngeal teeth will depend on particular signals during the cascade of epithelial-mesenchymal interactions leading to replacement tooth formation in the jaw medullary cavity. If, as is currently accepted, the hypermineralized layer on top of cichlid teeth is enameloid (Sasagawa, 1995), then the final 'crown' size and shape of the tooth is already established by the shape of the epithelial-mesenchymal boundary before the start of tooth matrix deposition (Shellis, 1978) and both cell layers must have been instructed by then. Instruction of tooth shape in mammals has long been thought to come from the mesenchymal population. However, according to the current view (e.g. Lumsden, 1988; challenged however by Ruch, 1995) tooth initiating properties (position and shape) reside in the epithelium. Increased loading on the teeth of the LPJ by crushing snails is probably registered by the pharyngeal epithelium that surrounds the functional teeth (including the reduced enamel epithelium of the functional teeth) and which may subsequently instruct a specific shape. However, forces are also transmitted to the mesenchyme in the medullary cavity through the pillars of attachment bone and the possibility that the mesenchyme instructs tooth shape cannot be entirely excluded. Evidence that the medullary cavity mesenchyme can respond to increased loading, is shown in this fish as bone deposition in the form of additional supportive bony trabeculae within the medullary cavity. This occurs in hard food specimens only (Huysseune *et al.*, 1994).

16.6.   Adjustment of the tooth replacement cycle

The replacement cycle (i.e. the interval between attachment of a tooth and attachment of its successor) can be divided into an inactive phase during which the functional tooth exists without a successor, and an active phase during which the successor develops from a young germ up to a tooth ready to replace its predecessor.

There is no experimental evidence (e.g. from labelling studies) as to whether the duration of the replacement cycle (cycle length) differs between both phenotypes. Yet the microradiographs reveal proportionally smaller numbers of replacement teeth among the central posteriormost teeth in soft food specimens (Huysseune, 1995). If, as one can reasonably assume, frequency of a given stage is indicative of its relative importance (length) in the replacement cycle, then we must assume that the active phase in these positions is relatively longer in hard food specimens. Replacement teeth in the hard food specimens must grow to a larger size. Assuming that rates of matrix deposition vary within narrow limits, growing to a larger size most likely needs more time. A combination of both an absolute and a relative increase of the active phase in hard food specimens necessarily implies a maintenance or increase, in absolute terms, of the inactive phase, resulting in a cycle length that is either maintained (possibly even shortened) or increased. If cycle length is increased, i.e. if cycle length is still sufficient to match the need for a dentition that is susceptible to heavy wear, then one must assume that teeth are replaced in soft food specimens even before there is a functional need to do so. On the other hand, if cycle length is maintained (or even shortened), then some mechanism must account for an earlier activation of replacement tooth formation. There is a close correspondence between the amount of wear of a functional tooth and the stage of development of its successor; both events are apparently synchronized. I hypothesize that wear of a functional tooth and initiation of its successor are the result of a common, mechanical, factor (pressure on the teeth by hard food particles). Length of the replacement cycle may furthermore depend on age. Both an increase and a decrease of cycle length have been observed with increasing age (in piranhas, Berkovitz, 1980; in trout, Berkovitz and Moore, 1974).

16.7.   **Conclusions**

The example of environmentally induced plasticity discussed in this paper leaves us with some intriguing questions on the developmental mechanisms involved in tooth formation in this polyphyodont species. The current data suggest that tooth number in this model is regulated by decisions made in the epithelium, concerning tooth initiation sites and their timing. Indeed, a new downgrowth of epithelium has to be produced every time a replacement tooth has to be formed. If the timing of this downgrowth can be influenced, as our data on durophagous forms seem to indicate, and in the absence of any overt accumulation of mesenchymal cells (no mesenchymal condensation is visible until the aboral end of the strand within the jaw cavity starts to fold), then the epithelium is the likely site of tooth initiation. This supports the notion of epithelial primacy in tooth initiation, as suggested by Lumsden (1988) for the mouse and also by Westergaard and Ferguson (1990) for alligator dentitions.

On the other hand, changes in shape and possibly also in size may be regulated by the mesenchymal cell population of the jaw medullary cavity providing the dental papillae. This population has been shown to be responsive to changes in the mechanical environment (cf. the deposition of reinforcing trabeculae in hard food specimens).

Clearly there are morphogenetic and architectonic constraints to the amount of plasticity that can be realized between successive tooth generations. We do not yet know, in quantitative terms, how much one tooth generation can change with regard to the previous one in terms of tooth size, shape and position, mainly because homologization of individual teeth between specimens of different stages is very difficult. However, it is likely that the changes in features observed here in conditions of altered function are of a magnitude which can only be accomplished through successive tooth generations. It is therefore logical to associate developmental plasticity

with polyphyodonty and developmental stability with diphyodonty.

In diphyodont mammals, anlagen of the permanent teeth are present during embryonic life as offshoots of the dental lamina of the deciduous teeth and are therefore very much less, if at all, susceptible to variation induced by environmental changes in postnatal life, despite possible interpopulation differences of feeding habits of the young. Altering the diet of the developing animal can nevertheless perturb coordinated development and growth of the jaws and result in malocclusion. Of particular interest are experiments by Beecher, Corruccini and co-workers (discussed in Miles and Grigson, 1990) who found that, in a number of mammals, skull measurements decreased and muscle weight was reduced in animals fed an unusually soft diet. Tooth positions were also affected. In such cases, effects on the dentition are usually assumed to be secondary and related to changes in growth of the skull and jaw.

There is no discussion about the fact that variation in meristic characters such as tooth number is common among fish, amphibians and reptiles. What I have shown here is that the environment, and in particular the diet, can significantly alter developmental programmes. The occurrence of diet-related plasticity, in my opinion, depends on two features, besides its relationship to polyphyodonty: whether the organism shows heterodonty (*sensu lato*) and whether tooth–tooth contacts are crucial to masticatory efficiency. This combination may limit the phenomenon to a small number of taxa, even among polyphyodonts. Cichlid pharyngeal jaws not only exhibit heterodonty but these jaws are actually involved in mastication and food is ground at the tooth surface, suggesting a feedback mechanism between diet, feeding mode and morphology. Meyer (1993) collected evidence suggesting that fishes in general and cichlids in particular may be among the most plastic vertebrates with regard to trophic morphology. In contrast, most polyphyodont vertebrates have homodont dentitions, often with numerous teeth, mostly involved in orthal movements like grasping and biting, rather than in mastication and thus there is hardly any food processing at the tooth surface. Further studies of variable species, comparison of wild and captive specimens, and controlled breeding experiments should reveal the commonality of intraspecific variation in the dentition and the involvement of genetic mechanisms.

This case study demonstrates conclusively adaptability in the dentition, or 'the ability to acquire an adaptive variation during an individual's lifetime' (Jablonka and Lamb, 1995). Yet, it also suggests that the extent to which the dentition may be regarded as an adaptable functional unit may be limited to instances showing the rare combination of polyphyodonty, heterodonty and functional involvement in mastication.

Is plasticity in the dentition then a marginal phenomenon of little significance to vertebrate biologists? In my opinion it is not. Taxonomists and palaeontologists may perhaps feel reassured, but to developmental biologists the dentition of *Astatoreochromis* presents some challenges for understanding the mechanisms of tooth formation. Despite the inconvenience that individual teeth, thus far, cannot be homologized between stages and between individuals (unlike, for example, in alligator dentitions: Westergaard and Ferguson, 1990), the (lower) pharyngeal jaw in phenotypically plastic cichlid species may provide us with a unique model to study regulation of individual tooth initiation in a polyphyodont. To further explore the mechanisms responsible for plasticity in tooth variables, we need careful morphological observations (serial semi-thin sections, three-dimensional reconstructions and transmission electron microscopy). In addition, experimental approaches, including *in vitro* culture (recently established for cichlid fish organ culture, Koumans and Sire, 1996) and molecular techniques, should provide essential information on the mechanisms that underly plasticity. The challenge will be to unravel how mechanical loading can modify gene expression patterns (homeobox genes, BMPs, other growth factors) to match the needs associated with a change of function.

may play a role in shape and size differences of the teeth among the two phenotypes. This case study demonstrates adaptability in the dentition, yet suggests that the extent to which the dentition may be regarded as an adaptable functional unit may be limited to instances showing the rare combination of polyphyodonty, heterodonty and functional involvement in mastication.

Acknowledgements

I am greatly indebted to Dr R. Slootweg, Dr J. Smits and Dr F. Witte (University of Leiden, the Netherlands) for kindly providing the *Astatoreochromis* material. Dr J. Smits generously made his tooth counts available. Mrs F. Allizard and Dr F. Meunier are gratefully acknowledged for help in preparing the ground sections and the microradiographs, respectively. Dr K. Desender (KBIN, Brussels) is gratefully acknowledged for statistical advise. I also wish to express my sincere thanks to Professor W. Verraes (Ghent), Dr F. Meunier and Dr J. -Y. Sire (Paris) and Dr J. Smits (Leiden) for their helpful comments on the manuscript. The work benefitted from a grant of the University of Ghent ('Verkennend Europees Onderzoek'). Part of the work was supported through a cooperation between the 'Ministerie van de Vlaamse Gemeenschap' (Belgium) and the CNRS (France) (project C94/009).

Summary

This chapter explores a case of environmentally induced phenotypic variation in the dentition. In particular it illustrates how dietary changes affect tooth variables in a polyphyodont species and discusses the developmental mechanisms possibly involved. The African cichlid fish *Astatoreochromis alluaudi* exhibits different phenotypes with respect to the dentigerous pharyngeal jaws (the major crushing apparatus): a form feeding on soft food, and a mollusc-crushing (durophagous) form. During growth, tooth number increases in soft food specimens, but is maintained or possibly even decreased after a switch to molluscivory. It is suggested that tooth number in this model is regulated by decisions made in the epithelium with respect to position and timing of the epithelial downgrowth produced anew for each germ. Data on the durophagous form seem to indicate that the timing of this downgrowth can be environmentally influenced. Molluscivorous specimens also produce larger teeth. The medullary cavity mesenchyme, providing the dental papillae, is responsive to increased loading and

References

Barel, C. D. N., Witte, F. and van Oijen, M. J. P. (1976). The shape of the skeletal elements in the head of a generalized *Haplochromis* species: *H. elegans* Trewavas 1933 (Pisces, Cichlidae). *Netherlands Journal of Zoology*, **26**, 163–265.

Barel, C. D. N., van Oijen, M. J. P., Witte, F. and Witte-Maas, E. (1977). An introduction to the taxonomy and morphology of the haplochromine Cichlidae from Lake Victoria. *Netherlands Journal of Zoology*, **27**, 333–389.

Berkovitz, B. K. B. (1980). The effect of age on tooth replacement patterns in piranhas (Pisces: Characidae). *Archives of Oral Biology*, **25**, 833–835.

Berkovitz, B. K. B. and Moore, M. H. (1974). A longitudinal study of replacement patterns of teeth on the lower jaw and tongue in the rainbow trout (*Salmo gairdneri*). *Archives of Oral Biology*, **19**, 1111–1119.

Brook, A. H. (1984). A unifying aetiological explanation for anomalies of human tooth number and size. *Archives of Oral Biology*, **29**, 373–378.

Casciotta, J. R. and Arratia, G. (1993). Jaws and teeth of American cichlids (Pisces: Labroidei). *Journal of Morphology*, **217**, 1–36.

Galis, F. (1992). A model for biting in the pharyngeal jaws of a cichlid fish: *Haplochromis piceatus*. *Journal of Theoretical Biology*, **155**, 343–368.

Greenwood, P. H. (1965). Environmental effects on the pharyngeal mill of a cichlid fish, *Astatoreochromis alluaudi* and their taxonomic implications. *Proceedings of the Linnean Society, London*, **176**, 1–10.

Hoogerhoud, R. J. C. (1984). A taxonomic reconsideration of the haplochromine genera *Gaurochromis* Greenwood, 1980 and *Labrochromis* Regan, 1920 (Pisces, Cichlidae). *Netherlands Journal of Zoology*, **34**, 539–565.

Hoogerhoud, R. J. C. (1986a). Taxonomic and ecological aspects of morphological plasticity in molluscivorous haplochromines (Pisces, Cichlidae). *Annales, Musée Royal de l'Afrique Centrale, Sciences Zoologique*, **251**, 131–134.

Hoogerhoud, R. J. C. (1986b). Ecological morphology of some cichlid fishes. PhD thesis, University of Leiden.

Huysseune, A. (1995). Phenotypic plasticity in the lower pharyngeal jaw dentition of *Astatoreochromis alluaudi* (Teleostei: Cichlidae). *Archives of Oral Biology*, **40**, 1005–1014.

Huysseune, A., Sire, J.-Y. and Meunier, F. J. (1994). Comparative study of the lower pharyngeal jaw structure in two phenotypes of *Astatoreochromis alluaudi* (Teleostei: Cichlidae). *Journal of Morphology*, **221**, 25–43.

Jablonka, E. and Lamb, M. J. (1995). *Epigenetic Inheritance and Evolution. The Lamarckian Dimension.* Oxford: Oxford University Press.

Koumans, J. T. M. and Sire, J.-Y. (1996). An *in vitro* serum-free organ culture technique for the study of development and growth of the dermal skeleton in fish. *In Vitro Cellular and Developmental Biology, Animal*, **32**, 612–626.

Kronmiller, J. E., Nguyen, T. and Berndt, W. (1995). Instruction by retinoic acid of incisor morphology in the mouse embryonic mandible. *Archives of Oral Biology*, **40**, 589–595.

Lauweryns, I., Carels, C. and Vlietinck, R. (1993). The use of twins in dentofacial genetic research. *American Journal of Orthodontics and Dentofacial Orthopedics*, **103**, 33–38.

Liem, K. F. and Kaufman, L. S. (1984). Intraspecific macroevolution: functional biology of the polymorphic cichlid species *Cichlasoma minckleyi*. In *Evolution of Species Flocks*, eds. A. A. Echelle and I. Kornfield, pp. 203–215. University of Maine: Orono Press.

Lumsden, A. G. S. (1988). Spatial organization of the epithelium and the role of neural crest cells in the initiation of the mammalian tooth germ. *Development*, **103** (Supplement), 155–169.

Melamed, Y., Barkai, G. and Frydman, M. (1994). Multiple supernumerary teeth (MSNT) and Ehlers–Danlos syndrome (EDS): a case report. *Journal of Oral Pathology and Medicine*, **23**, 88–91.

Meyer, A. (1990a). Ecological and evolutionary consequences of the trophic polymorphism in *Cichlasoma citrinellum* (Pisces: Cichlidae). *Biological Journal of the Linnean Society*, **39**, 279–299.

Meyer, A. (1990b). Morphometrics and allometry in the trophically polymorphic cichlid fish, *Cichlasoma citrinellum*: alternative adaptations and ontogenetic changes in shape. *Journal of Zoology, London*, **221**, 237–260.

Meyer, A. (1993). Trophic polymorphisms in cichlid fish: do they represent intermediate steps during sympatric speciation and explain their rapid adaptive radiation? In *New Trends in Ichthyology: an International Perspective*, eds J. H. Schröder, J. Bauer and M. Schartl, pp. 257–266, G. S. F. Bericht 7/92. Oxford: Blackwell.

Miles, A. E. W. and Grigson, C. (1990). *Colyer's Variations and Diseases of the Teeth of Animals.* Cambridge: Cambridge University Press.

Osborn, J. W. (1978). Morphogenetic gradients: fields versus clones. In *Development, Function and Evolution of Teeth*, eds. P. M. Butler and K. A. Joysey, pp. 171–201. London: Academic Press.

Osborn, J. W. (1984). From reptile to mammal: evolutionary considerations of the dentition with emphasis on tooth attachment. In *The Structure, Development and Evolution of Reptiles*, ed. M. W. J. Ferguson. *Symposia of the Zoological Society of London*, **52**, 549–574.

Reif, W.-E. (1982). Evolution of dermal skeleton and dentition in vertebrates. The odontode regulation theory. *Evolutionary Biology*, **15**, 287–368.

Reif, W.-E. (1984). Pattern regulation in shark dentitions. In *Pattern Formation. A Primer in Developmental Biology*, eds G. M. Malacinski and S. V. Bryant, pp. 603–621. New York: Macmillan.

Ruch, J. V. (1995). Tooth crown morphogenesis and cytodifferentiations: candid questions and critical comments. *Connective Tissue Research*, **32**, 1–8.

Sasagawa, I. (1995). Features of formation of the cap enameloid in the pharyngeal teeth of tilapia, a teleost. *ACBTE*, **4**, 7–14.

Schalk-van der Weide, Y., Steen, W. H. A., Beemer, F. A. and Bosman, F. (1994). Reductions in size and left-right asymmetry of teeth in human oligodontia. *Archives of Oral Biology*, **39**, 935–939.

Sharpe, P. T. (1995). Homeobox genes and orofacial development. *Connective Tissue Research*, **32**, 17–25.

Shellis, R. P. (1978). The role of the inner dental epithelium in the formation of the teeth in fish. In *Development, Function, and Evolution of Teeth*, eds. P. M. Butler and K. A. Joysey, pp. 31–42. London: Academic Press.

Slavkin, H. C. (1988). Editorial. Genetic and epigenetic challenges in tooth development. *Journal of Craniofacial Genetics and Developmental Biology*, **8**, 195–198.

Smith, M. M. and Miles, A. E. W. (1971). The ultrastructure of odontogenesis in larval and adult urodeles; differentiation of the dental epithelial cells. *Zeitschrift für Zellforschung und Mikroskopische Anatomie*, **121**, 470–498.

Smits, J. D. (1996). Trophic Flexibility through Spatial Reallocation of Anatomical Structures in the Cichlid Fish *Astatoreochromis alluaudi*. PhD thesis, University of Leiden.

Sofaer, J. A. (1978). Morphogenetic influences and patterns of developmental stability in the mouse vertebral column. In *Development, Function, and Evolution of Teeth*, eds. P. M. Butler and K. A. Joysey, pp. 215–227. London: Academic Press.

Stearns, S. C. (1989). The evolutionary significance of phenotypic plasticity. *BioScience*, **39**, 436–445.

Thesleff, I., Vaahtokari, A. and Partanen, A.-M. (1995). Regulation of organogenesis. Common molecular mechanisms regulating the development of teeth and other organs. *International Journal of Developmental Biology*, **39**, 35–50.

Tuisku, F. and Hildebrand, C. (1994). Evidence for a neural influence on tooth germ generation in a polyphyodont species. *Developmental Biology*, **165**, 1–9.

Vandewalle, P., Huysseune, A., Aerts, P. and Verraes, W. (1994). The pharyngeal apparatus in teleost feeding. *Advances in Comparative and Environmental Physiology*, **18**, 59–92.

Westergaard, B. and Ferguson, M. W. J. (1990). Development of the dentition in *Alligator mississippiensis*: upper jaw dental and craniofacial development in embryos, hatchlings, and young juveniles, with a comparison to lower jaw development. *American Journal of Anatomy*, **187**, 393–421.

Witte, F. and van Oijen, M. J. P. (1990). Taxonomy, ecology and fishery of Lake Victoria haplochromine trophic groups. *Zoologische Verhandelingen, Leiden*, **262**, 1–47.

Witte, F., Barel, C. D. N. and Hoogerhoud, R. J. C. (1990). Phenotypic plasticity of anatomical structures and its ecomorphological significance. *Netherlands Journal of Zoology*, **40**, 278–298.

17 Enamel microporosity and its functional implications

R. P. Shellis and G. H. Dibdin

17.1. Introduction

Despite its great hardness and density, enamel has an appreciable porosity. Pore structure affects the mechanical properties of enamel. It also influences the interaction between light and enamel, understanding of which is necessary for development of transillumination techniques for caries detection and for optical matching of restorative materials and tooth tissue. However, interest in enamel porosity has above all been stimulated by the important role played by diffusion of acids and mineral ions in caries formation. Because enamel mineral exists as very small crystals, organised in an elaborate structure, the internal pores are small and variable in form, orientation and distribution. Microscopical information on pore structure tends to be prone to artefact but much information has been obtained by a range of less direct methods. Most studies have employed human or bovine permanent enamel.

17.2. Pores in relation to enamel structure

17.2.1. Average porosity and the state of enamel water

Techniques used for estimating the mineral content (and hence porosity) of enamel include quantitative microradiography and measurements of enamel density. Both methods require a knowledge of the density of the apatite crystals of enamel, and in the past it has been assumed that they have the same density as hydroxyapatite (3.15 g/ml). However, recent work (Elliott, 1997), which concluded that enamel apatite has a lower density (2.99 g/ml), suggests that many previous estimates of

mineral content may be too low. For instance the much-quoted mineral estimate by Angmar et al. (1963) of 86.2% v/v (based on an apatite density of 3.15) has been revised upwards to 93.1% by Elliott (1997) using an apatite density of 2.99. Likewise, on the basis of overall enamel density (Weatherell et al., 1967), the mean mineral content was estimated by Elliott (1997) as 96% v/v. On the other hand, after similar revision, the microradiographic data of Wilson and Beynon (1989) still only give a mean mineral content of 87.8% v/v. Clearly, there remains some uncertainty about this important property of enamel. However, the new calculations reduce the disparity between (a) pore volume/water estimates obtained by difference from mineral-content measurements, and (b) the following direct measurements.

Some of the water in enamel pores is freely mobile, giving a narrow nuclear magnetic resonance (NMR) peak (Dibdin, 1972a), while some has a more restricted mobility, presumably because of binding interactions with the mineral or the matrix. Water released from enamel by heating to 100–200 °C is considered to include the free water, together with 'loosely bound' water. Permanent enamel dried in this way after prior equilibration at 100% RH at 37 °C loses about 2% w/w (c. 6% v/v) water (LeFevre and Manly, 1938; Bird et al., 1940; Burnett and Zenewitz, 1958) and the same fraction is exchangeable with ^3H_2O (Dibdin, 1993). The reduced water content of air-dried enamel (Burnett and Zenewitz, 1958; Bonte et al., 1988) indicates that much of the free or loosely bound fraction is quite labile. The remainder is at least in part associated with the organic matrix. Drying of enamel is fully reversible only up to 50 °C (Burnett and Zenewitz, 1958; Zahradnik and Moreno, 1975), probably because dehydration causes irreversible conformational changes in the organic matrix.

Studies with hydroxyapatite (Kibby and Hall, 1972; Rootare and Craig, 1977) suggest that some of the water in enamel pores will be more firmly bound to the mineral. From the crystal dimensions (Kérébel et al., 1979), density and mineral content of enamel (Elliott 1997), we calculate that enamel could accommodate about 0.85% w/w water as a monolayer adsorbed to the mineral. However, because of the complexity of the changes which occur in enamel at high temperatures (LeGeros et al., 1978; Holcomb and Young, 1980), the amount of firmly bound water has not been quantified exactly. Clearly, because of structural heterogeneities, the relative proportions of free, loosely and firmly bound water will vary considerably, with consequent effects on diffusion. In small pores, diffusion may be restricted because more of the pore water will have restricted mobility than in larger pores.

17.2.2. Histological variations

Quantitative microradiography and density mapping show that mineral content, and hence porosity, varies significantly within enamel (Angmar et al., 1963; Weatherell et al., 1967; Theuns et al., 1983; Wilson and Beynon, 1989). Mineral content falls from the outer surface to the enamel–dentine junction (EDJ), the volume not occupied by mineral being 50–100% greater in inner than in outer enamel. Both protein (Robinson et al., 1971) and water (Zahradnik and Moreno, 1975; Dibdin and Poole, 1982) are more abundant in inner enamel. The mineral content of outer enamel falls off from the cusps towards the cervical region (Weatherell et al., 1967; Crabb, 1968; Theuns et al., 1983; Wilson and Beynon, 1989), and is lower in enamel beneath fissures (Robinson et al., 1971). There may also be localised subsurface regions of low-density enamel, often near the cusps (Weatherell et al., 1967; Crabb, 1968).

17.2.3. Ultrastructure

The apatite crystals of enamel show considerable irregularity of outline but are roughly hexagonal in cross-section (Frazier, 1968; Selvig and Halse, 1972), with mean width 68.3 nm and mean thickness 26.3 nm (Kérébel et al., 1979). The crystals are too long for their true length to be determined (Boyde and Pawley, 1975; Orams et al., 1974, 1976).

The generally accepted model for the prismatic structure of human enamel (Poole and Brookes, 1961; Boyde, 1965; Meckel et al., 1965; Helmcke, 1967) implies three levels of perfection of crystal packing, associated with the prism bodies, prism tails and prism junctions (Shellis, 1984), and hence a heterogeneous pore structure. Knowledge of pore structure at the electron microscope level is based mainly on sections prepared by ion beam

thinning, a method which minimises mechanical damage.

The crystals are most closely packed within the prism 'bodies', which occupy 60–65% v/v of enamel (Shellis, 1984). Here the crystals are essentially parallel with each other and with the prism axis. Except in small groups, the crystals are randomly orientated with respect to their a-axes (Frazier, 1968; Hamilton et al., 1973; Selvig and Halse, 1972; Orams et al., 1974, 1976), but intercrystalline spaces are reduced by interdigitation of the irregular crystal profiles (Frazier, 1968; Selvig and Halse, 1972). While many crystals are tightly packed, pores with polygonal cross-sections and diameters of 1–10 nm occur as a result of the disorientation with respect to the crystal a-axes. These polygonal pores extend for some distance in three dimensions as minute 'tubules' (Orams et al., 1974, 1976).

In the 'tail' region, the crystals diverge from the prism axis, by about 15–45° in the cervical direction (Poole and Brookes, 1961). At the same time, they diverge laterally where, in the alternating prism arrangement typical of human enamel (Pattern 3), they fan out between the prism bodies in the next row (Poole and Brookes, 1961; Boyde, 1965; Meckel et al., 1965; Hamilton et al., 1973). This dispersion of crystal orientation increases porosity (Hamilton et al., 1973), although pore sizes have not been quantified.

The prism junctions, or boundaries, are sites where crystals belonging to the tail region of one prism meet those in the body of another, with an abrupt change in crystal orientation. Because of the imperfect crystal packing the junctions are regions of increased porosity (Hamilton et al., 1973; Orams et al., 1974, 1976). In human enamel, the prism junctions are typically incomplete in cross-section and so form laminar pores, with curved cross-sections, running from the EDJ to the outer surface. However, while, in outer enamel the prism junctions tend to be separate and thus exist as independent channels, those in inner enamel interconnect to form a three-dimensional network of laminar pores, especially in molars (Boyde, 1989; Shellis, 1996). The hypomineralization and greater porosity of cervical enamel are associated with a more or less aberrant prism structure (Poole et al., 1981).

Besides the major variations described above, there may be locally increased porosity at the striae of Retzius (Schmidt and Keil, 1971; Newman and Poole, 1974) and the cross-striations (Hardwick et al., 1965; Sundström and Zelander, 1968; Schmidt and Keil, 1971; Boyde, 1989). It is possible to form methacrylate replicas of some cross-striations in inner enamel (Shellis, 1996) but they have not been so demonstrated in outer enamel, which suggests that here the pores at the striations are either very small or inaccessible.

17.2.4. Matrix

Chromium sulphate demineralization (Sundström and Zelander, 1968) provides apparently reliable information on matrix distribution at the electron microscope level, although the specific surface areas obtained by Silness *et al.* (1973) suggest that some shrinkage occurs. Sometimes, the profiles produced by this technique are of individual crystals with a coating of matrix (Silness *et al.*, 1973) but often, their size, spacing and morphology (see Sundström and Zelander, 1968, figs. 9, 11) suggest that they are sections of the polygonal pores observed by Orams *et al.* (1974, 1976). These may be separated by matrix-free spaces in which the crystals are presumably so tightly packed that intervening matrix has been excluded. In the tail regions, matrix appears to be more abundant than in the head region. Prism junctions usually appear as regions where enlarged pores are filled with matrix, although often porosity seems to be not much greater than in the surrounding tissue.

The tufts in inner enamel are regions of high porosity, cutting across the prism structure, in which crystals are small and dispersed (Orams *et al.*, 1976) and protein is abundant (Weatherell *et al.*, 1968).

The matrix may influence the properties of enamel pores in several ways: (1) effective pore sizes may be reduced; (2) protein hydration may reduce water mobility within the pores; (3) diffusion of charged solutes may be altered by interactions with the negative charge of the protein; (4) the lipid component may influence diffusion of solutes according to their hydrophobicity.

17.3.   Pore sizes

Using polarising microscopy, Poole *et al.* (1961) and Darling *et al.* (1961) showed that enamel acts as a molecular sieve. Imbibition experiments indicated that methanol, which has a mean molecular radius (r) of 0.15 nm, seems to enter all pores accessible to water, but that higher alcohols, even ethanol ($r = 0.17$ nm), are excluded to some extent. Concentrated salt solutions, with an osmotic pressure > 500 atm, removed water from the small pores of cervical enamel (Poole *et al.*, 1981), which thus excludes ions as well as neutral organic molecules. The hydrated radii of ions used by Poole, Newman and Dibdin range from 0.33 nm (Cl^-) to 0.43 nm (Zn^{2+}). Although factors other than simple pore size may cause molecular exclusion (Poole *et al.*, 1981), it seems certain that some enamel pores are extremely small, possibly with radii as small as 0.2 nm. Poole and Stack (1965) pointed out that the molecular sieve properties of enamel could be due to either elongated, narrow pores or

Table 17.1. *Porosities and specific surface areas of human permanent enamel measured by sorption*

Porosity (% v/v)	Specific surface area, (m²/g)	Sorbate	Reference
0.25–0.28	0.37–0.45	Krypton	1
	3.5–4.0	Nitrogen	2, 3, 4
1.4–3.6	3.7–5.5	Water	5, 6, 7
	4.1	Chemisorption	4

References: 1, Dibdin (1969); 2, Huget *et al.* (1968); 3, Loebenstein (1973); 4, Misra *et al.* (1978); 5, Moreno and Zahradnik (1973); 6, Zahradnik and Moreno (1975); 7, Dibdin and Poole (1982).

wider 'ink-bottle' pores with constricted entrances, possibly 'plugged' by organic material.

Study of the adsorption and desorption of gases, including water vapour, can give much useful information about porosity, specific surface area and pore-size distributions. Data for enamel are presented in Table 17.1.

From the low porosity and specific surface area for krypton sorption, and its short equilibration time, it is clear that this sorbate penetrates only the larger pores. This may be due to the large size of the molecule, to kinetic factors or to exclusion from protein-filled pores. Calculations from an idealised model of enamel structure (Dibdin, 1972b) suggested that the krypton-accessible pores correspond to the prism junctions. This hypothesis is supported by evidence (Dibdin, 1969) that these pores have a laminar form, and that the pore-size distribution (mean width 6 nm) seems consistent with prism-junction ultrastructure. Morphometric data (Shellis, 1984) indicate that the prism junctions could account for about 60% of the krypton-specific surface area. Since these measurements did not take into account any irregularities in the pore walls, this is reasonably good agreement.

A much greater fraction of enamel pores is clearly accessible to nitrogen, water and, surprisingly, the large organic molecule used by Misra *et al.* (1978). Pore-size distributions from water-sorption data indicate peaks at about 2–3 and 4–6 nm radius (Moreno and Zahradnik, 1973; Zahradnik and Moreno, 1975; Dibdin and Poole, 1982). Although such analyses are subject to uncertainties (Dibdin and Poole, 1982) they do suggest, in conjunction with ultrastructural data, that these sorbates enter some intercrystalline pores.

From the desorption behaviour of water, which shows a sudden release at a relative vapour pressure of 40–50%, Moreno and Zahradnik (1973) deduced the presence of pores with constricted entrances. An increase in apparent water capacity on raising the temperature to 50 °C was attributed to activation of diffusion through the constrictions (Zahradnik and Moreno, 1975). Since removal of most of the organic material was found to reduce adsorption/desorption hysteresis, Zahradnik and Moreno (1975) further concurred with Poole and Stack (1965) that the pore constrictions were associated with organic 'plugs'.

However, the water desorption behaviour could also be consistent with the presence of elongated, very narrow pores (< 1–2 nm radius; Dibdin and Poole, 1982). At low enough pressures, suction forces may be enough to overcome the tensile strength of the water in such pores, so that the water molecules are pulled apart. The same mechanism could account for removal of water from enamel pores by high osmotic pressures (Poole et al., 1981). It might be expected that diffusion in such narrow pores would be an activated process, as found by Zahradnik and Moreno (1975). Evidence for ultramicroscopic pores comes from the fact that adsorption/desorption hysteresis extends down to the lowest vapour pressures (Moreno and Zahradnik, 1973; Zahradnik and Moreno, 1975; Dibdin and Poole, 1982) and from the finding that water continues to be lost indefinitely at low pressure, so that a perfectly dry state is probably not achievable (Dibdin and Poole, 1982). Moreover, a large proportion of pores is inaccessible to nitrogen, even after thorough removal of the organic component (Misra et al., 1978).

17.4.   Permeability and diffusion

Enamel is permeable to water (Bergmann and Siljestrand, 1963; Poole et al., 1963; Lindén, 1968), ions (Arwill et al., 1969), small organic solutes (Marthaler and Mühlemann, 1960; Poole et al., 1963) and dyes (Berggren, 1947; Jansen and Visser, 1950; Tarbet and Fosdick, 1971). Microscopical observations show that the prism junctions provide the main pathways (Jansen and Visser, 1950; Tarbet and Fosdick, 1971; Brudevold et al., 1977) although, in inner enamel, some transport was observed within the prisms (Lindén, 1968). Studies of fluid movement (Lindén, 1968) and electrical impedance (Hoppenbrouwers et al., 1986) show that permeability increases from the outer surface towards the EDJ.

Interestingly, Tarbet and Fosdick (1971) found that, 5 minutes after application to the outer tooth surface, acriflavine penetration into enamel was less in vivo than in vitro. This is an unexpected result since, in life, the positive pressure of the pulp (15–30 cm water: Vongsavan and Matthews 1991) would, if anything, tend to cause an outward flow of fluid which would oppose inward diffusion.

Enamel forms an imperfect semi-permeable membrane in that, while water is transported through the tissue under the influence of an osmotic gradient, solute also moves in the opposite direction (Klein and Amberson, 1929; Atkinson, 1947; Poole et al., 1963).

Permeability appears to be strongly influenced by matrix components. Boiling in KOH/ethylene glycol abolishes the osmotic properties of enamel (Atkinson, 1947). Permeability to ^{22}Na is increased considerably by treatment with concentrated solutions of urea or hydrogen peroxide but not by treatment with chloroform, suggesting that protein, but possibly not lipid, restricts ionic diffusion (Arwill et al., 1969). On the other hand, formation of artificial caries lesions is accelerated by extraction of enamel lipid (Featherstone and Rosenberg, 1984); this could involve enhancement of diffusion as well as of dissolution.

Diffusion through enamel has been measured for a variety of substances (Table 17.2) but the relevant coefficients can be defined in more than one way because of the porous, multiphase nature of the material. One type, called here the apparent diffusion coefficient (D_a), is obtained using Fick's first law (Dibdin, 1995), from the steady-state flux of diffusate through a slice of enamel under a known concentration gradient, using a two-chamber cell. The term 'apparent' indicates that the gradient assumed is based on concentrations in the bathing solutions rather than in the sample itself. It is therefore like a permeability coefficient, being influenced by the fractional volume occupied by the pores and their tortuosity, but not by binding interactions. In contrast, the effective diffusion coefficient (D_e) is a measure of the rate at which a front of diffusate moves through the material and is measured by following penetration of diffusate into the material or by clearance from a sample pre-equilibrated with the diffusate. D_e is influenced by the tortuosity of the pores (but not by porosity) and also by reversible binding with the pore surfaces, which retards movement of the front by immobilising some of the diffusate (Crank, 1975). Sometimes it is useful (Dibdin, 1995) to indicate this retarding effect of binding by the additional subscript r, as in D_{re}; D_e is then reserved to denote the value in the absence of binding. However, in all measurements it is essential to avoid interference from irreversible binding. This may be difficult in penetration studies but can often be achieved by extensive pre-equilibration of the sample with unlabelled diffusate. The penetration profile of radioactive tracer is then analysed at known time intervals (at least two) after addition to the bathing solution.

Enamel slices used for D_a measurements (Table 17.2A) have all been cut parallel with the outer surface, so that the prisms run obliquely across the thickness. The relatively low porosity and activation energy for D_a of water (Burke and Moreno, 1975) suggest that D_a probably

Table 17.2. *Diffusion coefficients (D) and activation energies (E$_{act}$) for diffusion in enamel*

A. Apparent diffusion coefficients

Diffusate	Temperature (°C)	D_a (cm^2/s × 10^8)	E_{act} (kJ/mol)	Remarks	Reference
Ca	4	0.76–0.95		Bovine, 13–25 mmol/l carrier	1
F	23	3.3		Deciduous human molar	2
	23	0.02		Permanent human premolar	2
F	23	0.46		Permanent human premolar	3
SiF$_6$	23	0.26		Permanent human premolar	3
PO$_3$F	23	0.17		Permanent human premolar	3
Cr-EDTA	23	2.8		Deciduous human molar	2
	23	0.07		Permanent human premolar	2
Chlohexidine	23	1.7		Deciduous human molar	2
	23	0.05		Permanent human premolar	2
H$_2$O	37	0.77–2.8	22–42	Permanent human incisor	4
	37	0.84–1.2	47–61	Permanent human canine	4
Rb	4	0.27–12.1		Bovine	5
Cl	4	0.18–7.8		Bovine	5
Sorbitol	4	0.04–2.5		Bovine	5
Glycerol	4	0.08–4.4		Bovine	5

B. Effective diffusion coefficients

Diffusate	Temperature (°C)	D_e (cm^2/s)	E_{act} (kJ/mol)	Remarks	Reference
Urea	37	~$1 \cdot 10^{-10}$		Human	6
Na	25	~$2 \cdot 10^{-10}$	21	Human	7
PO$_3$F	37	$1.7 \cdot 10^{-13}$		Interprismatic, bovine	8
	37	$1.8 \cdot 10^{-18}$		Intraprismatic, bovine	8
Phosphate	37	$1.4 \cdot 10^{-12}$	48	Interprismatic, bovine	9,10
	37	$3.5 \cdot 10^{-17}$	68	Intraprismatic, bovine	9,10
Ca	37	$1.6 \cdot 10^{-12}$	65	Interprismatic, bovine, 1 mmol/l Ca	11,10
	37	$1.9 \cdot 10^{-13}$	122	Intraprismatic, bovine, 1 mmol/l Ca	11,10
F	20	$3.0 \cdot 10^{-10}$		Interprismatic, bovine[a]	12
	20	$1.65 \cdot 10^{-13}$		Intraprismatic, bovine[a]	12
H$_2$O	35	$4.3 \cdot 10^{-7}$		Overall D_e, human	13
	35	$0.9 \cdot 10^{-5}$		Interprismatic, human	13

[a] Recalculated using structural parameters from reference 9.

References: 1, Borggreven *et al.* (1981); 2, Lindén *et al.* (1986); 3, Hattab (1986); 4, Burke and Moreno (1975); 5, Borggreven *et al.* (1977); 6, Arends *et al.* (1984); 7, Braden *et al.* (1971); 8, de Rooij *et al.* (1980); 9, de Rooij *et al.* (1980); 10, de Rooij and Arends (1981); 11, Flim and Arends (1977a, b); 12, Flim *et al.* (1978); 13. Dibdin (1993)

relates to 'through diffusion' in a small fraction of the total porosity, probably the larger pores at the prism junctions, where diffusion is essentially in free water. The work of Borggreven *et al.* (1980) on the effect on D_a of agents which alter water structure tends to support this.

Values of D_a are in general about $1 \cdot 10^{-9}$–$1 \cdot 10^{-7}$ cm^2/s. The very low values of Lindén *et al.* (1986), which conflict with those of Hattab (1986), obtained by the same method, may be due to inadequate pre-equilibration. Molecular size dependence of D_a (Borggreven *et al.*, 1977) provides direct evidence for the molecular-sieve properties of enamel.

One problem in applying apparent diffusion coefficients measured for low-porosity materials such as enamel is that these coefficients will, by their very nature, underestimate the real transport rate (as distinct from flux) through a sample. Burke and Moreno (1975) recognised this and allowed for it explicitly, whereas van Dijk *et al.* (1983) did not, although they did so when discussing membrane potentials (van Dijk *et al.*, 1980).

Properly, concentrations just inside the two faces of the sample should be used, in Fick's first law, to obtain the true diffusion coefficient. These can in principle be obtained, for a range of applied concentrations, by measuring the appropriate distribution coefficients between the bathing solution and both the pores and solid phase of the sample. If the distribution coefficient includes reversibly bound tracer, then the D thus derived is equivalent to D_{re} (if irreversible binding is absent). If it only includes the tracer in the pores, then it is equivalent to D_e.

Analysis of data for several diffusates suggests the existence of more than one diffusion process (Table 17.2B). Flim *et al.* (1977, 1978) suggested, in the case of fluoride, a fast diffusion process between the crystals and a slow process into the crystals. However, solid-state diffusion is negligible at ordinary temperatures and it was later concluded that, for calcium, phosphate and monofluorophosphate, the fast process (D_{fast}) was associated with interprismatic diffusion (i.e. along the prism junctions) and the slow process (D_{slow}) with diffusion in the small intraprismatic pores (Flim and Arends, 1977a, b; de Rooij et al., 1980; de Rooij and Arends, 1981, 1982; de Rooij et al., 1981). We have therefore revised the results of Flim et al. (1978) to obtain values for fluoride diffusion (Table 17.2B), using the same structural parameters l and d as used by de Rooij and co-workers. The D_{fast} values and their activation energies suggest that interprismatic diffusion takes place through free water in the prism-junction pores, but with some hindrance by the organic matrix.

Intraprismatic diffusion was much more restricted, and the nature of the restriction seemed to vary according to the diffusing species. Flim and Arends (1977a,b) suggested that intraprismatic calcium transport, which has a high activation energy, is by surface diffusion (from site to site on the crystal surfaces), rather than within the pore fluid. On the other hand, it appeared that intraprismatic diffusion of phosphate occurs within the pore fluid, because the activation energy was similar to that for interprismatic diffusion. For phosphate, both D_{fast} and D_{slow} increased with decreasing pH (de Rooij and Arends, 1982). Since, as pH is reduced, the negative charge of both the diffusing phosphate ions and of the pore walls will become smaller, it was concluded that the slowness of intraprismatic phosphate diffusion was due to electrostatic exclusion of the ions from the pores.

The hypothesis of two diffusion processes is clearly consistent with structural and other evidence. However, although the work of Arends and co-workers is valuable in suggesting possible models for diffusion in enamel, their analyses are open to some criticism. First, they adopted a simplified model of enamel structure as parallel, closely packed hexagonal prisms, whereas, in the bovine enamel they used, the prisms form rows separated by inter-row sheets (pattern 2) and the pore structure will be very different. The calculated length (l) of prism junctions per unit area in the plane normal to the prism direction ($4.62 \cdot 10^3$ cm/cm^2: de Rooij et al., 1980) was only about 50% higher than values determined directly for outer human permanent enamel ($3.4 \cdot 10^3$ cm/cm^2: Shellis, 1984), but the assumed mean width (d) of the prism junctions (300 nm) was much larger than ultrastructural or sorption studies indicate. Use of the value $d = 6$ nm (Dibdin, 1969) would increase D_{fast} 50-fold.

Second, in the mathematical model used to analyse calcium diffusion (Flim and Arends, 1977a, b), it is assumed that inter- and intraprismatic diffusion are two independent processes occuring in parallel, i.e. in separate compartments. However, it is highly unlikely that there is no exchange between the two pore sytems. Third, irreversible binding is a possible difficulty in these penetration studies. Calcium, phosphate and fluoride, being components of the enamel crystals, may precipitate if the solubility product is exceeded. Pre-incubating each sample for just 1 or 2 weeks with unlabelled solution may have been inadequate to achieve solubility equilibrium, since the tracer penetration profiles obtained after the same time periods showed gradients of 1-2 orders of magnitude. Further work is desirable to confirm that the reduction in D_{slow} for calcium with increasing calcium concentration is due to a surface-diffusion mechanism, as Flim and Arends (1977a,b) suggested, and not to irreversible binding.

Recently, Dibdin (1993) obtained D_e values for water by studying clearance of ^3H_2O from enamel slabs. The results were provisionally analysed in terms of a model which assumed interprismatic 'through pores' similar to those measured by krypton sorption (Dibdin, 1969), fed by an inner (intraprismatic) 'blind' porosity obtained from water-sorption studies (Moreno and Zahradnik, 1973; Dibdin and Poole, 1982). This model appeared reasonably consistent with the D_a data of Burke and Moreno (1975).

Diffusion of ionic solutes is affected, not only by pore size, but by interactions with the charge on the enamel solid (negative under quasi-physiological conditions: Neiders et al., 1970). Because these interactions occur within narrow pores, enamel has the properties of an ion-exchange membrane. Ions having the same charge as the pore walls are partly excluded, while ions of opposite charge can enter the pores more readily and thus diffuse faster: an effect demonstrated directly by Borggreven, et al. (1977). Consequently, an electric potential difference (membrane potential) develops between two salt solutions with different concentrations separated by enamel (Klein and Amberson, 1929; Klein, 1932; Chick and Waters, 1963, 1965; Waters, 1971, 1972, 1975). With gradients of salts such as KCl or NaCl (pH 6-7), the membrane potential has a negative sign, showing that the enamel has a negative charge. Phosphate, fluoride and various organic anions make the potential more negative, since these ions adsorb to the enamel and increase its negative charge (Chick and Waters, 1963, 1965; van Dijk et al., 1977). With a gradient of salts of divalent cations (e.g. Ca^{2+}, Mg^{2+}, Mn^{2+}), the sign of the potential is reversed, because cation adsorption reduces or reverses the charge on the enamel (Klein and Amberson, 1929; Klein, 1932; van Dijk et al., 1977; Borggreven et al., 1981).

Observations on the membrane potential and streaming potential at different pH values suggest that the isoelectric point of enamel lies at about pH 4 (Klein, 1932; Waters, 1972; Kambara et al., 1978). These values are of questionable validity, since the specimens were probably not in solubility equilibrium, but the isoelectric point of enamel seems at least to be lower than that of pure hydroxyapatite (6.5-8.5: Chander and Fuerstenau, 1982), possibly because of the presence of the organic matrix.

17.5.   **Variations with tooth type**

Among the permanent teeth, canine enamel appears to be significantly less permeable than molar or incisor enamel (Table 17.2). Enamel of deciduous teeth has a lower mineral content than that of permanent teeth, except near the EDJ (Bird *et al.*, 1940; Wilson and Beynon, 1989). Lindén *et al.* (1986) observed that D_a was about 30 times less for permanent enamel than for deciduous enamel (Table 17.2) but, in view of the discrepancies in D_a values for fluoride reported by these authors and Hattab (1986), further work is required.

17.6.   **Functional implications**

17.6.1.   Mechanical properties

Although the mechanical properties of enamel are discussed by Rensberger (Chapter 18), it is appropriate to stress here the fundamental role of pore structure in this respect. The orientated laminar pores provided by the prism junctions provide planes of weakness, so that fractures in enamel propagate preferentially along the junctions (Rasmussen *et al.*, 1976; Boyde, 1989). Haines *et al.* (1963) found that enamel is more compressible at low loads and suggested that this was due to displacement of free water from pores.

17.6.2.   Optical properties

In wet enamel, the prism junctions act as a diffraction grating (O'Brien, 1988). However, Vaarkamp *et al.* (1995) found that enamel showed less optical anisotropy than dentine and concluded that Rayleigh scattering by the crystals was more important than scattering by diffraction at the prism junctions.

Because of the molecular sieve properties of sound enamel, high-refractive-index liquids cannot penetrate all pores within the tissue. The resulting inability to eliminate form birefringence limits the quantitative application of polarising microscopy (Carlström and Glas, 1963; Theuns *et al.*, 1993).

17.6.3.   Caries

The material at the prism junctions has a raised solubility (Shellis, 1996). This, combined with faster diffusion there, accounts for the fact that, at the advancing front of a caries lesion, demineralisation occurs first at the junctions and then spreads laterally into the intraprismatic regions. The greater porosity of inner enamel, combined with a more interconnected prism-junction structure and a raised solubility, may enhance lateral spread of caries near the EDJ (Shellis, 1996). Enamel of deciduous teeth is more susceptible to *in vitro* caries formation and this may be linked with greater porosity, the higher prism-junction density in particular (Shellis, 1984). The thick, aprismatic surface layer commonly found on deciduous enamel does not seem to influence *in vitro* caries progression (Shellis, 1984), despite its lower porosity.

While the general pattern of diffusion is well understood, work is required to define more exactly the transport of mineral ions and to relate diffusion to the detailed structure of enamel and to its variations with depth and tooth type. Demineralisation increases porosity, so that the diffusion pathways in caries lesions will be very different from those in sound enamel. There is also scope for further work on electrochemistry of enamel. Van Dijk *et al.* (1977, 1979, 1980) showed that electrochemical and diffusion data can be integrated with other factors, such as solubility, to provide useful insights into the modes of action of caries-preventive agents.

Summary

The largest pores in enamel are associated with the prism junctions, but these constitute only a small fraction of the total porosity, most of which is associated with the prism bodies and tails. Here, most pores may exist as very narrow gaps between closely packed crystals but some, while still small, are elongated and tubule-like, and may well communicate with the prism-junction pores only through narrow intercrystalline pores. Organic matrix seems to be present within all pores and may alter pore sizes and modify diffusion. There are significant variations in porosity, apparently affecting both inter- and intraprismatic compartments, both within a single tooth (from outer to inner surface) and between tooth types. Pore structure affects the mechanical and optical properties of enamel. The formation of caries lesions is strongly influenced by the pathways for diffusion and by electrochemical effects arising from the charge on the pore walls.

References

Angmar, B., Carlström, D. and Glas, J-E. (1963). Studies on the ultrastructure of dental enamel IV. The mineralization of normal human enamel. *Journal of Ultrastructure Research*, **8**, 12–23.

Arends, J., Jongebloed, W. L. and Goldberg, M. (1984). Interactions of urea and human enamel. *Caries Research*, **18**, 17–24.

Arwill, T., Myrberg, N. and Söremark, R. (1969). Penetration of radioactive isotopes through enamel and dentine II. Transfer of ^{22}Na in fresh and chemically treated dental tissues. *Odontologisk Revy*, **20**, 47–54.

Atkinson, H. F. (1947). An investigation into the permeability of human enamel using osmotic methods. *British Dental Journal*, **83**, 205–214.

Berggren, H. (1947). Experimental studies on the permeability of enamel and dentine. *Swedish Dental Journal*, **40**, 6–109.

Bergman, G. and Siljestrand, B. (1963). Water evaporation in vitro from human dental enamel. *Archives of Oral Biology*, **8**, 37–38.

Bird, M. J., French, E. L., Woodside, M. R., Morrison, M. L. and Hodge, H. C. (1940). Chemical analyses of deciduous enamel and dentin. *Journal of Dental Research*, **19**, 413–423.

Bonte, E., Deschamps, N., Goldberg, M. and Vernois, V. (1988). Quantification of free water in human dental enamel. *Journal of Dental Research*, **19**, 880–882.

Borggreven, J. M. P. M., van Dijk, J. W. E. and Driessens, F. C. M. (1977). A quantitative radiochemical study of ionic and molecular transport in bovine dental enamel. *Archives of Oral Biology*, **22**, 467–472.

Borggreven, J. M. P. M., Driessens, F. C. M. and van Dijk, J. W. E. (1980). Diffusion through bovine enamel as related to the water structure in its pores. *Archives of Oral Biology*, **25**, 345–348.

Borggreven, J. M. P. M., van Dijk, J. W. E. and Driessens, F. C. M. (1981). Effect of mono- and divalent ions on diffusion and binding in bovine tooth enamel. *Archives of Oral Biology*, **26**, 663–669.

Boyde, A. (1965). The structure of developing mammalian dental enamel. In *Tooth Enamel*, ed. M. V. Stack and R. W. Fearnhead, pp. 163–167. Bristol: Wright.

Boyde, A. (1989). Enamel. In *Handbook of Microscopic Anatomy*, vol. V/6 *Teeth*, ed. A. Oksche and L. Vollrath, pp. 309–473. Berlin: Springer.

Boyde, A. and Pawley, J. B. (1975). Transmission electron microscopy of ion erosion thinned hard tissues. In *Calcified Tissues*, ed. S. Pors Nielsen and E. Hjorting-Hansen, pp. 117–123. Copenhagen: FADL's.

Braden, M., Duckworth, R. and Joyston-Bechal, S. (1971). The uptake of ^{24}Na by human dental enamel. *Archives of Oral Biology*, **16**, 367–374.

Brudevold, F., Srinavasan, B. N. and Skobe, Z. (1977). Separation of human tooth enamel microstructures by density fractionation. *Archives of Oral Biology*, **22**, 593–597.

Burke, E. J. and Moreno, E. C. (1975). Diffusion fluxes of tritiated water across human enamel membranes. *Archives of Oral Biology*, **20**, 327–332.

Burnett, G. W. and Zenewitz, J. (1958). Studies on the composition of teeth VII. The moisture content of calcified tooth tissues. *Journal of Dental Research*, **37**, 581–589.

Carlström, D. and Glas, J-E. (1963). Studies on the ultrastructure of dental enamel III. The birefringence of human enamel. *Journal of Ultrastructure Research*J, **8**, 1–11.

Chander, S. and Fuerstenau, D. W. (1982). On the dissolution and interfacial properties of hydroxyapatite. *Colloids and Surfaces*, **4**, 101–120.

Chick, A. O. and Waters, N. E. (1963). Membrane potentials in teeth: application of some common anions to enamel. *Journal of Dental Research*, **42**, 934–942.

Chick, A. O. and Waters, N. E. (1965). Membrane potentials in human teeth. The effect of some common organic anions. *Archives of Oral Biology*, **10**, 1–7.

Crabb, H. S. M. (1968). Structural patterns in human dental enamel revealed by the use of microradiography in conjunction with two dimensional microdensitometry. *Caries Research*, **2**, 235–252.

Crank, J. 1975. *The Mathematics of Diffusion*, second edition. Oxford: Clarendon Press.

Darling, A. I., Mortimer, K. V., Poole, D. F. G. and Ollis, W. D. (1961). Molecular sieve behaviour of normal and carious human enamel. *Archives of Oral Biology*, **5**, 251–273.

Dibdin, G. H. (1969). The internal surface and pore structure of enamel. *Journal of Dental Research*, **48**, 771–776.

Dibdin, G. H. (1972a). The stability of water in human dental enamel studied by proton nuclear magnetic resonance. *Archives of Oral Biology*, **17**, 433–437.

Dibdin, G. H. (1972b). Surface area of dental enamel: comparison of experimental values with values calculated for its structural components. *Journal of Dental Research*, **51**, 1256–1257.

Dibdin, G. H. (1993). The water in human dental enamel and its diffusional exchange measured by clearance of tritiated water from enamel slabs of varying thickness. *Caries Research*, **27**, 81–86.

Dibdin, G. H. (1995). Models of diffusion/reaction in dental plaque. *Microbial Ecology in Health and Disease*, **8**, 317–319.

Dibdin, G. H. and Poole, D. F. G. (1982). Surface area and pore size analysis for human enamel and dentine by water vapour sorption. *Archives of Oral Biology*, **27**, 235–241.

van Dijk, J. W. E., Waters, N. E., Borggreven, J. M. P. M. and Driessens, F. C. M. (1977). Some electrochemical characteristics of human tooth enamel. *Archives of Oral Biology*, **22**, 399–403.

van Dijk, J. W. E., Borggreven, J. M. P. M. and Driessens, F. C. M. (1979). The effect of fluoride and monofluorophosphate treatment on the electrochemical properties of bovine tooth enamel. *Archives of Oral Biology*, **24**, 753–758.

van Dijk, J. W. E., Borggreven, J. M. P. M. and Driessens, F. C. M. (1980). The effect of some phosphates and a phosphonate on the electrochemical properties of bovine tooth enamel. *Archives of Oral Biology*, **25**, 591–595.

van Dijk, J. W. E., Borggreven, J. M. P. M. and Driessens, F. C. M. (1983). Diffusion in mammalian tooth enamel in relation to the caries process. *Archives of Oral Biology*, **28**, 591–597.

Duckworth, R. and Braden, M. (1967). The uptake and release of fluorine-18 by human intact surface enamel in vitro. *Archives of Oral Biology*, **12**, 217–230.

Elliott, J. C. (1997). Structure, crystal chemistry and density of enamel apatites. In *Dental Enamel*, ed. D. Chadwick and G. Cardew, pp. 54–72. Chichester: Wiley (Ciba Foundation Symposium 205).

Featherstone, J. D. B. and Rosenberg, H. (1984). Lipid effect on the progress of artificial caries lesions in dental enamel. *Caries Research*, **18**, 52–55.

Flim, G. J. and Arends, J. (1977a). Diffusion of ^{45}Ca in bovine enamel. *Calcified Tissue Research*, **24**, 59–64.

Flim, G. J. and Arends, J. (1977b). The temperature dependency of ^{45}Ca diffusion in bovine enamel. *Calcified Tissue Research*, **24**, 173–177.

Flim, G. J., Kolar, Z. and Arends, J. (1977). Diffusion of fluoride ions in dental enamel at pH 7: a theoretical model. *Journal of Bioengineering*, **1**, 209–213.

Flim, G. H., Kolar, Z. and Arends, J. (1978). Diffusion of fluoride ions in dental enamel at pH 7. *Journal of Bioengineering*, **2**, 93–102.

Frazier, P. D. (1968). Adult human enamel: an electron microscopic study of crystallite size and morphology. *Journal of Ultrastructure Research*, **22**, 1–11.

Haines, D. J., Berry, D. C. and Poole, D. F. G. (1963). Behavior of tooth enamel under load. *Journal of Dental Research*, **42**, 885–888.

Hamilton, W. J., Judd, G. and Ansell, G. S. (1973). Ultrastructure of human enamel specimens prepared by ion micromilling. *Journal of Dental Research*, **52**, 703–710.

Hardwick, J. L., Martin, C. J. and Davies, T. G. H. (1965). The microstructure of mature dental enamel as observed under the optical microscope. In *Tooth Enamel*, ed. M. V. Stack and R. W. Fearnhead, pp. 168–171. Bristol: Wright.

Hattab, F. N. (1986). Diffusion of fluorides in human dental enamel in vitro. *Archives of Oral Biology*, **31**, 811–814.

Helmcke, J-G. (1967). Ultrastructure of enamel. In *Structural and Chemical Organization of Teeth*, Vol. 2, ed. A. E. W. Miles, pp. 135–164. London: Academic Press.

Holcomb, D. W. and Young, R. A. (1980). Thermal decomposition of human tooth enamel. *Calcified Tissue International*, **31**, 189–201.

Hoppenbrouwers, P. M. M., Scholberg, H. P. F. and Borggreven, J. M. P. M. (1986). Measurements of the permeability of dental enamel and its variation with depth using an electrochemical method. *Journal of Dental Research*, **65**, 154–157.

Huget, E. F., Brauer, G. M. and Loebenstein, W. V. (1968). Apparent heats of wetting and heats of reaction of the components of tooth structure and synthetic fluorapatite. *Journal of Dental Research*, **47**, 291–301.

Jansen, M. T. and Visser, J. B. (1950). Permeable structures in normal enamel. *Journal of Dental Research*, **29**, 622–632.

Kambara, M., Asai, T., Kumasaki, M. and Konishi, K. (1978). An electrochemical study on the human dental enamel with special reference to isoelectric point. *Journal of Dental Research*, **57**, 306–312.

Kérébel, B., Daculsi, G. and Kérébel, L. M. (1979). Ultrastructural studies of enamel crystallites. *Journal of Dental Research*, **58**, 844–850.

Kibby, C. L. and Hall, W. K. (1972). Surface properties of calcium phosphates. In *The Chemistry of Biosurfaces*, vol. 2, ed. M. L. Hair, pp. 663–730. New York: Marcel Dekker.

Klein, H. (1932). Physico-chemical studies on the structure of dental enamel IV–VIII. *Journal of Dental Research*, **12**, 79–98.

Klein, H. and Amberson, W. R. (1929). A physico-chemical study of the structure of dental enamel. *Journal of Dental Research*, **9**, 667–688.

LeFevre, M. L. and Manly, R. S. (1938). Moisture, inorganic and organic contents of enamel and dentin from carious teeth. *Journal of the American Dental Association*, **25**, 233–242.

LeGeros, R. Z., Bonel, G. and Legros, R. (1978). Types of 'H$_2$0' in human enamel and precipitated apatites. *Calcified Tissue Research*, **26**, 111–118.

Lindén, L.-A. (1968). Microscopic observations of fluid flow through enamel in vitro. *Odontologisk Revy*, **19**, 349–365.

Lindén, L-A., Björkman, W. and Hattab, F. (1986). The diffusion in vitro of fluoride and chlorhexidine in the enamel of human deciduous and permanent teeth. *Archives of Oral Biology*, **31**, 3–37.

Loebenstein, W. V. (1973). Adsorption of water on tooth components and related materials. *Journal of Dental Research*, **52**, 271–280.

Marthaler, T. M. and Mühlemann, H. R. (1960). Das Eindringen von radioaktiv markiertem Schwefelharnstoff in den Schmelz nach Natrium- und Kalziumchloridvorbehandlung. *Schweizerische Monatsschrift fuer Zahnheilkunde*, **70**, 10–17.

Meckel, A. H., Griebstein, W. J. and Neal, R. J. (1965). Ultrastructure of fully calcified human dental enamel In *Tooth Enamel*, ed. M. V. Stack and R. W. Fearnhead, pp. 160–162. Bristol: Wright.

Misra, D. N., Bowen, R. L. and Mattamal, G. J. (1978). Surface area of dental enamel, bone and hydroxyapatite: chemisorption from solution. *Calcified Tissue Research*, **26**, 139–142.

Moreno, E. C. and Burke, E. J. (1974). A diaphragm cell and the procedure for studying isothermal diffusion in dental enamel. *Archives of Oral Biology*, **19**, 417–420.

Moreno, E. C. and Zahradnik, R. T. (1973). The pore structure of human dental enamel. *Archives of Oral Biology*, **18**, 1063–1068.

Neiders, M. E., Weiss, L. and Cudney, T. L. (1970). An electrokinetic characterization of human tooth surfaces. *Archives of Oral Biology*, **15**, 135–151.

Newman, H. N. and Poole, D. F. G. (1974). Observations with scanning and transmission electron microscopy on the structure of human surface enamel. *Archives of Oral Biology*, **19**, 1135–1143.

O'Brien, W. J. (1988). Fraunhofer diffraction of light by human enamel. *Journal of Dental Research*, **67**, 484–486.

Orams, H. J., Phakey, P. P., Rachinger, W. A. and Zybert, J. J. (1974). Visualisation of micropore structure in human dental enamel. *Nature*, **252**, 584–585.

Orams, H. J., Zybert, J. J., Phakey, P. P. and Rachinger, W. A. (1976). Ultrastructural study of human dental enamel using selected-area argon-ion beam thinning. *Archives of Oral Biology*, **21**, 663–675.

Poole, D. F. G. and Brooks, A. W. (1961). The arrangement of crystallites in enamel prisms. *Archives of Oral Biology*, **5**, 14–26.

Poole, D. F. G. and Stack, M. V. (1965). The structure and physical properties of enamel. In *Tooth Enamel*, ed. M. V. Stack and R. W. Fearnhead, pp. 172–176. Bristol: Wright.

Poole, D. F. G., Mortimer, K. V., Darling, A. I. and Ollis, W. D. (1961). Molecular sieve behaviour of dental enamel. *Nature*, **189**, 998–1000.

Poole, D. F. G., Tailby, P. W. and Berry, D. C. (1963). The movement of water and other molecules through human enamel. *Archives of Oral Biology*, **8**, 771–772.

Poole, D. F. G., Newman, H. N. and Dibdin, G. H. (1981). Structure and porosity of human cervical enamel studied by polarizing microscopy and transmission electron microscopy. *Archives of Oral Biology*, **26**, 977–982.

Rasmussen, S. T., Patchin, R. E., Scott, D. B. and Heuer, A. H. (1976). Fracture properties of human enamel and dentine. *Journal of Dental Research*, **55**, 154–164.

Robinson, C., Weatherell, J. A. and Hallsworth, A. S. (1971). Variation in composition of dental enamel within thin ground tooth sections. *Caries Research*, **5**, 44–57.

de Rooij, J. F. and Arends, J. (1981). Phosphate diffusion in whole bovine enamel at pH 7. II. Temperature, time and concentration dependency. *Caries Research*, **15**, 353–362.

de Rooij, J. F. and Arends, J. (1982). Phosphate diffusion in whole bovine enamel. III. pH dependency. *Caries Research*, **16**, 211–216.

de Rooij, J. F., Kolar, Z. and Arends, J. (1980). Phosphate diffusion in whole bovine enamel at pH 7. *Caries Research*, **14**, 393–402.

de Rooij, J. F., Arends, J. and Kolar, Z. (1981). Diffusion of monofluorophosphate in whole bovine enamel at pH 7. *Caries Research*, **15**, 363–368.

Rootare, H. M. and Craig, R. G. (1977). Vapor phase adsorption of water on hydroxyapatite. *Journal of Dental Research*, **56**, 1437–1448.

Schmidt, W. J. and Keil, A. (1971). *Polarising Microscopy of Dental Tissues*, (translated by D. F. G. Poole and A. I. Darling). Oxford: Pergamon.

Selvig, K. A. and Halse, A. (1972). Crystal growth in rat incisor enamel. *Anatomical Record*, **173**, 453–468.

Shellis, R. P. (1984). Relationship between human enamel structure and the formation of caries-like lesions *in vitro*. *Archives of Oral Biology*, **29**, 975–981.

Shellis, R. P. (1996). Solubility variations in human enamel and dentine: a scanning electron microscope study. *Archives of Oral Biology*, **41**, 473–484.

Silness, J., Hegdahl, T. and Gustavsen, F. (1973). Area of the organic-inorganic interface of dental enamel. *Acta Odontologica Scandinavica*, **31**, 123–129.

Sundström, B. and Zelander, T. (1968). On the morphological organisation of the organic matrix of adult human enamel after decalcification by means of a basic chromium (III) sulphate solution. *Odontologisk Revy*, **19**, 1–15.

Tarbet, W. J. and Fosdick, L. S. (1971). Permeability of human dental enamel to acriflavine and potassium fluoride. *Archives of Oral Biology*, **16**, 951–961.

Theuns, H. M., van Dijk, J. W. E., Jongebloed, W. L. and Groeneveld, A. (1983). The mineral content of human enamel studied by polarizing microscopy, microradiography and scanning electron microscopy. *Archives of Oral Biology*, **28**, 797–803.

Theuns, H. M., Shellis, R. P., Groeneveld, A., van Dijk, J. W. E. and Poole, D. F. G. (1993). Relationships between birefringence and mineral content in artificial caries lesions of enamel. *Caries Research*, **27**, 9–14.

Vaarkamp, J., ten Bosch, J. J. and Verdonschot, E. H. (1995). Propagation of light through human dental enamel and dentine. *Caries Research*, **29**, 1–13.

Vongsavan, N. and Matthews, B. (1991). The permeability of cat dentine *in vivo* and *in vitro*. *Archives of Oral Biology*, **36**, 641–646.

Waters, N. E. (1971). The selectivity of human dental enamel to ionic transport. *Archives of Oral Biology*, **16**, 305–322.

Waters, N. E. (1972). Membrane potentials in teeth: the effect of pH. *Caries Research*, **6**, 346–354.

Waters, N. E. (1975). Electrochemistry of human enamel: selectivity to potassium in solutions containing calcium or phosphate ions. *Archives of Oral Biology*, **20**, 195–201.

Weatherell, J. A., Weidman, S. M. and Hamm, S. (1967). Density patterns in enamel. *Caries Research*, **1**, 42–51.

Weatherell, J. A., Weidman, S. M. and Eyre, D. R. (1968). Histological appearance and chemical composition of enamel protein from mature human molars. *Caries Research*, **2**, 281–293.

Wilson, P. R. and Beynon, A. D. (1989). Mineralization differences between human deciduous and permanent enamel measured by quantitative microradiography. *Archives of Oral Biology*, **34**, 85–88.

Zahradnik, R. T. and Moreno, E. C. (1975). Structural features of human dental enamel as revealed by isothermal water vapour sorption. *Archives of Oral Biology*, **20**, 317–325.

18 Pathways to functional differentiation in mammalian enamel

J. M. Rensberger

18.1. Introduction

Dental morphology has been the most important source of data for the interpretation of both phyletic relationships and dietary behaviors of extinct mammals. Differences in gross dental shape have traditionally contributed most of this information, although the complexity of enamel structure was recognized in the last century as subsequently were patterns of enamel microstructural differences among diverse mammalian taxa (Kawai, 1955).

However, it has only been in recent years that relationships of enamel microstructure to dental function have been identified (Walker *et al.*, 1978; Koenigswald, 1980; Rensberger and Koenigswald, 1980; Pfretzschner, 1988; Maas, 1991). The organization of the enamel microstructure is now known to have an adaptive relationship to the stresses generated during mastication and other uses of the teeth, indicating that these structures can be used to make inferences about differences in dietary behavior in extinct taxa.

An increased interest in enamel structure in the past two decades is contributing to a greatly expanded knowledge of the microstructure. As the structures in larger numbers of taxa have been clarified, the diversity of known microstructural conditions has increased. This has resulted in an increasingly complex picture of phyletic evolution that is clearly not explained by one-to-one relationships of structure and diet. For example, some highly specialized structural conditions occur in unrelated living taxa with quite different gross dental specializations and dietary behaviors. Nevertheless, even in such cases there may be functional components that are shared and may explain the microstructural similarities.

This chapter reviews functional bases for differentia-tion of enamel structure, examines several key morphologies that have appeared in independent lineages and attempts to integrate these structures and their functional attributes into a coherent set of relationships that may explain their pathways of selection.[1]

18.2. Dental structure and diet

The evidence that dental structures are predictors of diet in mammals falls into two categories: associations between dental morphology and dietary behaviors in living mammals, and existence of mechanical relationships that constrain a particular dental structure in its dietary function.

A classic example of a structure–diet association is the high cheek-tooth crowns that occur in bovoid artiodactyls, equine horses and arvicoline rodents, all of which feed on grasses. An association of structure and behavior does not alone assure that the structure cannot function effectively under circumstances differing from those observed. Identifying structural constraints that limit the functional utility of alternate morphologies (e.g., low-crowned teeth cannot endure the high rates of wear in grass-eating mammals) strengthens the functional association of a structure. This information becomes

---

[1] Throughout this chapter, the following abbreviations of museum collections will be used:

FMNH   Field Museum of Natural History, Chicago
UCMP   University of California Museum of Paleontology, Berkeley
UWBM   Burke Museum of Natural History and Culture, University of Washington, Seattle

increasingly important the farther back in geologic time that interpretations are applied, when environments were increasingly different and relationships to modern taxa more remote. In the example of high cheek-tooth crowns, the constraint validating the relationship is the increased rate of mechanical wear caused by the phytolith content of grasses (Walker *et al.*, 1978) coupled with the low nutrient yield of cellulose-rich foods resulting in greater volume requirements (Janis and Fortelius, 1988) and the high occlusal stresses (and consequent high rates of wear) required to fracture grass fibers (Rensberger, 1973).

18.3. Mechanical constraints

Mammalian teeth, in spite of their diversity, all function fundamentally by transmitting forces from the jaw muscles to food items. In fracturing food objects, the critical requirement is that the stresses generated exceed the strength of the food material. The physical properties of food materials vary enormously. For example, the fibers in leaves of different plants range over an order of magnitude in tensile strength (Gordon, 1976), and within an individual plant, the cellulose and lignin content increases with ontogenetic age (Lucas, 1979). Thus the critical stress that is required to fracture different materials will vary greatly. Tooth shapes and materials, together with the differences in shapes and physical properties of food materials, can be viewed as the mechanical environment in which teeth function. The evolutionary histories of enamel microstructures are at least, in part, responses to this complex environment, and the success of using these structures as predictors of diets in extinct forms depends upon the extent to which we can understand the relationships of the different structures to these environments and can identify the sequence of selective factors that have operated in a given lineage.

To reach critical occlusal stresses for tough foods, mammals have evolved dental shapes that focus the masticatory force in small areas of the contacting surfaces and thus elevate stresses to a level exceeding the strength of the structural components of the food (Rensberger, 1973). The diversity of dental shapes can be broadly classified as two types with different functional relationships to food materials: cusps or lophs with relatively sharp points or knifelike edges, and blunt projections or surfaces with low relief. Sharp structures have been selected for their ability to concentrate masticatory forces in small areas that transfer high stresses to tough food objects, such as insect carapaces or plant fibers, that can be fractured by cusp penetration or shearing movements. Blunt structures have been acquired for their ability to compress brittle objects that are easily fractured when deformed. For example, many seeds when compressed develop high tensile stresses in the stiff hull which cause it to fracture.

The physical properties of dental materials that are most closely related to their function, and differentiate them from other skeletal materials, are high stiffness and abrasion resistance. Dental structures are composed of the stiffest materials in vertebrate skeletons. Materials less stiff than enamel and dentine would compress proportionately more under high stresses, resulting in deformation during the milliseconds of high stress and a corresponding reduction in the level of the stresses being transmitted to the object.

Bones, including limb elements and antlers, have elastic moduli (in compression, GNm^{-2}) in the range of about 7 to 13, whereas elastic moduli in dentine are 11 to 16 and in enamel 20 to 90 (Waters, 1980; Currey, 1984). Whale tympanic bone has an elastic modulus of about 30 GNm^{-2} (Currey, 1984), but this high value emphasizes the functional role of stiffness in skeletal parts.

Dentine and enamel-like materials appeared early in vertebrate history. True enamel is present in the teeth of sarcopterygian fishes (Smith, 1989, 1992; see also Chapter 10). An outer surface layer of scales and dermal plates bearing similar materials were present in Paleozoic fishes as protective armor that apparently functioned to prevent penetration of external objects by distributing stresses away from the point of contact. As denticles began functioning as teeth, the stiffness of the dental materials served to concentrate stresses (rather than dissipate them), and transmit them to external objects.

Because scales and teeth function in contact with external objects, they are subject to abrasion. The hardest, as well as stiffest, of the dental and scale materials are the enamels, and these two factors probably explain the universal presence of these materials as thin outer coatings.

18.4. Basis for evolutionary modifications

18.4.1. Abrasion resistance
The high resistance of enamel to abrasion, as well as its high stiffness, results from its high mineral content (97% by weight: Waters, 1980). This abrasion resistance is not directionally uniform at the crystallite level and evolutionary responses of enamel to abrasion have been affected by this anisotropic characteristic.

That enamel prisms are anisotropic in their resistance

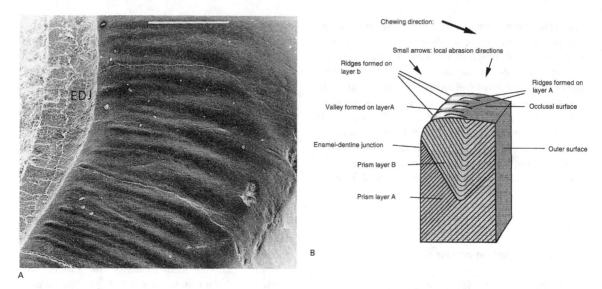

Figure 18.1. **Ridges and valleys on enamel surface caused by anisotropic wear rate of prisms with different inclinations in rhinocerotoids. A. Scanning electron micrograph of worn enamel surface in *Subhyracodon*, UWBM 31913. EDJ, enamel-dentin junction. Scale bar represents 1 mm. B. Enamel prism structure underlying occlusal ridges and valleys in rhinocerotoids. Prisms of the inner half of the enamel (left half of vertical section) have directions that differ by almost 90° in adjacent vertical layers, but in the outer enamel the directions of prisms converge on a single outward-occlusal direction. The greater resistance of prisms to abrasion parallel to their long axes than perpendicular to their long axes results in ridges and valleys corresponding respectively to the two sets of prism directions near the leading (left) edge of the enamel. The different direction of abrasion at the middle region of the enamel thickness results in another set of ridges corresponding to the prisms that form the valleys at the leading edge. See also Rensberger and Koenigswald (1980)**

to abrasion is not obvious because, in most mammals, prisms at the surface in any given region are inclined at about the same angle and the worn surfaces of the enamel are smooth. However, in rhinoceroses unique ridges and valleys are present on the occlusal surface (Figure 18.1A). The ridges coincide with regions in which prisms reach the abrading surface with their long axes aligned at angles parallel with the abrasion vector (direction in which the enamel is thinning under abrasion) and the valleys coincide with intervening positions where prisms are more nearly perpendicular to the abrasion vector (Figure 18.1B; Rensberger and Koenigswald, 1980). There are two sets of ridges, one on the leading edge and the other in the middle of the enamel thickness. At the leading edge, prisms aligned toward that edge (prisms in layer B) form a set of ridges because prisms wear more slowly parallel to their long axes. Prisms of layer A are perpendicular to the wear direction at the leading edge, wear more rapidly and form valleys there. At a distance away from the leading edge, the direction of wear is vertical (because material can separate most easily from a surface in a direction normal to the surface) and in that region prisms of layer A are locally more nearly parallel to the abrasion direction than prisms of layer B and thus form ridges.

That prisms offer greatest resistance to abrasion occurring parallel to their long axes results from a

general behavior of apatite. Apatite is most resistant to abrasion in the direction parallel to the C crystallographic axis, which is the long axis of the crystallite. The directions of the C axes of crystallites in enamel vary in a more or less regular pattern related to the domains of the ameloblasts that deposited the enamel. Within the region that we define as a prism, the crystallite C-axes are on average parallel to long axis of the prism.

18.4.2.   Chemical resistance

The direction of maximum resistance of enamel to corrosion by acid differs by 90° from the direction of maximum resistance to abrasion. Erosion of the enamel proceeds most rapidly in a direction parallel to the long axes of prisms. For example, application of an acidic solution to the occlusal surface of rhinoceros enamel, in which prism decussation planes intercept the chewing surface perpendicularly, produces ridges and valleys in reverse correspondence to the ridges and valleys that occur naturally due to abrasion (Rensberger and Koenigswald, 1980). Potentially, this weakness of prisms to acids could influence the evolution of prism attitudes near the surface in mammals that feed on acidic foods, but as yet such modifications have not been observed. Apparently, surface erosion due to mechanical abrasion and fracture in most cases is greater than erosion by chemical corrosion.

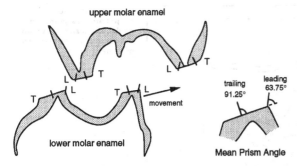

Figure 18.2. **Prism directions in leading and trailing enamel of** *Phascolarctos cinerus* **(koala) shown in vertical section (after Young et al., 1987). Arrow indicates direction of movement of lower molar during chewing. L, leading edge; T, trailing edge. Prism directions (angles) measured from occlusal surface.**

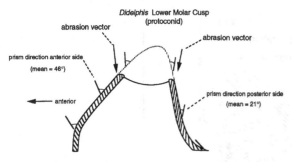

Figure 18.3. **Prism directions on anterior and posterior sides of protoconid in the opossum *Didelphis virginiana* (after Stern et al., 1989).**

18.5.   Microstructures associated with specific mechanical environments

18.5.1.   Prisms

Prisms are the result of columnar discordances in the directions of enamel crystallites. In most mammals the average direction of the crystallites tends to be aligned with the path of the ameloblast and the long axis of the prism. However, the pattern of discordance varies and, in some taxa, this results in a distinction between *interprismatic* crystallites and the bulk of the crystallites that more closely run parallel to the axis of the prism: *prismatic* crystallites.

The anisotropic resistance of prisms to abrasion has provided a basis for functional differentiation of prism directions in mammals. In the koala, *Phascolarctos cinerus*, Young et al., (1987) found an asymmetry in prism direction related to the chewing direction. Prisms in the leading edge of the occlusal enamel form an average angle of 64° with the flat occluding surface of the enamel, but in the trailing region the prism angle averages 91° (Figure 18.2). Young et al. noted that the prisms in the leading enamel are approximately parallel

with the abrasion vector in that region and maximally resist abrasion. We can extend that relationship to the enamel in the trailing region as well, because the direction in which enamel is being worn is being controlled by the flat contact surface between the opposing tooth surfaces. There the surface must wear in a vertical direction – normal to the plane of occlusal movement of the two surfaces. Thus, there are two environments of abrasion on the tooth, one controlled by contact with the opposing surface and the other on the leading edge, which is free of contact and is able to wear parallel to the movement of abrasive particles impacting obliquely against the surface. Selection has maximized the resistance to wear in the two regions by aligning the prism directions with the respective abrasion vectors.

This concept of two distinct factors controlling the direction of the abrasion vector on a single loph also explains the presence of two sets of ridges that are frequently distinguishable on occlusal surfaces of rhinocerotoid enamel (Rensberger and Koenigswald, 1980; Figure 18.1B). Ridges form on the leading edge from prism direction B, which is inclined toward the leading edge where the abrasion vector is oblique, whereas prisms of direction A form ridges in the trailing region where the abrasion vector is vertical and largely controlled by near contact with the opposing enamel surface. The two sets of prism directions here are not a phyletic response to these different abrasion vectors, but are a consequence of the vertical attitude of the prism decussation plane selected to resist fracture propagation (see section *18.5.2 on prism decussation*).

Stern et al. (1989) documented the prism directions in the molar cusps of the opossum *Didelphis virginiana*, and their data show a different alignment of prisms on anterior and posterior sides of the protoconid. Prism directions on the anterior side form an angle of about 46° with the surface whereas the prisms on the posterior side form an angle of about 21° with the surface (Figure 18.3). The prism directions on either side of the cusp are not very different from the orthal (vertical) chewing direction, except that, as in most mammals, they are slanted toward the outer surface. The prism direction on the anterior side intercepts the enamel surface at a more obtuse angle than on the posterior side, consistent with the less vertical attitude of the anterior dental surface with respect to the vertical chewing direction. As the occlusal surface wears down, the angle of the occlusal prisms on the anterior side slant more strongly anteriorly. This is consistent with the masticatory vector during chewing, which in primitive mammals has a significant anterior component that holds the shearing surfaces together. With wear, anterior movement of the lower dental surface is less constrained by interlocking

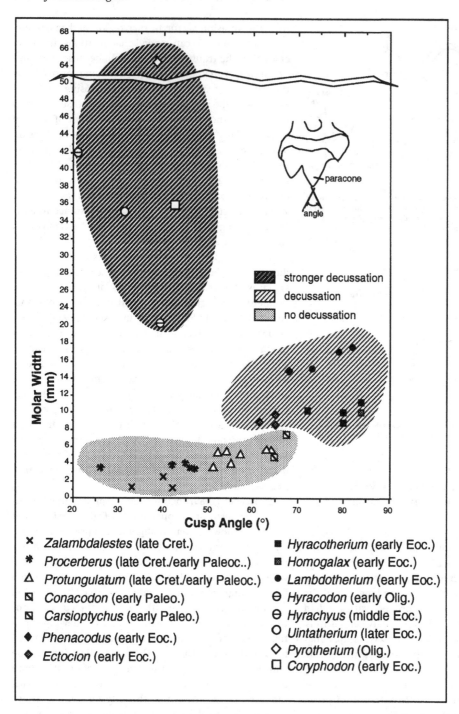

Figure 18.4. **Relationships of cusp sharpness (paracone angle), molar size and prism decussation in early Tertiary mammals.**

✕ *Zalambdalestes* (late Cret.)

✳ *Procerberus* (late Cret./early Paleoc..)

△ *Protungulatum* (late Cret./early Paleoc.)

◩ *Conacodon* (early Paleo.)

◪ *Carsioptychus* (early Paleo.)

◆ *Phenacodus* (early Eoc.)

◈ *Ectocion* (early Eoc.)

■ *Hyracotherium* (early Eoc.)

▨ *Homogalax* (early Eoc.)

● *Lambdotherium* (early Eoc.)

⊖ *Hyracodon* (early Olig.)

⊖ *Hyrachyus* (middle Eoc.)

○ *Uintatherium* (later Eoc.)

◇ *Pyrotherium* (Olig.)

☐ *Coryphodon* (early Eoc.)

cusps, and this would increase the anterior component of wear. Therefore the asymmetry in prism direction with respect to the enamel surface is consistent with a difference in the abrasion vector on the two surfaces.

In summary, because prisms are anisotropic in their resistance to abrasion, the directions of the prisms in both *Phascolarctos* and *Didelphis* have become asymmetric, that is, the ameloblasts have moved in different directions in their path away from the enamel–dentin junction (EDJ) and toward the outer surface in ways that locally maximize resistance to abrasion and help maintain edge sharpness during wear. However, neither of

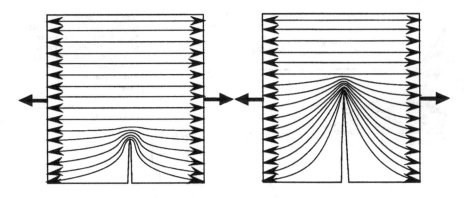

Figure 18.5. **Stress trajectories and concentration of tensile stresses around tips of cracks.** As a crack becomes longer, trajectories become increasingly concentrated at its tip, resulting in higher stresses there.

these taxa have modified prism directions as complexly as have the many mammals that have prism decussation.

18.5.2.   Prism decussation

Prism decussation in enamel (often visible as Hunter–Schreger bands) is associated with large tooth size, with a few notable exceptions. Mammals with molars more than 6 mm wide usually have Hunter–Schreger bands (HSB) and those of smaller size usually lack this structure (Koenigswald *et al.*, 1987). The main exceptions to this association occur in rodents and whales.

Rodents, even the smallest taxa, universally have prism decussation in their incisors. This exception to the association of size and decussation seems to identify stress as the specific factor in the association. Incisors of even the smallest of rodents must develop extreme stresses as a result of their thin profiles, cantilevered projection from the supporting bones and loading at the tip.

The association of decussation with high dental stresses is reinforced by early trends in the evolution of cusp shape in ungulates.

Cusp bluntness is correlated with size in early Tertiary mammals (Rensberger, 1988). Sharp cusps concentrate stresses generated by jaw musculature, and as the absolute size of the musculature increases, the stresses eventually become too large to be sustained by the strength of the enamel. Among *smaller* taxa (lower part of Figure 18.4), when the angle formed by the sides of the paracone is plotted with respect to tooth size, as size becomes larger the angle becomes larger, making the cusps blunter. However, when *larger*, more advanced taxa are included (upper part of Figure 18.4), the correlation of size and bluntness breaks down and we see large taxa with smaller cusp angles and sharper structures. The degree of prism decussation (shaded regions in Figure 18.4) appears to be the limiting factor because the advanced taxa that have sharper structures have increased degrees of prism decussation.

The notable exception – larger mammals having poor or no decussation – is provided by whales (Koenigswald *et al.*, 1988). Here the reduction or absence of decussation is associated with low dental stresses. Primitive whales have prism decussation (Sahni, 1981) and the reduction of decussation in later whales correlates with a reduction in dental stresses in an altered mechanical environment in which teeth no longer occlude with one another.

18.5.3.   Fracture resistance

Why might prism decussation be associated with body size? The answer must be that prism decussation helps resist the increased tendency for fracture as chewing loads increase. Materials that have high stiffness and hardness tend to fracture easily. The work-of-fracture in mammalian enamel is only 200 joules per square meter, compared with values of 550 joules for dentine and 1710 joules for bone (Rasmussen *et al.*, 1976; Currey, 1979).

Brittle materials fail largely in response to tensile stresses (Kingery *et al.*, 1976). Cracks in these materials propagate due to an increasing concentration of stress at the tip of the crack as the crack becomes longer (Figure 18.5). Other factors in the materials tend to inhibit the propagation of cracks, so that fracture actually occurs when the stresses normal to the fracture plane (the relationship shown in the inset diagram in Figure 18.6) exceed some critical magnitude specific to the particular material and are able to overcome the inhibiting factors. The critical magnitude (σ) depends upon the relationship (Griffith, 1921)

$$\sigma \cong \sqrt{\frac{2E\gamma}{\pi c}\left(\frac{\rho}{3a}\right)}$$

in which stress to fracture becomes smaller as the crack length (c) becomes larger.

On the other hand, as the radius (ρ) of the crack tip becomes wider or the amount of work required to form new crack surface (γ) becomes greater, the stress required for fracture increases.

That prism decussation can provide a mechanism that resists fracturing depends on an inherent weakness of

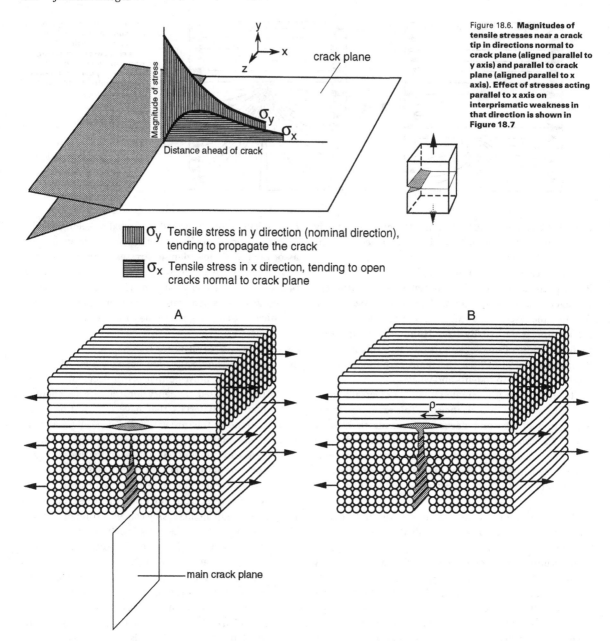

Figure 18.6. **Magnitudes of tensile stresses near a crack tip in directions normal to crack plane (aligned parallel to y axis) and parallel to crack plane (aligned parallel to x axis). Effect of stresses acting parallel to x axis on interprismatic weakness in that direction is shown in Figure 18.7**

σ_y Tensile stress in y direction (nominal direction), tending to propagate the crack

σ_x Tensile stress in x direction, tending to open cracks normal to crack plane

Figure 18.7A,B. **Inhibition of crack propagation by tensile stress ahead of crack tip acting in direction of crack plane. A. Crack appearing in weak interprismatic region ahead of and perpendicular to main crack. B. Tip of main crack reaching and being blunted by smaller crack normal to main crack plane; ρ, radius of main crack tip.**

prisms. Microscopic observations have shown that, in prismatic enamel, cracks occur more commonly along prism boundaries than through prism diameters (Boyde, 1976; Rasmussen *et al.*, 1976).

This anisotropic weakness of prisms allows certain modifications in prism directions to increase the magnitude of stress required for a crack to propagate (Rensberger, 1995a). Although there is a region of greatly elevated stresses at the crack tip that act *normal* to the crack plane (and thus favor extension of the crack), a short distance ahead of the crack tip there is another, though lesser, concentration of stresses (Figure 18.6), aligned *parallel* to the crack plane (Parker 1957; Cook and Gordon, 1964). If there are prisms in the region ahead of the crack tip running perpendicular to the crack plane, the oblique stresses there (Figure 18.7) will tend to

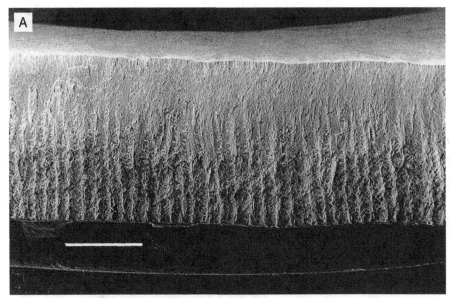

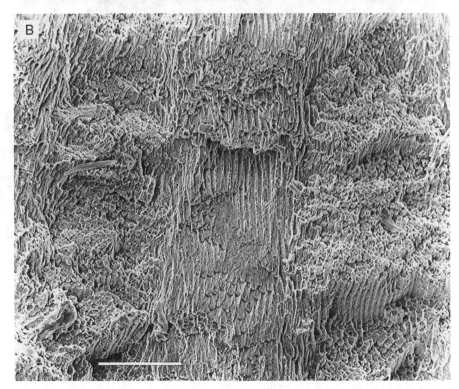

Figure 18.8A,B **Scanning electron micrographs (SEMs) of rough fracture surface in molar enamel of *Astrapotherium*, a later late Tertiary ungulate, UCMP 38825. A. Occlusal view of fracture paralleling the occlusal surface; scale bar represents 1 mm. B. Same but at higher magnification; note layer in the middle of the SEM in which prisms run parallel to the plane of the image, flanked by layers on either side composed of prisms running perpendicular to the plane of the image; scale bar represents 50 μm. Microscopic fracture planes follow prism boundaries, even in directions perpendicular to the main fracture plane, before fracturing through prism long axes, resulting in perpendicular steps in the fracture surface.**

produce small cracks along those prism boundaries. These cracks will be oblique or even perpendicular to the plane of the main crack and when the main crack reaches them they will have the effect of enlarging the crack tip radius (ρ), thus increasing the magnitude of the applied stress required to extend the crack beyond that point.

Prism decussation also raises the stress required for crack propagation by increasing γ, the energy required to form new crack surfaces. Cracks passing in most directions through enamel that has prism decussation can be extremely rough (Figure 18.8A,, B) and have more surface area than cracks in enamel lacking prism decussation. Prisms aligned normal to the crack plane do not all fracture precisely when the crack tip passes and will

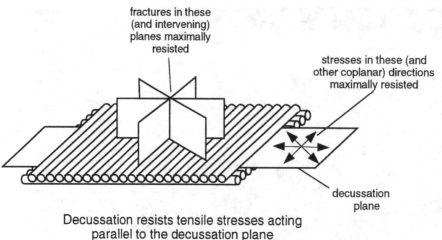

fractures in these
(and intervening)
planes maximally
resisted

stresses in these (and
other coplanar) directions
maximally resisted

decussation
plane

**Decussation resists tensile stresses acting
parallel to the decussation plane**

Figure 18.9. **Relationship of
resisted stresses to attitude of
decussation plane.**

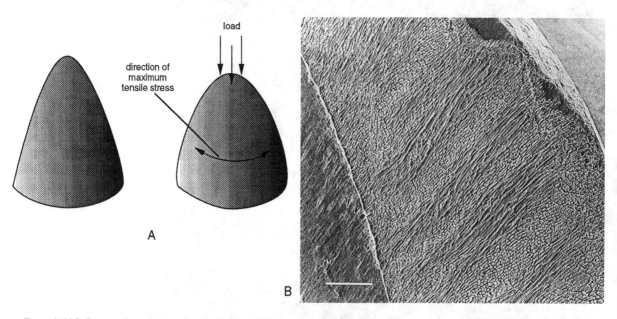

direction of
maximum
tensile stress

load

A

B

Figure 18.10A,B. **Decussation and stress direction in ellipsoidal cusps. A. Horizontal maximum tensile stresses in ellipsoidal cusp caused by
expansion under vertical load. B. Vertical enamel section with horizontal prism decussation: long axes of prisms aligned parallel to vertical
section in one set and normal or oblique to section in adjacent set; decussation planes normal to section plane. Anterior enamel of entoconid in
Hyracotherium (equid, early Eocene), UWBM 59782; enamel–dentine junction left; wear facet on outer margin of enamel at upper right. Scale
represents 100 μm.**

continue to pull out and consume energy in overcoming
friction after the crack tip passes; the small prism-sized
pits and prisms that extend above their neighbors in
Figure 18.8B are indications of prism pullout.

Where prisms do not actually decussate but bend
simultaneously (Koenigswald, 1988), cracks tend to bend
away from the plane in which they originated, and when
this happens the crack plane is no longer normal to the
direction of the maximum tensile stress, reducing the
energy available for propagation. Crack direction may be

similarly diverted when the crack penetrates a region of
enamel in which the prism directions are reversed; there
the crack may lose energy by following the new prism
directions for a short distance (Pfretzschner, 1988).

18.5.4.   Direction of decussation planes

The resistance of decussating enamel prisms to fracture
depends on the orientation of the decussation fabric. We
can define the decussation plane as a plane parallel to
the prism long axes in both sets of diverging prisms

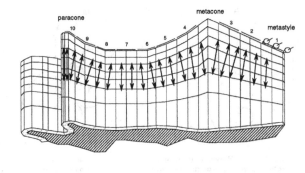

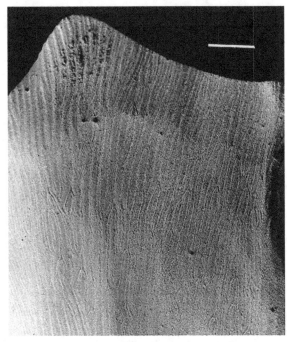

Figure 18.11A,B. **Stress directions and decussation planes in rhinocerotoid upper molar ectoloph (inverted, chewing surface at top). A. Predicted directions of maximum tensile stresses under loads normal to outer surface near occlusal edge. Static stress analysis of three-dimensional finite element model; directions shown are averages of separate local loadings (labeled 1–10) across entire edge. B. Tangential section through inner part of ectoloph enamel in Oligocene rhinocerotoid** Subhyracodon, **UWBM 32019. Scanning electron micrograph. Note divergence of decussation planes to left and right under metacone (cusp at top), as in model. Scale represents 1 mm.**

(Figure 18.9). Most prism decussation is tabular, with the decussation planes consistent in their attitudes.

The cusps in primitive mammals have circular cross-sections and ellipsoidal profiles. The chewing direction in these taxa is vertical and when such cusps are subjected to vertical compressive forces at the tip, the highest tensile stresses in the enamel are horizontal (Pfretzschner, 1988). This is true of low cusps of various profiles that have a circular cross-section and are loaded vertically (Rensberger, 1992). The compressive force causes the cusp to expand in all directions normal to the force, setting up circumferential tensile stresses acting horizontally and tending to open vertical cracks if unimpeded (Figure 18.10A).

In most such teeth, prism decussation planes are more or less horizontal (Figure 18.10B), that is, perpendicular to vertical fracture planes. Thus the planes of decussation tend to parallel the directions of maximum tensile stresses. This sets up the situation in which the propagating crack, after passing rather easily through a thin zone of unidirectional prisms, must encounter a zone of new prism directions where its progress is impeded by the mechanisms outlined above.

If this were the only attitude of decussation planes and planes of maximum tensile stresses in mammals, the geometry of the decussation could be regarded as possibly due to some developmental constraint rather than the result of selection for strength. However, a different, vertical arrangement of decussation planes occurs in the cheek-teeth of rhinocerotoids and several extinct groups of mammals (Rensberger and Koenigswald, 1980; Fortelius, 1985; Boyde and Fortelius, 1986), showing that the decussation plane attitudes are not developmentally bound to horizontal planes. The chewing direction in these taxa is strongly horizontal and the cusps have been phyletically replaced by thin lophs. Modeling the stresses in these structures shows that the combination of chewing direction and loph shape results in maximum tensile stresses that are vertical (Figure 18.11A), as they are in rhinocerotoid enamel (Figure 18.11B) and this is confirmed by the horizontal directions and arcuate shapes of the premortem cracks in rhino enamel (Rensberger, 1995b). Thus in these mammals, as in mammals with primitively circular cusp cross-sections, the planes of prism decussation are aligned in precisely the direction that maximizes resistance to fracture.

There is also evidence that the amount or degree of prism decussation is related to the magnitude of stresses that are developed in the enamel. Decussation is weakly developed in primates that eat soft foods but better developed in species that include hard foods in their diets (Maas, 1986). Decussation is also poorly developed

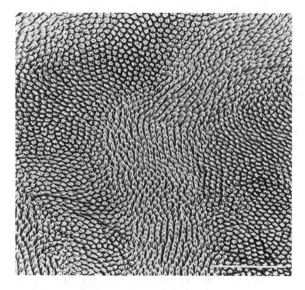

Figure 18.12. **Wavy decussation planes in Oligocene astrapothere canine tusk enamel. Scanning electron micrograph of tangential section in enamel fragment from *Parastrapotherium* FMNH P15050. Scale represents 50 μm.**

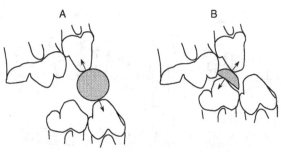

Figure 18.13A,B. **How hard materials can load cusps in different direction, depending on position of object. A. Bone cross-section loaded between P³ and P₃ in hyaenid dentition. B. Bone loaded between P³ and P₄; arrows indicate directions of loading of the teeth.**

in regions of the molars of the primitive perissodactyl *Hyracotherium* where finite element modeling indicates maximum tensile stresses are low, and decussation is better developed in areas where higher stresses are indicated (Rensberger, 1993). In North American Eocene notharctid primates, the presence of prism decussation is associated with taxa of larger body size in which higher dental stresses are expected (Maas, 1996).

The decussation planes in the dentitions discussed thus far are essentially flat. In the tusks of astrapotheres (Rensberger and Pfretzschner, 1992), the decussation planes undulate in a wavy pattern (Figure 18.12). Strictly flat decussation planes, though resistant to stresses acting parallel to the decussation plane, are weak to tensile stresses acting normal to the decussation plane. In theory, wavy decussation planes would minimize this weakness, yet retain much of the resistance of flat decussation planes to stresses parallel to the average attitude of the decussation planes. The wavy decussation in astrapothere tusk enamel is associated with sets of premortem cracks of distinctly different directions (Rensberger and Pfretzschner, 1992) and appears to have evolved in response to multiple loading directions. Varied loading is probably an attribute of elongate tusklike teeth, unlike the uniform loading that occurs in the cheek-teeth of most ungulates.

Peak loads in many of the teeth of carnivorous mammals are also expected to be highly variable. Carnivores, in capturing as well as in dismembering verte-

brate prey, frequently bring their teeth into contact with bone. The loading directions in bone–tooth contact can be highly variable, owing to different positions of impact on high-profile cusps (Figure 18.13) and small differences in loading position and direction cause large differences in stress directions within the cusps (Figure 18.14). Consistent with this variability of stress directions, decussation planes in many carnivorous mammals tend to be wavy.

A radically different geometry of decussation occurs in extant mammals in which bone eating has a significant role in their dietary behavior. Living hyenas, large felids and wolves display such bone-eating behaviors, and exhibit above average incidents of dental fracture as a correlate (Van Valkenburgh and Ruff, 1987). Among these three groups, the hyenas have the highest frequency of dental fracture, indicating exertion of large forces on the teeth, and have the most highly modified teeth. Almost all of the teeth in the spotted hyena (*Crocuta crocuta*) have a very specialized enamel microstructure (Figure 18.15A) in which the moderately wavy HSB typical of most carnivorans have been folded into high-amplitude waves (Rensberger, 1995a; Stefen, 1997). In addition, the crests and troughs are uniformly aligned with crests and troughs of the vertically adjacent layers. Furthermore, the reversed prism directions that characterize the adjacent layers of decussated enamel in antecedent taxa have disappeared in the crests and troughs of the aligned folds, and the prisms in adjacent crests are uniformly aligned in one direction and those of the troughs in the opposite direction, forming a higher level (Level II) decussation that is vertical (Figure 18.15B).

In the canines of *Crocuta*, which have regions of both Level I and Level II decussation, the distribution of the Level II decussation occurs in the predicted regions of highest tensile stresses when the tooth is loaded parallel with the tooth axis at the root (Rensberger, 1995a). The position of the most common premortem cracks in the

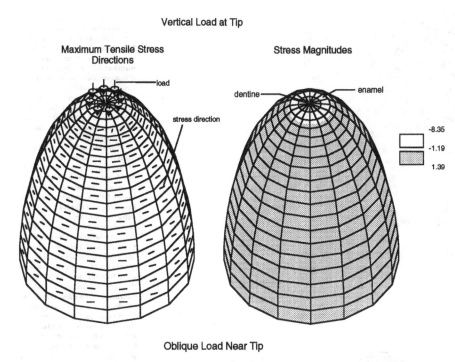

Vertical Load at Tip

Maximum Tensile Stress
Directions

Stress Magnitudes

Figure 18.14A,B. **Stresses in the enamel shell of an ellipsoidal cusp loaded in different directions. Based on static finite element calculations of maximum tensile stresses (directions of stresses indicated by short lines, relative magnitudes shown in adjacent diagrams). Polygonal units are hexagonal solid elements representing enamel surrounding solid pentagonal elements representing dentine. Maximum tensile stress directions and magnitudes: under vertical load (A); under oblique load of same magnitude (B).**

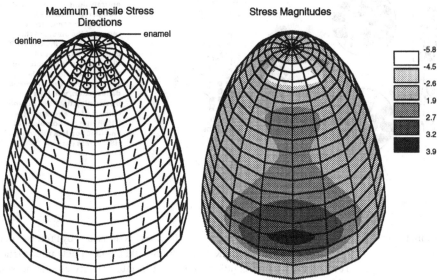

Oblique Load Near Tip

Maximum Tensile Stress
Directions

Stress Magnitudes

canine corresponds to the position of these highest stresses. The cougar (*Puma concolor*), a felid that has a similarly shaped canine but ranks low in bone-eating behavior, lacks Level II decussation altogether, indicating that Level II decussation is a correlate of high dental stresses.

Level II decussation seems to have appeared phyletically in response to the unusually high biting stresses that are generated when teeth bite bones with great force and cause the teeth to bend. The common pre-mortem cracks in the hyena canines are horizontal, occur on the convex side of the canine, and must have resulted from high vertical tensile stresses resulting from bending of the long axis of the tooth. But even in hyena premolars, where the axis is essentially straight, when the load is off-center the teeth will tend to bend and consequently the peak tensile stresses will be aligned vertically on the side that is loaded (Figure

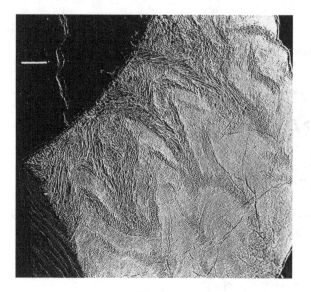

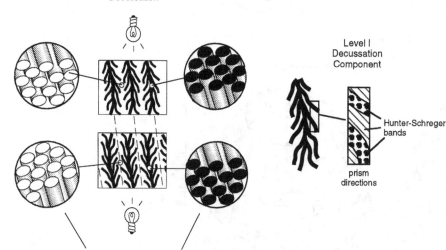

Level II
Decussation

Level I
Decussation
Component

Hunter-Schreger
bands

prism
directions

Hunter-Schreger bands

Figure 18.15A,B. **Prism decussation in the spotted hyena (*Crocuta crocuta*). A. Scanning electron micrograph of worn canine tip in specimen number A3650, Transvaal Museum, Zooarchaeology Collection. HSB appear as a tightly folded series of wave crests and troughs. Scale bar represents 100 μm. B. Diagram showing how differences in prism direction can be interpreted under reflected light microscopy from structures appearing as dark and light bands in hyena enamel. The Hunter–Schreger bands (HSB) representing the homologs of the HSB in most other mammals are represented here and in A as thinner dark/light bands on the flanks of the folds. The prisms in the centers of the crests and troughs of the folds have lost the primitive HSB differentiation and are all running in one direction (seen as either solid dark or solid light regions). The prism direction can be seen to alternate from crest to trough because under one lighting direction a crest region that is entirely light alternates with a trough region that is dark, and when the lighting direction is reversed the colors of the crests and troughs switch. The consistently reversed prism directions of the crests and troughs function as a large scale (Level II) vertical prism decussation, whereas the intervening flanks of the folds consist of smaller scale (Level I) HSB that are homologs of and function like the wavy HSB of less derived carnivores.**

18.14B). The Level II decussation, being aligned vertically, resists these stresses.

As seen in Figure 18.15A, B, the folded Level I HSB continue to be expressed on the flanks of the folds, where the decussation planes are oblique, so that this complex structure locally preserves a universal resistance to stresses acting in diverse directions. The fact that the Level II decussation occurs entirely around the premolars of hyenas, which are the main bone crushing teeth, is consistent with the variability of the loading directions that occur during bone cracking.

18.5.5.   Crystallite decussation

The early horses has horizontal prism decussation (Figure 18.10B) as a fracture-resisting mechanism

(Rensberger, 1993). However, as described by Pfretzschner (1993) and Kozawa and Suzuki (1995), a new microstructural condition emerged in late Tertiary equine horses. Pfretzschner (1993) showed that in the equines the interprismatic enamel matrix (IPM) and prismatic enamel are of equal thickness, in contrast to earlier horses in which the prisms are thick and the IPM is thin, and noted the correlation of this structure with hypsodonty in equids and other ungulates.

In the early Eocene equid *Hyracotherium*, the prisms run toward the outer surface in an occlusal direction and the IPM crystallites run toward the outer surface in a cervical direction. The prisms are subcylindrical and thick in comparison with the IPM matrix around the prisms and the IPM appears in discontinuous patches

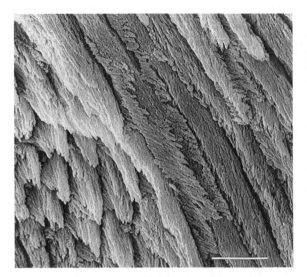

Figure 18.16. **Scanning electron micrograph of ground vertical section showing prisms and interprismatic enamel in inner enamel of early Eocene equid *Hyracotherium* (UWBM 31821). Occlusal direction toward top, enamel–dentin junction (EDJ) near right edge of image. Note prisms at left running oblique to plane of section, prisms at right parallel (dipping toward upper left). Scale bar represents 10 μm.**

(Figure 18.16). The subcylindrical shape of the prisms and the discontinuity of the IPM contrasts with the condition in the equine horses.

In vertically fractured *Equus* enamel, the structure is seen to be very different. The enamel has acquired an extremely plywoodlike microstructure (Figure 18.17A). The prisms appear as thin 'laths' separated from adjacent prism rows by equally thin and uniform sheets of interprismatic enamel. The directions of the crystallites in the prisms and the crystallites of the IPM differ by approximately 70°. The enamel has essentially become thin, parallel sheets of alternating crystallite directions, with the distinctiveness of the prisms almost disappearing. In fact, the prism and IPM sheets are difficult to distinguish from one another except for the difference in direction of their crystallites. In both functional and descriptive senses, this structure can be defined as crystallite decussation.

The crystallite decussation planes in this structure are aligned in a vertical and radial attitude, as are the prism decussation planes in rhinocerotoids and astrapotheres. When these high, parallel-sided teeth are loaded by the strong horizontal chewing movements that are distinctive of equine horses (Rensberger *et al.*, 1984), the maximum tensile stresses are vertical (Figure 18.17B), as in rhinocerotoids (Rensberger, 1992, 1995*b*) and the similarly adapted astrapotheres. The enamel microstructure accomplishes resistance to crack propagation in the equines through a mechanism that is functionally

similar to the vertical prism decussation in the rhinocerotoids and astrapotheres because the decussation planes in both cases are vertical and radial, but with a finer decussating fabric of crystallites instead of prisms.

Given the extent of horizontal translational movement among the diverse large ungulates living today, an important question is why only one family, the Rhinocerotidae, has vertical prism decussation.

Vertical crystallite decussation arose in equids after a long history involving animals of small to moderate size, modest development of prism decussation and slow emergence of lophodonty. Artiodactyla, the other modern ungulate group with crystallite decussation, also has an extended early history of small to moderate size and slow appearance of lophodonty.

In contrast, the earliest rhinocerotoids, astrapotheres and pyrotheres, all of which have vertical prism decussation, were very large mammals and were highly lophodont in the early Tertiary. One other group of fossil ungulates, the deperetellid tapiroids, were characterized by a comparable degree of vertical prism decussation, had a high degree of lophodonty and were larger than contemporaneous Eocene equids and artiodactyls. Thus, rates of evolution of dental morphology and body size may have been determining factors in the evolution of fracture-resisting mechanisms that arose in enamels

18.5.6.   Competition between fracture resistance and abrasion resistance

Decussation of all types must compete with simple parallel prisms for space in the enamel. On the one hand, enamel with decussating prisms cannot offer optimal resistance to abrasion because, when prisms are not parallel, all prisms can not be aligned parallel to the abrasion vector. On the other hand, parallel prism directions optimized for wear resistance weakens the fracture resistance of enamel. These tradeoffs explain much of the differential distribution of these types of structure through the enamel thickness. Decussation in most instances occurs in the inner part of the enamel. In astrapotheres (Rensberger and Pfretzschner, 1992) and rhinocerotoids (Rensberger, 1995*b*) vertical prism decussation occurs in the inner enamel and some degree of horizontal decussation in the inner part of the outer enamel, in correspondence to peak stresses locally aligned in those directions. However, the intensity of the horizontal decussation within the outer enamel is much weaker than the vertical decussation in the inner enamel, so that the prisms in the outer enamel, even though decussating, are predominantly running in an outward direction. This general restriction of decussation to the inner enamel of most mammals seems to be related to a higher incidence of abrasive wear at or near

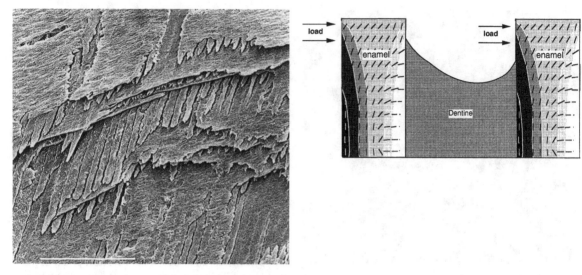

Figure 18.17A. **Scanning electron micrograph showing crystallite/crystallite decussation in equine enamel. Vertical fracture surface in inner enamel of Pleistocene** *Equus,* **UWBM 82572. Occlusal direction toward upper right corner of image, 45° from vertical, EDJ toward lower right corner. Narrow slat-like strips are prisms with crystallites nearly parallel to prism axis; thin continuous sheets intervening between layers of prisms are interprismatic (IP) enamel with crystallites (the fine, linear fabric) forming an angle of about 70° with prism crystallite axes. Scale bar represents 50 μm. B. Tensile stress intensity and direction in** *Equus* **enamel. Stress intensity contours (darkest represents highest magnitudes) and directions (small lines) based on a finite element model.**

the outer enamel surface, although quantitative data for wear rates across the enamel thickness have not yet been obtained. In other words, in regions in which wear resistance is of greatest importance, parallel prisms aligned toward the abrasion vector prevail. Away from the outer tooth surfaces where wear less frequently occurs, if stresses are high and would tend to fracture unreinforced enamel, decussation of prisms or crystallites prevails.

Summary

Identifying constraints that disallow selection for certain structures and favor others is important in interpreting dietary behaviors in extinct taxa because simple associations between diet and structure in extant taxa do not alone assure that the same dental structure may not have been used differently in earlier environments. Teeth function by transmitting stresses to food or other objects, and constraints on their function are imposed by limits to their resistance to fracture and abrasive wear. The better we understand those limits, the better we can interpret dental function in prehistoric animals.

Among the factors limiting fracture resistance are brittleness of enamel and shapes of the cusps. Ability to transmit stresses to food objects often correlates with cusp sharpness whereas fracture resistance often correlates with cusp bluntness. Consequently, early increases

in the size of jaw musculature were accompanied by decreasing cusp sharpness; attainment of sharper cusps and high occlusal stresses were made possible, however, by strengthening the fracture resistance through decussation, either at the prism level or at the crystallite level. Decussated enamel optimally resists fracture only when the decussation plane is parallel to the direction of the maximum tensile stress. Consequently, adequate dental strength in mammalian carnivores, in which the directions of peak stresses are highly variable due to the variability of tooth–bone contacts, was attained by acquiring decussation planes that bend in an undulating pattern. Ultimately, the highest tensile stresses attained in brittle materials are those resulting from bending of the central axis of the structure. However, most mammals with low-crowned teeth have relatively blunt teeth that minimize this type of deformation, and are constrained mainly by deformation resulting from vertical compression and horizontal tensile stresses in the outer part of the tooth. The brittleness of the enamel in this region created selective pressure for horizontal decussation, which allowed sharper tooth shapes to evolve and allowed larger mammals to tolerate higher occlusal stresses and to process tougher foods.

However, as very large size was attained, the general limits to bending that brittle materials can sustain without fracture became a constraint. This was overcome only by acquisition of some type of vertical decussation. Of the two types of vertical decussation that emerged, prism decussation may have been the easiest pathway for mammals that were increasing in size very rapidly. The most common vertical decussation in late Cenozoic mammals, however, is crystallite decussation, in which prisms have become flattened into sheets of crystallites with a common direction, and interprismatic crystallites have formed almost identical alternating sheets with an opposed crystallite direction.

Carnivorous mammals (hyenas, canids, including the extinct borophagines and many earlier Tertiary carnivorous taxa) that became adapted to extensive bone eating acquired a still different type of vertical decussation to withstand the high bending stresses that occur even in blunt cusps: vertical folding of primitively horizontal prism decussation planes. At the very centers of the folds and troughs, the original decussation planes have been lost. In those regions all prisms run in a single direction that is reversed from crest to trough, creating a second level of decussation with vertical decussation planes that resist bending. This Level II decussation is superimposed on the primitive Level I decussation that is characteristic of other mammals. Level II decussation is probably a reliable indication of bone eating, where it appears in Cenozoic taxa, because of its association with the high bending stresses in modern carnivorans that extensively consume bone.

Acknowledgements

I thank Mark Teaford for inviting me to participate in a symposium on teeth at the International Congress of Vertebrate Morphology in Chicago a few years ago and later to contribute an article for this volume. I am also extremely grateful to the following people who provided fossil and Recent dental tissues for my study: William Clemens, Laurence Frank, J. Howard Hutchinson and Donald E. Savage (University of California, Berkeley); Elizabeth de Wet (Transvaal Museum, Pretoria, South Africa; John Flynn, William Turnbull and Bruce Patterson (Field Museum of Natural History, Chicago); Blaire Van Valkenburgh (Biology, UCLA).

References

Boyde, A. (1976). Enamel structure and cavity margins. *Operative Dentistry*, 1, 13-28.

Boyde, A. and Fortelius, M. (1986). Development, structure and function of rhinoceros enamel. *Zoological Journal of the Linnean Society*, 87, 181-214.

Cook, J. and Gordon, J. E. (1964). A mechanism for the control of cracks in brittle systems. *Proceedings of the Royal Society, Series A*, 282, 508-520.

Currey, J. D. (1979). Mechanical properties of bone with greatly differing functions. *Journal of Biomechanics*, 12, 313-319.

Currey, J. D. (1984). *The Mechanical Adaptations of Bones*. Princeton, New Jersey: Princeton University Press.

Fortelius, M. (1985). Ungulate cheek teeth: developmental, functional, and evolutionary interrelations. *Acta Zoologica Fennica*, 180, 1-76.

Gordon, J. E. (1976). *The New Science of Strong Materials*, second edition. Princeton, New Jersey: Princeton University Press.

Griffith, A. A. (1921). The phenomena of rupture and flow in solids. *Philosophical Transactions of the Royal Society, Series A*, 221, 163.

Janis, C. M. and Fortelius, M. (1988). On the means whereby mammals achieve increased functional durability of their dentitions, with special reference to limiting factors. *Biological Reviews*, 63, 197-230.

Kawai, N. (1955). Comparative anatomy of the bands of Schreger. *Okijimas Folia Anatomica Japonica*, 27, 115-131.

Kingery, W. D., Bowen, H. K. and Uhlmann, D. R. (1976). *Introduction to Ceramics*. New York: Wiley.

Koenigswald, W. v. (1980). Schmelzstruktur und Morphologie in den Molaren der Arvicolidae (Rodentia). *Abhandlungen der Senckenbergischen Naturforschenden Gesellschaft*, 539, 1-129.

Koenigswald, W. v. (1988). Enamel modification in enlarged front teeth among mammals and the various possible reinforcements of the enamel. In *Teeth Revisited: Proceedings of the VIIth International Symposium on Dental Morphology, Paris*, ed. D. E. Russell, J.-P. Santoro and D. Sigogneau-Russell, pp. 147-167. Paris: Mémoires du Muséum National d'Histoire Naturelle, (Series C), 53.

Koenigswald, W. v., J. M. Rensberger and H. U. Pfretzschner. (1987). Changes in the tooth enamel of early Paleocene mammals allowing dietary diversity. *Nature*, 328, 150-152.

Kozawa, Y. and K. Suzuki. (1995). Appearance of new characteristic features of tooth structure in the evolution of molar teeth of Equoidea and Proboscidea. In *Aspects of Dental Biology: Palaeontology, Anthropology and Evolution*, ed. J. Moggi-Cecchi, pp. 27-31. Florence, Italy: International Institute for the Study of Man.

Lucas, P. W. (1979). The dental-dietary adaptations of mammals. *Neues Jahrbuch für Geologie und Paläontologie Abhandlungen*, 8, 486-512.

Maas, M. C. (1986). Function and variation of enamel prism decussation in ceboid primates. *American Journal of Physical Anthropology*, 69, 233-234.

Maas, M. C. (1991). Enamel structure and microwear: an experimental study of the response of enamel to shearing force. *American Journal of Physical Anthropology*, 85, 31-49.

Maas, M. C. (1996). Evolution of molar enamel microstructure in North American Notharctidae (primates). *Journal of Human Evolution*, 31, 293-310.

Parker, E. R. (1957). *Brittle Behavior of Engineering Structures*. New York: Wiley.

Pfretzschner, H. U. (1988). Structural reinforcement and crack propagation in enamel. In *Teeth Revisited: Proceedings of the VIIth International Symposium on Dental Morphology, Paris*, ed. D. E. Russell, J.-P. Santoro and D. Sigogneau-Russell, pp. 133–143. Paris: Mémoires du Muséum National d'Histoire Naturelle, (Series C), **53**.

Pfretzschner, H. U. (1992). Enamel microstructure and hypsodonty in large mammals. In: *Structure, Function and Evolution of Teeth*, eds. P. Smith and E. Tchernov, pp. 147–162. London: Freund Publishing House.

Pfretzschner, H. U. (1993). Enamel microstructure in the phylogeny of the Equidae. *Journal of Vertebrate Paleontology*, **13**, 342–349.

Rasmussen, S. T., Patchin, R. E., Scott, D. B. and Heuer, A. H. (1976). Fracture properties of human enamel and dentin. *Journal of Dental Research*, **55**, 154–164.

Rensberger, J. M. (1973). An occlusion model for mastication and dental wear in herbivorous mammals. *Journal of Paleontology*, **47**, 515–528.

Rensberger, J. M. (1988). The transition from insectivory to herbivory in mammalian teeth. In: *Teeth Revisited: Proceedings of the VIIth International Symposium on Dental Morphology*, ed. D. E. Russell, J.-P. Santoro and D. Sigogneau-Russell, pp. 355–369. Paris: Mémoires du Muséum National d'Histoire Naturelle, (Series C), **53**.

Rensberger, J. M. (1992). Relationship of chewing stress and enamel microstructure in rhinocerotoid cheek teeth. In: *Structure, Function and Evolution of Teeth*, eds. P. Smith and E. Tchernov, pp. 163–183. London: Freund Publishing House.

Rensberger, J. M. (1993). Adaptation of enamel microstructure to differences in stress intensity in the Eocene perissodactyl *Hyracotherium*. In *Structure, Formation and Evolution of Fossil Hard Tissues*, eds. I. Kobayashi, H. Mutvie and A. Sahni. pp. 131–145. Tokyo: Tokai University Press.

Rensberger, J. M. (1995a). Determination of stresses in mammalian dental enamel and their relevance to the interpretation of feeding behaviors in extinct taxa. In: *Functional Morphology in Vertebrate Paleontology*, ed. J. Thomason, pp. 151–172. Cambridge: Cambridge University Press.

Rensberger, J. M. (1995b). Relationship of chewing stresses to 3-dimensional geometry of enamel microstructure in rhinocerotoids In: *Aspects of Dental Biology: Palaeontology, Anthropology and Evolution*. ed. J. Moggi-Cecchi, pp. 129–146. Florence, Italy: International Institute for the Study of Man.

Rensberger, J. M. and Koenigswald, W. v. (1980). Functional and phylogenetic interpretation of enamel microstructure in rhinoceroses. *Paleobiology*, **6**, 477–495.

Rensberger, J. M. and Pfretzschner, H. U. (1992). Enamel structure in astrapotheres and its functional implications. *Scanning Microscopy*, **6**, 495–510.

Rensberger, J. M., Forstén, A. and Fortelius, M. (1984). Functional evolution of the cheek tooth pattern and chewing direction in Tertiary horses. *Paleobiology*, **10**, 439–452.

Sahni, A. (1981). Enamel ultrastructure of fossil Mammalia: Eocene Archaeoceti from Kutch. *Journal of the Palaeontological Society of India*, **25**, 33–37.

Smith, M. M. (1989). Distribution and variation in enamel structure in the oral teeth of sarcopterygians: its significance for the evolution of a protoprismatic enamel. *Historical Biology*, **3**, 97–126.

Smith, M. M. (1992). Microstructure and evolution of enamel amongst osteichthyan fishes and early tetrapods. In *Structure, Function and Evolution of Teeth*, eds. P. Smith and E. Tchernov, pp. 73–101, London: Freund Publishing House.

Stefen, C. (1997). Differentiations in Hunter-Schreger bands of carnivores. In *Teeth Enamel Microstructure*, eds. W. v. Koenigswald and P. M. Sander, pp. 123–136. Rotterdam: Balkema.

Stern, D., Crompton, A. W. and Skobe, Z. (1989). Enamel ultrastructure and masticatory function in molars of the American opossum, *Didelphis virginiana*. *Zoological Journal of the Linnean Society*, **95**, 311–334.

Van Valkenburgh, B. and Ruff, C. B. (1987). Canine tooth strength and killing behavior in large carnivores. *Journal of Zoology, London*, **212**, 379–397.

Walker, A. C., Hoeck, H. N. and Perez, L. (1978). Microwear of mammalian teeth as an indicator of diet. *Science*, **201**, 908–910.

Waters, N. E. (1980). Some mechanical and physical properties of teeth. In *The Mechanical Properties of Biologic Materials*, ed. J. F. V. Vincent and J. D. Currey. Cambridge: Cambridge University Press.

Young, W. G., McGowan, M. and Daley, T. J. (1987). Tooth enamel structure in the koala, *Phascolarctos cinereus*: some functional interpretations. *Scanning Microscopy*, **1**, 1925–1934.

19 Trends in the evolution of molar crown types in ungulate mammals: evidence from the northern hemisphere

J. Jernvall, J. P. Hunter and M. Fortelius

19.1. Food, fossils, form and function

Food is one of the main dimensions of an animal's environment. If we know what animals eat, then we know a great deal about them. If we can infer an animal's food from its anatomy, then we can reconstruct what extinct animals ate. And if we sum this sort of knowledge over the entire fossil record of a group of animals, we might obtain a reasonable idea of what changes, if any, have occurred in the group's ecology. In this way, mammalian paleontology, largely through the study of tooth evolution, has provided a detailed and dynamic picture of the biotic and abiotic changes that have shaped the terrestrial world during the past 65 million years (Archibald, 1983; Janis, 1989, 1993; Janis and Damuth, 1990; Stucky, 1990, 1992, 1995; Maas and Krause, 1994; Gunnell et al., 1995).

Despite this wealth of knowledge, the ties between dental morphology and food remain loose and methods for obtaining more precise information (such as the study of 'microwear'; see Teaford, 1994; Teaford et al., 1996) are currently far too laborious to be used on entire faunas, guilds, or clades over long intervals. Samples consisting of hundreds of species are required for the study of large-scale patterns of *biotal evolution* (*sensu* Van Valen, 1991). For this sort of study, we must identify different ecological groups of mammals – which can, but do not have to be, taxa – rather than phylogenetic lineages. Comparisons of extinct mammalian faunas and their dietary specializations require broadly applicable grouping criteria for teeth. Mammalogy textbooks (e.g., Vaughan, 1986) do give the impression that a sound morphological grouping of molar crown shapes exists: bunodont, selenodont, lophodont and a few odd types like lamellar grinders of elephants or the slicing carnas-

sials of carnivores. But a closer look reveals that this system leaves vast volumes of morphospace unaccounted for, and that its basis is a composite of form, size and inferred relationships. For example, few would describe the teeth of a non-artiodactyl as 'selenodont,' yet the molars of the koala, a marsupial, are perfectly selenodont in that they possess cusps compressed into lophs shaped like crescent moons (see Young and Robson, 1987).

Ideally, a dental classification should allow a neutral, functional classification of a wide range of tooth shapes. Moreover, morphological classifications focusing on characters that are useful in determining genealogical relationships might not be useful in determining ecological associations; in other words, these classifications 'cannot see the tooth for the cusps.' To correct this, we have recently developed a more neutral classification system for all kinds of teeth (Hunter and Jernvall, 1995; Jernvall, 1995; Jernvall et al., 1996), and in this chapter we apply this method to take a broad look at molar evolution using ungulate mammals as a case study.

That this kind of approach might work, or be worthwhile, was recently argued by Hunter and Jernvall (1995) who showed that squaring off the mammalian upper molar (by the hypocone cusp) fulfills the criteria of a key innovation – an evolutionary change important for new adaptations. The Cenozoic mammalian radiations have had a dichotomous nature. The lineages possessing the hypocone speciated greatly while those without did not. As the hypocone is associated with herbivory in modern mammals, the trend toward greater species diversity among mammals with hypocones is a signal of *ecological evolution*. This demonstration raised a number of questions. What is the morphological nature of the diversification process? How does species diversity relate to morphological-ecological diversity? Was the process similar on different continents? How does diversification result in the origination of higher taxa, that is, in the emergence over time of the differences that characterize biological groups? And what happens during evolutionary radiations anyway?

19.2.   The method

Addressing the kind of questions outlined above could easily consume many full-time careers. To make a modest start, we have confined ourselves to the two most diverse ungulate orders, both today and in the geologic past, the Artiodactyla and Perissodactyla, as well as an order of archaic ungulates, the Condylarthra, during the Cenozoic. We included in this study the condylarthran families Arctocyonidae, Hyopsodontidae, Meniscotheriidae, Mioclaenidae, Periptychidae, Phenacodontidae and Tricuspiodontidae. The artiodactyls and perissodactyls are each usually acknowledged to be holophyletic (i.e., taxa containing the most recent common ancestor and all descendants of that ancestor; *sensu* Ashlock, 1971). The condylarths, while being monophyletic (i.e., derived from an ancestor that was also a condylarth; *sensu* Ashlock, 1971) and forming an adaptively unified (Van Valen, 1978) stage in ungulate phylogeny, do not contain all the descendants of their most recent common ancestor (e.g., artiodactyls and perissodactyls) and are thus paraphyletic (Prothero *et al.*, 1988). Therefore, these groups are continuous both genealogically and adaptively, although admittedly they represent only part of ungulate phylogeny (nevertheless the part responsible for the lion's share of ungulate species diversity through the Cenozoic). The taxonomic data is predominantly derived from Savage and Russell (1983), the only such compilation treating each continent separately. Notable changes involve classification of the archaic ungulates (following Stucky and McKenna, 1993; Hunter, 1997) and time scale (following Woodbourne, 1987; Prothero, 1995). Because using genera is more conservative than using species in estimating the number of evolutionary lineages, we report the results of taxonomic trends using genera. However, analyses using species produced similar results.

The hypocone classification was based on a three-step tabulation of the relative development of a single cusp (Hunter and Jernvall, 1995). The present, more comprehensive classification of tooth shapes involves a new scheme of so-called crown types (Jernvall, 1995; Jernvall *et al.*, 1996). Like the hypocone classification, this system is purely topographical. No phylogenetic relations (i.e., homologies) are implied. Rather, all that is required is to characterize tooth shape based on its 'look.' This pro-

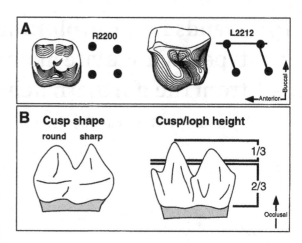

Figure 19.1A. **Examples of different molar shapes, their cusps and lophs (dots and lines) and their crown types. The crown type consists of five entries (12345): (1) cusp shape (R, round; S, sharp; L, loph), number of buccal (2) and lingual (3) cusps and the number of longitudinal (4) and transverse (5) lophs. The examples are (left) a pig molar and (right) a rhinoceros molar. B. Schematic drawing illustrating the criteria used to determine cusp shape (convex or concave slopes) and the cut-off point used to determine which features were tabulated as cusps and lophs. Note that, for example, three lingual cusps can be any combination of protocone, hypocone, metaconule, paraconule, or protostyle (epicone).**

cedure is simple and this simplicity makes it possible to obtain neutral descriptions of vast numbers of teeth. Six variables are defined, and the crown types are in the form R2200B or L2212H, for example (Figure 19.1A). The first letter stands for cusp shape. The general shape of the cusps on a tooth can be round (R), sharp (S) or 'lophy' (i.e., bladey; L). The next two numbers stand for the number of buccal and lingual cusps, respectively. Thus, our two examples both have two buccal and two lingual cusps (Figure 19.1A). The last two numbers stand for the number of longitudinal and transverse lophs, respectively. This time, the first example has no lophs, and the second example has one longitudinal and two transverse lophs. Finally, the last letter stands for crown height: B is for brachydont, and H is for hypsodont. We actually omitted crown height from this study for the sake of simplicity and because crown height does not necessarily affect the size and position of cusps and lophs (Jernvall, 1995).

Our two examples (Figure 19.1A) are therefore a bunodont, four-cusped and low-crowned tooth (e.g., pig) and a trilophodont, four-cusped and high-crowned tooth (e.g., black rhinoceros). We have limited counts of the number of cusps and lophs to three. If there are four or more cusps or lophs, then we tabulated the number to be 'many' or simply 'M' in the code, as in RM200B. Also, this system does not differentiate where a single loph on a crown is. Tabulation of this is certainly possible (e.g., separately for buccal and lingual longitudinal lophs), but this complication has been left out to keep the system simple. Interestingly, all upper molars with single longitudinal lophs encountered so far are buccal (ectolophs), and all single transverse lophs are in the distal part of the crown.

We have one critical piece of advice for anyone undertaking a crown type classification: it should be done on actual teeth or casts (we have made extensive use of several fossil collections including those at the American Museum of Natural History). We have found that published descriptions and many figures often fail to provide two crown type variables. First, the shape of cusps, whether round or sharp, can be rather elusive when many taxa are compared. Therefore, when in doubt, we classified a cusp as round if it had generally convex slopes, sharp if the slopes look more concave (Figure 19.1B). The second problem is the number of lophs. Lophs can be very variable in height. As an arbitrary cut-off point in determining what is and what is not a loph, we used one third of the overall crown height. That is, when in doubt, say, about a longitudinal loph on a selenodont tooth, we examined the crown from its buccal aspect. If the depth of the valley between the cusps is less that one third of the height of the whole crown, then we said that the cusps are joined by a loph (Figure 19.1B). And if the valley is relatively deeper, there is not a loph in our scheme. Note that a loph can consist of a single cusp or of several cusps.

A more general issue is the actual number of cusps. In our usage, cusps can be conules or stylar cones, but whether they are actually counted depends on the relative size of the cusps on the crown. As for lophs, we counted a cusp if it stood above two-thirds the crown height. Indeed, here we did not count many distinct, but relatively small hypocones among the early ungulates. This is not to say that we refute their existence; rather their exclusion from our counting is a byproduct of the same robust criteria applied to all teeth, which tries to minimize any teleological influence on our characterization of tooth shape. On that note, it is worth mentioning that *Hyracotherium*, the earliest perissodactyl, shares its crown type (R2200) mainly with condylarths, not with other perissodactyls. Our observation provides empirical support for Gould's (1991) claim, echoing Simpson (1953, p. 345), that a traditional (i.e. 'evolutionary') systematist living in the early Eocene would not have erected an Order Perissodactyla on the basis of *Hyracotherium* alone. Likewise, *Diacodexis*, the earliest artiodactyl (Rose, 1996), not unexpectedly shares its crown type (R2100) mainly with condylarths confirming Van Valen's (1971a) suspicion that the origin of artiodactyls was unaccompanied by major change in their resource use – at least not immediately.

19.3.   Paleontological patterns and biological inferences

Only a small fraction of all theoretically possible crown types are realized in actual teeth among extant mammals (Jernvall, 1995). Our initial study of crown types in extinct ungulates suggests that the same was true in the geologic past. Figure 19.2 depicts the 28 crown types that we encountered in the course of this study of which only seven (25%) are found among extant members of Artiodactyla and Perissodactyla. For convenience, Figure 19.2 displays the distribution of ungulate families across crown types. We carried out the survey of crown types, however, at the level of genera and species.

The critical assumption in the analysis below is that crown type diversity reflects ecological diversity more closely than does taxonomic diversity. Is this assumption justified? While a fossil taxon is a combination of (preferably) many characters, a crown type manifests only a few components of a mammal. Therefore, the ecological correlates of crown types can be more easily demonstrated, and indeed different dietary specializations have been shown to associate with different crown types in extant mammals (Jernvall, 1995; and section 19.3.3). Another concern with taxonomic diversity is the comparability of different species-level classifications across higher taxa (Gould, 1991; Foote, 1993). For example, some extinct mammalian groups might appear more species rich as a result of more thorough studies of their evolution irrespective of the amount of ecological diversification within the group. This should not be a problem with crown types.

19.3.1.   Diversity in crown types and genera

Let us begin by considering trends in relative crown type diversity (Figure 19.3). Relative crown type diversity appears to have reached a peak in ungulates in the later part of the Eocene, declined during the Oligocene and rose to – more or less – modern levels in the Miocene. This pattern is repeated on North America, Europe and Asia (Figure 19.3), and we interpret this parallel pattern to indicate that the crown types are picking up similar, large-scale evolutionary processes on each continent. To demonstrate these patterns on the southern continents, our crown typing method must be applied to the endemic South American litopterns and notoungulates as well as the Australasian Diprotodontia. A similar Eocene peak in crown type diversity in Africa might also be found among hyraxes once their fossil record is better known (Rasmussen et al., 1992).

Trends in relative generic diversity roughly parallel trends in crown type diversity on each continent (Figure 19.3) as is the general expectation (see Foote, 1996). Both

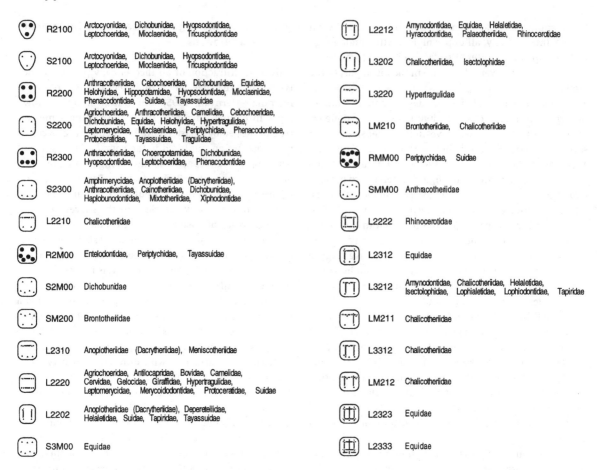

Figure 19.2. Crown types encountered in this study and families possessing them. The caricatures on the left represent occlusal views of each simplified crown type. Buccal side is toward the top of the figure. The survey of crown types was carried out at the level of species, and therefore some families appear under more than one crown type.

crown type diversity and generic diversity experienced a steep increase in the late Eocene and decline in the Oligocene. Eocene ungulates appear to have steadily invaded more and more morphospace as they diversified in number of taxa, and the Eocene can be interpreted as the time of the greatest number of ways for an ungulate to make a living (see Prothero, 1994, for a similar view) followed by the Oligocene as a time of crisis. Nevertheless, there are discordances between crown type and taxonomic diversity. Although numbers of ungulate genera rose in the Miocene, the number of crown types increased much less markedly (100–200% increase in number of genera from the Oligocene to the Miocene compared with less than 50% increase in number of crown types, Figure 19.3). Differences in sampling intensity have not been taken into account in our Figure 19.3, and poor sampling in the Oligocene may account for some of the decline in species diversity in the Oligocene, thus exaggerating the Miocene increase in numbers of

species, at least in North America (Alroy, 1998). Nevertheless, similarity among the patterns on each continent, as well as the decline in crown type diversity (crown types should be less sensitive to sampling artifacts than numbers of taxa), suggests that the Oligocene crisis strongly affected the adaptive diversity of ungulates regardless of any effect on species diversity *per se*.

At least in North America and Europe, crown type diversity declined only *after* the decline in taxonomic diversity in the Oligocene. Foote (1993) interpreted a similar delay (in Ordovician trilobites and echinoderms) as due to morphologically nonselective extinction. In our case, the extinction may have been nonselective on crown types, but perhaps not on some other biological (or biogeographical) attributes of the taxa. A closer look, however, reveals that a great deal of species loss occurred in lineages possessing crown types with one loph (brontotheres (LM210) and anoplotheres (L2310); see Figure 19.5 and section 19..3.3), suggesting that the late Eocene–

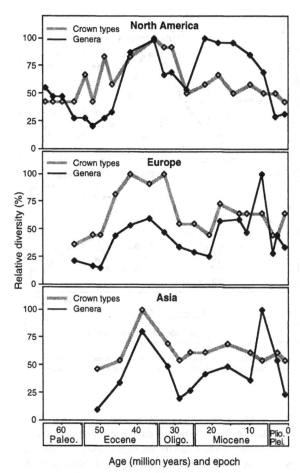

Figure 19.3. **Relative (percentage of maximum) diversity in crown types and genera through the Cenozoic in North America, Europe and Asia. Note the concordant peaks in crown type and generic diversity in the Eocene and the peak in generic diversity at low crown type diversity in the Miocene. Spearman rank correlations (r_s) between genus and crown type diversity are for the Paleogene (Paleocene, Eocene and Oligocene); North America: r_s=0.58 (P<0.05), Europe: r_s=0.89 (P<0.05) and Asia: r_s=0.90 (P<0.05). For the Neogene (Miocene, Pliocene and Pleistocene); North America: r_s=0.66 (NS), Europe: r_s=0.69 (NS) and Asia: r_s=0.06 (NS).**

Table 19.1. *Number of shared crown types among Orders and total number of crown types*

Mammal age	C-P-A	C-P	C-A	P-A	N
North America					
Graybullian (l. Eocene)	–	2	1	–	8
Lysitean (l. Eocene)	1	–	1	–	5
Lostcabinia (l. Eocene)	–	1	1	–	10
Bridgerian (m. Eocene)	–	–	1	1	7
Uintan (u. Eocene)	–	–	–	1	10
Europe					
Sparnacian (l. Eocene)	–	1	1	–	5
Cuisan (l. Eocene)	–	1	–	–	5
Luteian (m. Eocene)	1	–	–	–	9
Robiacian (u. Eocene)	–	–	–	1	11
Headonian (u. Eocene)	–	–	–	1	10
Asia					
Middle Eocene	–	–	–	1	7
Upper Eocene	–	–	–	1	13

Eocene is the only epoch when crown types are shared among orders (C, Condylarthra; P, Perissodactyla; A, Artiodactyla; *N*, Total number of crown types).

dietary specializations, and, hence, several crown types. One additional point to remember is that because we have analyzed only a portion of the mammalian Cenozoic radiations (ungulates), we obviously cannot test for ecological replacement by other mammalian groups. Some of the crown type morphospace vacated by ungulates in the late Eocene–Oligocene may have been filled in by rodents, for example.

19.3.2.   **What groups are responsible for the patterns?**
How have these three higher taxa of ungulates (condylarths, artiodactyls and perissodactyls) each contributed to the overall patterns? Figure 19.4 depicts diversity in genera (Figure 19.4A) and crown types (Figure 19.4B) within each of these taxa through the Cenozoic of North America and Europe. The patterns for Asia would look generally similar, but the fossil record of condylarths in the Paleocene of Asia remains spotty (see Lucas and Williamson, 1995, for a reassessment of previously described forms). In both North America and Europe, condylarths began to decline in taxonomic diversity *before* the appearance of artiodactyls and perissodactyls in the Eocene, a trend which continues until their final extinction (Figure 19.4A). However, condylarths vacate crown types only with the appearance and radiation of artiodactyls and perissodactyls (Figure 19.4B). Whatever caused the crash in number of taxa of condylarths – obviously not competition with artiodactyls or perissodactyls – acted across crown types rather than clearing selected regions of morphospace (see section 19.3.1 above; Foote, 1993). Moreover, the early Eocene was the time of greatest overlap in crown types among the three orders (Table 19.1) in keeping with an origin of both

Oligocene decline in taxa may not, after all, be completely independent of crown types. Stucky (1990, 1992, 1995) has suggested that a decline in within-habitat species diversity (i.e., alpha diversity) during the late Eocene–Oligocene in North America occurred as a result of a reduction in the number of ecological niches. We suggest that this degradation of environment may have been responsible for the extinction of crown types as well. That is, decline in crown type diversity shows that the Oligocene was also a time of *ecological* (niche) impoverishment, with an ensuing recovery in taxonomic diversity, but not in ecological diversity, in the Miocene. In general, the loss of niches may have resulted from reduction of tropical-like forests and thus affected several

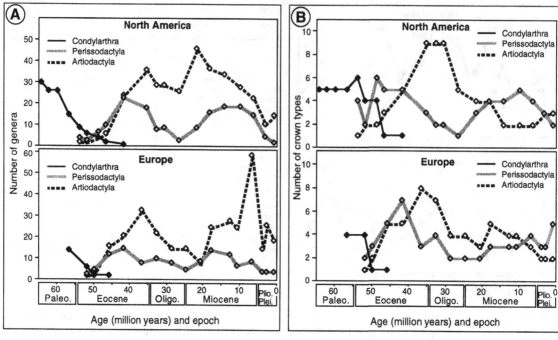

Figure 19.4. **Generic (A) and crown type (B) diversity profiles of the three ungulate orders studied through the Cenozoic. Note that crown type diversity peaks earlier in perissodactyls than artiodactyls. Although generic diversity of condylarths declines before the appearance of the modern orders, crown type diversity does not.**

artiodactyls and perissodactyls from condylarths (Radinsky, 1966; Van Valen, 1971b, 1978; Thewissen and Domning, 1992; Archibald, 1996; Rose, 1996) or from condylarth-like mammals (McKenna *et al.*, 1989).

In both North America and Europe, crown type diversity peaks first for perissodactyls during the Eocene and later for artiodactyls at the Eocene–Oligocene boundary (Figure 19.4B) suggesting greater initial morphological diversification among perissodactyls than artiodactyls. Temporal displacement between the two ordinal peaks is less apparent in generic diversity (Figure 19.4A). The largest contributors to generic diversity in the post-Oligocene world are the artiodactyls but with little increase in crown type diversity (Figure 19.4). Indeed, the Neogene taxonomic and crown type diversity trends appear more concordant in perissodactyls than in artiodactyls (Figure 19.4), a difference that may reflect the distinct physiological strategies of these two ungulate groups toward herbivory (hindgut fermentation versus rumination, see Janis, 1993 and references therein). On the other hand, closer concordance in perissodactyls than in artiodactyls may simply reflect the greater importance of teeth in perissodactyl systematics (species of ruminant artiodactyls are often distinguished more by horn and antler morphology than by teeth). The Eocene and post-Oligocene peaks in generic diversity are furthermore comprised of different families. For perissodactyls, these were

the brontotheres, chalicotheres and rhinoceratoids in the Eocene, but primarily equids after the Oligocene (mainly North American forms; Figure 19.4). For artiodactyls, these were the anthracotheres, entelodonts, dichobunids and oreodonts in the Eocene, but primarily pecoran ruminants and camelids (in North America) after the Oligocene. The relatively high artiodactyl generic diversity in the North American Oligocene is made of oreodonts, which have low crown type diversity. The taxonomic diversity of oreodonts is, however, probably overestimated compared with other fossil taxa (Lander, 1977; cited in Janis, 1989).

19.3.3.   Ecological-evolutionary trends in crown types
Ecological inferences need not be limited to the level of crown type diversity. We can also investigate trends in functionally significant characters that comprise the crown types. A convenient and informative grouping of ungulate crown types that reflects increasing structural complexity is loph number (Figure 19.5). In general, an increasing number of lophs means increasingly derived and complex teeth. Moreover, lophs can be related directly to resource use: lophs are best developed in herbivores consuming fibrous and tough plant foods (e.g., leaves and grass; see also Lucas *et al.*, 1995; Chapter 20) and less developed in consumers of less fibrous plant foods (e.g., fruit). Looking at crown types in extant

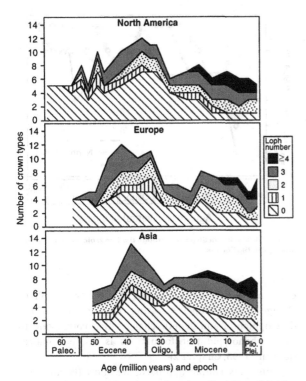

Figure 19.5. **Crown type diversity as a function of loph (dental blade) number through the Cenozoic. Lophed crown types are associated with specialized herbivory (Jernvall 1995). Note that the peak in crown type diversity in the Eocene is due to the appearance of lophodont crown types and that the lower crown type diversity in the Neogene is brought about mainly by a steady decline in the number of nonlophodont crown types.**

mammals, 89% of families having a diet of only fibrous vegetation have two or more lophs on their upper molars. This contrasts with 64% of less-specialized herbivorous families (eating in addition fruits, nectar or seeds) and 45% of generalist families (data and dietary classification in Hunter and Jernvall, 1995; Jernvall, 1995). Therefore, loph number is clearly associated with specialized herbivory (see also Fortelius, 1985; Janis and Fortelius, 1988).

From inspection of Figure 19.5, we find that the Eocene peak in number of crown types was largely due to the appearance and proliferation of lophodont crown types, whereas the Oligocene crash in number of crown types was due mainly to the steady loss – continuing into the present – of nonlophodont crown types. The addition during the Miocene of forms with many (i.e., four or more) lophs is mainly due to a radiation of hypsodont horses with complex occlusal morphology (Figure 19.5). Although the trends in North America, Europe and Asia are quite similar, they are not made up of the same crown types. Only a minimum of 44% (in the Oligocene) and a maximum of 54% (in the Miocene) of crown types

are shared among all three continents. Similar lophedness trends among biotas reflects ecological similarity rather than morphological resemblance *per se*. Thus, ecological factors probably played a central role in the evolution of ungulate dental morphology.

The first lophed crown types appear already in the Paleocene, when condylarths, but not artiodactyls or perissodactyls, were present. This may explain why we observed no drop in the total number of condylarthran crown types in spite of declines in the number of condylarthran genera and of nonlophodont crown types (Figure 19.4). The appearance of ungulates with lophodont crown types in the Paleocene thus preceded their diversification in numbers of taxa in the Eocene. Radiation in the Eocene of ungulates with lophs indicates that the ungulate biota became increasingly adapted to fibrous diets. This effect is also represented in our data as the steep rise in lophedness in the early Eocene (Figure 19.5). Interestingly, lophedness does not appear to change in the late Eocene and Oligocene (the average remains around one loph per crown type, Figure 19.5; Jernvall *et al.*, 1996), in face of apparent fluctuations in generic and crown type diversity (Figure 19.3). Whatever the environmental changes were during that period, the dietary specializations of ungulates do not appear to have experienced marked change, at least on the level of average lophedness. Thus, the Oligocene crisis may have been so broad (severe) that all crown types suffered more or less equally. However, as mentioned above (section 19.3.1), an interesting detail is the disappearance of forms possessing only one loph (Figure 19.4). This disappearance is real because different families comprise this class of crown types before and after the Oligocene–Miocene boundary in North America. This class of crown type seems particularly vulnerable as it never reappeared in Asia or Europe, and because they possessed only one loph, their extinction did not alter the average lophedness (which was around one). Furthermore, the delay in the change of average lophedness in the Oligocene may be an artifact of our crown typing system: we counted only well-developed lophs, thus ignoring early stages in loph evolution. Perhaps half a loph may indeed have been better than none (M. Cartmill, personal communication).

We interpret loss of the nonlophodont and one-lophed crown types to indicate that these forms may have been poorly adapted to post-Eocene cool, dry and seasonal climates with more open-forest environments (Upchurch and Wolfe, 1987; Knobloch *et al.*, 1993; Wing and Greenwood, 1993; Greenwood and Wing, 1995). These environments supported a greater proportion of fibrous vegetation, which, in turn, required herbivorous taxa to evolve more complex, bladed teeth or else go extinct (see Janis, 1989 and Prothero, 1994, for similar views). Thus, the world (food) may have turned literally too tough for the nonlophodont ungulates (see Chapter 20 for a discussion of food toughness). The fact that maximum lophedness did not develop until the Miocene most probably reflects environmental effects of the progressive decrease in global temperature during the Neogene and, perhaps more importantly, the increasing seasonality (Potts and Behrensmeyer, 1992; Bernor *et al.*, 1996*b*). The change, as seen in our study, appears to have been gradual and unrelated to any single global event. Indeed, it seems to have been somewhat delayed in Europe, where a major change from predominantly closed to predominantly open habitats is well documented as late as the beginning of the late Miocene, about 10 Ma (Fortelius *et al.*, 1996).

19.3.4. Disparity among crown types

So far we have discussed trends in the number of crown types and other measures of morphological diversity. We also investigated morphological disparity (*sensu* Gould, 1989) of the ungulate biota through time. Morphological disparity as conceived by Gould (1989, 1991) and implemented by Foote (1993, 1995) is different conceptually from simply the number of morphological types (e.g., our crown type diversity). In general, diversity is the number of different kinds of things, and disparity is the magnitude of the differences between or among things. In morphological evolution, disparity is 'range of anatomical design' (Gould, 1991, p. 412) or span of morphospace filled, and it is usually approached either as the distance among samples in morphospace or as some proxy for this quantity such as *total variance* (Van Valen, 1974*a*) in all of the dimensions of morphospace (Foote, 1993). In Figure 19.6 we show trends in maximum disparity as the maximum 'distance' between any two crown types present at a given time. Distance between any pair of crown types was determined in the following way. Consider our two original examples, R2200B (bunodont pig molar) and L2212H (trilophodont rhinoceros molar). First, the crown types must be converted into a completely numeric code. In cusp type, R (round) is coded as 0 and S (sharp) as 1. To avoid overestimating the morphological distance between lophed and nonlophed teeth, cusp shape L (loph) was tabulated as having a zero distance to R or S (difference in number of lophs is added later). As above, crown height was dropped from this study. The two converted crown types examples are 02200 and 02212. Distance is calculated as the sum of the absolute values of the differences between the crown types in each variable and thus the distance between these two crown types is $0 + 0 + 0 + 1 + 2 = 3$. It is worth pointing out that even a difference of one is morphologi-

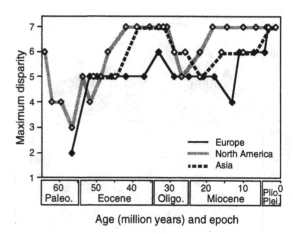

Figure 19.6. **Maximum disparity through the Cenozoic. Note the steep increase in the Eocene.**

cally quite large (addition or loss of one cusp or loph, or change in cusp shape). Assessing the magnitude of the developmental and genetic changes that correspond to these phenotypic transformations is another matter (Jernvall, 1995; Chapter 2)

Using this simple approach, we show (Figure 19.6) that maximum disparity among crown types rose quickly in the early Eocene, in keeping with this being a time of expansion in the number of crown types. The curves show a rise in maximum disparity to modern levels in the late Eocene (except in Europe), and after a decline in the late Oligocene–early Miocene, the values reach a maximum of seven on all three continents in the Pliocene. The initial peak in the North American record in the early Paleocene is largely due to the morphologically diverse and distinct (from other condylarths) family Periptychidae. This observation supports Van Valen's (1978) impression that the Periptychidae already warranted familial distinction in the early Paleocene on the basis of genera then extant irrespective of their descendants.

Given that maximum crown type disparity and thus maximum span of morphospace, increased quickly in the Eocene, how densely packed has this total space been? This is illustrated in Figure 19.7 using average disparity. This was tabulated by simply adding all distances between all pairs of crown types and dividing this sum by the number of crown type comparisons (*n* number of crown types generate $n \cdot (n-1) / 2$ crown type comparisons). Average disparity not only takes into account the distances between pairs of crown types, but also the number of crown types present. Average disparity increased through the Cenozoic (Figure 19.7). This could have been predicted because the total (and slightly expanding) morphospace (maximum disparity; Figure 19.6) has been shared among fewer and fewer crown types

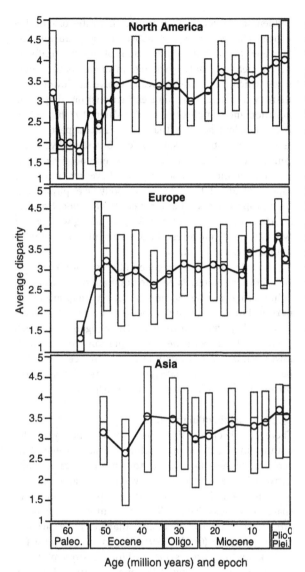

Figure 19.7. **Average disparity through the Cenozoic. Circles, mean (average) disparity; bars. quartiles; horizontal lines subdividing bars, median disparity.**

crown type disparity mainly as a result of the appearance of animals with lophodont and multicusped molar teeth. The Neogene radiations, on the other hand, were accompanied by increased average crown type disparity partially as a result of a decline in the number of crown types. These contrasting patterns of ungulate radiations bring us right into the problem of what constitutes an evolutionary radiation.

Was ungulate diversity generated by different processes in the Paleogene versus the Neogene? As has been discussed by several workers (e.g., Jablonski, 1980; Benton, 1988; Erwin, 1992; Heard and Hauser, 1995; Patzkowsky, 1995), evolutionary radiations are a complex group of phenomena, indeed perhaps as diverse as mammalian crown types themselves. Can the crown types inform us about the ecological processes underlying the ungulate radiations? Presumably, yes, assuming that teeth are useful indicators of an animal's food and that this has been the case throughout the Cenozoic.

The Paleogene radiations, comprised of parallel increases in taxonomic diversity, crown type diversity and crown type disparity, fit what Van Valen (1985) described as the 'bloom phase' of an adaptive radiation. That is, assuming that crown types do tell something about an animal's food (see section 19.3.3 above; Jernvall, 1995), the Paleogene (particularly the Eocene) saw a rapid increase in the number of ways (increase in crown type diversity and lophedness) for an ungulate to make a living (see also Prothero, 1994). Rosenzweig (1995, pp. 97–110) proposed that radiations resulting in new morphologies (*via* increase in phenotypic variance) could arise by a process he calls *competitive speciation*. While it is likely that all radiations involve competition at some level (Schluter and McPhail, 1993; Schluter, 1996), the critical point in this scenario is that both reproductive isolation and eco-phenotypic divergence can be promoted simultaneously when environmental variance (environmental patchiness) is larger than population variance (i.e., there is niche vacancy; Rosenzweig, 1995, p. 101). This disparity in variance will result in the rise of new adaptations as well as competitive speciation that is 'sympatric' at the level of several populations forming a metapopulation (basically parapatric speciation, i.e., on the level of populations, speciation would result from decreased successful immigration among individual populations within a metapopulation; see Hanski and Gilpin, 1991, on metapopulations). The Paleogene ungulate radiations involved the expansion of occupied dental morphospace accommodating more taxa utilizing more energy through time (see Van Valen, 1985, 1991), a process that may not have necessarily required geographical barriers *per se*. Rather, competitive ungulate radiations merely required enough resources (food) to

(Figure 19.3). After the peak in crown type diversity in the late Eocene, morphospace thinned through the rest of the Cenozoic leaving only islands of occupied morphospace.

19.3.5.   Generation of diversity in the Paleogene versus Neogene

To summarize the trends shown above, the superficially similar Paleogene (mainly Eocene) and Neogene (mainly Miocene) ungulate radiations are very different from each other. The Paleogene saw the rapid increase in ungulate taxonomic diversity, crown type diversity and sustain populations that were large enough to be isolated without going extinct (see Van Valen, 1973, for the importance of food in limiting the number of kinds of herbivores).

That new resources allowing this kind of radiation existed, is indirectly supported by the inferred Cenozoic temperature maximum for the early Eocene at high latitudes (e.g., Raymo and Ruddiman, 1992). High mean annual temperature and low mean annual range in temperature at high latitudes in the Paleogene (evidenced by leaf size and morphology; Upchurch and Wolfe, 1987; Knobloch *et al.*, 1993; Wing and Greenwood, 1993; Greenwood and Wing, 1995) suggest that tropical types of vegetation were widespread during this time. Also, plant diversity in Eocene assemblages appears to have been comparatively high (Wing, *et al.*, 1995; Knoll, 1986), which could be an indication of greater local productivity (see also Stucky, 1990). Moreover, high floral diversity could have created more 'crown type niches' by increasing environmental variance.

Similarly, steep climatic cooling in the late Eocene may have been responsible for the apparent 'crunch' in diversity (e.g., Van Valen, 1985; Janis, 1989; Prothero, 1994). The ensuing Neogene recovery was comprised of increased taxonomic diversity and crown type disparity but not of comparable increase in crown type diversity. Because the Neogene recovery (manifested by Miocene appearances) happened without marked increase in the number of crown types, we cannot assume new ways of living (diet) for most of the new Miocene taxa. Therefore, most of the diversifications during the Miocene could have been a result of *geographical speciation* (*sensu* Rosenzweig, 1995, pp. 108–110), that is, new species arose in isolation without marked changes in their resource use. Increased isolation, evidenced by provinciality, has been hypothesized for North America (Webb, 1977; Stucky, 1990, 1992, 1995) but an opposite trend has also been suggested for the late Cenozoic of North America (Van Valkenburgh and Janis, 1993). In western Eurasia, at least, provinciality decreased during the late Miocene (Fortelius *et al.*, 1996). One explanation is that the taxonomic proliferation of species-rich ungulates (bovids and horses) was a simple species/area effect, due to the spread of low-biomass/open woodland biomes over vast continental areas. Thus, the effect of increase in the area of open woodlands would have surpassed the effect of declining provinciality on continental taxonomic diversity.

This does not mean that the Neogene was totally devoid of diversifications involving invasions of new adaptive zones. For example, the evolution of hypsodont horses was certainly a new way of living *per se*, but even this did not result in a large number of new crown types (or diet; see MacFadden and Cerling, 1994). Indeed, Mac-

Fadden and Hulbert (1988) argued that rates of morphological change were not above average during the Miocene taxonomic proliferation of horses (unfortunately measured only by change in linear tooth dimensions which, like 'size,' typically evolve at slower rates than taxa (originations and extinctions) or 'shape' (innovation and complexity); see Van Valen, 1974*b*, 1985). It is nonetheless possible that this result is part of the general trend toward species-rich, but dentally homogeneous diversifications in the Neogene observed here. This taxonomic diversification may have involved other aspects of ungulate ecology (e.g., body size, physiology, limb proportions, see Bernor *et al.*, 1989, 1996*a*, Hayek *et al.*, 1991) and analysis of those aspects along with crown types are needed to assess the degree of mosaic evolution in ungulates radiations.

The Paleogene and Neogene radiations produced two very different biotas. Paleogene ungulate specializations, culminating in the Eocene, involved adaptations to narrow, but 'high quality' niches, reflected in the diverse and tightly packed crown types. The Neogene radiations created a fauna with dental adaptations to deal with lower primary productivity and thus we see the disappearance of intermediate crown types as the world became harsher and ungulates more disparate.

19.3.6.   Origin of higher taxa

There has been considerable debate concerning the manner in which the morphological changes that characterize higher taxa (and the morphological gaps between higher taxa) originate, whether by large changes early in a radiation – analogous to dispersal in morphospace – or by selective weeding out of intermediate forms after a period of diversification – analogous to vicariance in morphospace (Gould, 1991). We found evidence for both kinds of patterns in the biotic history of these ungulates. In the early Paleocene, periptychid condylarths expand in number of taxa and crown types in a way that sets them apart from other, contemporaneous condylarths and would warrant their recognition as a higher taxon to a traditional ('evolutionary') systematist (see also Van Valen, 1978; Hunter, 1997). Through the Cenozoic, artiodactyls and perissodactyls have paralleled each other in many ways, but much initial similarity between them has been destroyed by extinction. We believe that apparent morphological discontinuities between artiodactyls and perissodactyls, on the one hand, and the condylarths on the other (Table 19.1), are a byproduct of this thinning out of morphological space, largely (as we have already seen in Figure 19.5) selecting against nonlophodont forms, the condylarths and the more primitive artiodactyls and perissodactyls. Thus, most artiodactyls and perissodactyls today occupy a different adaptive zone – judging from their teeth – from that of their condylarthran ancestors (Van Valen, 1971a). And, of course, this process that gave rise to the adaptively distinct artiodactyls and perissodactyls, also converted condylarths from being holophyletic to being paraphyletic.

19.4.   What are crown types good for?

In this chapter, we have demonstrated the following patterns in the evolutionary history of condylarths, artiodactyls and perissodactyls:

1.   Trends in molar crown type (i.e., morphological) diversity generally have paralleled trends in taxonomic diversity, but not always. For example, generic and crown type diversity seem to have coevolved during the Paleogene, but have become somewhat decoupled in the Neogene.

2.   Perissodactyls and artiodactyls may appear synchronously in the fossil record (Krause and Maas, 1990), but perissodactyls appear to have diversified earlier than artiodactyls in tooth morphology.

3.   Some ungulate families are diverse in morphology, others less so.

4.   As the world has become colder, harsher, more open and more latitudinally stratified in the Neogene (as compared with the Paleogene), ungulates have become more disparate and more 'lophy.' We have interpreted this to mean that different processes underlie the Paleogene and Neogene ungulate radiations.

5.   Nevertheless, despite the different evolutionary processes implicated, higher taxa of ungulates have originated (i.e., large changes in morphology have occurred) both early and late in their radiation.

We have demonstrated these and other patterns in this paper using an exceedingly simple approach that anyone with access to skeletal collections and collections of fossil mammals can utilize. We have also shown how this approach can enable one to determine the ecological significance of morphological and taxonomic evolution and thus arrive at a new, dynamic and exciting view of mammalian evolution. We hope that others will adopt our approach and that we will soon witness a 'bloom phase' of morphological studies of tooth evolution.

Acknowledgements

We thank the staff of the American Museum of Natural History (New York) for help and access to their collections of fossil ungulates. Comments from Christine Janis and Mark Teaford greatly improved this chapter. Mathew Cartmill suggested to us the importance of half a loph. This is a contribution from the Valio Armas Korvenkontio Unit of Dental Anatomy in Relation to Evolutionary Theory.

References

Alroy, J. (1998). Long-term equilibrium in North American mammalian diversity. In *Biodiversity Dynamics: Turnover of Populations, Taxa and Communities*, ed. M. L. McKinney, pp. 232–287. New York: Columbia University Press.

Archibald, J. D. (1983). Structure of the K-T mammal radiation in North America: speculations on turnover rates and trophic structure. *Acta Palaeontologica Polonica*, **28**, 7–17.

Archibald, J. D. (1996). Fossil evidence for a Late Cretaceous origin of 'hoofed' mammals. *Science*, **272**, 1150–1153.

Ashlock, P. D. (1971). Monophyly and associated terms. *Systematic Zoology*, **20**, 63–69.

Benton, M. J. (1988). The nature of an adaptive radiation. *Trends in Ecology and Evolution*, **3**, 127–128.

Bernor, R. L., Tobien, H. and Woodburne, M. O. (1989). Patterns of Old World hipparionine evolutionary diversification. In *European Neogene Mammal Chronology*, eds. E.H. Lindsay et al., pp. 263–319. New York: Plenum Press.

Bernor, R. L., Koufos, G. D., Woodburne, M. O. and Fortelius, M. (1996a). The evolutionary history and biochronology of European and Southwest Asian Late Miocene and Pliocene hipparionine horses. In *The Evolution of Western Eurasian Neogene Mammal Faunas*, eds. R. L. Bernor, V. Fahlbusch and H.-W. Mittmann, pp. 307–338. New York: Columbia University Press.

Bernor, R. L., Fahlbusch, V., Andrews, P., de Bruijn, H., Fortelius, M., Rögl, F., Steininger, F. F. and Werdelin, L. (1996b). The evolution of western Eurasian Neogene mammal faunas: a chronologic, systematic, biogeographic, and palaeoenvironmental synthesis. In *The Evolution of Western Eurasian Neogene Mammal Faunas*, eds. R. L. Bernor, V. Fahlbusch and H.-W. Mittmann, pp. 449–469. New York: Columbia University Press.

Erwin, D. H. (1992). A preliminary classification of evolutionary radiations. *Historical Biology*, **6**, 133–147.

Foote, M. (1993). Discordance and concordance between morphological and taxonomic diversity. *Paleobiology*, **19**, 185–204.

Foote, M. (1995). Morphological diversification of Paleozoic crinoids. *Paleobiology*, **21**, 273–299.

Foote, M. (1996). Perspective: evolutionary patterns in the fossil record. *Evolution*, **50**, 1–11.

Fortelius, M. (1985). Ungulate cheek teeth: developmental, functional and evolutionary interrelations. *Acta Zoologica Fennica*, **180**, 1–76.

Fortelius, M., Werdelin, L., Andrews, P., Bernor, R. L., Gentry, A., Humphrey, L., Mittmann, H.-W. and Viranta, S. (1996). Provinciality, diversity, turnover and paleoecology in land mammal faunas of the later Miocene of Western Eurasia. In *The Evolution of Western Eurasian Neogene Mammal Faunas*, eds. R. L. Bernor, V. Fahlbusch and H.-W. Mittmann, pp. 414–448. New York: Columbia University Press.

Gould, S. J. (1989). *Wonderful Life: the Burgess Shale and the Nature of History*. New York: Norton.

Gould, S. J. (1991). The disparity of the Burgess Shale arthropod fauna and the limits of cladistic analysis: why we must strive to quantify morphospace. *Paleobiology*, 17, 41–423.

Greenwood, D. R. and Wing, S. L. (1995). Eocene climate and latitudinal temperature gradients on land. *Geology*, 23, 1044–1048.

Gunnell, G. F., Morgan, M. E., Maas, M. C. and Gingerich, P. D. (1995). Comparative paleoecology of Paleogene and Neogene mammalian faunas: trophic structure and composition. *Palaeogeography, Palaeoclimatology, Palaeoecology*, 115, 265–286.

Hanski, I. and Gilpin, M. (1991). Metapopulation dynamics: brief history and conceptual domain. In *Metapopulation Dynamics: Empirical and Theoretical Investigations*, eds. M. Gilpin and I. Hanski, pp. 3–16. London: Academic Press (for the Linnean Society of London).

Hayek, L.-A. C., Bernor, R. L., Solounias, N. and Steigerwald, P. (1991). Preliminary studies of hipparionine horse diet as measured by tooth microwear. *Annales Zoologici Fennici*, 3–4, 187–200.

Heard, S. B. and Hauser, D. L. (1995). Key evolutionary innovations and their ecological mechanisms. *Historical Biology*, 10, 151–173.

Hunter, J. P. (1997). Adaptive radiation of early Paleocene ungulates (Mammalia, Condylarthra). PhD dissertation. State University of New York at Stony Brook.

Hunter, J. P. and Jernvall, J. (1995). The hypocone as a key innovation in mammalian evolution. *Proceedings of the National Academy of Sciences, USA*, 92, 10718–10722.

Jablonski, D. (1980). Adaptive radiations: fossil evidence for two modes [abstract]. *Second International Congress of Systematic and Evolutionary Biology*, 243.

Janis, C. M. (1989). A climatic explanation for patterns of evolutionary diversity in ungulate mammals. *Palaeontology*, 32, 463–481.

Janis, C.M. (1993). Tertiary mammal evolution in the context of changing climates, vegetation and tectonic events. *Annual Review of Ecology and Systematics*, 24, 467–500.

Janis, C. M. and Damuth, J. (1990). Mammals. In *Evolutionary Trends*, ed. K. J. McNamara, pp. 301–345. London: Belhaven Press.

Janis, C. M. and Fortelius, M. (1988). On the means whereby mammals achieve increased functional durability of their dentitions, with special reference to limiting factors. *Biological Reviews*, 63, 197–230.

Jernvall, J. (1995). Mammalian molar cusp patterns: developmental mechanisms of diversity. *Acta Zoologica Fennica*, 198, 1–61.

Jernvall, J., Hunter, J. P. and Fortelius, M. (1996). Molar tooth diversity, disparity and ecology in Cenozoic ungulate radiations. *Science*, 274, 1489–1492.

Knobloch, E., Kvacek, Z., Buzek, C., Mai, D. H. and Batten, D. J. (1993). Evolutionary significance of floristic changes in the Northern Hemisphere. *Review of Palaeobotany Palynology*, 78, 41–54.

Knoll, A. H. (1986). Patterns of change in plant communities through geological time. In *Community Ecology*, eds. J. Diamond and T. J. Case, pp. 126–141. New York: Harper and Row.

Krause, D. W. and Maas, M. C. (1990). The biogeographic origins of late Paleocene–early Eocene mammalian immigrants to the Western Interior of North America. *Geological Society of America, Special Paper*, 243, 71–105.

Lander, E. B. (1977). A review of the Oreodonta (Mammalia, Artiodactyla), parts I, II and III. PhD dissertation. University of California, Berkeley.

Lucas, P. W., Darvell, B. W., Lee, P. K. D., Yuen, T. D. B. and Choong, M. F. (1995). The toughness of plant cell walls. *Philosophical Transactions of the Royal Society of London Series B*, 348, 363–372.

Lucas, S. G. and Williamson, T. E. (1995). Systematic position and biochronological significance of *Yuodon* and *Palasiodon*, supposed Paleocene 'condylarths' from China. *Neues Jahrbuch für Geologie und Paläontologie Abhandlungen*, 196, 9 3–107.

Maas, M. C. and Krause, D. W. (1994). Mammalian turnover and community structure in the Paleocene of North America. *Historical Biology*, 8, 91–128.

MacFadden, B. J. and Cerling, T. E. (1994). Fossil horses, carbon isotopes and global change. *Trends in Ecology and Evolution*, 9, 481–486.

MacFadden, B. J. and Hulbert, R. C. Jr. (1988). Explosive speciation at the base of the adaptive radiation of Miocene grazing horses. *Nature*, 336, 466–468.

McKenna, M. C., Chow, M., Ting, S. and Luo, Z. (1989). *Radinskya yupingae*, a perissodactyl-like mammal from the late Paleocene of China. In *The Evolution of Perissodactyls*, eds. D. R. Prothero and R. M. Schoch, pp. 24–36. Oxford: Clarendon Press.

Patzkowsky, M. E. (1995). A hierarchical branching model of evolutionary radiations. *Paleobiology*, 21, 440–460.

Potts, R. and Behrensmeyer, A. K. (1992). Late Cenozoic terrestrial ecosystems. In *Terrestrial Ecosystems Through Time*, eds. A. K. Behrensmeyer, J. D. Damuth, W. A. DiMichele, R. Potts, H. Sues and S. L. Wing, pp. 419–541. Chicago: University of Chicago Press.

Prothero, D. R. (1994). *The Eocene–Oligocene Transition: Paradise Lost*. New York: Columbia University Press.

Prothero, D. R. (1995). Geochronology and magnetostratigraphy of Paleogene North American land mammal 'ages': an update. In *Geochronology, Time Scales and Global Stratigraphic Correlation*, eds. W. A. Berggren, D. V. Kent, M.-P. Aubry and J. Hardenbol, pp. 305–315. Tulsa: SEPM (Society for Sedimentary Geology).

Prothero, D. R., Manning, E. M. and Fischer, M. (1988). The phylogeny of the ungulates. In *The Phylogeny and Classification of the Tetrapods*, ed. M. J. Benton, pp. 201–234. Oxford: Clarendon Press.

Radinsky, L. B. (1966). The adaptive radiation of the phenacodontid condylarths and the origin of the Perissodactyla. *Evolution*, 20, 408–417.

Rasmussen, D. T., Bown, T. M. and Simons, E. L. (1992). The Eocene–Oligocene transition in continental Africa. In *Eocene–Oligocene Climatic and Biotic Evolution*; eds. D. R. Prothero and

W. A. Berggren, pp. 548–566. Princeton: Princeton University Press.

Raymo, M. E. and Ruddiman, W. F. (1992). Tectonic forcing of late Cenozoic climate. *Nature*, **359**, 117–122.

Rose, K. D. (1996). On the origin of the order Artiodactyla. *Proceedings of the National Academy of Sciences, USA*, **93**, 1705–1709.

Rosenzweig, M. L. (1995). *Species Diversity in Space and Time*. Cambridge: Cambridge University Press.

Savage, D. E. and Russell, D. E. (1983). *Mammalian Paleofaunas of the World*. Reading, MA: Addison-Wesley.

Schluter, D. (1996). Ecological causes of adaptive radiation. *American Naturalist*, **148**, S40– S64.

Schluter, D. and McPhail, J. D. (1993). Character displacement and replicate adaptive radiation. *Trends in Ecology and Evolution*, **8**, 197–200.

Simpson, G. G. (1953). *The Major Features of Evolution*. New York: Columbia University Press.

Stucky, R. K. (1990). Evolution of land mammal diversity in North America during the Cenozoic. In *Current Mammalogy*, ed. H. H. Genoways, pp. 375–429. New York: Plenum Press.

Stucky, R. K. (1992). Mammalian faunas in North America of Bridgerian to early Arikareean 'ages' (Eocene and Oligocene). In *Eocene–Oligocene Climatic and Biotic Evolution*, eds. D. R. Prothero and W. A. Berggren, pp. 464–493. Princeton: Princeton University Press.

Stucky, R. K. (1995). Problems and speculations in mammalian paleoecology and evolution: western North America during the Paleogene. In *Proceedings of the Third Annual Fossils of Arizona Symposium*, eds. D. Boaz, S. Bolander, P. Dierking, M. Dornan and B. J. Tegowski, pp. 97–103 Mesa: Mesa Southwest Museum.

Stucky, R. K. and McKenna, M. C. (1993). Mammalia. In *The Fossil Record 2*, ed. M. J. Benton, pp. 739–771. London: Chapman and Hall.

Teaford, M. F. (1994). Dental microwear and dental fuction. *Evolution and Anthropology*, **3**, 17–30.

Teaford, M. F., Maas, M. C. and Simons, E. L. (1996). Dental microwear and microstructure in early Oligocene primates from the Fayum, Egypt: implications for diet. *American Journal of Physical Anthropology*, **101**, 527–543.

Thewissen, J. G. M. and Domning, D. P. (1992). The role of phenacodontids in the origin of the modern orders of ungulate mammals. *Journal of Vertebrate Paleontology*, **12**, 494– 504.

Upchurch, G. R., Jr. and Wolfe, J. A. (1987). Mid-Cretaceous to early Tertiary vegetation and climate: evidence from fossil leaves and woods. In *The Origin of Angiosperms and their Biological Consequences*, eds. E. M. Friis, W. G. Chaloner and P. R. Crane, pp. 75–105. Cambridge: Cambridge University Press.

Van Valen, L. M. (1971a). Adaptive zones and the orders of mammals. *Evolution*, **25**, 420–428.

Van Valen, L. M. (1971b). Toward the origin of the artiodactyls. *Evolution*, **25**, 523–529.

Van Valen, L. M. (1973). Pattern and the balance of nature. *Evolutionary Theory*, **1**, 31–49.

Van Valen, L. M. (1974a). Multivariate structural statistics in natural history. *Journal of Theoretical Biology*, **45**, 235–247.

Van Valen, L. M. (1974b). Two modes of evolution. *Nature*, **252**, 298–300.

Van Valen, L. M. (1978). The beginning of the age of mammals. *Evolutionary Theory*, **4**, 45–80.

Van Valen, L. M. (1985). Why and how do mammals evolve unusually rapidly? *Evolutionary Theory*, **7**, 127–132.

Van Valen, L. M. (1991). Biotal evolution: a manifesto. *Evolutionary Theory*, **10**, 1–13.

Van Valkenburgh, B. and Janis, C. M. (1993). Historical diversity patterns in North American large herbivores and carnivores. In *Species Diversity in Ecological Communities: Historical and Geographical Perspectives*, eds. R. E. Ricklefs and D. Schluter, pp. 330– 340. Chicago: The University of Chicago Press.

Vaughn, T. A. (1986). *Mammalogy*. Fort Worth: Saunders College Publishing.

Webb, S. D. (1977). A history of savannah vertebrates in the New World. I. North America. *Annual Review of Ecology and Systematics*, **8**, 355–380.

Wing, S. L. and Greenwood, D. R. (1993). Fossils and fossil climate: The case for equable continental interiors in the Eocene. *Philosophical Transactions of the Royal Society of London, Series B*, **341**, 243–252.

Wing, S. L., Alroy, J. and Hickey, L. J. (1995). Plant and mammal diversity in the Paleocene to Early Eocene of the Bighorn Basin. *Palaeogeography, Palaeoclimatology, Palaeoecology*, **115**, 117–155.

Woodbourne, M. O. (1987). *Cenozoic Mammals of North America: Geochronology and Biostratigraphy*. Berkeley: University of California Press.

Young, W. G. and Robson, S. K. (1987). Jaw movements from microwear on the molar teeth of the koala *Phascolarctos cinereus*. *Journal of Zoology, London*, **213**, 51–61.

20 Function of postcanine tooth crown shape in mammals

P. W. Lucas and C. R. Peters

20.1. Introduction

The postcanine tooth crowns of mammals have complex shapes adapted for mastication. Mastication is a process that reduces the particle sizes of foods. Once particle sizes are sufficiently small for saliva to bind particles together into a bolus, mastication stops and the food is swallowed. The extra food surface area exposed by particle size reduction enables a mammal to meet its high metabolic requirements by more rapid enzymic digestion than if a higher intake of large particles were more slowly digested. Mastication is an essential process if mammals are to have locomotor stamina because it allows them to be lightly built.

The repeated fracture of food particles by mastication is a mechanical process. The rate at which it takes place depends, of course, on the chewing rate of any given mammal. However, for a given chewing rate, the rate of breakdown depends on the physical characteristics of food particles, e.g. the size and shape of particles and, most particularly, on their material properties. The shapes and sizes of postcanine teeth in different mammalian species reflect adaptations to differing physical properties of foods in their diets. The tailoring of tooth form to diet allows mammals to meet their energy requirements.

For about 15 years, there has been an analysis available of mastication that allows a distinction to be made between two aspects of the process: on the one hand, the probability of food particles being fractured, and on the other, the number and size of fragments formed when any particle actually fractures. In this chapter, we will distinguish between:

(1)   *The probability that a particle has of being fractured*

within the mouth. Despite the abilities of the tongue to organise the intra-oral transport of food particles and the extensive sensory feedback that is available, food particles in the mouth are not certain to be fractured in any given chew. The greater the volume of food particles in the mouth (the mouthful) and the smaller their sizes, the lower their individual probabilities of being fractured. The approximate measure of this probability per chew is termed the *selection function*. There are other features of foods than the mouthful and particle size that influence the value of the selection function, such as particle shape, the stickiness of particles and their abrasiveness, but they are united by being characteristics of the food surface. The dental variable that adapts to varying surface features of foods so as to optimise the rate of food breakdown is the description of the surface of the teeth, which we will refer to in general as *tooth size*.

(2)   *The size distribution of the fragments of particles that are broken.* When defined per chew, the measurement of this is called the *breakage function*. Some foods may break down into a minimum of only two fragments; others may break into many. The characteristics of food particles that influence their fragmentation are their material properties which act variably to resist the formation of new food surface by cracking. The dental variable that adapts to try to produce the maximum number of fragments from each particle is *tooth shape*, in particular, the shape of the tooth cusps – the working (occlusal) surface of the tooth crown.

This chapter restricts its discussion to the breakage

function and, therefore, to tooth crown shape. Until recently, all analyses of this subject have been qualitative. The most general analysis, based on the ease of crack propagation within food particles, is that of Lucas (1979). He proposed that soft tough foods probably require sharp-bladed teeth to break them down efficiently. The sharpness of the blade concentrates stress in the food, which is otherwise wrapping itself around the blade's edges. Even so, cracks in these foods will not travel far ahead of the blade because their toughness resists this. A blade needs to have a length greater than the size of any food particle, otherwise there is no separation of fragments and the blade simply forms a cleft in the food particle (this is why a sharp pointed tooth will not do). A long enough blade subdivides the particle into two. A more broken food product, that provides more than two fragments when it is fractured, requires more blades acting simultaneously. This is, however, a feature of tooth size and, therefore, the selection function and so beyond the scope of this chapter. Fracture in the mouth involves reciprocal features on teeth mounted in upper and lower jaws. The only logical pairing is that of two blades, oriented such that they pass each other during the final part of jaw closure. This analysis implies the working features of a pair of scissors.

In contrast, hard (stiff) brittle (i.e. not tough) foods are probably most efficiently comminuted by blunt pointed teeth. Cracks in such foods propagate rapidly well in advance of the tooth part that initiates the fracture and so blades are unnecessary. If instead of blunt cusps, cusps were sharp, the small contact areas between two such stiff materials might cause fracture of the cusp because of the stress concentrations involved. Also, acting as an indenter, the sharper the cusp is, the higher the peak stresses in the food (Lawn and Wilshaw, 1975). This encourages plastic behaviour which will actually tend to stop cracking because plasticity is the most potent toughening (i.e. fracture-resisting) mechanism available. So, a blunt cusp would be best. There are two possible ways in which blunt cusped teeth could be arranged in upper and lower jaws so that they act reciprocally. Together with another blunt cusp (usually two cusps in one jaw and one in the other work together), there is the potential for only limited size reduction. A greater capacity involves a cusp fitting loosely, to avoid friction, into a basin (a 'pestle and mortar' system). An uppermost size limit is defined by the diameter of the basin but, below this size, the bluntness of the cusp has another benefit. The rapidity of cracking in these foods is normally associated with multiple fragments forming well ahead of the cusp. Later, during the same jaw closure (the same chew), these fragments may be reduced in size further if they lie in the basin close to the cusp tip (Lucas and Luke, 1984).

For any field worker face-to-face (hopefully) with a mammal, or for a museum worker looking at an array of jaws and teeth from living and fossil forms, the above has to appear very limited and unspecific. Would the physical properties of the diet of a range of mammals help explain their dental morphologies? Will further investigation of the actual behaviour of foods under load help a field worker to understand the feeding behaviour of a particular mammalian species? We argue yes in both cases. To that end, this chapter addresses itself to the following concerns:

1. What exactly are foods with either soft tough or hard brittle behaviours and can such properties be measured?

2. Does the information obtained from this cast any more light on why the postcanine teeth of some mammals are sharp while others are blunt?

3. What can we say from this about the human dentition and diet and our recent evolution?

20.2.   **The material properties of foods**

If material properties of foods act variably to obstruct particle size reduction, then knowledge of these material properties is clearly a necessary prerequisite for the analysis of the dentition. There are three main properties: *strength, toughness* and *Young's modulus*.

Strength is the critical stress (stress being the force normalised to the cross-sectional area over which it is acting) at which some aspect of the food structure fails. This failure may be defined as the onset of permanent deformation (the yield strength) or the initiation of a crack (the fracture stress). Once a structure has cracked, we need to define the resistance to the crack growing. *Toughness* is the property that does this and is the work required to generate the crack area formed. Toughness is the resistance to failure after the fracture stress has been achieved, i.e. while a crack is propagating. Crack growth is not associated with any particular (non-zero) stress but is controlled energetically – hence toughness is defined in terms of requisite work or energy. The *Young's (or elastic) modulus* is the ratio of stress to strain, the latter being the change in proportions in the linear dimension parallel to the stress. It defines the deformability of a material at low strains when the stress-strain curve is (usually) linear. Both strength and Young's modulus have the same units because strain is dimensionless. The Pascal is a small unit equal to 1 Nm^{-2}. Strength has units of kiloPascals (kPa) or MegaPascals (MPa) while Young's modulus is either in MPa or Gigapascals (GPa). Toughness is measured in Joules per square metre (J m^{-2}) or kJ m^{-2}.

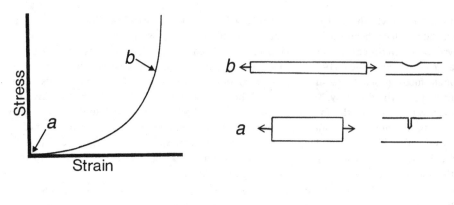

Figure 20.1A. **The stress–strain curve in tension for a highly extensible material such as mammalian skin. The curve resembles the letter 'J'. B. The appearance of the central section of a specimen at positions *a* (lower diagram) and *b* (upper diagram) on the curve. At *b*, the material has not only elongated greatly, it has also reduced its lateral dimensions to a similar degree. C. The appearance of the central section of a notched specimen at the same strains as in B. The notch is soon blunted. It is reversible on unloading and greatly increases the toughness of the tissue.**

The resistance that foods have to being fractured during chewing would appear to be captured simply by a measure of fracture resistance, such as strength. However, strength only measures the initial point of failure of a material. As such, strength is perfectly sufficient for many engineering applications; for example, a buckled, plastically deformed or cracked structural element of a tall building has failed and it is the onset of such events, which strength defines, that needs to be designed against. The engineer does not generally need to design for or against an extremely large primate wading among such buildings, breaking them apart with its teeth. However, suspending judgement for a moment, suppose engineers did have to design against a chewed (fragmented) building. They would then certainly have to take into account the propagation of cracks through its structural elements. The fracture stress does not describe resistance to this crack growth. The fracture stress is, in fact, size dependent – on the size of the building, that is, not on the size of the primate – and is not a fundamental property of the material of the building at all (Atkins and Mai, 1985). This can be a critical issue in situations where the size of chewed buildings varies enormously. Going back now to primates and their regular diet, chewing is a comminution process that can easily generate several orders of magnitude of particle size in one chewing sequence. So, if we are to characterise the process properly, we will want to measure mechanical parameters that control this process, not those that are affected by it.

From fracture mechanics, we know that the stress at which cracks grow from small flaws involves toughness (Atkins & Mai, 1985). Toughness appears, at first sight, to describe crack resistance with satisfaction but, in fact, it only begins to do so. As shown clearly by Ashby (1989),

the resistance of any structure to propagating cracks is also a function of its deformability (Young's or elastic modulus) *if* either the displacement through which the element must fracture is limited or else the strain at which the group of materials break is stereotyped. Then, the controlling parameter is the square root of the ratio of toughness to Young's modulus (Sharp *et al.*, 1993). There are some difficulties in applying this, particularly with foods that have non-linear stress-strain curves (such as the material shown in Figure 20.1), but there is already reason for confidence that this ratio, which we call the *fragmentation index*, does indeed apply to mastication: Agrawal *et al.* (1997) describe a strong negative correlation ($r^2 = 0.77$) between this index and the degree of fragmentation produced by human subjects on 28 different foods (nut, meats, cheeses and raw vegetables).

The analysis of Ashby (1989) assumes that loading by the teeth initially stores elastic strain energy in food particles. This strain energy then later pays for crack growth. However, as food particles become smaller during mastication, or are ingested in the form of thin sheets (such as leaves, for example), the work done by the teeth on the particles becomes used directly to produce cracks with negligible intervening energy storage. For many plant foods, the ease of breakdown of particles smaller than about 1 mm in thickness may depend purely on their toughness, not on their fragmentation index (Lucas *et al.*, 1995, 1997). Despite these complications, this information relates very simply to the qualitative model of Lucas (1979). Soft tough foods, e.g. ratskin, are usually those with a very low Young's modulus but high toughness while hard brittle foods (e.g. dental enamel, bone and seed shells) exhibit the converse: high Young's modulus but low toughness (see Lucas and Teaford, 1994, for a recent survey).

20.3.   Measurement of material properties of foods

It stands to reason that if we could measure the material properties of the foods of mammals in the field, we could discover whether mammals generally consume foods characterised by a particular range of the fragmentation index or toughness or, alternatively, have a dietary range circumscribed by certain limiting values of these parameters. It does not follow, however, that though the rate at which food breaks down is controlled by these indices that these are the sole criteria for the adaptation of tooth crown. Selection pressures on cusp height, for example, may depend more on the ratio of strength to Young's modulus (Lucas and Peters, in preparation).

Field tests are now possible, and toughness, strength and Young's modulus can be measured on small portable testers (e.g. Darvell et al., 1996). A general inventory of tests which give comparable results is provided in Vincent (1992). Toughness is usually obtained by wedging or scissoring. Young's modulus and strength could be measured in either tension or compression. Small testers have limitations on the loads that they can support. The stiffness of the tester is also critical; the specimen should deform, not the tester. These restrictions can often be overcome by making small enough specimens. Those foods that cannot be tested in the field are likely to be low in moisture content and will probably survive storage until a laboratory is reached when they can be tested in less trying circumstances. The remaining part of the field test kit that needs to be developed is, in fact, a consistent set of methods for preparing specimens. Softer foods are usually easy to shape with a scalpel, razor blade or a cork borer, though for many toughness tests (e.g. scissoring leaves or pods: Hill and Lucas, 1996; Lucas et al., 1991a, 1995), it may not be necessary for much preparation at all.

With due caution, it is now possible to construct texture maps (i.e. bivariate graphs depicting the mechanical properties that contribute towards the ease or difficulty of food breakdown; Lucas and Teaford, 1994) for mammalian species which can be used to provide insights into their dental morphology. In combination with behavioural data on the rate of acquisition and processing, this will help answer the question of just how sharp or how blunt tooth features have to be for a given diet. Prior to the acquisition of this data, we can make some fairly specific predictions. The manner by which foods are toughened will be important. It is tempting to view toughness in the same way as Young's modulus, as something immutable. Toughness is not like this, however. First, toughness is achieved by various mechanisms – elastic/plastic crack blunting, fibre (or fibril)

Table 20.1. *Properties of ratskin (kJ m^{-2})*

Trouser-tearing[a]	Scissoring[b]
14–20	0.59

[a] Purslow (1983).
[b] Pereira et al. (1997), with blades of 16 μm sharpness.

pullout and plastic buckling (Atkins & Mai, 1985) – which certain loadings can suppress, as we suggest in the next section. Second, real foods are heterogeneous and may have a very different toughness for alternative crack paths within them. The imposition of a particular crack path by a sharp dental feature may involve a much higher cost than if the crack were allowed to wander along the cheapest route through the material. Both these factors appear to be important in explaining the form of some dentitions. We use two examples to illustrate this.

20.3.1.   Foods of high fragmentation index: mammalian soft tissues

The most important and best investigated of mammalian soft tissues is their integument. The skin protects mammals by virtue of its 'J- shaped' stress-strain curve in tension (Figure 20.1). The curve has a very low initial slope which becomes steeper with increasing stress. At low strains, the behaviour of skin appears to be dominated by the extracellular connective tissue matrix, the ground substance (Gordon, 1978). As the strain increases collagen fibres start to be loaded, as a result of which the material starts to get much stiffer. Individual collagen fibres, or groups of them, may then begin to fail. However, the very low shear modulus of the extracellular matrix obstructs the communication of strain energy between these groups of fibres. The effective energetic isolation of these tensile elements, the collagen fibres, results in a notch-insensitive structure that is very resistant to crack propagation (Gordon, 1978, 1980). The toughness of ratskin in a standard trouser tear test, a test which probably mixes out-of-plane shear and tension fracture, is between 14 and 20 kJ m^{-2} (Table 20.1). The relatively high toughness value is produced largely by elastic crack-blunting. Whenever a solid object is cracked and loaded so as to open the crack up, the stress near the crack tip is going to be much higher than that remote from it. However, the low slope of the stress-strain curve at low strain relieves this sharpness by elastically stretching the tissue greatly while the stress is still low; thus, the stress at the crack tip becomes very little different from elsewhere in the tissue. This toughening mechanism is thwarted by fracture with a very sharp blade, because the skin gets no opportunity to stretch and thus

Table 20.2. *Plant toughness (kJ m^{-2})*

Between cells[a]	Across cell[b]	
	Cell wall	'Woody toughness'
0.35	3.5	35.0

[a] Gibson and Ashby (1988).
[b] Lucas *et al.* (1995, 1997).

absorb energy. The data in Table 20.1 for ratskin show that toughness is reduced by a factor of more than 20 when it is scissored. We argue that this is precisely why any vertebrate that breaks such tissues inside the mouth must have very sharp teeth. The sharpness subverts the major toughening mechanism and so effectively reduces the fragmentation index.

20.3.2.   Foods of low fragmentation index: seed shells

Seed shells are extremely dense woody tissues, denser than wood itself because theirs is a purely protective function, not compromised by any other need such as water transport. Wood itself is a prodigiously tough material which would seem, from common experience, to require a bladed type of instrument to break down. Why then are seed eaters that eat thick-shelled seeds almost always depicted/predicted to have blunt cusped molars? The answer appears to lie in the discrepancy between high toughness across fibres versus low toughness between them and in the arrangement of the fibres in seed shells.

All plant tissues are cellular in a mechanical sense in that they have multiple air/fluid-filled spaces surrounded by solid cell walls (Gibson & Ashby, 1988). The toughness of non-woody tissue is affected principally by the fraction of tissue volume that is occupied by cell wall in the tissue (though wall composition is also important). The data available suggest the wall itself may have an intrinsic toughness of about 3.5 kJ m^{-2} (Table 20.2). It is relatively easy to calculate toughness of most plant tissues with isodiametric cells from this normalising factor. For example, watermelon flesh, with 0.3% cell wall by volume as determined by histological measurements, has a predicted – and actual – toughness of about 10 J m^{-2} (Lucas *et al.*, 1995). A simpler alternative to microscopic assessment of cell wall content is to measure fibre content. To do this, fibre content should be expressed as a proportion of the wet weight of the tissue and compensated for the difference in density between moisture and cell wall. Neutral detergent fibre is the best measure of fibre content for this purpose because this includes all components of the cell wall (Choong, 1996; Hill and Lucas, 1996).

The Young's modulus of *non-woody* plant tissues depends on turgor pressure. There is no adequate mathe-

matical relation for this dependence at present. However, even if we cannot put it on a firm quantitative basis, it is clear that turgid cells are very much stiffer than flaccid cells. Toughness does not seem to depend on turgor to anything like the same extent as the modulus, and so the fragmentation index of flaccid cells is very much higher than for turgid cells, i.e. flaccid tissues will be much more resistant to breaking down in the mouth (even without considering that they may crack-blunt in the human mouth as described for mammalian soft tissues).

Woody tissues, by contrast, possess a toughness that is well out of proportion to their fibre content. For example, calculating in the same way as for watermelon flesh, the across-the-grain (i.e. across the cell wall) toughness of a sample of teak wood with about 55% cell wall by volume would be predicted to be 1.925 kJ m^{-2}. However, this is an order of magnitude lower than experimental results (e.g. Jeronimidis, 1980). The extra work comes not from a difference in cell wall composition, as is sometimes thought (e.g. lignification), but from a preferred orientation of the cellulose. Helical windings of cellulosic microfibrils in the S2 layer of the secondary cell wall, the thickest layer, of wood fibres cause them to buckle plastically when they are loaded (Gordon and Jeronimidis, 1974, 1980; Jeronimidis, 1980). Perpendicular to the long axis of the cell, this can provide a potential toughness of up to 35 kJ m^{-2} (Table 20.2), if a wood cell were completely solid (Lucas *et al.*, 1997). This is 10 times that possible in non-woody tissues. The alignment of the cellulose causes anisotrophy of the Young's modulus of the cell wall. This is about 35 GPa along the long axis of cells but only 15 GPa across it (Gibson and Ashby, 1988). The result though is a highly crack-resistant structure.

Fibrous seed shells exhibit woody toughness. As shown in Figure 20.2, the toughness of their wall (the intrinsic toughness) is about that expected from their high density (Figure 20.2b). When cracks are forced across cells, they display a lot of plastic work which is expended in buckling (Figure 20.2a). The work involved is slightly below the value for the highest woods but this can be explained by a very narrow cellular lumen (due to their high density) which obstructs buckling.

The toughness of the outer zone of a *Mezzettia* shell directly across fibres as tested with scissors is about 25 kJ m^{-2}. That for *Macadamia* or *Schinziophyton* shells is about half of this value because their fibres are arranged randomly in the shell. If the shell were to be broken with bladed teeth, this is what fracture would cost. However, the toughness of these seed shells with free-running cracks, such as would be produced by blunt cusped teeth, is very much lower, about 1–2 kJ m^{-2} (Jennings and Macmillan, 1986; Lucas *et al.*, 1991b; Williamson and

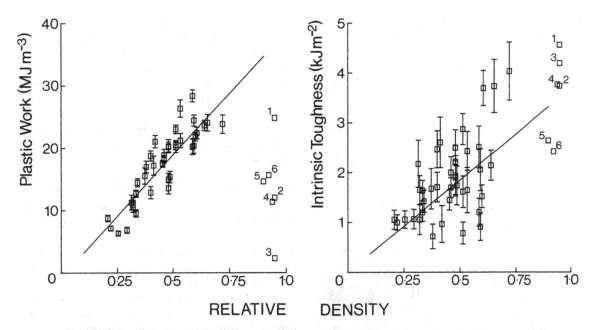

Figure 20.2A,B. **The toughness of seed shells (indicated by numbers) and woods (mean ± S.E.) in relation to their relative densities (i.e. the volume fraction of tissue that is cell wall). A. The ability of tissues to display plastic deformation is a function both of cell wall structure and cellularity: the wall buckles and collapses into the lumen. Seed shells show less ability to buckle plastically because they are so dense, they have insufficient lumen. B. shows intrinsic toughness, a function purely of cell wall composition and structure. Woods and seed shells all lie roughly on the same regression line. When relative density is greater than 0.1–0.2, a free-running crack will deflect between cells in seed shells because the energy required to do this is less than across cells (see Table 20.2). This is the major reason why seed shells propagate cracks so rapidly. Key to seed shells: 1, 2 and 3 are three zones in the *Mezzettia parviflora* seed shell (Annonaceae): 1 is cracking at right-angles to fibres; 2 is cracking across randomly directed fibres while 3 is cracking across isodiametric cells of the same relative density as 1 and 2 but without much woody toughness (i.e. little plastic work). 4 is *Schinziophyton rautenii* (Anacardiaceae) with randomly-directed fibres and toughness similar to 2. 5 and 6 are cracking across fibres of *Scheelea* sp. (Palmae) and *Aleurites moluccana* (Euphorbiaceae) shells.**

Lucas, 1995). The reason is that free-running cracks naturally follow the cheapest path that they can, given the direction of loading. As far as possible, they track between fibres. This seems to cost about 0.35 kJ m^{-2} in any woody tissue (Table 20.2), though it can rise to several times this value depending on the geometry of loading and the entanglement of fibres. When fibres are crossed by cracks, it is only at their weakest points. The random ordering of fibres in a seed shell does not help much. Toughness is low, with that of *Schinziophyton rautenii* seed shell is about 1 kJ m^{-2} (Williamson and Lucas, 1995) while that of *Mezzettia parviflora* is about double this (Lucas *et al.*, 1991*b*).

At the cost of a large force at initial loading, the fracture of these seed shells is an order of magnitude cheaper, and much faster, with a blunt loading regime than with blades. The cost of fracturing seed shells (and possibly also mollusc shells and bone) is greatly reduced by the use of blunt teeth. There is, however, the qualification that fast cheap fracture of seed shells comes at the cost of high initial loads. For example, orang-utans may have to exert between 1 and 2 kN in order to indent *Macadamia* or *Mezzettia* seed shells so as to achieve fast fracture (Lucas *et al.*, 1994). Smaller mammals are not

capable of exerting these bite forces. Are such foods unavailable to them? Not in principle, because we know that rodents often eat these resources. Rats, for example, with very small maximum bite forces are major predators of the *Macadamia* crop in Australia. How do they do open these seeds to get at the tissue inside? The answer is that they use their sharp bladed incisor teeth, constantly sacrificing tooth tissue, to open the shells by crossing the material slowly in almost its toughest direction. The only energetic respite that they can get is to orientate bites at a very shallow angle to fibres (Figure 20.3). Woody fibres are toughest at right-angles to their long axis but toughness falls off rapidly as the sine of the angle decreases. At about five degrees to the cellular axis, plastic buckling is almost completely suppressed and the toughness falls to the intrinsic toughness due to the amount of cell wall present. The randomly oriented fibres of many seed shells may foil such a loading but rodents that gnaw through the wood of tree trunks may benefit from this strategy. It is very difficult to generalise about the evolution of the rodent incisor given the wide variety of diets seen in modern forms (Landry, 1970), the tooth form seems to have great value for coping with woody tissues when available bite forces are small.

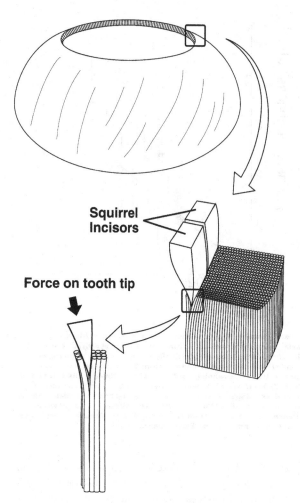

Squirrel Incisors

Force on tooth tip

Figure 20.3. **The presumed mode of entry of a squirrel (*Callosciureus* sp.) into an *Aleurites* shell, as assessed from observation of feeding and seed damage. The initial loading of the seed shell requires that the incisors of the squirrel be sharp so that fracture will be predominantly between fibres (very cheap) and not across them. (Nailing of wood follows very similar principles: Atkins and Mai, 1985). However, microfractures of the tip of the incisors must be expected. During propagation of the crack through the shell, it is likely that the teeth act as wedges between the fibres. The very shallow angle of attack by the incisors to the long axis of the fibres ensures that even if the latter are damaged, expensive plastic buckling will be minimised.**

20.4.   The diet and dentition of the modern human

As yet, we do not know the ideal design for the dentition of a large-bodied omnivore. It could hinge upon food resources that are critical at certain times of the year or on those dietary items that are eaten most frequently. The unaided dentition of humans seems able to cope efficiently only with foods of low fragmentation index (i.e. low toughness and/or high Young's modulus). In

Homo sapiens, extra-oral processing has allowed an emphasis on selecting a great variety of high-quality foods that are outside the range of dental quality. The increased dietary breadth of modern humans is associated with the recent evolution of tools and cooking. These technological developments have permitted the mastication of food objects that are otherwise beyond the capacity of the hominid dentition, by modifying them before ingestion. The carnivorous behaviour of early *Homo* is associated with the use of stone flakes to help, for example, to de-flesh antelope carcasses, fracturing the skin and reducing the lean meat to small pieces. Early *Homo* could not otherwise eat mammalian soft tissues very efficiently because the blunt cusped postcanine teeth would face the full toughness of the softest tissues due to the elastic crack-blunting described above. Once in the mouth, small pieces of meat may not be reduced in size further before being swallowed.

In contrast, our ancient near-relatives, the robust australopithecines of the Plio-Pleistocene, exhibited a massive heavily-molarised, postcanine dentition. Hypothetical diets for these hominids might range from carnivory to specialised herbivorous diets, such as seeds, to more generalised herbivory, to omnivory. The relatively blunt cusps of their unworn postcanine teeth and their apparent capacity for producing high bite forces suggests a diet of foods of low fragmentation index – very likely, thick-shelled seeds (Jolly, 1970; Peters, 1981, 1993) and/or hard fruits (Walker, 1981).

Cultural developments have produced a mismatch between the diet of modern humans and our postcanine dentition. The increased dietary breadth of modern humans is associated with the evolution of high-energy technologies. Although our postcanine teeth are similar in basic shape to those of early *Homo* (although smaller), these cultural developments probably act to reduce the fragmentation index or toughness, or both, of foods to make these break down faster in the mouth. The result is something akin, in a broad sense, to 'baby food'.

20.5.   Conclusion

Most mammals eat a wide variety of foods. Therefore, it is going to take quite some time to obtain enough information on the properties of foods to test and refine our hypotheses. It is going to take even longer to understand particular features of food behaviour in relation to the detailed morphology of dentitions. We hope, however, that the above discussion shows clearly why at least some postcanine dentitions are sharp, but others very blunt. To break mammalian soft tissues, it pays for tooth features to be very sharp. To break seed shells, it pays for tooth features to be very blunt. Nothing could highlight better the advantages of diversity in tooth types.

Acknowledgements

This research was supported by the Research Grants Council of Hong Kong to PWL and CRP and by the CRCG of the University of Hong Kong.

References

Agrawal, K. R., Lucas, P. W., Prinz, J. F. and Bruce, I. C. (1997). Mechanical properties of foods responsible for resisting their breakdown in the human mouth. *Archives of Oral Biology*, **42**, 1–9.

Ashby, M. F. (1989). Overview no. 80. On the engineering properties of materials. *Acta Metallurgica*, **37**, 1273–1293.

Atkins, A. G. and Mai, Y. W. (1985). *Elastic and Plastic Fracture*. Chichester: Ellis Horwood.

Choong, M. F. (1996). What makes a *Castanopsis fissa* leaf tough and how this affects herbivore behaviour. *Functional Ecology*, **10**, 668–674.

Darvell, B. W., Lee, P. K. D., Yuen, T. D. B. and Lucas, P. W. (1996) A portable fracture toughness tester for biological materials. *Measurement Science and Technology*, **7**, 954–962.

Gibson, L. J. and Ashby, M. F. (1988). *Cellular Solids*. Oxford: Pergamon Press.

Gordon, J. E. (1978). *Structures*. London: Penguin.

Gordon, J. E. (1980). The last stronghold of vitalism. In *The Mechanical Properties of Biological Materials*, eds. J. F. V. Vincent and J. D. Currey, pp. 1–11. Cambridge: Cambridge University Press.

Gordon, J. E. and Jeronimidis, G. (1974). Work of fracture of natural cellulose. *Nature* **252**, 116.

Gordon, J. E. and Jeronimidis, G. (1980). Composites with high work of fracture. *Philosophical Transactions of the Royal Society of London, Series A*, **294**, 545–550.

Hill, D. A. and Lucas, P. W. (1996). Toughness and fibre content of major leaf foods of Japanese macaques (*Macaca fuscata yakui*) in Yakushima. *American Journal of Primatology*, **38**, 221–231.

Jennings, J. S. and Macmillan, N. H. (1986). A tough nut to crack. *Journal of Materials Science*, **21**, 1517–1524.

Jeronimidis, G. (1980). The fracture behaviour of wood and the relations between toughness and morphology. *Philosophical Transactions of the Royal Society of London, Series B*, **208**, 447–460.

Jolly, C. J. (1970). The seed-eaters: a new model of hominid differentiation. *Man* (n.s.), **5**, 1–26.

Landry, S. O. (1970). The Rodentia as omnivores. *Quarterly Review of Biology*, **45**, 351–372.

Lawn, B. and Wilshaw, T. (1975) Indentation fracture: principles and applications. *Journal of Materials Science*, **10**, 45–51.

Lucas, P. W. (1979). The dental-dietary adaptations of mammals. *Neues Jahrbuch für Geologie und Paläontologie* **8**, 486–512.

Lucas, P. W. and Luke, D. A. (1984). Chewing it over – basic principles of food breakdown. In *Food Acquisition and Processing in Primates*, eds. D. J. Chivers, B. A. Wood and A. Bilsborough, pp. 283–302. New York: Plenum Press.

Lucas, P. W. and Teaford, M. F. (1994). Functional morphology of colobine teeth. In *Colobine Monkeys: their Ecology, Behaviour and Evolution*, eds. A. G. Davies and J. F. Oates, pp. 171–203. Cambridge: Cambridge University Press.

Lucas, P. W., Choong, M. F., Tan, H. T. W., Turner, I. M. and Berrick, A. J. (1991a). Fracture toughness of the leaf of the dicotyledonous angiosperm, *Calophyllum inophyllum* L. *Philosophical Transactions of the Royal Society of London, Series B* **334**, 95–106.

Lucas, P. W., Lowrey, T. K., Pereira, B., Sarafis, V. and Kuhn, W. (1991b). The ecology of *Mezzettia leptopoda* Hk. f. et Thoms. (Annonaceae) seeds as viewed from a mechanical perspective. *Functional Ecology*, **5**, 345–353.

Lucas, P. W., Peters, C. R. and Arrandale, S. (1994). Seed-breaking forces exerted by orang-utans with their teeth in captivity and a new technique for estimating forces produced in the wild. *American Journal of Physical Anthropology*, **94**, 365–378.

Lucas, P. W., Darvell, B. W., Lee, P. K. D., Yuen, T. D. B. and Choong, M. F. (1995). The toughness of plant cell walls. *Philosophical Transactions of the Royal Society of London Series B*, **348**, 363–372.

Lucas, P. W., Tan, H. T. W. and Cheng, P. Y. (1997). The toughness of secondary cell wall and woody tissue. *Philosophical Transactions of the Royal Society of London, Series B*, **352**, 341–352.

Pereira, B. P., Lucas, P. W. and Teoh, S. H. (1997) Ranking the fracture toughness of mammalian soft tissues using the scissors cutting test. *Journal of Biomechanics*, **30**, 91–94.

Peters, C. R. (1981). Gracile vs. robust early-hominid masticatory capabilities: the advantages of the megadonts. In *The Perception of Evolution: Essays Honoring Joseph B. Birdsell*, eds. L. L. Mai, E. Shanklin and R. W. Sussman, vol. 7, pp. 161–181. Los Angeles: Anthropology UCLA.

Peters, C. R. (1993). Shell strength and primate seed predation of nontoxic species in eastern and southern Africa. *International Journal of Primatology*, **14**, 315–344.

Purslow, P. P. (1983). Measurement of the fracture toughness of extensible connective tissues. *Journal of Materials Science*, **18**, 3591–3598.

Sharp, S. J., Ashby, M. F. and Fleck, N. A. (1993). Material response under static and sliding indentation loads. *Acta Metallurgica et Materialia*, **41**, 685–692.

Vincent, J. F. V. (1992). *Biomechanics – Materials*. Oxford: IRL Press.

Walker, A. (1981). Diet and teeth: dietary hypotheses and human evolution. *Philosophical Transactions of the Royal Society of London, Series B*, **292**, 57–64.

Williamson, L. and Lucas, P. W. (1995). The effect of moisture content on the mechanical properties of a seed shell. *Journal of Materials Science*, **30**, 162–166.

21 Primate dental functional morphology revisited

M. F. Teaford

We will never understand organisms without an integrative approach Schmidt-Kittler and Vogel, 1990

21.1. Introduction

It is no secret that over the past 20 years there has been a veritable explosion of information generated by scientific research. One of the beneficiaries of this information explosion has been the study of morphology, where new techniques and analyses have led to new insights into a wide range of topics. Teeth have been no exception to this, as advances in genetics, histology, microstructure, biomechanics, microscopy, and morphometrics have allowed researchers to study teeth from new perspectives. For instance, advances in genetics and histology have given researchers a much clearer perspective on the embryological origins of tooth morphology, as we now move from correlations between multiple gene expression patterns (see Chapter 11) to an understanding of very specific cause-and-effect relationships between genes and tooth cusp formation (Chapters 1 and 2). As another example, work by Chris Dean, David Beynon, and others has shown that variations in dental microstructure can be related to periodic fluctuations in tooth development which can, in turn, lead to new insights into the growth and development of our ancestors (see Chapter 9).

This new range of material requires two changes in perspective by all parties involved. First, given the technical complexities of many of the new approaches, communication and collaboration are the keys to the future. In short, we *must* work together if we are to successfully harness the techniques available to us. Second, in doing so, we will have no choice but to broaden our perspectives of "dental functional morphology'. Researchers have often made the distinction between 'functional biology' and 'evolutionary biology', or 'how does it work?' and 'how did it come into being?' (Bock, 1988,

1990; Dullemeijer, 1980; Mayr, 1982). Other investigators (e.g., Wake, 1992) have subdivided the study of morphology into a variety of components, including ecomorphology (Wainwright, 1991; Wainwright and Reilly, 1993), biomechanics (Gans, 1974; Vogel, 1988), and transformation morphology (Galis, 1996). The present volume has given but a brief glimpse of the range of perspectives that can be brought to bear on the study of teeth. Obviously, every possible perspective could not be explored in one volume, but we still cannot help but broaden our perspectives of dental functional morphology: it isn't just the study of tooth shape anymore.

The purpose of this chapter is to provide a brief sketch of some of the future possibilities for the study of primate dental functional morphology. In short, 'What new questions might be asked of primate teeth; and how might microscopic and macroscopic evidence yield a better picture of how teeth work?'

21.2. Microscopic evidence

21.2.1. Introduction
Teeth have been the subject of study in many scientific disciplines ranging from paleontology to dentistry. In many of those disciplines, teeth have been treated as homogeneous structures. For example, studies of the 'evolution' of teeth, or the 'reconstruction' of teeth, have usually been presented in terms of cusps, fossae, and fissures, with no mention of microscopic underpinnings. The microscopic evidence presented in this volume (e.g., Chapters 18, 17) reiterates quite emphatically that teeth are complex, heterogeneous structures. Studies of dental functional morphology must never lose sight of that fact.

For instance, the physical properties of teeth or enamel are often presented in summary form ('the hardness of enamel is . . .'). But what is the functional significance of such measurements if dental enamel is composed of a complex arrangement of hydroxyapatite crystals that varies over the tooth surface? Similarly, if enamel and dentin are said to be susceptible to the effects of acids, how might that occur? Recent investigations are helping to answer these, and other questions. In the process, they are showing us new possibilities for future work.

21.2.2. Dental microstructure

Recent improvements in microscopy (e.g., environmental electron microscopy, confocal microscopy, tandem scanning reflected light microscopy) have fueled a rapid growth in our understanding of dental microstructure. Much of the recent work on nonhuman primates has focused on the development of enamel structure and the insights it might yield into patterns of growth and development in prehistoric primates (Bromage and Dean 1985; Beynon and Wood, 1987; Beynon and Dean, 1988). Essentially, if certain features in enamel, most notably cross-striations and perikymata, can be shown to have a specific developmental periodicity, then they can be 'counted' in fossil teeth to determine the amount of time necessary to form tooth crowns and entire dentitions (see Chapter 9).

Additional microscopic work on enamel has focused on enamel prism patterns and their potential to decipher phylogenetic relationships among species (Gantt et al., 1977; Shellis and Poole, 1977; Vrba and Grine, 1978; Boyde and Martin, 1982; Gantt, 1983; Martin et al., 1988). The complexity of enamel microstructure, however, has made interpretations of that work extremely difficult. However, researchers have now begun to use that complexity to their advantage to gain new insights into the relationship between enamel microstructure and dental function.

As emphasized by Koenigswald and Clemens (1992), enamel microstructure is best organized on a hierarchical scale, from the crystallite level on up to variations between teeth in the dentition. A key point that should now be obvious is that each step in that hierarchy is of functional importance and should fall under the rubric of 'dental functional morphology' As investigators begin to make intuitive/functional leaps between steps in that hierarchy, exciting possibilities for future work arise.

It has long been known that hydroxyapatite crystallites form the building-blocks of enamel, and that these crystallites are bundled together in mammalian teeth to form prisms (Boyde, 1964; Carlson, 1990). (See Chapter 7 for a discussion of the evolutionary origins of prismatic enamel). What many people in anthropology and paleon-

tology fail to realize, however, is that this leaves enamel as a very anisotropic materials. In other words, despite the fact that enamel is the hardest and stiffest material in the body (Waters, 1980), its properties vary depending on how they are measured relative to the organization of its component parts. For instance, as Rensberger has noted in Chapter 18, the resistance of enamel to abrasion varies depending on the arrangement of its prisms and crystallites relative to the direction of abrasion. This is presumably reflected in variations in prism/crystallite orientation over the surface of teeth (Rensberger and Koenigswald, 1980; Pfretzschner, 1988; Maas, 1993, 1994; Rensberger 1997; Chapter 18). However, researchers are only beginning to understand the subtle variations in material properties (such as hardness and deformability) that might be underlying these patterns.

For instance, studies of hardness have 'evolved' from summary characterizations of measurements for specific teeth (Craig and Peyton, 1958), to analyses of finer resolution where the influence of prism orientation can be taken into account (e.g., Willems et al., 1993; Xu et al., 1998). Still, the microhardness testers used in recent studies are 10–12 times the size of enamel prisms. Thus, they cannot reach the level of resolution necessary to measure the hardness of specific prisms, not to mention the smaller crystallites within them. Recent advances demonstrate that measurements of much finer resolution are possible (Kinney et al., 1996a, b; T. P. Weihs and A. B. Mann, personal communication). When done, this has the potential to document patterns of hardness and deformability throughout the tooth, leading to a quantum leap, not only in our understanding of the structure of teeth, but also in clinical implications of structural differences.

For instance, there may well be age-related changes in physical properties that could affect such things as wear-resistance, dental reconstruction, and caries susceptibility. If the mineral content of enamel increases as one moves from the dentin–enamel junction to the outer occlusal surface (Angmar et al., 1963; Theuns et al., 1983), would measures of hardness show a parallel increase? If so, as enamel is lost due to tooth wear, would the abrasive wear resistance of enamel decrease with age? For that matter, if there are interspecific differences in the secretory rate of ameloblasts (Chapter 9), might the pattern and degree of mineralization also vary between species?

Moving to fracture properties of enamel, it has long been known that fractures of enamel propagate preferentially along prism boundaries (Rasmussen al., 1976); Boyde, 1989). However, as crystallite packing and porosity also vary within prisms, and between prism patterns (Shellis, 1984; Chapter 17), the high fracture resistance of enamel in different species might be due to such ultra-

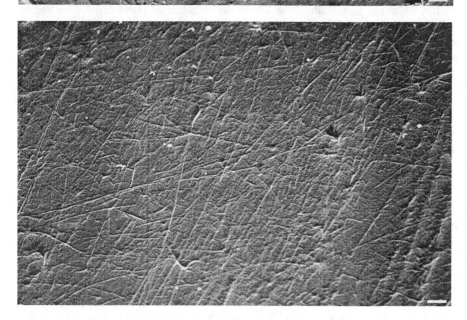

Figure 21.1A,B. **Molar microwear patterns in primates with different diets. A. Scanning electron micrograph of second molar of a patas monkey (a semi-terrestrial, cercopitheince monkey with a varied diet focusing on fruits). B. Scanning electron micrograph of second molar of a howling monkey, an arboreal New World monkey with a varied diet focusing on leaves. Scale bar represents 10 μm.**

structural differences, in addition to prism decussation (Pfretzschner, 1988; Rensberger, 1977). From another perspective, enamel is more compressible at low loads (Haines *et al.*, 1963), perhaps due to water displacement within the enamel. As body size increased through the early evolution of primates, higher loads could be applied to the teeth. To avoid fractures, might enamel porosity, and thus increased compressibility, be increased before the development of prism decussation?

As emphasized by Shellis and Dibdin in Chapter 17, a great deal of microstructural work has been spawned by interest in the diffusion of acids and mineral ions in caries formation. Such information might also be useful for other purposes, however. For instance, if inner enamel is more porous than outer enamel (Shellis, 1996), there might well be age-related changes in porosity as enamel is lost due to wear. In nonhuman primates, some species spend a significant portion of their time feeding on acidic fruits, and some species show a higher incidence of caries than do others (Colyer, 1936). If prism boundaries are the sites of increased porosity and solubility (Hamilton *et al.*, 1973; Orams *et al.*, 1976; Shellis, 1996), the combination of age-related changes in porosity and subtle differences in enamel microstructure might explain differences in the chemical erosion of enamel or in the incidence of caries. Clearly, analyses of dental microstructure can yield innumerable new perspectives on how teeth are used.

21.2.3. Dental microwear

One of the nagging problems that has always confronted paleontologists is that they have little direct evidence of the behaviors of extinct creatures. In other words, most functional studies of prehistoric morphology ultimately hinge upon assumptions of the usefulness and selective advantages of the structures in question. Thus dental microwear analysis, or the counting and measuring of microscopic scratches and pits on the surface of teeth, provides a true rarity in paleontology – a glimpse directly into the past – evidence of what the animal *did*, not what it was *capable* of doing.

Over the past two decades, improvements in microscopy and dental replication techniques have allowed researchers to document differences in dental microwear between modern animals with different diets (see Teaford, 1988, 1994; Rose and Ungar, in press; Ungar, in press, for reviews; see Figure 21.1). These data have then been used to make inferences about the diets of prehistoric animals including human ancestors (e.g., Puech and Albertini, 1984; Teaford and Walker, 1984; Grine, 1986; Lukacs and Pastor, 1988; Ryan and Johanson, 1989; Molleson and Jones, 1991; Strait, 1991; Ungar and Grine, 1991; Lalueza Fox and Perez-Perez, 1993; Solounias and Hayak, 1993; Ungar, 1996, in press). While similar work is continuing on many fronts today, there are two areas in which new advancements are most likely to occur.

The first involves methodological improvements. The measurement and analysis of dental microwear is a difficult and time-consuming task (Gordon, 1984, 1988). Most analyses to date have involved the use of scanning electron microscopy (SEM). Thus methodological improvements may involve refinements of the use of SEM images (Grine and Kay, 1988; Ungar et al., 1991), *or* changes to new methods of data collection altogether (e.g., Walker and Hagen, 1994). If methods can be standardized and simplified, larger samples can be analyzed and broader questions asked (e.g., those asked by Jernvall et al., in Chapter 19).

The second area of new research involves work on live animals and dental patients. The process of dental microwear formation is extremely complicated, and the only way to understand it is to examine it in creatures with known diets. Fortunately, high-resolution casts of teeth can now be used in dental microwear analyses (Beynon, 1988; Teaford and Oyen, 1989a), and work with live animals is now proceeding in a number of areas (e.g., Teaford and Glander, 1991; Strait and Overdorff, 1994).

Unfortunately, while investigators now have a far better understanding of certain causes of dental microwear (e.g., Ungar, 1992; Lucas and Teaford, 1995; Ungar et al., 1995), all of this work has focused on causes of *abrasive* wear. What might be the effects of acidic fruits on microwear patterns, and how might those patterns be related to differences in dental microstructure?

Finally, another type of microwear study involving living animals focuses on the *rates* of microscopic tooth wear rather than the *patterns* of microscopic tooth wear. As new scratches and pits are usually added to the tooth surface daily, the appearance (and disappearance) of microwear has been used to monitor day-to-day changes in rates of tooth wear (Teaford and Glander, 1991, 1996; Teaford and Tylenda, 1991; see Figure 21.2). This has certainly led to insights into dental microwear formation, but it is also leading unexpected ways. For instance, it has been used to demonstrate that pregnant and lactating females show a faster rate of microscopic tooth wear than do other adult females (Figure 21.3; Teaford and Glander, 1996). Accompanying behavioral observations suggest that this may sometimes be due to increased time feeding by the pregnant and lactating females, and other times due to differences in food items eaten by the different groups of monkeys. If these sorts of behavioral subtleties can be detected by analyses of rates of microscopic tooth wear, what kind of clinical questions might be answered through similar techniques? Can the progress of treatment for anorexia and bulimia be monitored through changes in rates of tooth wear? Can the functional effects of oral surgery or orthodontics be documented by changes in rates of tooth wear? Again, there is immense potential for new discoveries of how teeth are used through these microscopic analyses.

21.3   Macroscopic evidence

21.3.1   Introduction

The size and shape of teeth have long been recognized as important indicators of diet and dental function. For many years, only the most obvious correlations were noted, for example that meat-eaters had sharper teeth than did other animals. However, 25–30 years ago, this perspective began to change, as researchers began to notice subtler correlations between variations in tooth shape and variations in diet. In primates, with their varied diets, much of this work focused on differences between leaf-eaters and fruit-eaters. The former tended to have more occlusal relief on their molars and relatively long molar shearing crests while the latter tended to have larger incisors but flatter molars and relatively

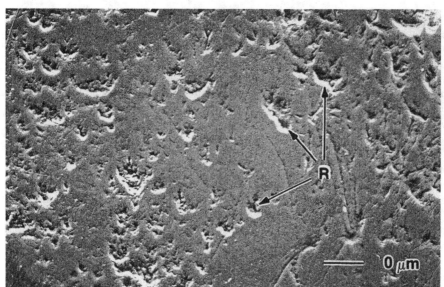

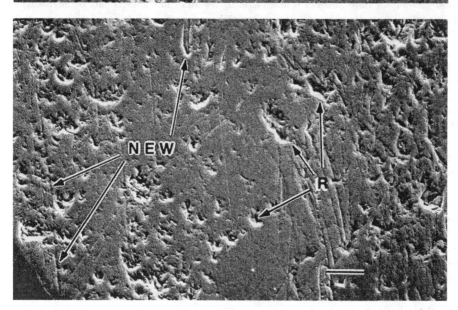

Figure 21.2A,B. **Two scanning electron micrographs from the same location on a molar of the same dental patient. A. Baseline cast: features marked R are reference features visible in each micrograph. B. Follow-up cast taken one week after A. Reference features are still visible, but new features have appeared within the 1-week period. This rate of turnover is fairly typical of that of humans. However, wild-caught nonhuman primates generally show a much faster rate of microscopic wear. Scale bars represent 50 μm.**

small molar shearing crests (Kay, 1973, 1975; Kay and Hiiemae, 1974; Hylander, 1975; Rosenberger and Kinzey, 1976; Swindler, 1976; Maier, 1977; Maier and Schneck 1981; see Figure 21.4, macaque versus langur scanning electron micrograph). When differences in body size were taken into account, primate insect-eaters were also shown to have relatively high molar shearing capacities (Kay and Hiiemae, 1974; Kay et al. 1978). Theoretical discussions suggested reasons for these differences, e.g., leaves might be tougher than most fruits and thus require precise, scissor-like cutting or shearing to be efficiently processed (Kay and Hiiemae, 1974; Kay, 1975;

Osborn and Lumsden, 1978; Lucas, 1979; Lucas and Luke, 1984). At the same time, laboratory studies began to show that the ability to process certain insects or plant-foods could be directly tied to differences in dental morphology (Walker and Murray, 1975; Sheine and Kay, 1977; Kay and Sheine, 1979).

Much of what has been done in the past decade has been an extension of all of the above as investigators strove to gain new insights into the various functions of fossil teeth. However, three other areas of research have brought new perspectives which should move future analyses even further.

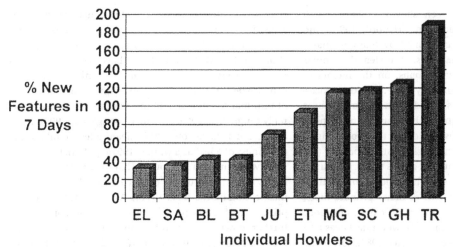

Figure 21.3. **Rates of molar wear in howling monkeys. Individuals JU, ET, MG, TR are pregnant of lactating females. Individuals SC and GH are adult males who were in the midst of territorial disputes at the time their dental impressions were taken. (From Teaford and Glander, 1996.)**

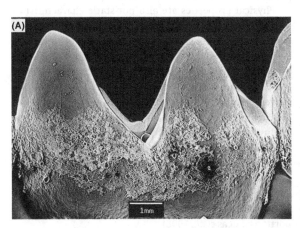

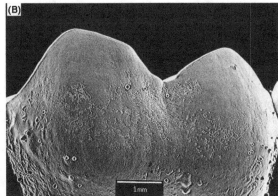

Figure 21.4A,B. **Scanning electron micrographs of colobine and cercopithecine molars. A. *Procolobus badius* (a colobine monkey). B. *Cercocebus albigena* (a cercopithecine monkey). Note difference in cusp height. (From Lucas and Teaford, 1994.)**

21.3.2. Physical properties of foods

The first of these new studies is unusual in that it doesn't even involve the study of teeth. Instead, it involves the physical properties of foods, as investigators try to determine the properties that are confronting teeth as they process food. Early workers used general terms such as 'shearing' and 'crushing' to describe the actions of teeth on food. These terms were easy to conceptualize, but almost impossible to measure consistently. Thus one of the big advances in the past decade has been a standardization of terminology (largely stimulated by the work of Lucas and his colleagues: Lucas and Pereira, 1990; Lucas et al., 1991a, b; Choong et al., 1992; Hill and Lucas, 1996; see also Kinzey and Norconk, 1990, 1993; Peters, 1993; Strait, 1993a, 1997; Yamashita, 1996) which has forced workers to realize that they need to measure physically meaningful properties, such as fracture toughness, if they are to compare and contrast results successfully.

With this new-found understanding of physical property testing, two major points have become clear.

First, interspecific differences in dental microstructure and tooth shape can indeed be correlated with interspecific differences in the physical properties of food items, at least at certain levels. For instance, primates that process foods of high fracture toughness tend to be primates that have relatively high molar shearing capacity. This might, at first glance, seem to tell us nothing more than we already knew, as the work of Kay and others (Kay, 1973, 1975; Kay and Hiiemae, 1974; Rosenberger and Kinzey, 1976; Maier, 1977; Kay et al., 1978; Maier and Schneck, 1981; Strait, 1993a, b) had previously shown that primate leaf- and insect-eaters tend to have relatively high molar shearing capacity. However, one way in which studies of food properties have broadened these perspectives is by documenting the physical properties of a wider range of foods. For instance, seeds with thin flexible coats have been shown to have tough storage tissues inside (Lucas and Teaford, 1994), and these storage tissues are consumed by many primates (e.g., colobine monkeys). Thus, the lengthy molar shearing crests, and tall cross-lophs, on the molars of colobine monkeys are eminently suited for the processing of tough, pliant seeds, in addition to leaves (Lucas and Teaford, 1994). By contrast, the short, blunt cusps of many primate teeth are best suited for the processing of foods of low toughness and high elastic modulus (see Chapter 20 for further discussion of this topic). Of course, the question still remains, 'What about soft fruits?' (which are a major component of many primate diets; Milton, 1987, 1988). Naively, one might expect *any* tooth shape to be able to process them. However, as Lucas and Luke have noted (1984; Luke and Lucas, 1983), soft fruits can be relatively easy to compress, but difficult to cut, in other words, soft but tough. In essence, they tend to 'ooze' out from between the occluding teeth. Thus the molars of many soft-fruit eaters tend to have peripheral shearing blades with concave edges, to trap and cut the fruit between them, and to retain fruit within the tooth's central basin to aid in the extraction of fruit juice (Kay and Hiiemae, 1974; Lucas and Luke, 1984; Anthony and Kay, 1993; Luke and Lucas, 1983; Strait, 1993b). From this perspective, the large dentin exposures often found on the molars of some primate fruit-eaters, such as chimpanzees, probably aid in the processing of fruit by effectively increasing basin depth and thus fruit retention between the teeth.

Of course another hallmark of primate diets is that they are often extremely variable (Hladik and Hladik, 1969; Milton, 1987). Consideration of this fact raises an age-old question: if variations in tooth shape can confer a selective advantage on certain individuals by allowing them to process certain foods better than can their conspecifics, which foods are having the most important influence on morphology, those that are eaten most frequently (Kay, 1973, 1975), or those that are of critical importance to the survival of the individual (Rosenberger and Kinzey, 1976; Kinzey, 1978)? For some primates, this may be a false dichotomy as their most frequent food items may be their most critical items. But for some primates (e.g., those existing in seasonal environments in which resource availability fluctuates dramatically), it may be a major consideration. Only with a more realistic understanding of primate diets and the physical properties of primate foods will we be able to answer the question, and make inferences about its importance in the fossil record.

The second major point that has become obvious in the past decade is that the physical properties of foods are extremely complex. As Lucas and Peters have empha-sized in Chapter 20, foods are heterogeneous structures. Thus, for example, some fruits eaten by primates have tough but soft outer shells and acidic flesh inside (Lucas et al., 1991b), while other fruits have hard outer shells and soft seeds inside (Kinzey and Norconk, 1993). Leaves can also be surprisingly complex, with spongy mesophyll of minimal toughness, and major veins which are extremely tough (Lucas et al., 1991a; Choong et al., 1992). Thus, as Lucas et al. have noted, 'it is meaningless to talk of the toughness of a leaf as a whole' (1991a, p. 104), for it is the veins which make a leaf tough. That toughness effectively limits the number of leaf cells which can be damaged in mastication (Lucas et al., 1991a). As a result, the teeth of primate leaf-eaters prepare leaves for microbial action further down the gastrointestinal tract, whereas the teeth of primate fruit-eaters may yield a faster return for their activity (Milton, 1981, 1984, 1987).

Physical properties are also not static characteristics of foods. For instance, the toughness of seed shells can be significantly decreased by the presence of moisture (Williamson and Lucas, 1995). Likewise, the veins in a leaf may grow rapidly tougher with age (Kursar and Coley, 1991), and measures of toughness often decrease as one moves from the base of a leaf to its apex (Lucas et al., 1998). State-of-the-art behavioral observations are now beginning to tie these physical subtleties to primate feeding patterns, as primates may preferentially feed on younger leaves (which are less tough) than on more mature leaves on the same tree (Hill and Lucas, 1996; Lucas et al., 1998), and they may also nip off the apical portions of leaves, seemingly ignoring the tougher basal portions (Lucas et al., 1998).

These complexities of food properties once again raise the question of the evolutionary importance of 'critical items' versus 'most frequent items' in the diet. For instance, the orang-utan is one of the vertebrates that routinely feeds on the seeds of *Mezzettia*. In fact, some investigators have suggested that the combination of thick, decussating molar enamel, short blunt cusps, and craniofacial biomechanics leaves the orang-utan admirably equipped to process hard, tough foods (Kay, 1981). Yet *Mezzettia* seeds form but a small portion of the orang-utan's diet. Is this a classic case of masticatory features being adapted to a critical item in the diet? Or is the orang-utan merely making the best use of a set of evolutionary constraints that fortuitously allow it to exploit a novel food source? Continued analyses of the properties of its foods will help us to answer these questions.

21.3.3.   3-D morphometrics

Some of the most dramatic changes in macroscopic ana-lyses of teeth have occurred in the *measurement* of teeth.

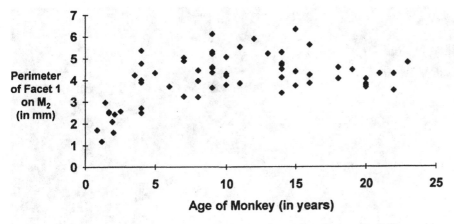

Figure 21.5. **Changes in molar shearing facet size with age in howling monkeys. Perimeter of shearing facets measured with 3-D measuring microscope (Reflex microscope). (From Teaford and Glander, 1996.)**

Early dental analyses (e.g., Biggerstaff, 1975; Kay, 1975; Lavelle, 1976; Swindler, 1976) were restricted to the use of two-dimensional measures, such as the length and width of teeth, or the length of occlusal features. Standard dial calipers provided the usual method of data collection, and calibrated eye-pieces on microscopes provided similar measures for small specimens. Methods of three-dimensional data collection were available (e.g., photogrammetrics; Savara, 1965; Clark et al., 1971; Jonason et al., 1974) but they were tedious to use and generally not of sufficient resolution to yield useful measures of the intricacies of tooth shape.

The past two decades, with the advent of personal computers, new measurement devices, and the accompanying software to run them, have changed all of this, and in the process made possible easy and accurate three-dimensional measurement of teeth.[1] Now researchers can choose among a variety of techniques (measuring microscopes, laser scanners, etc.: Hartman, 1989; MacLarnon, 1989; Teaford and Oyen, 1989b; Teaford, 1991; Zuccotti et al., 1998) to obtain detailed 3-D measurements of teeth.

Of course, 3-D data collection has the potential to generate an almost infinite amount of data. Thus an even more important question is 'What to do with it all?' Here the answers have gone in two different directions. Some investigators have effectively extended traditional 2-D analyses to 3-D, by developing new landmark-based analyses of coordinate data (e.g., Euclidean Distance Matrix Analysis or EDMA: Lele and Richtsmeier, 1991; Lele, 1993). The advantage of such analyses is that they can yield results which are roughly comparable to those of previous work. For instance, instead of measuring the

length of molar shearing crests on unworn teeth, as in the work of Kay and others e.g., Kay and Hiiemae, 1974; Kay, 1975; Kay and Hylander, 1978), one might measure shearing *facet* size on an ontogenetic series of worn teeth to chart the functional course of tooth wear (Teaford, 1991; Teaford and Glander, 1996; Figure 21.5). The disadvantage of such work is that it is still ultimately tied to a series of landmarks on the tooth surface. If those landmarks change dramatically, or vanish, during ontogeny or evolution, the analyses will generally be inappropriate.

The other new approach to 3-D analyses is a break with tradition, a move away from landmarks, to the analysis of *surfaces* or *shapes*. What is required is a change of perspective: from a focus on cusps and crests, as isolated, functional units, to an emphasis on surfaces or combinations of shapes. For instance, Ungar and co-workers (Zuccotti et al., 1998) have recently used a laser-scanner to map the occlusal surfaces of primate molar teeth, and then used special software (the Geographic Resources Analysis Support System or GRASS) to plot differences in drainage patterns between teeth of different species (Figure 21.6). As one of the functions of teeth is to help retain food in position while chewing (as noted above for soft fruits), such information may yield new insights into dental function. Moreover, similar analyses may also allow more effective comparison of teeth of different ages (in ontogenetic series of casts), or teeth of potential ancestors and descendants in evolution.

21.3.4. Tooth function and wear

The past two decades have also witnessed a steady increase in our knowledge of how teeth are used. Earlier work (e.g., Butler and Mills, 1959; Mills, 1955, 1973; see also Chapter 14) had used analyses of dental morphology, specifically the size and location of molar wear facets, to show that primates use more than simple up-and-down

---

[1] Technological advances have *also* made possible quicker analyses of 2-D measurements (e.g., of molar cusp areas) (Wood and Abbott, 1983; Wood and Engleman, 1988; Uchida, 1998a, b)

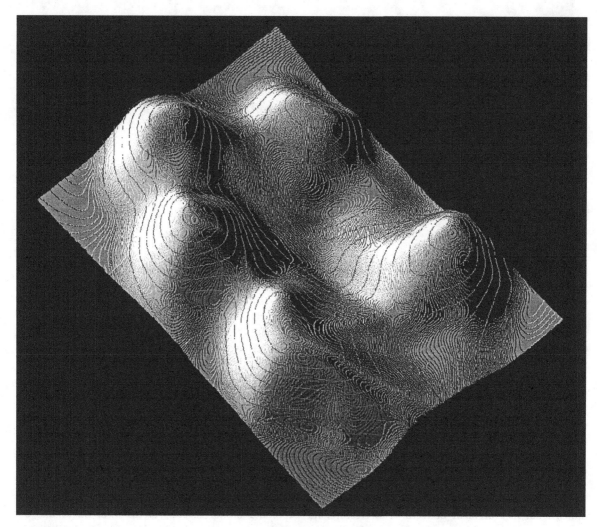

Figure 21.6. **Plot of tooth shape based on laser-scanned image of a howling monkey molar. (Provided by Peter Ungar.)**

jaw movements in chewing. Building on that work, Hiiemae and Kay (1973; Kay and Hiiemae, 1974) monitored the basic movements of mastication in laboratory primates fed known diets. This work was then followed by that of Hylander *et al.* (1987, 1992) in which more subtle differences of jaw movements during the various stages of the power stroke of chewing were noted.

With a basic understanding of chewing, researchers have now begun to take a more realistic view of dental function and dental wear. Essentially, molar shearing crests and molar wear facets can only take us so far in analyses of dental function. All primates, with the notable exception of Western, industrialized humans, spend most of their lifetimes with worn teeth. In fact, tooth wear usually proceeds so rapidly that molar wear facets are quickly obliterated (Figure 21.7), and molar shearing crests change dramatically in size and shape.

Again, if tooth shape is somehow to confer advantages in the processing of certain foods, is that advantage maintained throughout the lifetime of an individual? Lanyon and Sanson's classic paper on koalas (1986) suggested that the advantage is only maintained to a certain age, after which the cumulative effects of tooth wear cause a decrease in chewing efficiency. But similar work is just starting in primates, where ongoing analyses of wild-caught howling monkeys suggest that there are age-related changes in molar shearing capacity (Teaford and Glander, 1996; Figure 21.8). If these changes in shearing capacity can be tied to changes in chewing efficiency in different social groups (Figure 21.9) or in different species, it will make possible a whole new series of functional analyses of fossil teeth, where more precise inferences of prehistoric primate diets might then be made.

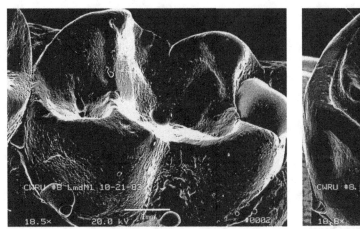
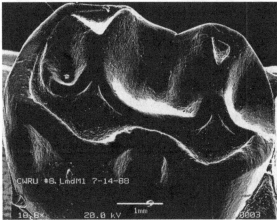

Figure 21.7. **Changes in molar shape with wear. A. Scanning electron micrograph of first molar of a vervet monkey (mesial, left; buccal, bottom). B. Scanning electron micrograph of the same tooth 56 months later. Note obliteration of tip facets on buccal cusps.**

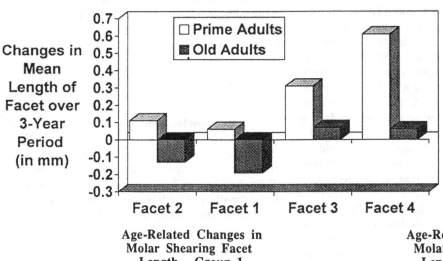

Changes in Mean Length of Facet over 3-Year Period (in mm)

Figure 21.8. **Changes in molar shearing facet length in old and prime adult howling monkeys over a 3-year period. Note that old adults show a decrease in the size of certain shearing facets. (From Teaford and Glander, 1996.)**

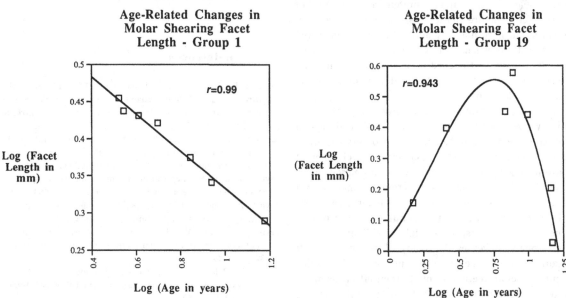

Figure 21.9A,B. **Age-related differences in molar shearing facet length in howling monkeys from different social groups. Note that group 1 (A) shows a progressive decrease in facet size, whereas group 19 (B) shows an initial increase in facet length followed by a decrease in facet length.**

21.4.   Broader perspectives

In this day and age of new techniques and complicated analyses, it is very easy to remain narrowly focused on a topic and, as the saying goes, 'lose sight of the forest among the trees'. This book is an attempt to show how varied that forest really is, and yet how intricately intertwined different 'forests' can be. A couple of examples will give a small taste of how multiple perspectives might yield insights into broader questions.

A great deal has been written in the literature about evolutionary novelties, i.e., new structures that confer functional, and thus selective, advantages in certain organisms (see Müller and Wagner, 1991, for a review). Jernvall *et al.* have discussed one such novelty in Chapter 19, the hypocone on mammalian teeth. Another example might be the enamel prism (Chapter 7; see also Grine *et al.*, 1979; Grine and Vrba, 1980). From a different perspective, that of evolutionary theory, researchers have noted (e.g., Roth and Wake, 1989) that structural novelties are often 'de-couplings', as structures that were formerly intimately linked become free of each other's influence to proceed on separate paths. Some of the work outlined in this volume (e.g., Chapters 1, 2 and 11) shows how close we are to understanding the genetics and development of dental tissue. Other chapters (e.g., 7, 10 and 5) make concerted efforts to pull together fossil evidence with developmental evidence. The combination of these perspectives cannot help but lead to insights. For instance, if reptilian amelogenesis is truly viewed as 'enamel surface morphogenesis' (Chapter 7), could the origin of prismatic enamel represent a de-coupling of tooth microstructure and tooth shape? If so, the enamel prism might have not only allowed a reduction in the number of tooth generations (Grine *et al.*, 1979: Grine and Vrba, 1980), but opened up a myriad of functional possibilities as well, through the development of the heterodont dentition.

As another example, the incorporation of evolutionary novelties into a lineage of animals will depend on a complex mosaic of factors. Some workers (e.g., Galis *et al.*, 1994) have turned to analyses of developmental patterns to gain insights into how evolutionary changes take place. For instance, the functional demands on a structure are not constant during ontogeny, i.e., there are crises sometimes and accumulated insurance at others. The same perspective might shed light on dental evolutionary novelties in many different ways. If weaning and old-age are different forms of developmental crises for teeth, how might their effects be manifested on evolutionary novelties? How might their effects be selected for (or against) in the evolution of such complex traits (see Chapter 11)? In an evolutionary perspective, there might also be periods of crisis and periods of relative insurance for a given species. Variable environments might leave a species with greater capacity to adapt in periods of crisis (Lowell, 1985). How would 'old age' affect it?

21.5.   Postscript

Clearly, the amount of work involved in bringing together dramatically different perspectives is immense. Thus, a single volume such as this one cannot *begin* to cover everything. What it can do, however, is broaden perspectives on dental functional morphology, and act as a catalyst to bring more people together into collaborative research. We cannot predict when or where our next major advance in dental research will appear. However, one thing is certain, if we don't take the time to communicate with each other, those advances will be few and far between. ·

Acknowledgements

This work was supported by NSF grants 8904327, 9118876, 9601176. Special thanks to go Ken Glander for all his help on the howler project. Thanks also go to Albert Capati and Rose Weinstein for taking some of the scanning electron micrographs used here.

References

Angmar, B., Carlström, D. and Glas, J.-E. (1963). Studies on the ultrastructure of dental enamel IV. The mineralization of normal human enamel. *Journal of Ultrastructure Research*, **8**, 12–23.

Anthony, M. R. L. and Kay, R. F. (1993). Tooth form and diet in ateline and alouattine primates: reflections on the comparative method. *American Journal of Science*, **293**-A, 356–382.

Beynon, A. D. (1988). Replication technique for studying microstructure in fossil enamel. *Scanning Microscopy* **1**(2), 663–669.

Beynon, A. D. and Dean, M. C. (1988). Distinct dental development patterns in early fossil hominids. *Nature*, **335**, 509–514.

Beynon, A. D. and Wood, B. A. (1987). Patterns and rates of enamel growth in the molar teeth of early hominids. *Nature*, **326**, 493–496.

Biggerstaff, R. H. (1975). Cusp size, sexual dimorphism, and heritability of cusp size in twins. *American Journal of Physical Anthropology*, **42**, 127–140.

Bock, W. J. (1988). The nature of explanations in morphology. *American Zoologist*, **28**, 205–215.

Bock, W. J. (1990). Explanations in Konstruktionsmorphologie and evolutionary morphology. In *Constructional Morphology and Evolution*, eds. N. Schmidt-Kittler and K. Vogel, pp. 9–29. Berlin: Springer.

Boyde, A. (1964). The Structure and Development of Mammalian Enamel. PhD thesis, University of London.

Boyde, A. (1989). Enamel. In *Handbook of Microscopic Anatomy*, vol. V/6, *Teeth*, eds. A. Oksche and L. Vollrath, pp. 309–473. Berlin: Springer.

Boyde, A. and Martin, L. B. (1982). Enamel microstructure determination in hominoid and cercopiothecoid primates. *Anatomy and Embryology*, **165**, 193–212.

Bromage, T. G. and Dean, M. C. (1985). Re-evaluation of the age at death of immature fossil hominids. *Nature*, **317**, 525–527.

Butler, P. M. and Mills, J. R. E. (1959). A contribution to the odontology of *Oreopithecus*. *Bulletin of the British Museum (Natural History) Geology*, **4**, 1–26.

Carlson, S. J. (1990). Vertebrate dental structures. In *Skeletal Biomineralization: Patterns Processes and Evolutionary Trends*, vol. 1, ed. J. G. Carter, pp. 531–556. New York: Van Nostrand Reinhold.

Choong, M. F., Lucas, P. W., Ong, J. S. Y., Pereira, B., Tan, H. T. W. and Turner, I. M. (1992). Leaf fracture toughness and sclerophylly: their correlations and ecological implications. *New Phytologist*, **121**, 597–610.

Clark, C. E., Lavelle, C. L. B. and Flinn, R. M. (1971). Definition of tooth shape by means of photogrammetry. *Journal of Dental Research*, **50**, 1178.

Colyer, F. (1936). *Variations and Diseases of the Teeth of Animals*. London: John Bale & Sons & Danielson.

Craig, R. G. and Peyton, F. A. (1958). The microhardness of enamel and dentin. *Journal of Dental Research*, **37**, 661–668.

Dullemaijer, P. (1980). Functional morphology and evolutionary biology. *Acta Biotheoretica*, **29**, 151–250.

Galis, F. (1996). The application of functional morphology to evolutionary studies. *Trends in Ecology and Evolution*, **11**, 124–129.

Galis, F., Terlouw, A. and Osse, J. W. M. (1994). The relation between morphology and behaviour during ontogenetic and evolutionary changes. *Journal of Fish Biology* **45** (Suppl. A), 13–26.

Gans, C. (1974). *Biomechanics, an Approach to Vertebrate Biology*. Philadelphia: Lippincott.

Gantt, D. G. (1983). The enamel of Neogene hominoids: structural and phyletic implications. In *New Interpretations of Ape and Human Ancestry*, eds. R. L. Ciochon and R. S. Corruccini, pp. 249–298. New York: Plenum Press.

Gantt, D. G., Pilbeam, D. R. and Steward, G. (1977). Hominoid enamel prism patterns. *Science*, **198**, 1155–1157.

Gordon, K. D. (1984). Hominoid dental microwear: complications in the use of microwear analysis to detect diet. *Journal of Dental Research*, **63**, 1043–1046.

Gordon, K. D. (1988). A review of methodology and quantification in dental microwear analysis. *Scanning Microscopy*, **2**, 1139–1147.

Grine, F. E. (1986). Dental evidence for dietary differences in *Australopithecus* and *Parathropus*: a quantitative analysis of permanent molar microwear. *Journal of Human Evolution*, **15**, 783–822.

Grine, F. E. and Kay, R. F. (1988). Early hominid diets from quantitative image analysis of dental microwear. *Nature*, **333**, 765–768.

Grine, F. E. and Vrba, E. S. (1980). Prismatic enamel: a pre-adaptation for mammalian diphyodonty? *South African Journal of Science*, **76**, 139–141.

Grine, F. E., Vrba, E. S. and Cruickshank, A. R. I. (1979). Enamel prisms and diphyodonty: linked apomorphies of Mammalia. *South African Journal of Science*, **75**, 114–120.

Haines, D. J., Berry, D. C. and Poole, D. F. G. (1963). Behavior of tooth enamel under load. *Journal of Dental Research*, **42**, 885–888.

Hamilton, W. J., Judd, G. and Ansell, G. S. (1973). Ultrastructure of human enamel specimens prepared by ion micromilling. *Journal of Dental Research*, **52**, 703–710.

Hartman, S. E. (1989). Stereophotogrammetric analysis of occlusal morphology of extant hominoid molars: Phenetics and function. *American Journal of Physical Anthropology*, **80**, 145–166.

Hiiemae, K. M. and Kay, R. F. (1973). Evolutionary trends in the dynamics of primate mastication. In *Symposium of the Fourth International Congress of Primatology*, vol. 3, *Craniofacial Biology of Primates*, ed. M. R. Zingeser, pp. 28–64. Basel: Karger.

Hill, D. A. and Lucas, P. W. (1996). Toughness and fiber content of major leaf foods of Japanese macaques (*Macaca fuscata yakui*) in Yakushima. *American Journal of Primatology*, **38**, 221–231.

Hladik, A and Hladik, C. M. (1969). Rapports tropiques entre vegetation et primates dans la foret de Barro Colorado (Panama). *La Terre et la Vie*, **23**, 25–117.

Hylander, W. L. (1975). Incisor size and diet in anthropoids with special reference to Cercopithecoidea. *Science*, **189**, 1095–1098,

Hylander, W. L., Johnson, K. R. and Crompton, A. W. (1987). Loading patterns and jaw movements during mastication in *Macaca fascicularis*: a bone-strain, electromyographic, and cineradiographic analysis. *American Journal of Physical Anthropology*, **72**, 287–314.

Hylander, W. L., Johnson, K. R. and Crompton, A. W. (1992). Muscle force recruitment and biomechanical modeling: an analysis of masseter muscle function in *Macaca fascicularis*. *American Journal of Physical Anthropology*, **88**, 365–387.

Jonason, C., Frykholm, K. O. and Frykholm, A. (1974). Three-dimensional measurement of tooth impression of criminological investigation. *International Journal of Forensic Dentistry*, **2**, 70–78.

Kay, R. F. (1973). Mastication, Molar Tooth Structure, and Diet in Primates. PhD dissertation, Yale University.

Kay, R. F. (1975). The functional adaptations of primate molar teeth. *American Journal of Physical Anthropology*, **43**, 195–215.

Kay, R. F. (1981). The nut-crackers: a new theory of the adaptations of the Ramapithecinae. *American Journal of Physical Anthropology*, **55**, 141–151.

Kay, R. F. and Hiiemae, K. M. (1974). Jaw movement and tooth use in recent and fossil primates. *American Journal of Physical Anthropology*, **40**, 227–256.

Kay, R. F. and Hylander, W. L. (1978). The dental structure of mammalian folivores with special reference to Primates and Phalangeroidea. In *The Ecology of Arboreal Folivores*, ed. G. G. Montgomery, pp. 173–191. Washington, D.C. : Smithsonian Institution Press.

Kay, R. F. and Sheine, W. S. (1979). On the relationship between chitin particle size and digestibility in the primate *Galago senegalensis*. *American Journal of Physical Anthropology*, **50**, 301–308.

Kinney, J. H., Balooch, M., Marshall, S. H., Marshall, Jr., G. W. and Weihs, T. P. (1996*a*). Hardness and Young's Modulus of peritubular and intertubular dentine: dependence of mechanical properties on intra-tooth position. *Archives of Oral Biology*, **41**, 9–13.

Kinney, J. H., Balooch, M., Marshall, S. H., Marshall, Jr., G. W. and Weihs, T. P. (1996*b*). Atomic force microscope measurements of the hardness and elasticity of peritubular and intertubular human dentin. *Journal of Biomechanical Engineering*, **118**, 133–135.

Kinzey, W. G. (1978). Feeding behavior and molar features in two species of titi monkey. In *Recent Advances in Primatology*, vol. 1, *Behaviour*, ed. D. J. Chivers and J. Herbert, pp. 373–385. New York: Academic Press.

Kinzey, W. G. and Norconk, M. A. (1990). Hardness as a basis of fruit choice in two sympatric primates. *American Journal of Physical Anthropology*, **81**, 5–15.

Kinzey, W. H. and Norconk, M. A. (1993). Physical and chemical properties of fruit and seeds eaten by *Pithecia* and *Chiropotes* in Surinam and Venezuela. *International Journal of Primatology*, **14**, 207–226.

Koenigswald, W. v. and Clemens, W. A. (1992). Levels of complexity in the microstructure of mammalian enamel and their application in studies of systematics., *Scanning Microscopy*, **6**, 195–218.

Koenigswald, W. v. and Pfretzschner, H. U. (1991). Biomechanics in the enamel of mammalian teeth. In *Constructional Morphology and Evolution*, eds. N. Schmidt-Kittler and K. Vogel, pp. 113–125. Berlin: Springer.

Koenigswald, W. v., Rensberger, J. M. and Pfretzschner, H. U. (1987). Changes in the tooth enamel of early Paleocene mammals allowing increased diet diversity. *Nature*, **328**, 150–52.

Kursar, T. A. and Coley, P. D. (1991). Nitrogen content and expansion rate of young leaves of rain forest species: implications for herbivory. *Biotropica*, **23**, 141–150.

Lalueza Fox, C. Perez-Perez, A. (1993). The diet of the Neanderthal child Gibralter 2 (Devil's Tower) through the study of the vestibular striation pattern. *Journal of Human Evolution*, **24**, 29–41.

Lanyon, J. M. and Sanson, G. D. (1986). Koala (*Phascolarctos cinereus*) dentition and nutrition. II. Implications of tooth wear in nutrition. *Journal of Zoology, London*, **209A**, 169–181.

Lavelle, C. L. B. (1976). Odontometric study of African monkey teeth. *Acta Anatomica*, **96**, 115–127.

Lele, S. (1993). Euclidean Distance Matrix Analysis (EDMA): estimation of mean form and mean form difference. *Mathematical Geology*, **25**, 573–602.

Lele, S. and Richtsmeier, J. T. (1991). Euclidean Distance Matrix Analysis: a coordinate-free approach for comparing biological shapes using landmark data. *American Journal of Physical Anthropology*, **86**, 415–427.

Lowell, R. B. (1985). Selection for increased safety factors of biological structures as environmental unpredictably increases. *Science*, **221**, 1009–1011.

Lucas, P. W. (1979). The dental-dietary adaptations of mammals. *Neues Jahrbuch für Geologie und Paläontologie Monatshefte*, **8**, 486–512.

Lucas, P. W. and Luke, D. A. (1984). Chewing it over: basic principles of food breakdown. In *Food Acquisition and Processing in Primates*, eds. D. J. Chivers, B. A. Wood and A. Bilsborough, pp. 283–301. New York: Plenum Press.

Lucas, P. W. and Pereira, B. (1990). Estimation of the fracture toughness of leaves. *Functional Ecology*, **4**, 819–822.

Lucas, P. W. and Teaford, M. F. (1994). Functional morphology of colobine teeth. In *Colobine Monkeys: Their Ecology, Behavior, and Evolution*, eds. A. G. Davies and J. F. Oates, pp. 173–203. Cambridge: Cambridge University Press.

Lucas, P. W. and Teaford, M. F. (1995). Significance of silica in leaves eaten by long-tailed macaques (*Macaca fascicularis*). *Folia Primatologica*, **64**, 30–36.

Lucas, P. W., Choong, M. F., Tan, H. T. W., Turner, I. M. and Berrick A. J. (1991*a*). The fracture toughness of the leaf of the dicotyledon *Callophyllum inophyllym* L. (Guttiferae). *Philosophical Transactions of the Royal Society of London, B.*, **334**, 95–106.

Lucas, P. W., Lowrey, T. K., Pereira, B. P., Sarafis, B. and Kuhn, W. (1991*b*). The ecology of *Mezzettia leptopoda* (Hk. f. et Thoms.) Oliv. (Annonaceae) seeds as viewed from a mechanical perspective. *Functional Ecology*, **5**, 545–553.

Lucas, P. W., Teaford, M. F., Ungar, P. S. and Glander, K. E. (1998). Physical properties of foods in *Alouatta palliata*. *American Journal of Physical Anthropology*, **Suppl. 26**, 152–153.

Lukacs, J. R. and Pastor, R. F. (1988). Activity induced patterns of dental abrasion in prehistoric Pakistan: evidence from Mehrgarh and Harappa. *American Journal of Physical Anthropology*, **76**, 377–398.

Luke, D. A. and Lucas, P. W. (1983). The significance of cusps. *Journal of Oral Rehabilitation*, **10**, 197–206.

Maas, M. C. (1993). Enamel microstructure and molar wear in the greater galago, *Otolemur crassicaudatus* (Mammalia, Primates). *American Journal of Physical Anthropology*, **92**, 217–233.

Maas, M. C. (1994). Enamel microstructure in Lemuridae (Mammalia, Primates): assessment of variability. *American Journal of Physical Anthropology*, **95**, 221–241.

MacLarnon, A. M. (1989). Applications of the Reflex instruments in quantitative morphology. *Folia Primatologica*, **53**, 33–49.

Maier, W. (1977). Die Evolution der bilophodonten Molaren der Cercopithecoidea. *Zeitschrift für Morphologie und Anthropologie*, **68**, 26–56.

Maier, W. and Schneck, G. (1981). Konstruktionsmorphologische Untersuchungen am Gebiss der hominoiden Primaten. *Zeitschrift für Morphologie und Anthropologie*, **72**, 127–169.

Martin, L. B., Boyde, A. and Grine, F. E. (1988). Enamel structure in primates: a review of scanning electron microscope studies. *Scanning Microscopy*, **2(3)**, 1503–1526.

Mayr, E. (1982). *The Growth of Biological Thought. Diversity, Evolution and Inheritance*. Cambridge, MA: Harvard University Press.

Mills, J. R. E. (1955). Ideal dental occlusion in the primates. *Dental Practitioner*, **6**, 47–61.

Mills, J. R. E. (1973). Evolution of mastication in primates. In *Symposia of the Fourth International Congress of Primatology*, vol. 3, eds. W. Montagna and M. R. Zingeser, pp. 65–81. Basel: Karger.

Milton, K. (1981). Food choice and digestive strategies of two sympatric primate species. *American Naturalist*, **117**, 496–505.

Milton, K. (1984). The role of food processing factors in primate food choice. In *Adaptations for Foraging in Nonhuman Primates*, eds. P. S. Rodman and J. G. H. Cant, pp. 249–279. New York: Columbia University Press.

Milton, K. (1987). Primate diets and gut morphology: Implications for hominid evolution. In *Food and Evolution: Toward a Theory of Human Food Habits*, eds. M. Harris and E. B. Ross, pp. 93–115. Philadelphia: Temple University Press.

Milton, K. (1988). Foraging behavior and the evolution of primate intelligence. In *Machiavellian Intelligence*, eds., R. W. Byrne and A. Whiten, pp. 285–305. Oxford: Clarendon Press.

Molleson, T. and Jones, K. (1991). Dental evidence for dietary changes at Abu Hureyra. *Journal of Archaeological Science*, 18, 525–539.

Müller, G. B. and Wagner, G. P. (1991). Novelty in evolution: restructuring the concept. *Annual Review of Ecology and Systematics*, 22, 229–256.

Orams, H. J., Zybert, J. J., Phakey, P. P. and Rachinger, W. A. (1976). Ultrastructural study of human dental enamel using selected-area argon-ion beam thinning. *Archives of Oral Biology*, 21, 663–675.

Osborn, J. W. and Lumsden, A. G. S. (1978). An alternative to 'thegosis' and a re-examination of the ways in which mammalian molars work. *Neues Jahrbuch für Geologie und Paläontologie Abhandlungen*, 156, 371–392.

Peters, C. R. (1993). Shell strength and primate seed predation of nontoxic species of Eastern and Southern Africa. *International Journal of Primatology*, 14, 315–344.

Pfretzschner, H. U. (1988). Structural reinforcement and crack propagation in enamel. In *Teeth Revisited: Proceedings of the VIIth International Symposium on Dental Morphology*, eds. D. E. Russell, J. P. Santoro, and D. Sigogneau-Russell, pp. 133–143. Paris: Mémoirs du Musée national d'Histoire naturelle C53.

Puech, P.-F. and Albertini, H. (1984). Dental microwear and mechanisms in early hominids from Laetoli and Hadar. *American Journal of Physical Anthropology*, 65, 87–91.

Rasmussen, S. T., Patchin, R. E., Scott, D. B. and Heuer, A. H. (1976). Fracture properties of human enamel and dentin. *Journal of Dental Research*, 55, 154–164.

Rensberger, J. M. (1997). Mechanical adaptation in enamel. In *Tooth Enamel Microstructure*, eds. W. v. Koenigswald and P. M. Sander, pp. 237–258. Rotterdam: A. A. Balkema.

Rensberger, J. M. and Koenigswald, W. v. (1980). Functional and phylogenetic interpretation of enamel microstructure in rhinoceroses. *Paleobiology*, 6, 477–495.

Rose, J. C. and Ungar, P. S. (in press). Gross dental wear and dental microwear in historical perspective. In *Dental Anthropology: Fundamentals, Limits, Prospects*, eds. K. W. Alt, F. W. Rosing, M. Teschler-Nicola. Stuttgart: Gustav-Fischer.

Rosenberger, A. L. and Calvao, C. M. (1998). Digital morphology: The Smithsonian's biovisualization lab. *Evolutionary Anthropology*, 94, 77–78.

Rosenberger, A. L. and Kinzey, W. G. (1976). Functional patterns of molar occlusion in platyrrhine primates. *American Journal of Physical Anthropology*, 45, 281–298.

Roth, G. and Wake, D. B. (1989). Conservatism and innovation in the evolution of feeding in vertebrates. In *Complex Organismal Functions: Integration and Evolution in Vertebrates*, eds. D. B. Wake and G. Roth, pp. 7–21. New York: Wiley.

Ryan, A. S. and Johanson, D. C. (1989). Anterior dental microwear in *Australopithecus afarensis*. *Journal of Human Evolution*, 18, 235–268.

Savara, B. S. (1965). Application of photogrammetry for quantitative study of tooth and face morphology. *American Journal of Physical Anthropology*, 23, 427–434.

Schmidt-Kittler, N. and Vogel, S. (1990). *Constructional Morphology and Evolution*. Berlin: Springer.

Sheine, W. S. and Kay, R. F. (1977). An analysis of chewed food particle size and its relationship to molar structure in the primates *Cheirogaleus medius* and *Galago senegalensis*. *American Journal of Physical Anthropology*, 47, 15–20.

Sheine, W. S. and Kay, R. F. (1982). A model for comparison of masticatory effectiveness in primates. *Journal of Morphology*, 172, 139–149.

Shellis, R. P. (1984). Relationship between human enamel structure and the formation of caries-like lesions in vitro. *Archives of Oral Biology*, 29, 975–981.

Shellis, R. P. (1996). Solubility variations in human enamel and dentine: a scanning electron microscopic study. *Archives of Oral Biology*, 41, 473–484.

Shellis, R. P. and Poole, D. F. G. (1977). The calcified dental tissues of primates. In *Evolutionary Changes to the Primate Skull and Dentition*, eds. C. L. B. Lavelle, R. P. Shellis and D. F. G. Poole, pp. 197–279. Springfield, IL: Thomas.

Solounias, N. and Hayak, L.-A. C. (1993). New methods of tooth microwear analysis and application to dietary determination of two extinct antelopes. *Journal of Zoology, London*, 229, 421–445.

Spears, I. R. (1997). A three-dimensional finite element model of prismatic enamel: a re-appraisal of the data on the Young's Modulus of enamel. *Journal of Dental Research*, 76, 1690–1697.

Strait, S. G. (1991). Dietary Reconstruction in Small-Bodied Fossil Primates. PhD thesis, S. U. N. Y. Stony Brook.

Strait, S. G. (1993a). Molar morphology and food texture among small-bodied insectivorous mammals. *Journal of Mammalogy*, 74, 391–402.

Strait, S. G. (1993b). Differences in occlusal morphology and molar size in frugivores and faunivores. *Journal of Human Evolution*, 25, 471–484.

Strait, S. G. (1997). Tooth use and the physical properties of food. *Evolutionary Anthropology*, 5, 199–211.

Strait, S. G. and Overdorff, D. J. (1994). A preliminary examination of molar microwear in Strepsirhine primates. *American Journal of Physical Anthropology*, Suppl. 18, 190.

Swindler, D. R. (1976). *Dentition of Living Primates*. New York: Academic Press.

Teaford, M. F. (1988). A review of dental microwear and diet in modern mammals. *Scanning Microscopy*, 2, 1149–1166.

Teaford, M. F. (1991). Measurements of teeth using the Reflex Microscope. *Proceedings of the International Society for Optical Engineering, SPIE*, 1380, 33–43.

Teaford, M. F. (1994). Dental microwear and dental function. *Evolutionary Anthropology*, 3, 17–30.

Teaford, M. F. and Glander, K. E. (1991). Dental microwear in live, wild-trapped *Alouatta* from Costa Rica. *American Journal of Physical Anthropology*, 85, 313–319.

Teaford, M. F. and Glander, K. E. (1996). Dental microwear and diet in a wild population of mantled howling monkeys (*Alouatta palliata*). In *Adaptive Radiations of Neotropical Primates*, eds. M. A. Norconk, A. L. Rosenberger and P. A. Garber, pp. 433–449. New York: Plenum Press.

Teaford, M. F. and Oyen, O. J. (1989a). Live primates and dental replication: new problems and new techniques. *American Journal of Physical Anthropology*, **72**, 255–261.

Teaford, M. F. and Oyen, O. J. (1989*b*). Differences in the rate of molar wear between monkeys raised on different diets. *Journal of Dental Research*, **68**, 1513–1518.

Teaford, M. F. and Tylenda, C. A. (1991). A new approach to the study of tooth wear. *Journal of Dental Research*, **70**, 204–207.

Teaford, M. F. and Walker, A. C. (1984). Quantitative differences in dental microwear between primate species with different diets and a comment on the presumed diet of *Sivapithecus*. *American Journal of Physical Anthropology*, **64**, 191–200.

Theuns, H. M., van Dijk, J. W. E., Jongebloed, W. L. and Groeneveld, A. (1983). The mineral content of human enamel studied by polarizing microscopy, microradiography and scanning electron microscopy. *Archives of Oral Biology*, **28**, 797–803.

Uchida, A. (1998*a*). Variation in tooth morphology of *Gorilla gorilla*. *Journal of Human Evolution*, **34**, 55–70.

Uchida, A. (1998*b*). Variation in tooth morphology of *Pongo pygmaeus*. *Journal of Human Evolution*, **34**, 71–79.

Ungar, P. S. (1992). Incisor Microwear and Feeding Behavior of Four Sumatran Anthropoids. PhD thesis, S. U. N. Y. Stony Brook.

Ungar, P. S. (1996). Dental microwear of European Miocene catarrhines: evidence for diets and tooth use. *Journal of Human Evolution*, **31**, 335–366.

Ungar, P. S. (in press). Dental allometry, morphology and wear as evidence for diet in fossil primates. *Evolutionary Anthropology*.

Ungar, P. S. and Grine, F. E. (1991). Incisor size and wear in *Australopithecus africanus* and *Paranthropus robustus*. *Journal of Human Evolution*, **20**, 313–340.

Ungar, P. S. and Kay, R. F. (1995). The dietary adaptations of European Miocene catarrhines. *Proceedings of the National Academy of Sciences, USA*, **92**, 5479–5481.

Ungar, P. S., Simon, J.-C. and Cooper, J. W. (1991). A semiautomated image analysis procedure for the quantification of dental microwear. *Scanning Microscopy*, **13**, 31–36.

Ungar, P. S., Teaford, M. F., Pastor, R. F. and Glander, K. W. (1995). Dust accumulation in the canopy: implications for the study of dental microwear in primates. *American Journal of Physical Anthropology*, **97**, 93–99.

Vogel, S. (1988). *Life's Devices: the Physical World of Animals and Plants*. Princeton: Princeton University Press.

Vrba, E. S. and Grine, F. E. (1978). Australopithecine enamel prism patterns. *Science*, **202**, 890–892.

Wainwright, P. C. (1991). Ecomorphology: experimental functional anatomy for ecological problems. *American Zoologist*, **31**, 680–693.

Wainwright, P. C. and Reilly, S. M. (eds.) (1993). *Ecological Morphology*. Chicago: University of Chicago Press.

Wake, M. H. (1992). Morphology, the study of form and function, in modern evolutionary biology. In *Oxford Surveys in Evolutionary Biology*, vol. 8, pp. 289–346. Oxford: Oxford University Press.

Walker, P. L. and Hagen, E. H. (1994). A topographical approach to dental microwear analysis. *American Journal of Physical Anthropology*, **Suppl. 18**, 203.

Walker, P. and Murray, P. (1975). An assessment of masticatory effectiveness in a series of anthropoid primates with special reference to the Colobinae and Cercopithecinae. In *Primate Functional Morphology and Evolution*, ed. R. H. Tuttle, pp. 135–150. The Hague: Mouton.

Waters, N. E. (1980). Some mechanical and physical properties of teeth. In *The Mechanical Properties of Biological Materials*, eds. J. F. V. Vincent and J. D. Currey, Cambridge: Cambridge University Press.

Willems, G., Celis, J. P., Lambrechts, P., Braem, M. and Vanherle, G. (1993). Hardness and Young's modulus determined by nanoindentation technique of filler particles of dental restorative materials, compared with human enamel. *Journal of Biomedical Materials Research*, **27**, 747–755.

Williamson, L. and Lucas, P. (1995). The effect of moisture content on the mechanical properties of a seed shell. *Journal of Materials Science*, **30**, 162–166.

Wood, B. A. and Abbott, S. A. (1983). Analysis of the dental morphology of Plio-Pleistocene hominids. I. Mandibular molars, crown area measurements and measurements and morphological traits. *Journal of Anatomy*, **136**, 197–219.

Wood, B. A. and Engleman, C. A. (1988). Analysis of the dental morphology of Plio-Pleistocene hominids. IV. Maxillary postcanine tooth morphology. *Journal of Anatomy*, **151**, 1–35.

Xu, H. H. K., Smith, D. T., Jahanmir, S., Romberg, E., Kelly, J. R., Thompson, V. P. and Rekow, E. D. (1998). Indentation damage and mechanical properties of human enamel and dentin. *Journal of Dental Research*, **77**, 472–480.

Yamashita, N. (1996). Seasonal and site specificity of mechanical dietary patterns in two Malagasy lemur families (Lemuridae and Indriidae). *International Journal of Primatology*, **17**, 355–387.

Zuccotti, L. F., Williamson, M. D., Limp, W. F. and Ungar, P. S. (1998). Modeling primate occlusal morphology in three dimensions using Geographic Resources Analysis Support System software. *American Journal of Physical Anthropology*, Suppl. 26, 238–239.

Index